INTERNATIONAL SERIES IN MATHEMATICS

ESSENTIALS OF MATHEMATICAL STATISTICS

The Jones & Bartlett Learning Series in Mathematics

Geometry

Euclidean and Non-Euclidean Geometries, Second Edition
Libeskind (978-0-7637-9334-0) © 2015

Geometry with an Introduction to Cosmic Topology
Hitchman (978-0-7637-5457-0) © 2009

Euclidean and Transformational Geometry: A Deductive Inquiry
Libeskind (978-0-7637-4366-6) © 2008

A Gateway to Modern Geometry: The Poincaré Half-Plane, Second Edition
Stahl (978-0-7637-5381-8) © 2008

Lebesgue Integration on Euclidean Space, Revised Edition
Jones (978-0-7637-1708-7) © 2001

Liberal Arts Mathematics

Exploring Mathematics: Investigations with Functions
Johnson (978-1-4496-8854-7) © 2015

Understanding Modern Mathematics
Stahl (978-0-7637-3401-5) © 2007

Precalculus

Precalculus: A Functional Approach to Graphing and Problem Solving, Sixth Edition
Smith (978-1-4496-4916-6) © 2013

Precalculus with Calculus Previews, Fifth Edition
Zill/Dewar (978-1-4496-4912-8) © 2013

Essentials of Precalculus with Calculus Previews, Fifth Edition
Zill/Dewar (978-1-4496-1497-3) © 2012

Algebra and Trigonometry, Third Edition
Zill/Dewar (978-0-7637-5461-7) © 2012

College Algebra, Third Edition
Zill/Dewar (978-1-4496-0602-2) © 2012

Trigonometry, Third Edition
Zill/Dewar (978-1-4496-0604-6) © 2012

Calculus

Brief Calculus for the Business, Social, and Life Sciences, Third Edition
Armstrong/Davis (978-1-4496-9516-3) © 2014

Multivariable Calculus
Damiano/Freije (978-0-7637-8247-4) © 2012

Single Variable Calculus: Early Transcendentals, Fourth Edition
Zill/Wright (978-0-7637-4965-1) © 2011

Multivariable Calculus, Fourth Edition
Zill/Wright (978-0-7637-4966-8) © 2011

Calculus: Early Transcendentals, Fourth Edition
Zill/Wright (978-0-7637-5995-7) © 2011

Calculus: The Language of Change
Cohen/Henle (978-0-7637-2947-9) © 2005

Applied Calculus for Scientists and Engineers
Blume (978-0-7637-2877-9) © 2005

Calculus: Labs for Mathematica
O'Connor (978-0-7637-3425-1) © 2005

Calculus: Labs for MATLAB®
O'Connor (978-0-7637-3426-8) © 2005

Linear Algebra

Linear Algebra with Applications, Eighth Edition
Williams (978-1-4496-7954-5) © 2014

Linear Algebra with Applications, Alternate Eighth Edition
Williams (978-1-4496-7956-9) © 2014

Linear Algebra: Theory and Applications, Second Edition
Cheney/Kincaid (978-1-4496-1352-5) © 2012

Advanced Engineering Mathematics

Advanced Engineering Mathematics, Fifth Edition
Zill/Wright (978-1-4496-9172-1) © 2014

A Journey into Partial Differential Equations
Bray (978-0-7637-7256-7) © 2012

An Elementary Course in Partial Differential Equations, Second Edition
Amaranath (978-0-7637-6244-5) © 2009

Complex Analysis

Complex Analysis for Mathematics and Engineering, Sixth Edition
Mathews/Howell (978-1-4496-0445-5) © 2012

A First Course in Complex Analysis with Applications, Second Edition
Zill/Shanahan (978-0-7637-5772-4) © 2009

Classical Complex Analysis
Hahn (978-0-8672-0494-0) © 1996

Real Analysis

Elements of Real Analysis
Denlinger (978-0-7637-7947-4) © 2011

An Introduction to Analysis, Second Edition
Bilodeau/Thie/Keough (978-0-7637-7492-9) © 2010

Basic Real Analysis
Howland (978-0-7637-7318-2) © 2010

Closer and Closer: Introducing Real Analysis
Schumacher (978-0-7637-3593-7) © 2008

The Way of Analysis, Revised Edition
Strichartz (978-0-7637-1497-0) © 2000

Statistics

Essentials of Mathematical Statistics
Albright (978-1-4496-8534-8) ©2014

Topology

Foundations of Topology, Second Edition
Patty (978-0-7637-4234-8) © 2009

The Jones & Bartlett Learning Series in Mathematics

Discrete Mathematics and Logic

Essentials of Discrete Mathematics, Second Edition
Hunter (978-1-4496-0442-4) © 2012

Discrete Structures, Logic, and Computability, Third Edition
Hein (978-0-7637-7206-2) © 2010

Logic, Sets, and Recursion, Second Edition
Causey (978-0-7637-3784-9) © 2006

Numerical Methods

Numerical Mathematics
Grasselli/Pelinovsky (978-0-7637-3767-2) © 2008

Exploring Numerical Methods: An Introduction to Scientific Computing Using MATLAB®
Linz (978-0-7637-1499-4) © 2003

Advanced Mathematics

Basic Modern Algebra
Turner (978-1-4496-5232-6) © 2015

A Transition to Mathematics with Proofs
Cullinane (978-1-4496-2778-2) © 2013

Mathematical Modeling with Excel®
Albright (978-0-7637-6566-8) © 2010

Clinical Statistics: Introducing Clinical Trials, Survival Analysis, and Longitudinal Data Analysis
Korosteleva (978-0-7637-5850-9) © 2009

Harmonic Analysis: A Gentle Introduction
DeVito (978-0-7637-3893-8) © 2007

Beginning Number Theory, Second Edition
Robbins (978-0-7637-3768-9) © 2006

A Gateway to Higher Mathematics
Goodfriend (978-0-7637-2733-8) © 2006

For more information on this series and its titles, please visit us online at http://www.jblearning.com. Qualified instructors, contact your account representative at 1-800-832-0034 or info@jblearning.com to request review copies for course consideration.

The Jones & Bartlett Learning International Series in Mathematics

Basic Modern Algebra
Turner (978-1-4496-5232-6) © 2015

Essentials of Mathematical Statistics
Albright (978-1-4496-8534-8) © 2014

A Transition to Mathematics with Proofs
Cullinane (978-1-4496-2778-2) © 2013

Linear Algebra: Theory and Applications, Second Edition, International Version
Cheney/Kincaid (978-1-4496-2731-7) © 2012

Multivariable Calculus
Damiano/Freije (978-0-7637-8247-4) © 2012

Complex Analysis for Mathematics and Engineering, Sixth Edition, International Version
Mathews/Howell (978-1-4496-2870-3) © 2012

A Journey into Partial Differential Equations
Bray (978-0-7637-7256-7) © 2012

Association Schemes of Matrices
Wang/Huo/Ma (978-0-7637-8505-5) © 2011

Advanced Engineering Mathematics, Fifth Edition, International Version
Zill/Wright (978-1-4496-8980-3) © 2014
Canada/UK (978-1-4496-9302-2)

Calculus: Early Transcendentals, Fourth Edition, International Version
Zill/Wright (978-0-7637-8652-6) © 2011

Real Analysis
Denlinger (979-0-7637-7947-4) © 2011

Mathematical Modeling for the Scientific Method
Pravica/Spurr (978-0-7637-7946-7) © 2011

Mathematical Modeling with Excel®
Albright (978-0-7637-6566-8) © 2010

An Introduction to Analysis, Second Edition
Bilodeau/Thie/Keough (978-0-7637-7492-9) © 2010

Basic Real Analysis
Howland (978-0-7637-7318-2) © 2010

For more information on this series and its titles, please visit us online at http://www.jblearning.com. Qualified instructors, contact your account representative at 1-800-832-0034 or info@jblearning.com to request review copies for course consideration.

INTERNATIONAL SERIES IN MATHEMATICS

ESSENTIALS OF MATHEMATICAL STATISTICS

BRIAN ALBRIGHT, DSc
Concordia University, Nebraska

JONES & BARTLETT
LEARNING

World Headquarters
Jones & Bartlett Learning
5 Wall Street
Burlington, MA 01803
978-443-5000
info@jblearning.com
www.jblearning.com

Jones & Bartlett Learning books and products are available through most bookstores and online booksellers. To contact Jones & Bartlett Learning directly, call 800-832-0034, fax 978-443-8000, or visit our website, www.jblearning.com.

Substantial discounts on bulk quantities of Jones & Bartlett Learning publications are available to corporations, professional associations, and other qualified organizations. For details and specific discount information, contact the special sales department at Jones & Bartlett Learning via the above contact information or send an email to specialsales@jblearning.com.

Copyright © 2014 by Jones & Bartlett Learning, LLC, an Ascend Learning Company

All rights reserved. No part of the material protected by this copyright may be reproduced or utilized in any form, electronic or mechanical, including photocopying, recording, or by any information storage and retrieval system, without written permission from the copyright owner.

Essentials of Mathematical Statistics is an independent publication and has not been authorized, sponsored, or otherwise approved by the owners of the trademarks or service marks referenced in this product.

Some images in this book feature models. These models do not necessarily endorse, represent, or participate in the activities represented in the images.

Production Credits
Executive Publisher: Kevin Sullivan
Senior Developmental Editor: Amy Bloom
Director of Production: Amy Rose
Production Editor: Tiffany Sliter
Production Assistant: Eileen Worthley
Senior Marketing Manager: Andrea DeFronzo
V.P., Manufacturing and Inventory Control: Therese Connell
Composition: Northeast Compositors, Inc.
Cover & Title Page Design: Kristin E. Parker
Director of Photo Research and Permissions: Amy Wrynn
Cover & Title Page Image: Courtesy of Micah Hollenbeck
Printing and Binding: Edwards Brothers Malloy
Cover Printing: Edwards Brothers Malloy

Library of Congress Cataloging-in-Publication Data
Albright, Brian, 1977-
  Essentials of mathematical statistics / Brian Albright.
    pages cm
  Includes index.
  ISBN 978-1-4496-8534-8 (casebound) — ISBN 1-4496-8534-X (casebound)
  1. Mathematical statistics—Textbooks. I. Title.
  QA276.12.A39824 2014
  519.5—dc23
                            2012029915

6048

Printed in the United States of America
17 16 15 14 13   10 9 8 7 6 5 4 3 2

Contents

	Preface	xi
1	**Basics of Probability**	**1**
1.1	Introduction	2
1.2	Basic Concepts and Definitions	2
1.3	Counting Problems	13
1.4	Axioms of Probability and the Addition Rule	26
1.5	Conditional Probability and the Multiplication Rule	37
1.6	Bayes' Theorem	44
1.7	Independent Events	54
2	**Discrete Random Variables**	**65**
2.1	Introduction	66
2.2	Probability Mass Functions	67
2.3	The Hypergeometric and Binomial Distributions	78
2.4	The Poisson Distribution	92
2.5	Mean and Variance	101
2.6	Functions of a Random Variable	111
2.7	The Moment-Generating Function	117

3 Continuous Random Variables — 125

- 3.1 Introduction — 126
- 3.2 Definitions — 128
- 3.3 The Uniform and Exponential Distributions — 139
- 3.4 The Normal Distribution — 150
- 3.5 Functions of Continuous Random Variables — 161
- 3.6 Joint Distributions — 168
- 3.7 Functions of Independent Random Variables — 179
- 3.8 The Central Limit Theorem — 190
- 3.9 The Gamma and Related Distributions — 198
- 3.10 Approximating the Binomial Distribution — 212

4 Statistics — 219

- 4.1 What Is Statistics? — 220
- 4.2 Summarizing Data — 232
- 4.3 Maximum Likelihood Estimates — 249
- 4.4 Sampling Distributions — 259
- 4.5 Confidence Intervals for a Proportion — 270
- 4.6 Confidence Intervals for a Mean — 281
- 4.7 Confidence Intervals for a Variance — 293
- 4.8 Confidence Intervals for Differences — 300
- 4.9 Sample Size — 315
- 4.10 Assessing Normality — 324

5 Hypothesis Testing   333

- 5.1 Introduction 334
- 5.2 Testing Claims About a Proportion 342
- 5.3 Testing Claims About a Mean 356
- 5.4 Comparing Two Proportions 365
- 5.5 Comparing Two Variances 374
- 5.6 Comparing Two Means 383
- 5.7 Goodness-of-Fit Tests 397
- 5.8 Test of Independence 408
- 5.9 One-Way ANOVA 418
- 5.10 Two-Way ANOVA 431

6 Simple Regression   443

- 6.1 Introduction 444
- 6.2 Covariance and Correlation 446
- 6.3 Method of Least Squares 459
- 6.4 The Simple Linear Model 468
- 6.5 Sums of Squares and ANOVA 478
- 6.6 Nonlinear Regression 484
- 6.7 Multiple Regression 492

7 Nonparametric Statistics   509

- 7.1 Introduction 510
- 7.2 The Sign Test 510
- 7.3 The Wilcoxon Signed-Rank Test 518
- 7.4 The Wilcoxon Rank-Sum Test 524
- 7.5 The Runs Test for Randomness 532

A Proofs of Selected Theorems — 541

- A.1 A Proof of Theorem 3.7.5 — 541
- A.2 A Proof of the Central Limit Theorem — 542
- A.3 A Proof of the Limit Theorem of De Moivre and Laplace — 545
- A.4 A Proof of Theorem 4.6.1 — 547
- A.5 Confidence Intervals for the Difference of Two Means — 547
- A.6 Coefficients in the Linear Regression Equation — 550
- A.7 Wilcoxon Signed-Rank Test Distribution — 551

B Software Basics — 555

- B.1 Minitab — 555
- B.2 R — 555
- B.3 Excel — 558
- B.4 TI-83/84 Calculators — 560

C Tables — 561

D Answers to Selected Exercises — 577

Index — 587

Preface

Most statistics books fall into one of two categories: elementary statistics or mathematical statistics. Elementary statistics books focus on conceptual understanding, provide a superficial treatment of probability, and discuss very little theory. Mathematical statistics books, on the other hand, focus on the theory behind the statistics and prove most results in great detail and generality.

In this text we combine elements of both types of books. *Essentials of Mathematical Statistics* began as notes for a class for math majors who needed something more than an elementary statistics class, but not as much theory as a traditional mathematical statistics class. This text covers many of the same topics as an elementary book, but also discusses much of the underlying theory. However, we do not present every theorem in its full generality in order to avoid "losing the forest for the trees." Proofs of relatively simple theorems are given in the text. Proofs of some of the more complex theorems are given in Appendix A, while others are justified with the results of simulations.

This text is appropriate for math, engineering, and science majors who need an introduction to probability and statistics. No prior knowledge of probability or statistics is assumed, but knowledge of Calculus I and II is needed. Double and triple integrals are used in only one section (Section 3.6).

The text begins with an axiomatic treatment of elementary probability theory. We then introduce discrete and continuous random variables in Chapters 2 and 3 and include common types of distributions. In Chapter 4, we introduce statistics and stress the connection with random variables. In Chapter 5, we discuss many types of hypothesis tests, and in Chapter 6, we introduce topics in regression. We end with a brief survey of nonparametric hypothesis tests in Chapter 7.

This text contains numerous exercises (732 to be exact, an average of about 14.4 per section, many of which have multiple parts). These exercises range from computational to theoretical. Some exercises introduce new concepts that challenge the students to go beyond what is discussed in the section.

We end each section with a description of how to perform the calculations described therein with the software Minitab, R, and Excel, and the TI-83/84 calculators. Appendix B provides a brief introduction to using these software.

A detailed solutions manual, PowerPoint lecture outlines, and a Test Bank are available for instructors. To request access, visit go.jblearning.com/Albright or contact your account representative.

Acknowledgments

The author would like to thank his students who allowed their data to be used for examples and exercises. Credit is given to them all. The author would also like to thank his students who were the "guinea pigs" for early versions of the manuscript. Thanks for finding mistakes and suggesting improvements. Special thanks go to Kristen Shavlik, who helped write the solutions manual, and Micah Hollenbeck who designed the cover.

The author would also like to thank the following reviewers who provided invaluable feedback throughout the preparation of the manuscript: Michelle Quinlan, PhD, Senior Biostatistician, Novartis Oncology; Leszek Gabvarecki, Kettering University; Scott Gilbert, Southern Illinois University, Carbondale; and Bogdan Doytchinov, Elizabethtown College.

Above all, the author would like to thank his wife KaiPing who provided support through the countless hours of writing and rewriting. Wuine.

CHAPTER 1

Basics of Probability

Chapter Objectives

- Introduce fundamental concepts of probability
- Define basic terms
- Introduce counting techniques including permutations and combinations
- Provide an axiomatic foundation for the study of probability
- Introduce addition and multiplication rules
- Introduce Bayes' theorem

1.1 Introduction

Probability deals with describing *random experiments*. A random experiment is an activity in which the result is not known until the activity is performed. Most everything in life is a random experiment. For example, when you take a test, you don't know your grade until the test is graded. If you invest money in the stock market, you don't know exactly how much you will gain or lose until you sell your stock.

Because random experiments are so prevalent in life, probability has a wide range of applications. Informally, probability is a measure of how likely something is to occur. This is certainly not a rigorous mathematical definition, but it gives us a conceptual starting point for understanding the ideas behind probability. In this chapter, we discuss basic terminology and concepts and use an axiomatic approach to discover the rules and properties of probability.

1.2 Basic Concepts and Definitions

To illustrate the basic concepts and definitions associated with probability, consider a bag containing 10 cubes, of which 4 are blue, 3 are red, 2 are green, and 1 is yellow. The cubes are well mixed, and we randomly choose one cube. The activity of choosing one cube is an example of a *random experiment* because we don't know exactly what cube we will get until we choose it. By *randomly*, we mean that each cube has an equally likely chance of being selected. In other words, we don't try to choose any one particular cube, or any one color, over any other.

One question we might ask about this experiment is, How likely is it that we get a green cube? To begin to answer this question, we need to define some terms.

> **Definition 1.2.1** An *outcome* is a result of a random experiment. The *sample space* of a random experiment is the set of all possible outcomes. An *event* is a subset of the sample space.

In this random experiment, we have 10 possible outcomes because there are 10 different cubes (some of them are the same color, but they are all different cubes). The same space is

$$S = \{B_1, B_2, B_3, B_4, R_1, R_2, R_3, G_1, G_2, Y_1\}$$

where B_1 denotes the first blue cube, R_2 denotes the second red cube, etc. There are many events we could consider. Almost any subset of S (stated another way, almost any collection of outcomes) could be an event. Events can be described using formal set notation, or they can be described in words. For example,

$$A = \{B_2,\ B_4,\ R_2,\ Y_1\}$$

is the event that we choose the second blue cube, the fourth blue cube, the second red cube, or the yellow cube. In this experiment, there are four rather obvious events to consider. Let

$$B = \{B_1,\ B_2,\ B_3,\ B_4\},\ R = \{R_1,\ R_2,\ R_3\},\ G = \{G_1,\ G_2\},\ \text{and}\ Y = \{Y_1\}.$$

Then, in words, B, R, G, and Y are the events that we get a blue, red, green, and yellow cube, respectively.

Informally, probabilities are measures of how likely events are to occur. If A is an event, then the probability of A is denoted $P(A)$. We give probabilities values between 0 and 1, inclusive, with $P(A) = 1$ meaning event A is "certain" to occur and $P(A) = 0$ meaning it is "impossible" for A to occur. A probability close to 0 means the event is "unlikely," and a probability close to 1 means it is "likely."

The question we want to answer is, How likely is it that we get a green cube? In terms of probabilities and events, this question can be stated as, What is the value of $P(G)$? In this section, we introduce two different ways of calculating probabilities, both of which will be justified in later sections or chapters. In our first approach, we use relative frequencies.

Relative Frequency Approximation: To *estimate* the probability of an event A, repeat the random experiment several times (each repetition is called a *trial*) and count the number of times the event occurred. Then

$$P(A) \approx \frac{\text{number of times } A \text{ occurred}}{\text{number of trials}}.$$

The fraction on the right is called a *relative frequency*.

Suppose a person chooses a cube from the bag, records its color, places it back in the bag, mixes up the contents of the bag, and repeats this sequence 50 times. This constitutes 50 trials of the random experiment. The results are summarized in the first two columns of Table 1.1.

Color	Freq	Rel Freq
Blue	19	0.38
Red	15	0.30
Green	12	0.24
Yellow	4	0.08

Table 1.1

The relative frequencies in the third column of Table 1.1 are calculated by dividing each frequency by 50, the number of trials. The relative frequencies are approximations of the probabilities of events B, R, G, and Y. From the table, we see that the answer to our question is $P(G) \approx 0.24$.

To understand why this approach only gives *approximations* of probabilities, we must realize that if we were to do another 50 trials, we would not necessarily get the same results. For example, another student does 50 trials in the same way and gets the results shown in Table 1.2. Notice that these relative frequencies are different from, but similar to, the results that the first student got. The second student estimates $P(G) \approx 0.14$.

Color	Freq	Rel Freq
Blue	21	0.42
Red	16	0.32
Green	7	0.14
Yellow	6	0.12

Table 1.2

The fact that the two students got different frequencies is referred to as *sampling variability*. Because of the random nature of random events, we may not get the same thing every time we perform a set of trials. Therefore, relative frequencies can, at best, only be considered approximations of probabilities.

Because relative frequencies are subject to sampling variability, we want a more theoretical way of calculating probabilities. The following way of doing so seems natural:

> **Theoretical Approach**: If all outcomes of a random experiment are equally likely, S is the finite sample space, and A is an event, then
>
> $$P(A) = \frac{n(A)}{n(S)}$$
>
> where $n(A)$ is the number of elements in set A, and $n(S)$ is the number of elements in set S.

The qualifier that all outcomes are equally likely is *extremely* important and should not be ignored. In this example, all outcomes are equally likely because the choices are made randomly, so we can apply this approach to find probabilities. To find $P(G)$, we calculate

$$P(G) = \frac{n(G)}{n(S)} = \frac{2}{10} = 0.20.$$

Note that the first student's approximation was a bit higher than this theoretical value and the second student's was a bit lower. This approach gives an exact value because it's based on theory and not repeating the experiment, so it is not subject to sampling variability.

A rule that relates these two approaches to calculating probabilities is known as the *law of large numbers*.

> **Law of Large Numbers**: As the number of trials gets larger, the relative frequency approximation of $P(A)$ gets closer to the theoretical value.

Informally, the law of large numbers says that to get a better approximation, perform more trials. To illustrate this, we combine the results of the two students. Combined, they have a total of 100 trials, and in 19 of those trials, they got a green cube. Using these results, we calculate

$$P(G) \approx \frac{19}{100} = 0.19,$$

which is a closer approximation of the theoretical value than either student got individually.

The law of large numbers helps us to understand what the theoretical probability $P(G) = 0.20$ means in practical terms. It means that if we were to perform *many* trials of the random experiment and count the number of times we got a green cube, then

$$\frac{\text{Total number of green cubes}}{\text{Number of trials}} \approx 0.20,$$

and the larger the number of trials, the closer this fraction will (probably) be to 0.20. In other words, we get a green cube about 20% of the time. It is important to understand that a probability does not tell us what will happen on any one trial. It does, however, tell us what will happen in the long run. Stated another way:

A probability is an average in the long run.

Example 1.2.1 **Probability of an Individual Outcome** An event can consist of one outcome. For example, let $A = \{R_3\}$. This is the event that we get the third red cube. We can calculate $P(A)$, using the theoretical approach, as

$$P(A) = \frac{n(A)}{n(S)} = \frac{1}{10} = 0.10.$$

This example illustrates that for a random experiment with a finite sample space in which every outcome is equally likely, the probability of obtaining any one particular outcome is $1/n(S)$. □

Example 1.2.2 **Children in a Family** Suppose we randomly choose a family with two children and we want to know the probability of selecting a family with at least one boy. This activity of choosing a family with two children is a random experiment because we don't know what the genders of the children will be until the family is chosen. The sample space is shown in Table 1.3.

	Second Child	
First Child	B	G
B	BB	BG
G	GB	GG

Table 1.3

We see that $S = \{BB, BG, GB, GG\}$. We assume that in a single birth, each gender is equally likely (this may not be technically true, but it's a reasonable assumption) so that each outcome is equally likely. If we let $A = \{BB, BG, GB\}$ be the event that we select a family with at least one boy, then we have

$$P(A) = \frac{n(A)}{n(S)} = \frac{3}{4} = 0.75.$$

Similarly, we see that the probability of selecting a family with exactly one boy and one girl is 0.50. □

Example 1.2.3 Rolling Two Dice Suppose we roll a pair of fair six-sided dice. By *fair* we mean that each face of the die is equally likely to occur on a single roll. One question we might ask is, What is the probability that the sum of the dice is greater than 7? To answer this, we construct the sample space in Table 1.4.

	Second Die					
First Die	1	2	3	4	5	6
1	2	3	4	5	6	7
2	3	4	5	6	7	8
3	4	5	6	7	8	9
4	5	6	7	8	9	10
5	6	7	8	9	10	11
6	7	8	9	10	11	12

Table 1.4

From Table 1.4, we see that the sample space consists of 36 outcomes ranging in value from 2 to 12. In 15 of them, the sum is greater than 7. All of these outcomes are equally likely because the dice are fair. If A is the event that the sum is greater than 7, we can use the theoretical approach to calculate $P(A)$:

$$P(A) = \frac{15}{36} = \frac{5}{12} \approx 0.417.$$

We can also use the sample space to determine the probability that the sum equals any one specific value. If we let the variable X denote the sum of the dice (such a variable is called a *random variable*), we see that X has values between 2 and 12. Table 1.5 shows the values of X, denoted x; the number of outcomes in which that value is obtained; and the probability that value occurs, denoted $P(x)$.

x	2	3	4	5	6	7	8	9	10	11	12
Num Outcomes	1	2	3	4	5	6	5	4	3	2	1
$P(x)$	$\frac{1}{36}$	$\frac{2}{36}$	$\frac{3}{36}$	$\frac{4}{36}$	$\frac{5}{36}$	$\frac{6}{36}$	$\frac{5}{36}$	$\frac{4}{36}$	$\frac{3}{36}$	$\frac{2}{36}$	$\frac{1}{36}$

Table 1.5

A table such as Table 1.5 that gives the values of a random variable and the associated probabilities is called a *distribution*. We will study random variables and distributions in greater depth starting in Chapter 2. □

Many random experiments cannot be repeated many times, so relative frequencies cannot be used to estimate probabilities. Nor can we always assume that all the outcomes are equally likely, so the theoretical approach does not apply either. In cases such as this, we use subjective estimates of probabilities.

> **Subjective Probabilities**: $P(A)$ is *estimated* by using knowledge of pertinent aspects of the situation.

Example 1.2.4 illustrates the use of subjective probabilities.

Example 1.2.4 Predicting the Weather If you're planning a picnic for tomorrow, you might want to know the probability that it will rain. Tomorrow's weather is a random experiment because you don't know what it will be until it happens. Here our sample space is $S = \{\text{rain, not rain}\}$. If $A = \{\text{rain}\}$ is the event that it rains tomorrow, we might be tempted to say that $P(A) = \frac{1}{2}$ using the theoretical approach. However, the theoretical approach does not apply to this random experiment because our outcomes are not equally likely and it is not reasonable to assume they are so.

Nor can we use the relative frequency approach to estimate this probability because we cannot repeat the random experiment (the weather tomorrow) several times. Instead, weather forecasters use knowledge of weather and information from radar, satellites, etc., to estimate the likelihood that it will rain tomorrow. This number is not calculated using techniques as in the previous examples. □

Example 1.2.5 Choosing a Number Suppose we randomly choose an integer between 1 and 100. What is the probability that the number is greater than or equal to 75?

In this experiment, $S = \{1, \ldots, 100\}$. If we let $A = \{75, \ldots, 100\}$ be the event that the number is greater than or equal to 75, then we calculate

$$P(A) = \frac{n(A)}{n(S)} = \frac{26}{100} = 0.26.$$

Now suppose we do not restrict ourselves to integers. Suppose we can choose *any* number—integer, rational, or irrational—between 1 and 100. In this case, $S = [1, 100]$ is an *interval* of values, and $n(S)$ is not finite, so the theoretical approach as used above does not apply.

If we let $B = [75, 100]$ be the event that the number is greater than or equal to 75, then it seems reasonable to assign a probability to B as

$$P(B) = \frac{\text{length of interval } B}{\text{length of interval } S} = \frac{25}{99} \approx 0.253.$$

□

Example 1.2.6 Dartboard Suppose a circular dartboard has a diameter of 18 in and the circular bull's-eye has a diameter of $\frac{5}{8}$ in. Also suppose that we randomly throw a dart at the board (by *randomly* we mean that any point on the board has an equal chance of being hit). Further suppose that we are guaranteed to hit the board. Find the probability that we will hit the bull's-eye.

In this experiment, the sample space is all points on the board, which is infinite. Therefore, the theoretical approach does not apply. If we let A be the event that we hit the bull's-eye, it seems reasonable to assign the following probability to A:

$$P(A) = \frac{\text{area of the bull's-eye}}{\text{area of the dartboard}} = \frac{\pi\left(\frac{5}{16}\right)^2}{\pi(9)^2} \approx 0.0012.$$

In practical terms, this probability means that if we were to randomly throw a dart at the dartboard *many* times and hit the board each time, we would hit the bull's-eye approximately 0.12% of the time. Note that this probability does not apply if we were aiming at the bull's-eye as a real dart player would. □

Example 1.2.7 The Monty Hall Problem In the famous game show *Let's Make a Deal*, hosted by Monty Hall, one game required a contestant to choose one of three doors. Behind one of the doors was a prize (such as money or a car), and behind the other two doors were dummy prizes (such as a donkey). Once the choice was made, Monty Hall would open one of the unchosen doors, revealing one of the dummy prizes. The contestant was then given a choice to either switch to the remaining unopened door or keep the door that was already chosen. The contestant would get whatever "prize" was behind the door.

To illustrate this game, consider the scenario in Figure 1.1, where the real prize is behind door 2 (unbeknownst to the contestant) and the contestant chooses door 1. Monty would then open door 3, revealing a dummy prize. The contestant next had to decide whether to switch to door 2 (and consequently win the real prize) or not switch (and consequently not win the real prize).

Suppose we are playing this game. The initial choice of the door is random. The difficult part is to decide whether to switch. At first glance, it may seem as if it doesn't matter. There are two doors; one contains the real prize, and the other contains the dummy prize. Therefore, the probability of getting the real prize is $\frac{1}{2}$ (in other words, there is a 50-50 chance of winning). However, this reasoning assumes that there is an equal chance the real prize is behind either door. This is not obviously true.

To determine if we should switch, we break down the situation to two cases. In the first case, suppose that before the game begins, we decide to *not* switch. In this case, we win only if we initially choose the door containing the real prize. Because there are three equally

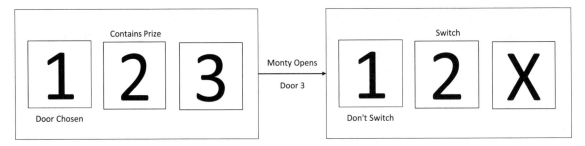

Figure 1.1

likely outcomes to the experiment of initially choosing the door (the three doors) and in one of them we win, we have

$$P(\text{winning if we don't switch}) = P(\text{initially choosing door with real prize}) = \frac{1}{3}.$$

In the second case, suppose that before the game begins, we decide to switch. To win in this case, we must switch *to* the door containing the real prize. This can be done only if we initially choose a door containing a dummy prize. Because two of the doors contain a dummy prize, we have

$$P(\text{winning if we do switch}) = P(\text{initially choosing a door with a dummy prize}) = \frac{2}{3}.$$

Thus we see that the probability of winning if we do switch is twice as high as if we don't switch. It is important to remember that these probabilities don't tell us what will happen on a single play of the game. They do tell us what will happen if we play the game *many* times. They tell us that if we play the game many times and switch *every* time, we will win twice as often as if we don't switch every time. Therefore, it is advantageous to switch. □

Exercises

1. Suppose we flip a fair coin three times. If the notation HTH denotes the outcome that the first coin is heads, the second is tails, and the third is heads, list the other seven outcomes in the sample space. If A is the event that at least one flip is tails, list the outcomes in this event and find its probability.

2. Four cars arrive at an intersection where each can turn either left or right. If the notation RRRL denotes the outcome that the first three cars turn right and the fourth car turns left, list the other 15 outcomes in the sample space. Assuming that all these outcomes are equally likely, find the probability that exactly two cars turn left. How reasonable do you think this assumption is? Explain.

3. Suppose you guess a person's birthday.

   a. Find the probability that you guess the correct month and day. Assume that all 365 days of the year are equally likely to be a birthday.

   b. Find the probability that you guess the correct month only. Assume that all 12 months of the year are equally likely.

4. A simple casino game is played by rolling a fair six-sided die and then flipping a coin.

   a. Describe the sample space of this random experiment, using a table such as that in Example 1.2.2.

   b. Describe the event of getting a tail and at least a 5 as a collection of outcomes. Find the probability of this event.

5. A number, not necessarily an integer, is randomly chosen from the interval $(-2, 5)$. Call the number X. Calculate each of the following probabilities:

   a. $P(X \leq 3)$ b. $P(X \geq -1.5)$ c. $P(-1 \leq X \leq 4)$
   d. $P(0 \leq X \leq 6)$ e. $P(X \geq 4)$ f. $P(X = -0.75)$

6. Two fair six-sided dice are rolled, and the *product* of the dice is calculated.

   a. Describe the sample space, using a table similar to Table 1.4.

   b. Describe the distribution of the product, using a table similar to Table 1.5.

   c. Find the probability that the product is between 5 and 25, inclusive.

7. A statistics professor has three sections of *Introduction to Statistics* with a total of 115 students.

   a. In one particular class, 9 students received an A, 15 got a B, 8 got a C, 5 got a D, and 2 got an F. Consider the experiment of randomly selecting a student *from this class*. Let B denote the event of selecting a student *from this class* who got a B. Find $P(B)$. What approach are you using to calculate this probability: the relative frequency approach or the theoretical approach? Why?

   b. Suppose the professor randomly selects 20 students from the group of 115 students, with replacement, and notes that 4 got an A, 8 got a B, 4 got a C, 3 got a D, and 1 got an F. Consider the experiment of randomly selecting a student *from the group of 115 students*. Let B denote the event of selecting a student *from the group of 115 students* who got a B. Use the given information to estimate $P(B)$. What approach are you using to calculate this probability? Why?

   c. Why is the probability in part a an exact value and the probability in part b is only an estimate?

   d. Suppose we randomly select a statistics student *from all Introduction to Statistics classes in the United States*. Let B denote the event of selecting a student *from the United States* who got a B. Would either probability from part a or part b apply to estimating $P(B)$ in this case? Why or why not?

8. Use a subjective probability approach to give reasonable estimates of the probability of each of the following events. We often consider an event to be "unusual" if its probability is less than 0.05. Using this criterion, identify each event as unusual or not.

   a. A randomly chosen elevator gets stuck between floors.

   b. You see a blue car while driving to school.

   c. A randomly selected person recognizes the brand Coca-Cola.

   d. You win the lottery next week.

   e. Someone in the United States wins the lottery next week.

9. A student explains that he likes math class because every answer to a homework problem is either right or wrong. Therefore, if you guess an answer, there is a 50-50 chance of getting it right. Explain what is wrong with this reasoning.

10. Consider a certain casino game in which the probability of winning is 0.10. A novice gambler claims that this means that if she plays 10 times, she is guaranteed to win exactly once. Explain what is wrong with this reasoning.

11. A dartboard has a radius of 9 in and is composed of a bull's-eye with a radius of 1 in surrounded by four concentric rings, each of which has a width of 2 in. If you randomly throw a dart at the board, find the probability of hitting each individual ring as well as the bull's-eye.

1.3 Counting Problems

To calculate probabilities with the theoretical approach, we need to know only the *number* of elements in an event and in the sample space. It is not necessary to list all the elements in either set. In this section, we discuss some techniques for counting the number of elements in these sets without having to list them.

The *fundamental counting principle* forms the foundation for all counting techniques.

> **Fundamental Counting Principle**: Suppose a choice has to be made that consists of a sequence of two subchoices. If the first choice has n_1 options and the second choice has n_2 options, then the total number of options for the overall choice is $n_1 \times n_2$.

Example 1.3.1 Choosing an Outfit Suppose an outfit consists of a shirt and a pair of pants. Choosing an outfit requires us to choose first a shirt and then a pair of pants (the two subchoices). If we have 6 shirts from which to choose and 4 pairs of pants, then we have $6 \times 4 = 24$ different outfits.

We can easily extend the fundamental counting principle to a sequence of more than two choices. Suppose an outfit consists of a shirt, a pair of pants, and a pair of shoes. If we have three pairs of shoes, then we have a total of $6 \times 4 \times 3 = 72$ different outfits. □

Example 1.3.2 Choosing a Batting Lineup Suppose a baseball manager has chosen the 9 starting players for a game. In how many ways can the batting lineup be chosen? This choice consists of 9 subchoices (the first position, second position, etc.). There are 9 options for the first subchoice. For the second subchoice, there are only 8 options because one player has already been chosen and cannot be chosen again. Likewise, there are 7 options for the third subchoice. Therefore, the total number of batting lineups is

$$9 \times 8 \times 7 \times 6 \times 5 \times 4 \times 3 \times 2 \times 1 = 362{,}880. \tag{1.1}$$

□

Products such as that in Equation (1.1) occur frequently in counting problems, so we define a special symbol to denote them.

> **Definition 1.3.1**   Let $n > 0$ be an integer. The symbol $n!$ (read "n factorial") is
> $$n! = n \times (n-1) \times (n-2) \times \cdots \times 2 \times 1.$$
> For convenience, we define $0! = 1$.

Example 1.3.3   Boat Race   Suppose 7 boats enter a race. How many possibilities are there for the first- through third-place finishers?

We could think of this scenario as though we are "choosing" the three finishers by choosing the first-place finisher, then the second-place finisher, and finally the third-place finisher. So we can apply the fundamental counting principle. There are 7 options for the first choice, 6 for the second, and 5 for the third, so there are

$$7 \times 6 \times 5 = 210$$

different possibilities. □

We make a few observations about this example:

1. We are interested in the number of *arrangements* of boats.

2. There are 7 *different* boats.

3. We are choosing 3 of the 7 boats *without replacement* (meaning we choose one boat from the 7, then we choose another from the remaining 6, and so on).

4. The order in which we choose the boats *is important* because the first boat we choose gets first place, the second boat gets second place, and the third boat gets third place.

Also note that the expression $7 \times 6 \times 5$ looks almost like a factorial, but not quite. However, note that

$$7 \times 6 \times 5 = \frac{7 \times 6 \times 5 \times 4 \times 3 \times 2 \times 1}{4 \times 3 \times 2 \times 1} = \frac{7!}{4!} = \frac{7!}{(7-3)!}$$

We generalize these observations in the following definition:

Definition 1.3.2 Suppose we are choosing r objects from a set of n objects and these requirements are met:

1. The n objects are all *different*.
2. We are choosing the r objects *without replacement*.
3. The order in which the choices are made *is important*.

Then the number of ways the overall choice can be made (also described as the number of *arrangements*) is called the *number of permutations of n objects chosen r at a time*. This number is denoted $_nP_r$ and

$$_nP_r = \frac{n!}{(n-r)!}.$$

Example 1.3.4 Members of a Committee If there are 15 members of a town council and they have to choose a committee consisting of a chair, a cochair, and a secretary, how many different committees are possible?

We can treat this scenario as though the first person chosen is the chair, the second person is the cochair, and the third is the secretary so that the order is important. Thus all the requirements for a permutation are met, and the solution is

$$_{15}P_3 = \frac{15!}{12!} = 2730.$$

□

Example 1.3.5 Choosing Roommates Suppose Martin is going to choose 3 friends from a group of 9 friends to be his roommates for next year. In how many ways can he make the selection?

At first glance, this appears to be a permutation problem. However, the order is not important because there is no indication that the roommates will have different roles as in Example 1.3.3. We are not interested in the number of arrangements (or permutations) of friends. Instead, we say that we are interested in the number of *combinations* of friends that can be chosen.

To calculate the number of combinations of friends, suppose that three of the friends are named A, B, and C. It is possible to make the following choices:

ABC, ACB, BAC, BCA, CAB, and CBA.

These $6 = 3!$ different possibilities are all different permutations; however, each one is the same *combination* of friends. Thus for each unique combination, there are $3!$ different permutations. So we have

$$(\text{Number of unique combinations}) \times 3! = {}_9P_3$$

$$\Rightarrow \text{Number of unique combinations} = \frac{{}_9P_3}{3!} = \frac{9!/(9-3)!}{3!} = \frac{9!}{3!\,(9-3)!} = 84.$$

Thus there are 84 different ways to choose the roommates. □

We generalize Example 1.3.5 in the following definition:

Definition 1.3.3 Suppose we are choosing r objects from a set of n objects and these requirements are met:

1. The n objects are all *different*.

2. We are choosing the r objects *without replacement*.

3. The order in which the choices are made is *not important*.

Then the number of ways the overall choice can be made is called the *number of combinations of n objects chosen r at a time*. This number is denoted ${}_nC_r = \binom{n}{r}$ (read "n choose r"), and

$$_nC_r = \binom{n}{r} = \frac{n!}{r!(n-r)!}.$$

Example 1.3.6 Public Safety Committee A town is going to form an 8-person committee to study ways to improve public safety. The committee will consist of 3 representatives from the 8-member town council, 2 members of the 6-person citizens advisory board, and 3 of the 10 police officers on the force. How many different committees could be formed?

To solve this problem, we use both the fundamental counting principle and combinations. Although we are choosing 8 people, we can break this down into three choices: (1) Choose 3 from the council; (2) choose 2 from the advisory board; and (3) choose 3 from the police force.

Because the order in which the council members are chosen does not matter, there are $\binom{8}{3}$ options for the first choice. Likewise, there are $\binom{6}{2}$ options for the second choice and $\binom{10}{3}$ options for the third choice. Thus, by the fundamental counting principle, there are

$$\binom{8}{3}\binom{6}{2}\binom{10}{3} = 56 \times 15 \times 120 = 100{,}800$$

different possible committees. □

Example 1.3.7 Student Activities Committee A local college is investigating ways to improve the scheduling of student activities. A 15-person committee consisting of 5 administrators, 5 faculty members, and 5 students is being formed. A 5-person subcommittee is to be formed from this larger committee. The chair and cochair of the subcommittee must be administrators, and the remainder will consist of faculty and students. How many different subcommittees could be formed?

As in Example 1.3.6, we can break this down into two subchoices: (1) Choose 2 administrators and (2) choose 3 faculty and students. Because the 2 administrators have different roles, the order in which we choose them *is* important. So there are $_5P_2$ ways of making this choice. We have 10 people to choose from for the remaining 3 members, and their roles are no different, so order is *not important*. Thus there are $\binom{10}{3}$ ways of making this choice.

Therefore, overall there are

$$_5P_2 \times \binom{10}{3} = 20 \times 120 = 2400$$

different possible subcommittees. □

Example 1.3.8 Arranging Boys and Girls Suppose we have a group of two boys and three girls, and we want to seat all five of them in a row of chairs. In how many ways can their genders be arranged?

At first glance, this situation appears to be a simple permutation problem. We have five objects to choose from, we are choosing five of them, and the order is important. Therefore, the number of permutations is

$$_5P_5 = \frac{5!}{(5-5)!} = \frac{5!}{0!} = 5!.$$

However, this situation does not meet the first requirement of a permutation problem because the objects are not all different. The objects (the boys and girls) are all different people, but they are not all different genders. Two of them are identical, and three of them are identical. To determine the number of different arrangements, note that $2 = 2!$ different permutations of the five students are

$$B_1 B_2 G_1 G_2 G_3 \quad \text{and} \quad B_2 B_1 G_1 G_2 G_3$$

where the boys are denoted B_1 and B_2 and the girls are denoted G_1, G_2, and G_3. These $2!$ permutations are identical to each other in terms of the arrangements of genders. So for each unique arrangement of genders, there are $2!$ different permutations of the boys. By a similar argument, there are $3!$ different permutations of the girls for each unique arrangement of genders. Therefore,

$$(\text{Number of unique arrangements}) \times 2! \times 3! = 5!$$

$$\Rightarrow \text{Number of unique arrangements} = \frac{5!}{2!\, 3!} = 10.$$

So there are 10 different ways of arranging the students in a row of five chairs according to gender. □

We generalize this example in the following definition:

Definition 1.3.4 Suppose we are arranging n objects, where n_1 are identical, n_2 are identical, ..., n_r are identical ($n_1 + \cdots + n_r = n$). Then the *number of unique arrangements of the n objects* is

$$\frac{n!}{n_1!\, n_2! \cdots n_r!}.$$

Note that in the case where $n_1 = \cdots = n_r = 1$, the situation meets the requirements for a permutation problem where we are choosing all n objects and the number of unique arrangements equals the number of permutations. For this reason, the number of unique arrangements is also referred to as the *number of permutations*.

Example 1.3.9 Arranging Letters How many "words" can be formed with the letters in the word MISSISSIPPI? By *word* we mean a unique arrangement of the letters, not necessarily a word that can be found in the dictionary.

Notice that we are arranging 11 letters. There are four S's, four I's, two P's, and one M. Therefore, the number of words is

$$\frac{11!}{4!\,4!\,2!\,1!} = 34{,}650. \qquad \square$$

Example 1.3.10 Lotto Game A lotto game is won by selecting the five winning numbers from $1, 2, \ldots, 35$. Each number can be selected at most once, and the order does not matter. If the winning numbers are randomly selected, find the probability of winning.

The random experiment in this situation is to select five numbers. Because the winning numbers are randomly selected, we can use the theoretical approach to calculate the probability of selecting the winning numbers. The sample space S consists of all possible combinations of 5 numbers chosen from a set of 35, so $n(S) = \binom{35}{5} = 324{,}632$. If A is the event that the winning numbers are selected, then $n(A) = 1$, so

$$P(A) = \frac{1}{324{,}632} \approx 3.08 \times 10^{-6}.$$

Therefore, we have an extremely small probability of winning. $\qquad \square$

Example 1.3.11 Poker Hands Each card in a standard deck of 52 cards is described by its rank (also called the *face value*), ace, 2, 3, $\ldots$, 10, jack, queen, or king; and its suit, hearts, diamonds, clubs, or spades. Assume that 5 cards are randomly dealt from the deck to form a poker hand. Because the order in which the cards are dealt does not matter, there are $\binom{52}{5} = 2{,}598{,}960$ possible poker hands.

Let A be the event that a hand consisting of 5 hearts is dealt. Because there are 13 different hearts, there are 5 hearts in the hand, and the order does not matter, $n(A) = \binom{13}{5} = 1287$ so that

$$P(A) = \frac{1287}{2{,}598{,}960} \approx 4.95 \times 10^{-4}.$$

A hand is called a "two-pair" if it contains two cards of the same rank, two cards of another rank that match each other but are different from the first pair, and a fifth card of a rank different from the two pairs. Let B be the event that a two-pair is dealt. When dealing (or choosing) a two-pair hand, there are $\binom{13}{2}$ ways of choosing the two ranks of the two different pairs, $\binom{4}{2}$ ways of choosing the suits of the first pair, $\binom{4}{2}$ ways of choosing the suits of the second pair, and 44 ways of choosing the fifth card. So by the fundamental counting principle,

$$n(B) = \binom{13}{2}\binom{4}{2}\binom{4}{2} 44 = 123{,}552.$$

Thus
$$P(B) = \frac{123{,}552}{2{,}598{,}960} \approx 0.0475.$$
□

Binomial Theorem

Combinations arise in expanding binomial expressions. To illustrate this, consider the following expansion:

$$\begin{aligned}(a+b)^3 &= (a+b)(a+b)(a+b) = aaa + aab + aba + abb + baa + bab + bba + bbb \\ &= a^3 + a^2b + a^2b + ab^2 + a^2b + ab^2 + ab^2 + b^3 \\ &= a^3 + 3a^2b + 3ab^2 + b^3.\end{aligned}$$

Consider the ab^2 term. Its coefficient of 3 comes from adding the terms

$$abb, \, bab, \, \text{and} \, bba.$$

Note that the difference between these terms lies in the positions of the b's. To count the number of such terms, we can think of it as though we are selecting two positions for the b's from the three available positions. There are $\binom{3}{2}$ ways of doing this.

To generalize this calculation, let n denote the exponent. Observe that each term in the above expansion has powers of a and b whose sum is 3, which is n in this case. Also observe that the powers of b increase from $k = 0$ to $k = 3 = n$. The ab^2 term is then the $k = 2$ term. Thus, using these variables, the coefficient of the ab^2 term is $\binom{n}{k}$.

In general, we get

$$(a+b)^n = \sum_{k=0}^{n} \binom{n}{k} a^{n-k} b^k$$

which is called the *binomial theorem*.

Example 1.3.12 **Using the Binomial Theorem** Use the binomial theorem to expand $(x+y)^4$.

Solution:

$$\begin{aligned}(x+y)^4 &= \sum_{k=0}^{4} \binom{4}{k} x^{4-k} y^k \\ &= \binom{4}{0} x^{4-0} y^0 + \binom{4}{1} x^{4-1} y^1 + \binom{4}{2} x^{4-2} y^2 + \binom{4}{3} x^{4-3} y^3 + \binom{4}{4} x^{4-4} y^4 \\ &= x^4 + 4x^3 y + 6x^2 y^2 + 4xy^3 + y^4\end{aligned}$$

This result can easily be verified by direct expansion. □

> ## Software Calculations
>
> **Excel**: Factorials, permutations, and combinations are calculated with the following functions:
> - $n!$: FACT(n)
> - $_nP_r$: PERMUT(n,r)
> - $_nC_r$: COMBIN(n,r)
>
> **TI-83/84 Calculators**: The commands for factorials, permutations, and combinations are accessed in the MATH $\to$ PRB menu. To calculate $_5P_3$, for instance, enter 5 at the home menu, next press MATH $\to$ PRB $\to$ 2:nPr, then press 3, and finally press ENTER.

Exercises

1. A license plate in a certain state consists of three letters followed by three digits. Find the total number of such license plates in each of the following cases:

   a. The letters and digits can be repeated.

   b. No letter or digit can be repeated.

   c. The letters and digits can be repeated, but the first letter cannot be A and the first digit cannot be 0.

2. A password for a computer consists of four characters. Each character may be a digit, an uppercase (or capital) letter or a lowercase (small) letter, and characters may be repeated.

   a. Find the number of possible passwords.

   b. If you forget the password and make a random guess, find the probability that you guess the correct password.

   c. If you forget the last character of the password, but remember the first three, find the probability that you guess the correct password.

3. A sandwich shop offers a choice of four types of bread, five types of meat, six types of vegetables, and a choice of mayonnaise and/or mustard.

a. If a customer does not want vegetables and will choose exactly one type of bread and one type of meat (along with the option of mayonnaise and/or mustard), how many different sandwiches are possible?

b. If a customer wants exactly three types of vegetables, in how many ways can the vegetables be chosen?

c. If a customer can choose any number of types of vegetables, in how many ways can the vegetables be chosen?

d. If a sandwich consists of exactly one type of bread, one type of meat, any number of types of vegetables, and a choice of mayonnaise and/or mustard, how many different sandwiches are possible?

e. The owner of the sandwich shop is considering offering additional sauces. If a sandwich is as described in part d, with the additional option of any number of these sauces, find the minimum number of sauces the owner should offer so that there are more than 40,000 different sandwich possibilities.

4. Three boys and two girls want to sit in a row of five chairs so that the boys are all to the right of the girls. Find the number of such arrangements.

5. Fifteen students are competing for three awards in a mathematics competition. No student can receive more than one award. In how many ways can the awards be given if the awards are all different? What if they are all identical?

6. Four Germans, five Canadians, and three Italians are to be seated in a row of 12 chairs so that those of the same nationality sit together. If the order in which those of the same nationality sit is important, in how many ways can this be done? (**Hint**: There are four subchoices: (1) the order of the nationalities, (2) the order of the Germans, (3) the order of the Canadians, and (4) the order of the Italians.)

7. A coding system consists of hanging five identical blue flags and three identical green flags on a vertical pole. How many such codes are possible? Solve this problem in two ways: (a) as an arrangement problem where some items are identical and (b) as a combination problem.

8. The owner of a sporting goods store observes that there are five customers in the store. If two of them purchase shoes, one purchases other equipment, and two simply browse, find the number of orders in which these three types of customers could have entered the store (assume they entered one at a time).

9. The dean of the college of arts and sciences needs to form a committee consisting of two of the three math faculty members, three of the five theology faculty members, and one of the three chemistry faculty members. How many such committees are possible?

10. A student must answer seven of the nine questions on a statistics final exam. Find out how many choices she has in these situations:

   a. She can select any seven of the nine.

   b. She must answer the first three questions.

   c. She must answer at least two of the first three questions. (**Hint**: Consider two cases: (1) she answers all three of the first three and (2) she answers exactly two of the first three. Calculate the number of choices in each case and add them.)

11. Expand and simplify $(x+3)^9$.

12. Find the coefficient of the $x^{14}y^7$ term in the expansion of $(x+y)^{21}$.

13. Consider a set A with n elements. Find the number of possible subsets of A. (**Hint**: The choice of a subset consists of n subchoices. For each element of A, we must choose if it is in the subset or not. Use the fundamental counting principle to determine the overall number of possible choices.)

14. A professor must choose 5 or 6 students from a class of 10 students to participate in a study. In how many ways can this be done? (**Hint**: Find the number of ways to choose 5 students and the number of ways to choose 6 students, and add the results.)

15. If we randomly deal 5 cards from a standard deck of 52, find the probability of obtaining each of the following poker hands:

   a. Four of a kind: four cards of one rank and a fifth card. (**Hint**: There are two subchoices: (1) the rank of the four and (2) the fifth card.)

   b. Full house: three cards of one rank and two cards of another rank. (**Hint**: There are four subchoices: (1) the rank of the three, (2) the suits of the three, (3) the rank of the two, and (4) the suits of the two.)

   c. Three of a kind: three cards of one rank and two cards of different rank from each other and from the first three. (**Hint**: There are five subchoices: (1) the rank of the three, (2) the suit of the three, (3) the ranks of the two, (4) the suit of the fourth card, and (5) the suit of the fifth card.)

16. A bag contains 25 cubes, 15 of which are red, 6 of which are blue, and 4 of which are green. Suppose you randomly select 8 cubes. Find the probability that exactly 3 of them are red. (**Hint**: When describing the number of elements in the event, there are two subchoices: (1) the three red cubes and (2) the five other cubes.)

17. A statistics class consists of five football players, six basketball players, three tennis players, and two soccer players. The professor randomly selects a group of four students to work together on a project. Find the probability that the group consists of these different types of players:

   a. One of each type.

   b. Only basketball players.

   c. Only tennis and soccer players.

   d. Three football players and one tennis player.

18. There are five roads between cities A and B and seven roads between cities B and C. Find the number of possible routes in these situations:

   a. You drive from city A to city C through city B.

   b. Same as in part a, except you also return to city A by going through city B.

   c. Same as in part b, except you cannot use any road twice.

19. Consider making a four-digit number with the digits 1, 3, 8, and 9 where each digit must be used exactly once.

   a. How many numbers can be made?

   b. What if you used the digits 1, 3, 3, and 9?

   c. What about 1, 3, 3, and 3?

20. How many different six-letter "words" can be formed using four different consonants and two different vowels (do not count y as a vowel)? (**Hint**: There are three subchoices: (1) the choice of the four consonants, (2) the choice of the two vowels, and (3) the arrangement of the six letters.)

21. A state lottery game is played by selecting three numbers between 0 and 9, inclusive, with replacement. The winning numbers are selected in the same way. You win if your numbers match the winning numbers *in any order*.

a. Find the total number of ways the winning numbers can be selected.

b. If you select the numbers 8, 4, and 6, find the probability that you win. (**Hint**: You win if the winning numbers 8, 4, and 6 are chosen in any order. Find the number of ways this can be done.)

c. If you select the numbers 1, 1, and 5, find the probability that you win.

d. If you select the numbers 9, 9, and 9, find the probability that you win.

e. Generalize your results. If you want to maximize your probability of winning, should you select the same number more than once? Do the actual values of the numbers matter? Explain.

22. A restaurant offers four different types of hamburgers. If three customers order hamburgers, find the probability that they all order the same type. State the assumptions you make.

23. A group of n students, including Cory and Katie, are to be randomly seated in a single row of n chairs. Find the probability that Cory and Katie are seated next to each other. (**Hint**: When an arrangement in which Katie and Cory sit next to each other is chosen, there are three subchoices: (1) who sits on the left, (2) the chair in which the person on the left sits, and (3) the chairs in which the other $(n-2)$ students sit.)

24. Consider a bag with n blue cubes.

a. Suppose that the blue cubes are arranged in a row, and then r red cubes ($r < n$) are placed between the blue cubes in such a way that there is a least one blue cube between each pair of red cubes and no red cube is at either end of the row. Show that the number of ways this can be done is $\binom{n-1}{r}$.

b. Now suppose there are $r+1$ empty bins where $r < n$, and the blue cubes are placed in the bins in such a way that there is there is at least one cube in each bin and each blue cube is placed in exactly one bin. Show that the number of ways this can be done is also $\binom{n-1}{r}$.

25. Consider the experiment of randomly selecting a number X from the interval $[0, 1]$. Following Example 1.2.5, we calculate $P(X = x) = 0$ where x is any given number between 0 and 1. This means that it is impossible to choose any one given number. This does not seem to agree with our intuition. It seems as if there should be some positive, albeit small, probability of choosing a given number. To help explain this disconnect, note that in this theoretical experiment, we are allowed to choose a number with *any* number of decimal places, so that there is an infinite sample space. In reality, if we were to perform this experiment, we would unconsciously place an upper limit on the number of decimal places. This results in a finite sample space. Suppose we limit ourselves to five decimal places. Find the size of the sample space (do not forget to include the possibility of choosing exactly 0 or 1), and calculate $P(X = x)$, where x is any given number between 0 and 1.

26. Show that $\binom{n_1 + n_2}{n_1} = \binom{n_1 + n_2}{n_2}$ for any integers $n_1, n_2 > 0$.

27. Prove the following identity, called *Pascal's equation*:

$$\binom{n}{r} = \binom{n-1}{r} + \binom{n-1}{r-1}.$$

1.4 Axioms of Probability and the Addition Rule

Probability is really all about describing uncertainty. Uncertainty is a philosophical concept, and thus people throughout the ages have had philosophical differences about what probability is and how to study it mathematically. There is no universally accepted "definition" of the idea of probability. There is even greater disagreement as to the interpretation and application of the results of probability. The interpretation of probability we present in this book—as an "average in the long run"—is known as a *frequentist* view of probability, but it is not the only view.

In the 1650s, mathematicians began studying probability in the context of games of chance and betting strategies. In the late 1800s, ideas of probability were used in the study of particle physics, and mathematicians became more seriously interested in formally studying probability. They attempted to axiomatize probability to form a solid mathematical foundation for its study. Many different attempts were made, and in 1933, the Russian mathematician Andrey Kolmogorov published the axioms that are the most widely used to this day. The axioms presented below are a simplified version of Kolmogorov's.

This axiomatization of probability consists of two parts. First, we state properties, or *axioms*, that the set of all events must satisfy. The purpose of these axioms is to limit ourselves to only "nice" events (in more technical terms, we limit ourselves to only *measurable* subsets of the sample space) to which we can assign probabilities. Second, we state three relatively simple properties that seem reasonable for every measure of probability to satisfy. We then use these axioms to prove other, more complex properties of probability.

Axioms of Events Let S be the sample space of a random experiment. An *event* is a subset of S. Let E be the set of all events, and assume that E satisfies the following three properties:

1. $S \in E$
2. If $A \in E$, then $\bar{A} \in E$ (where $\bar{A}$ is the complement of A).
3. If $A_1, A_2, \ldots \in E$, then $A_1 \cup A_2 \cup \cdots \in E$.

Axioms of Probability Assume that for each event $A \in E$, a number $P(A)$ is defined in such a way that the following three properties hold:

1. $0 \leq P(A) \leq 1$
2. $P(S) = 1$
3. If $A_1, A_2, \ldots$ are events for which $A_i \cap A_j = \emptyset$ whenever $i \neq j$, then
$$P(A_1 \cup A_2 \cup \cdots \cup A_n) = P(A_1) + P(A_2) + \cdots + P(A_n)$$
for each integer $n > 0$ and
$$P(A_1 \cup A_2 \cup \cdots) = P(A_1) + P(A_2) + \cdots$$

The triple (S, E, P) is called a *probability space*. Note that in these axioms of probability, we are not defining the *idea* of probability. Nor are we specifying how to calculate probabilities. Rather, we are stating properties that probability as a *set function* satisfies. Some comments on the practical implications of these axioms of probability are warranted:

1. The first axiom simply states that probabilities must take values between 0 and 1, inclusively. We could use numbers of any size to measure probability, but by convention we use numbers between 0 and 1. This is consistent with the fact that probabilities are related to relative frequencies.

2. The second axiom simply says that in any trial of a random experiment, *something* will happen.

3. If $A_i \cap A_j = \emptyset$, we say A_i and A_j are *disjoint*. Formally, this means that the two events have no element in common. In practical terms, this means that both events

cannot occur in one trial of the random experiment. The third axiom states that the probability of their union is the sum of their probabilities.

As in set theory, the union symbol $\cup$ and the word *or* are used almost synonymously. Likewise, the intersection symbol $\cap$ and the word *and* are often interchanged. The first two of these three axioms are very simple, but the third one is a bit more complex. Example 1.4.1 helps to justify this axiom.

Example 1.4.1 Cubes in a Bag Again consider the experiment of randomly selecting a cube from a bag that contains four blue, three red, two green, and one yellow cube, as discussed in Section 1.2. If G and Y are the events that we get a green and yellow cube, respectively, then by the theoretical approach

$$P(G) = 0.2 \text{ and } P(Y) = 0.1.$$

The set

$$G \cup Y = \{G_1, G_2, Y_1\}$$

is the event that we get a green *or* a yellow cube. Likewise,

$$G \cap Y = \emptyset$$

is the event that we get a green *and* a yellow cube. By the theoretical approach,

$$P(G \cup Y) = 0.3,$$

and by axiom 3,

$$P(G \cup Y) = P(G) + P(Y) = 0.2 + 0.1 = 0.3.$$

Thus axiom 3 agrees with the theoretical approach. Because it is impossible to get a green *and* a yellow cube, it seems reasonable to let $P(G \cap Y) = 0$. □

In Section 1.2, we stated the "theoretical approach" for calculating probabilities in a random experiment where all outcomes are equally likely and the sample space is finite. We can use the axioms of probability to justify this approach. Suppose a random experiment has a finite sample space $S = \{1, 2, \ldots, k\}$ where each outcome is equally likely. By *equally likely* we mean that

$$P(\{1\}) = P(\{2\}) = \cdots = P(\{k\}).$$

Because $S = \{1\} \cup \{2\} \cup \cdots \cup \{k\}$ and these k events are mutually disjoint (meaning $\{i\} \cap \{j\} = \emptyset$ whenever $i \neq j$), we have from axioms 2 and 3

$$\begin{aligned} 1 &= P(S) \\ &= P(\{1\} \cup \{2\} \cup \cdots \cup \{k\}) \\ &= P(\{1\}) + P(\{2\}) + \cdots + P(\{k\}) \\ &= P(\{1\}) + P(\{1\}) + \cdots + P(\{1\}) \\ &= kP(\{1\}). \end{aligned}$$

Thus we have $P(\{1\}) = P(\{2\}) = \cdots = P(\{k\}) = 1/k$ as claimed in Example 1.2.1.

Now, using this result, we can prove the theoretical approach.

Theorem 1.4.1 *If all outcomes of a random experiment are equally likely, S is the finite sample space, and A is an event, then*

$$P(A) = \frac{n(A)}{n(S)}$$

where $n(A)$ is the number of elements in set A and $n(S)$ is the number of elements in set S.

Proof. Without loss of generality, assume that $S = \{1, 2, \ldots, k\}$ and $A = \{1, 2, \ldots, j\}$ where $0 < j \leq k$ are integers. By an argument similar to the above,

$$\begin{aligned} P(A) &= P(\{1\} \cup \{2\} \cup \cdots \cup \{j\}) \\ &= P(\{1\}) + P(\{2\}) + \cdots + P(\{j\}) \\ &= j \cdot \left(\frac{1}{k}\right) \\ &= \frac{n(A)}{n(S)} \end{aligned}$$

thus proving the result. $\square$

If A is an event, then the *complement* of A, denoted $\bar{A}$ or A', is the set of everything in the sample space S that is not in A. Formally, $\bar{A} = \{s \in S : s \notin A\}$. Informally, $\bar{A}$ is the event that is the opposite of A. The probabilities of an event and its complement are related by the following theorem.

Theorem 1.4.2 $P(A) + P(\bar{A}) = 1$

Proof. Note that $S = A \cup \bar{A}$ and that $A \cap \bar{A} = \varnothing$. So, using axioms 2 and 3, we have
$$1 = P(S) = P(A \cup \bar{A}) = P(A) + P(\bar{A})$$
as desired. $\square$

Corollary 1.4.1 $P(\varnothing) = 0$

Proof. Note that $\bar{\varnothing} = S$. So, by Theorem 1.4.2,
$$1 = P(\varnothing) + P(S) = P(\varnothing) + 1 \Rightarrow P(\varnothing) = 0$$
as claimed. $\square$

Examples 1.4.2 and 1.4.3 illustrate the usefulness of complements.

Example 1.4.2 **Rolling a Die** Consider the random experiment of rolling a fair six-sided die successively until the sum of all the rolls is at least 11. Find the probability that it takes more than two rolls.

To calculate this probability directly, we would need to somehow count all the ways the sum could be at least 11 in exactly three rolls, then four rolls, etc., up to 11 rolls, and then divide by the total number of possibilities. This would be a daunting task. An easier way is to use complements.

If A is the event that it takes more than two rolls, then $\bar{A}$ is the event that it takes two or fewer rolls. Since it is impossible to get a sum of at least 11 in zero rolls or one roll, we only need to consider two rolls. Looking at Table 1.4 on page 7, we see that there are three ways to get a sum of at least 11 in two rolls of a die. Thus $P(\bar{A}) = \frac{3}{36} = \frac{1}{12}$, so
$$1 = P(A) + P(\bar{A}) = P(A) + \frac{1}{12} \Rightarrow P(A) = \frac{11}{12}.$$
$\square$

Example 1.4.3 **The Birthday Problem** In a class of n students, what is the probability that at least two students share a birthday (month and day)? This famous problem is known as the *birthday problem*. We ignore leap years and assume that each of the 365 days of the year is equally likely to be a birthday.

Consider the random experiment of selecting $n \leq 365$ students and noting their birthdays. If S is the sample space, then
$$n(S) = 365^n$$
because there are 365 different possibilities for the birthday of each student. Let A be the event that at least two students share a birthday. Then $\bar{A}$ is the event that no two students share a birthday and
$$n(\bar{A}) = 365 \cdot 364 \cdots (365 - n + 1)$$
because there are 365 possibilities for the birthday of the first student, 364 possibilities for the second student, etc. Thus
$$P(A) = 1 - P(\bar{A}) = 1 - \frac{365 \cdot 364 \cdots (365 - n + 1)}{365^n} = 1 - \frac{365!}{365^n (365 - n)!}.$$
A graph of $P(A)$ for values of n between 1 and 92 is shown in Figure 1.2. Notice that for n larger than about 60, $P(A) \approx 1$, meaning that it is almost certain that at least two students will share a birthday in a class of 60 or more students. Also note that for $n = 23$, $P(A) \approx 0.5$.

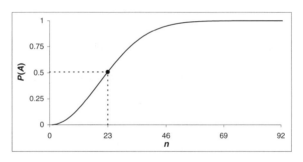

Figure 1.2

□

Theorem 1.4.3   Let A and B be events. If $A \subset B$, then $P(A) \leq P(B)$.

Proof. By elementary set theory, we have
$$B = A \cup (B \cap \bar{A}) \quad \text{and} \quad \varnothing = A \cap (B \cap \bar{A}).$$
So, by axiom 3,
$$P(B) = P(A \cup (B \cap \bar{A})) = P(A) + P(B \cap \bar{A}) \geq P(A)$$
because $P(B \cap \bar{A}) \geq 0$ by axiom 1.

□

Theorem 1.4.3 says that if event B contains the same outcomes as event A plus some more, then $P(B)$ will be at least as large as $P(A)$. This agrees with our intuitive understanding of probability. Theorem 1.4.4 is a generalization of axiom 3 that deals with calculating the probability of the union of nondisjoint events.

Theorem 1.4.4 The Addition Rule

$$P(A \cup B) = P(A) + P(B) - P(A \cap B)$$

Proof. By elementary set theory, we have

$$A \cup B = A \cup (\bar{A} \cap B) \quad \text{and} \quad \varnothing = A \cap (\bar{A} \cap B).$$

So, by axiom 3,

$$P(A \cup B) = P(A) + P(\bar{A} \cap B). \tag{1.2}$$

Also note that

$$B = (A \cap B) \cup (\bar{A} \cap B) \quad \text{and} \quad \varnothing = (A \cap B) \cap (\bar{A} \cap B).$$

Thus, by axiom 3,

$$P(B) = P(A \cap B) + P(\bar{A} \cap B) \Rightarrow P(\bar{A} \cap B) = P(B) - P(A \cap B).$$

Now, substituting this value for $P(\bar{A} \cap B)$ into Equation (1.2) yields

$$P(A \cup B) = P(A) + P(B) - P(A \cap B)$$

as desired. □

If A and B are disjoint, then $P(A \cap B) = P(\varnothing) = 0$ and Theorem 1.4.4 gives $P(A \cup B) = P(A) + P(B)$, which is exactly what axiom 3 says. Informally, Theorem 1.4.4 says that we must subtract $P(A \cap B)$ so that we don't "double-count" any outcomes. Example 1.4.4 illustrates this point.

Example 1.4.4 Exam Review A statistics professor offers her students a set of optional review problems before an exam. She records how many students completed the problems and their grades on the exam, as shown in Table 1.6.

Consider the experiment of randomly selecting one of the 80 students. If E is the event that we select a student who did not complete the problems, then

$$P(E) = \frac{38}{80} = 0.475$$

1.4 Axioms of Probability and the Addition Rule

	Grade		
	A or B	C or below	Totals
Completed	34	8	42
Did Not Complete	12	26	38
Totals	46	34	80

Table 1.6

because there are 80 elements in the sample space (the 80 students) and a total of 38 students who did not complete the problems. If F is the event that we select a student who got a C or below, then

$$P(F) = \frac{34}{80} = 0.425.$$

In words, $E \cap F$ is the event that we select a student who did not complete the problems and got a C or below. The probability of this event is

$$P(E \cap F) = \frac{26}{80} = 0.325.$$

Likewise, $E \cup F$ is the event that we select a student who did not complete the problems or got a C or below. To calculate probability of this event, we might be tempted to simply add $P(E) + P(F) = 0.475 + 0.425 = 0.900$. However, this double-counts the 26 students who did not complete the problems and got a C or below. To properly calculate the probability, we need to subtract the probability of the intersection:

$$P(E \cup F) = 0.475 + 0.425 - 0.325 = 0.575.$$

Exercises

1. In each part below, a random experiment and two events are described. Determine if it is reasonable to assume that the two events are disjoint. Briefly explain your reasoning.

  a. A student is worried about oversleeping and missing an important exam, so she sets two different alarm clocks to ring at 6:30 AM. Let A be the event that the first alarm clock fails to ring and B be the event that the second does not ring.

  b. A man asks two women for a date. Let A be the event that the first woman says no and B be the event that the second says no.

c. A professor randomly selects a student from a class. Let A be the event that the student gets a C or better and B be the event that the student gets a D or an F.

d. A father takes his four-year-old son to the park. Let A be the event that the boy plays on the swings and B be the event that he plays on the slide.

e. A flower is randomly selected from a garden. Let A be the event that the flower is red and B be the event that it is blue.

f. A truck is randomly selected from an assembly line. Let A be the event that the truck is free of defects and B be the event that the truck has a damaged taillight.

2. Consider the experiment of randomly selecting a single card from a standard deck of 52 cards. Let R be the event that the card is red and K be the event that the card is a king. Find $P(R)$, $P(K)$, $P(R \cap K)$, and $P(R \cup K)$.

3. Use the given information to find the indicated probabilities.

   a. If the probability that you win a lottery game is 1/1987, find the probability that you lose the game.

   b. A man asks a woman for a date and figures the probability that she says yes is 0.25. Find the probability that she says no.

   c. A student begins an introduction to statistics class and believes the probability that she will get a C or better is 0.8. Find the probability that she gets a D or an F.

4. A random experiment consists of selecting a positive integer n in such a way that $P(\{n\}) = \left(\frac{1}{2}\right)^n$ for each $n = 1, 2, \ldots$. Let $A = \{1, 2, 3\}$, $B = \{n : 5 \leq n \leq 10\}$, and $C = \{n : n \geq 5\}$. Find $P(A)$, $P(B)$, $P(A \cap B)$, $P(A \cup B)$, and $P(C)$. (**Hint:** Consider $P(\bar{C})$.)

5. A programmer claims to have written a computer program that can select an integer n between 1 and 10, inclusive, in such at way that $P(\{n\}) = \left(\frac{1}{3}\right)^n$ for each $n = 1, 2, \ldots, 10$. Explain why this cannot happen.

6. A horseracing bookie claims that for a certain horse, the probability that it finishes in fifth place or better is 0.35 and the probability that it finishes in the top three is 0.45. Explain why these probabilities are not consistent with the axioms of probability.

7. A weather forecaster makes the following claim regarding the chance of rain today and tomorrow:

$$P(\text{rain today}) = 0.25, \quad P(\text{rain tomorrow}) = 0.50,$$

$$P(\text{rain today and tomorrow}) = 0.10, \quad \text{and} \quad P(\text{rain today or tomorrow}) = 0.70.$$

Explain why these probabilities are not consistent with the axioms of probability. (**Hint**: Let A be the event that it rains today and B be the event that it rains tomorrow. Try plugging the given probabilities into the addition rule.)

8. In Example 1.2.5, we considered the random experiment of choosing any number from the interval $S = [1, 100]$. If B is an interval contained in interval S, then we assigned the probability of choosing a number in B as

$$P(B) = \frac{\text{length of interval } B}{\text{length of interval } S}.$$

   a. Show that this assignment of probability is consistent with probability axioms 1 and 2.

   b. Let A and B be nonoverlapping intervals both contained in interval S. Suppose we randomly choose a number from S and consider the event that the number is in interval A or B. Describe how to assign a probability to this event in a way that is consistent with axiom 3.

9. Let S be the sample space of a random experiment and A and B be two events. If $S = A \cup B$, $P(A) = 0.6$, and $P(B) = 0.8$, find $P(A \cap B)$.

10. Two students agree to meet at 7:30 PM in the library to study together. One student says the probability that he will be late is 0.35, and the other says the probability that he will be late is 0.65. If the probability that they are *both* late is 0.25, find the probability that at least one is late.

11. A very simple casino game is played by rolling two fair six-sided dice. If either die results in a number greater than 3, the game is won. A novice gambler reasons that because the probability that one die is greater than 3 is $\frac{3}{6} = 0.5$ and $0.5 + 0.5 = 1$, he is guaranteed to win. Explain what is wrong with this reasoning.

12. A marijuana drug test was given to a group of 581 people. Each person tested either positive or negative. At the same time the test was given, each person was asked how much

he or she used marijuana. The response was frequently, occasionally, or never. The results are summarized in the table below.

	Frequently	Occasionally	Never	Totals
Positive	58	148	15	221
Negative	4	98	258	360
Totals	62	246	273	581

Suppose we randomly choose one person from the group of 581. Let A be the event that the person tested negative, B be the event that the person tested positive, C be the event that the person frequently used marijuana, and D be the event that the person said he or she never used marijuana. Find $P(A)$, $P(B)$, $P(A \cap B)$, $P(A \cup B)$, $P(A \cup C)$, and $P(B \cap D)$. Are events A and B disjoint? What about B and D? Why or why not?

13. Find the probability that in a class of 5 students, at least 2 students share a birth *month*. Find the same probability for a class of 15 students.

14. The *odds against* event A occurring are the ratio $P(\bar{A})/P(A)$. For example, if $P(A) = \frac{1}{4}$,k then the odds against A are

$$\frac{1-\frac{1}{4}}{\frac{1}{4}} = \frac{\frac{3}{4}}{\frac{1}{4}} = \frac{3}{1}.$$

This is often expressed in the form of $3:1$ and read "3 to 1." Likewise, the *odds in favor* of A occurring are $P(A)/P(\bar{A})$.

   a. A roulette wheel has 38 slots numbered 0 though 36 and one numbered 00. Suppose you place a bet that the outcome is an odd number. Assuming all outcomes are equally likely, find the probability of winning and the odds against winning.

   b. If A is an event from a random experiment with a finite sample space where all outcomes are equally likely, show that the odds against A equal $n(\bar{A})/n(A)$.

   c. If the odds against A are $a:b$, find $P(A)$ in terms of a and b.

15. Consider the experiment of randomly choosing a number n from the set $S = \{1, 2, \ldots\}$. Show that not all the outcomes can have the same positive probability. That is, show that it is not possible for $P(\{n\}) = p$ for all $n = 1, 2, \ldots$ and some $p > 0$.

16. Let A and B be events of a sample space S. Prove the following relationships:

   a. $P(A \cap B) \leq P(A)$

   b. If $P(A) \leq P(B)$, then $P(\bar{B}) \leq P(\bar{A})$.

   c. $P(A \cap \bar{B}) = P(A) - P(A \cap B)$. (**Hint**: Use the fact that S is the union of the disjoint sets $(A \cap \bar{B})$, $\bar{A}$, and $(A \cap B)$ and apply axioms 2 and 3.)

   d. $P(A \cap B) \geq P(A) + P(B) - 1$. (**Hint**: Use part c.)

17. Two people agree to meet at a restaurant. Suppose the probability that person A is on time is 0.35, the probability that person B is on time is 0.65, and the probability that they are both on time is 0.10. Apply the results of Exercise 16c to find the probability that person A is on time, but person B is not.

18. Prove the following relationship: If A, B, and C are events, then

$$P(A \cup B \cup C) = P(A) + P(B) + P(C) - P(A \cap B) - P(A \cap C) - P(B \cap C) + P(A \cap B \cap C).$$

(**Hint**: By associativity, $A \cup B \cup C = A \cup (B \cup C)$. Apply the addition rule.)

1.5 Conditional Probability and the Multiplication Rule

A conditional probability involves finding the probability of an event where some information about the outcome is already known. We motivate this idea with an example.

Example 1.5.1 **Statistics Exam** The data in Table 1.7 give the results of a statistics exam and its review problems as described in Example 1.4.4.

	Grade		
	A or B	C or below	Totals
Completed	34	8	42
Did Not Complete	12	26	38
Totals	46	34	80

Table 1.7

Consider the random experiment of selecting a student at random; let E be the event that we select a student who got a C or below, and let F be the event that we select a student who completed the problems. Suppose we select a student and we find out the student completed the problems. Find the probability that the student got a C or below.

The probability in question is denoted $P(E\,|\,F)$ (read "probability of E given F"). Because we already know the student completed the problems, our sample space has been effectively reduced from 80 to 42. Out of those 42 students who completed the problems, 8 got a C or below. Thus our probability is

$$P(E\,|\,F) = \frac{8}{42} \approx 0.190.$$ □

To generalize this example, note that

$$P(E\,|\,F) = \frac{8}{42} = \frac{n(E \cap F)}{n(F)} = \frac{n(E \cap F)/80}{n(F)/80} = \frac{P(E \cap F)}{P(F)}.$$

This leads us to our definition of *conditional probability*.

Definition 1.5.1 The *conditional probability* of event A given that event B has occurred is

$$P(A\,|\,B) = \frac{P(A \cap B)}{P(B)}$$

provided that $P(B) \neq 0$.

Example 1.5.2 Two Boys in a Family If we randomly choose a family with two children and find out that at least one child is a boy, find the probability that both children are boys.

When we choose a family with two children, the sample space is

$$S = \{\text{boy boy, boy girl, girl boy, girl girl}\},$$

and it is reasonable to assume that these four outcomes are all equally likely. If A is the event that both children are boys and B is the event that at least one is a boy, then we want to find $P(A\,|\,B)$. From the sample space, we see that

$$P(B) = \frac{3}{4} \quad \text{and} \quad P(A \cap B) = \frac{1}{4}$$

so that

$$P(A\,|\,B) = \frac{P(A \cap B)}{P(B)} = \frac{\frac{1}{4}}{\frac{3}{4}} = \frac{1}{3}.$$ □

Example 1.5.3 Rolling Two Dice Consider the random experiment of rolling two fair six-sided dice and finding the sum. Find the conditional probability that the sum is at least 7 given that the first die is even. The sample space is given in Table 1.8.

First Die \ Second Die	1	2	3	4	5	6
1	2	3	4	5	6	7
2	3	4	5	6	7	8
3	4	5	6	7	8	9
4	5	6	7	8	9	10
5	6	7	8	9	10	11
6	7	8	9	10	11	12

Table 1.8

To solve this, let A be the event that the sum is at least 7 and B be the event that the first die is even. Using this notation, we want to find $P(A\,|\,B)$. Note that

$$P(B) = \frac{18}{36} = \frac{3}{6} = \frac{1}{2}$$

and that

$$P(A \cap B) = \frac{12}{36} = \frac{1}{3}.$$

Hence,

$$P(A\,|\,B) = \frac{P(A \cap B)}{P(B)} = \frac{\frac{1}{3}}{\frac{1}{2}} = \frac{2}{3}.$$

Now consider the probability $P(B\,|\,A)$. Note that

$$P(A) = \frac{21}{36} = \frac{7}{12}$$

so that

$$P(B\,|\,A) = \frac{P(B \cap A)}{P(A)} = \frac{\frac{1}{3}}{\frac{7}{12}} = \frac{4}{7}.$$

This example illustrates that $P(A\,|\,B)$ does not necessarily equal $P(B\,|\,A)$. □

The definition of a conditional probability can be rewritten as the following very useful formula called the *multiplication rule*.

Definition 1.5.2 Multiplication Rule The probability that events A and B both occur in one trial of a random experiment is

$$P(A \cap B) = P(A)P(B \,|\, A).$$

The multiplication rule says, informally, to find the probability that two events occur, multiply the probability of the first event by the probability of the second event, but make sure to take into account the occurrence of the first event when calculating the probability of the second event.

Example 1.5.4 Choosing Students Refer to the data regarding the statistics exam given in Table 1.7. Suppose we successively choose two different students at random. Find the probability that they *both* completed the problems.

Let A be the event that the first student completed the problems and B be the event that the second student completed them. We want to find $P(A \cap B) = P(A)P(B \,|\, A)$. For the first choice, there are 80 students, 42 of which completed the problems, so

$$P(A) = \frac{42}{80} = \frac{21}{40}.$$

For the second choice, there are only 79 students left, 41 of whom completed the problems, so it seems reasonable to assign

$$P(B \,|\, A) = \frac{41}{79}.$$

Thus we have

$$P(A \cap B) = P(A)P(B \,|\, A) = \frac{21}{40} \cdot \frac{41}{79} \approx 0.272. \qquad \square$$

We can extend the multiplication rule to more than two events, as illustrated in the next example.

Example 1.5.5 Dealing Cards If 5 cards are randomly dealt from a standard 52-card deck, find the probability of dealing exactly 5 hearts.

If we let H_1 be the event that the first card is a heart, H_2 be the event that the second is a heart, etc., then we want to find $P(H_1 \cap H_2 \cap H_3 \cap H_4 \cap H_5)$. To calculate this, we need to find the probability of each event, taking into account the occurrence of the previous event(s), and multiply the probabilities. Because there are 13 hearts,

$$P(H_1) = \frac{13}{52}.$$

When the second card is dealt, there are only 12 hearts in the remaining 51 cards, so

$$P(H_2 \mid H_1) = \frac{12}{51}.$$

The probability that the third card is a heart given that the first two are hearts is

$$P(H_3 \mid (H_1 \cap H_2)) = \frac{11}{50}.$$

Continuing in a similar fashion, we have

$$P(H_1 \cap H_2 \cap H_3 \cap H_4 \cap H_5) = \frac{13}{52} \cdot \frac{12}{51} \cdot \frac{11}{50} \cdot \frac{10}{49} \cdot \frac{9}{48} \approx 4.95 \times 10^{-4}.$$

This agrees with what we found in Example 1.3.11. □

Example 1.5.6 Selecting Cubes Suppose we have a bag containing three red cubes and two blue cubes, and we select two cubes at random *without replacement*. Find the probability that we get a red cube and a blue cube (not necessarily in that order).

First note that there are two ways of getting a red cube and a blue cube: we could select red and then blue or blue and then red. Let RB be the event that we get a red cube and then a blue cube, and let BR be the event that we get a blue cube and then a red cube. We want to find $P(RB \cup BR)$. Because the events RB and BR cannot occur at the same time, they are disjoint, so $P(RB \cup BR) = P(RB) + P(BR)$.

To find $P(RB)$, let R_1 be the event that we get a red cube on the first selection and B_2 be the event that we get a blue cube on the second. Then

$$P(RB) = P(R_1 \cap B_2) = P(R_1)\,P(B_2 \mid R_1) = \frac{3}{5} \cdot \frac{2}{4} = \frac{3}{10}.$$

By a similar argument,

$$P(BR) = \frac{2}{5} \cdot \frac{3}{4} = \frac{3}{10}.$$

Thus we have

$$P(RB \cup BR) = \frac{3}{10} + \frac{3}{10} = \frac{3}{5}.$$

□

Exercises

1. A field contains 390 flowers, each of which is red, blue, or yellow in color and short or tall in height, as summarized in the table below.

	Red	Blue	Yellow	Totals
Short	85	125	22	232
Tall	25	53	80	158
Totals	110	178	102	390

a. If one flower is picked at random, find the probability that (1) it is blue given that it is tall, (2) it is short given that it is blue or yellow, and (3) it is red or blue given that it is tall.

b. Now suppose two flowers are chosen without replacement. Find the probability that (1) both are blue and (2) the first is red and the second is not yellow.

c. Suppose two flowers are chosen without replacement. Find the probability that the first is short and the second is blue. (**Hint**: This can occur in two different ways: (1) The first is short and blue, and the second is blue; or (2) the first is short and not blue, and the second is blue.)

2. In a small town, 30% of the cars have a broken taillight, 15% have a broken headlight, and 10% have both a broken taillight and a broken headlight. Suppose we randomly choose one car. Find the probability that

a. the car has a broken headlight given that it has a broken taillight,

b. the car has a broken taillight given that it has a broken headlight, and

c. the car has either a broken headlight or a broken taillight.

3. A grocery store has 15 packages of shredded cheddar cheese in its dairy cooler, 3 of which are bad. Assuming that packages are sold in a random order, find the probability that

a. the first 3 packages sold are all good,

b. the first 3 packages sold are all bad,

c. the third package sold is the first bad package sold (**Hint**: This can happen only when the first two are good and the third is bad.), and

d. exactly one of the first 3 packages sold is bad.

4. One name is randomly selected from a list of motorcycle owners. Let A be the event that the person is male and B be the event that the person is named Pat. Which conditional probability do you think is higher, $P(A|B)$ or $P(B|A)$? Explain your reasoning.

5. A poker player is being dealt 5 cards from a standard deck of 52. The first two cards are kings. Find the probability that two of the next three cards are also kings.

6. A lone blackjack player is dealt a 2 and a 5 from a full standard deck of 52. He decides to "hit" and gets a 10, so that his total is 17. Find the probability that if he hits again, his total will be more than 21.

7. A bag contains 12 different pieces of candy, 5 of which you like. Suppose you reach into the bag and randomly select 2 pieces of candy. Find the probability that

a. you like both of them,

b. you like exactly one of them, and

c. you like neither of them.

8. A bag contains five red, six blue, and three green cubes. Two cubes are randomly selected without replacement.

a. Find the probability that they are the same color. (**Hint**: This can happen in three different ways. Find the probability of each way and add the results.)

b. Find the probability that they are both blue given that they are the same color.

9. A bag contains five cubes labeled 1, 2, 3, 4, and 5. The five cubes are randomly selected without replacement.

a. Find the probability that the first cube selected is labeled 1.

b. Find the probability that the third cube selected is labeled 3. (**Hint**: This can occur only if the first two cubes are not labeled 3.)

c. Find the probability that the second and third cubes selected are labeled 2 and 3, respectively.

10. Suppose a fair six-sided die is rolled five times. Find the probability of 3 coming up at least once. (**Hint**: Consider using complements.)

11. Suppose n people are randomly selected from a group of N people $(n < N)$, of which José is member.

a. If the selections are done *with replacement*, show that

$$P(\text{José is selected}) = 1 - \left(1 - \frac{1}{N}\right)^n.$$

(**Hint**: Use complements. If José is not selected, then n of the other $N - 1$ people are selected.)

b. If the selections are done *without replacement*, show that

$$P(\text{José is selected}) = \frac{n}{N}.$$

12. Let S be the sample space of a random experiment and $E \subset S$ be an event such that $P(E) > 0$. Show that the following properties are true:

a. For any event $A \subset S$, $0 \leq P(A\,|\,E) \leq 1$.

b. $P(S\,|\,E) = 1$.

c. If A and B are events such that $A \cap B = \varnothing$, then $P(A \cup B\,|\,E) = P(A\,|\,E) + P(B\,|\,E)$.

1.6 Bayes' Theorem

In this section, we present a relatively simple result involving conditional probabilities, called *Bayes' theorem*, which has many interesting applications. We motivate Bayes' theorem with an example.

Example 1.6.1 Drug Test Medical tests for the presence of drugs are not perfect. They often give *false positives*, where the test indicates the presence of the drug in a person who has not used the drug, and *false negatives*, where the test does not indicate the presence of the drug in a person who has actually used the drug.

Suppose a certain marijuana drug test gives 13.5% false positives and 2.5% false negatives and that in the general population, 0.5% of people actually use marijuana. If a randomly selected person from the population tests positive, find the conditional probability that the person actually used marijuana.

To find this, we define the following events:

$$U = \text{used marijuana}, \quad NU = \text{not used marijuana},$$
$$T^+ = \text{tested positive}, \quad \text{and} \quad T^- = \text{tested negative}.$$

Using this notation, we have

$$P(T^+ \mid NU) = 0.135, \quad P(T^- \mid NU) = 0.865,$$

$$P(T^- \mid U) = 0.025, \quad P(T^+ \mid U) = 0.975,$$

$$P(U) = 0.005, \quad \text{and} \quad P(NU) = 0.995.$$

We want to find $P(U \mid T^+)$. We might be tempted to erroneously assume that $P(U \mid T^+) = P(T^+ \mid U) = 0.975$. To make such an error is called *confusion of the inverse*. To correctly calculate the probability, we use the definition of a conditional probability:

$$P(U \mid T^+) = \frac{P(U \cap T^+)}{P(T^+)}.$$

To calculate $P(U \cap T^+)$, we use the multiplication rule:

$$P(U \cap T^+) = P(U)P(T^+ \mid U) = (0.005)(0.975) = 0.004875.$$

To calculate $P(T^+)$, we note that there are two ways a person can test positive: use and test positive or not use and test positive. In formal notation, we have

$$P(T^+) = P\left[(U \cap T^+) \cup (NU \cap T^+)\right].$$

Further note that the events $(U \cap T^+)$ and $(NU \cap T^+)$ cannot happen at the same time, so they are disjoint and we have

$$P(T^+) = P(U \cap T^+) + P(NU \cap T^+)$$
$$= P(U \cap T^+) + P(NU)\,P(T^+ \mid NU)$$
$$= 0.004875 + (0.995)(0.135)$$
$$= 0.1392.$$

Putting things together, we have

$$P(U\,|\,T^+) = \frac{0.004875}{0.1392} \approx 0.035.$$

In practical terms, this means that about 3.5% of people from the population who test positive have actually used marijuana. This very small probability means that we cannot be certain at all that a person has used marijuana in spite of a positive result.

Now suppose that instead of selecting a person from the general population, we select one from a subset of the population in which 80% use marijuana. Find $P(U\,|\,T^+)$. In this case,

$$P(U) = 0.80 \quad \text{and} \quad P(NU) = 0.20.$$

Redoing the calculations from above, we have

$$P(U\,|\,T^+) = \frac{(0.80)(0.975)}{(0.80)(0.975) + (0.20)(0.135)} \approx 0.967.$$

In this case, we can be much more confident that the person has used marijuana. These calculations illustrate why drug tests such as this are not given unless there is good reason to suspect that the person has used the drug. □

The probabilities in this example can be organized with a "tree diagram" such as that in Figure 1.3. The node on the left represents a person randomly selected from this population. The node on the top right represents the event that the person has used marijuana and tested positive. The probability of each node on the right is the product of the probabilities of the branches on the path from that node to the one on the left.

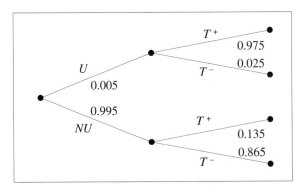

Figure 1.3

To find, for instance, $P(T^+)$, we note that there are two nodes in which a person tested positive:

$$P(T^+) = P(U \cap T^+) + P(NU \cap T^+) = (0.005)(0.975) + (0.995)(0.135) = 0.1392.$$

Thus we can calculate

$$P(U \mid T^+) = \frac{P(U \cap T^+)}{P(T^+)} = \frac{(0.005)(0.975)}{(0.005)(0.975) + (0.995)(0.135)} \approx 0.035.$$

To generalize this example, we introduce the following definition.

Definition 1.6.1 Let S be the sample space of a random experiment, and let A_1 and A_2 be two events such that

$$A_1 \cup A_2 = S \quad \text{and} \quad A_1 \cap A_2 = \emptyset.$$

The collection of sets $\{A_1, A_2\}$ is called a *partition* of S.

In Example 1.6.1, $\{U, NU\}$ forms a partition. Now let $B \subset S$ be any event; then as illustrated in Figure 1.4,

$$B = (A_1 \cap B) \cup (A_2 \cap B) \quad \text{and} \quad (A_1 \cap B) \cap (A_2 \cap B) = \emptyset.$$

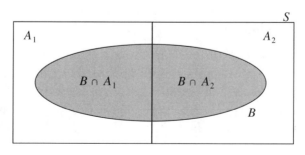

Figure 1.4

Using the definition of conditional expectation, for $i = 1, 2$,

$$P(A_i \mid B) = \frac{P(A_i \cap B)}{P(B)} = \frac{P(A_i)\,P(B \mid A_i)}{P((A_1 \cap B) \cup (A_2 \cap B))}$$

$$= \frac{P(A_i)\,P(B \mid A_i)}{P(A_1 \cap B) + P(A_2 \cap B)}$$

$$= \frac{P(A_i)\,P(B \mid A_i)}{P(A_1)\,P(B \mid A_1) + P(A_2)\,P(B \mid A_2)}.$$

This result is called *Bayes' theorem* (or *Bayes' rule*).

Theorem 1.6.1 Bayes' Theorem *If events A_1 and A_2 form a partition of the sample space S, and $B \subset S$ is any event, then for $i = 1, 2$,*

$$P(A_i \mid B) = \frac{P(A_i)\,P(B \mid A_i)}{P(A_1)\,P(B \mid A_1) + P(A_2)\,P(B \mid A_2)}.$$

A partition of S may consist of more than two events. In general, $\{A_1, \ldots, A_n\}$ is a partition of S if

$$A_i \cap A_j = \varnothing \quad \text{whenever } i \neq j$$

and $A_1 \cup \cdots \cup A_n = S$. Bayes' theorem applies to partitions such as those as illustrated in the next example.

Example 1.6.2 Missing Plane A plane has crashed in one of three equally likely, different regions. Region 1 is a wooded area, region 2 is a relatively flat farming area, and region 3 is a hilly area. Searchers choose to start looking in region 2 because it is the easiest to search and they believe that if a plane has crashed in this region, the probability that they will find it is 0.9. If they search region 2 and do not locate the plane, find the conditional probability that the plane is in region 3.

The random experiment in this scenario is the plane crashing, an outcome is the location of the crash, and the sample space consists of all possible locations. For $i = 1, 2, 3$, let A_i be the event that the plane crashed in region i. Because these regions do not overlap (we assume this is the case), then $\{A_1, A_2, A_3\}$ forms a partition of the sample space. Because the three regions are equally likely, we have

$$P(A_1) = P(A_2) = P(A_3) = \frac{1}{3}.$$

Let B be the event that the plane is *not* found in region 2. We want to know $P(A_3 \mid B)$. By Bayes' theorem,

$$P(A_3 \mid B) = \frac{P(A_3)\, P(B \mid A_3)}{P(A_1)\, P(B \mid A_1) + P(A_2)\, P(B \mid A_2) + P(A_3)\, P(B \mid A_3)}.$$

Now $P(B \mid A_3)$ is the probability that the plane is *not* found in region 2 given that it crashed in region 3. If it crashed in region 3, then it is impossible to find it in region 2. That is, if A_3 occurred, then B must occur. Thus $P(B \mid A_3) = 1$. By a similar argument, $P(B \mid A_1) = 1$.

Also, $P(B \mid A_2)$ is the probability that the plane is *not* found in region 2 given that it crashed in region 2. Because the researchers believe that if it crashed in region 2 and they search in region 2, then the probability they *do* find it is 0.9, and we have $P(B \mid A_2) = 0.1$. Thus we have

$$P(A_3 \mid B) = \frac{P(A_3)}{P(A_1) + P(A_2)\,(0.1) + P(A_3)} = \frac{\frac{1}{3}}{\frac{1}{3} + \frac{1}{3}(0.1) + \frac{1}{3}} = \frac{1}{2.1} \approx 0.476.$$

Notice that this probability is larger than $\frac{1}{3}$, but not quite $\frac{1}{2}$, as we might expect. □

In the next example, we present an alternate way of calculating conditional probabilities such as those in Example 1.6.1 by using *natural frequencies*.

Example 1.6.3 Cancer Screening Suppose that in the population of women, 0.9% have breast cancer, that 90% of women with cancer get a positive mammogram result, and that 6% of women without cancer get a false positive. Find the probability that a randomly chosen woman who gets a positive result actually has cancer.

This problem could be solved using Bayes' theorem. Instead of this approach, consider a hypothetical group of 1000 women, each of whom has a mammogram. In this group, we would expect

- $1000(0.009) = 9$ to have breast cancer and 991 to not have cancer,
- $9(0.90) \approx 8$ of those with cancer will test positive, and
- $991(0.06) \approx 59$ of those without cancer will test positive.

These expected frequencies are called *natural frequencies*. Based on these frequencies, we would expect a total of $8 + 59 = 67$ women to test positive. Out of these 67, only 8 have cancer. Therefore, the probability of having cancer given a positive result is

$$\frac{8}{67} \approx 0.12.$$

Note that as in Example 1.6.1, this probability is relatively small. It is much smaller than the probability of getting a positive result given that the woman has cancer. □

The view of probability we have presented in this text as an average in the long run is often called a *frequentist* approach to probability (because relative frequencies get closer to the theoretical probability as the number of trials increases). Another view of probability, called a *subjectivist* approach, is that it is an informal "measure of belief." This view of probability forms the basis of a field of statistics called *Bayesian statistics*.

As you might imagine, Bayes' theorem plays an important role in Bayesian statistics. In this field, the probabilities $P(A_i)$ are called the *prior probabilities* and $P(A_i | B)$ are called the *posterior probabilities*. The next example illustrates an application of the subjective probabilities and the meaning of prior and posterior probabilities.

Example 1.6.4 Colors of Cubes A bag contains a mixture of 10 red and blue cubes. We are told that 9 cubes are of one color and 1 is of the other color. We believe the probability that there are 9 red cubes and 1 blue cube in the bag is 0.9. To test this belief, we randomly select a cube from the bag and obtain a blue cube. Based on this belief, find the conditional probability that there are 9 red cubes and 1 blue cube in the bag given that we obtained a blue cube. Does this cause us to change our belief?

Define the events

$$A_1 = \text{event that there are 9 red cubes and 1 blue cube in the bag,}$$
$$A_2 = \text{event that there are 1 red cube and 9 blue cubes in the bag.}$$

Then based on our belief, we have

$$P(A_1) = 0.9 \quad \text{and} \quad P(A_2) = 0.1.$$

Observe that these are subjective probabilities because there is no average in the long run. Because the number of each color of cubes is already determined (albeit unknown), only one of these events can occur. Also note that these probabilities are given *before* the cube is selected from the bag. Hence, they are the *prior* probabilities.

Let B denote the event that we select a blue cube from the bag. Using this notation, we want to know

$$P(A_1 | B) = \frac{P(A_1) P(B | A_1)}{P(A_1) P(B | A_1) + P(A_2) P(B | A_2)}.$$

Now, $P(B \mid A_1)$ is the probability of selecting a blue cube given that there is one blue cube in the bag. This probability equals $\frac{1}{10}$. By a similar argument, $P(B \mid A_2) = \frac{9}{10}$. Thus we have

$$P(A_1 \mid B) = \frac{0.9\left(\frac{1}{10}\right)}{0.9\left(\frac{1}{10}\right) + 0.1\left(\frac{1}{10}\right)} = \frac{1}{2}.$$

This probability is calculated *after* the cube was selected, so it is the *posterior* probability. Note that this probability is much less than the originally believed value of $P(A_1) = 0.9$, so we would want to lower our believed value. Based on the calculations, a reasonable value would be $\frac{1}{2}$. □

Exercises

Directions: In each problem where Bayes' theorem is used, (1) define the events A_i and B, (2) give the values of $P(A_i)$ and $P(B \mid A_i)$, and (3) calculate the required conditional probability.

1. In each part below, a random experiment and two events are described. Determine if it is reasonable to treat the events as a partition of the sample space. Explain your reasoning.

    a. A biased six-sided die is rolled. Let A_1 be the event that the result is even and A_2 be the event that the result is odd.

    b. A boy gets injured on the playground. Let A_1 be the event that the injury is on his leg and A_2 be the event that the injury is on his head.

    c. A customer randomly chooses a package of fresh vegetables. Let A_1 be the event that the package weighs less than 16 oz and A_2 be the event that the vegetables are carrots.

    d. A satellite crash-lands on Earth. Let A_1 be the event that it lands on the ground and A_2 be the event that it lands in the water.

    e. A student purchases a new computer. Let A_1 be the event that the computer is overpriced and A_2 be the event that it is underpriced.

    f. A political pollster surveys a voter. Let A_1 be the event that the voter is a Republican and A_2 be the event that the voter is a Democrat.

    g. A customer randomly chooses a package of potato chips. Let A_1 be the event that the chips are nacho cheese flavored and A_2 be the event that the chips are barbecue flavored.

2. A toy factory has two machines that produce airplanes; call them machine A and machine B. Suppose machines A and B produce 60% and 40% of the total airplanes, respectively, and that they have a 2% and a 3% rate of defects, respectively. If an airplane is randomly selected from this factory and found to be defective, find the conditional probability that it was produced by machine B.

3. A small college campus consists of 60% females and 40% males. Of the females, 70% support increased enforcement of curfew in the dorms, while only 20% of males support it. One anonymous person was quoted in the school newspaper as being in support of it. Assuming the person was randomly selected from this campus, find the conditional probability that the person is male.

4. At a shopping mall, 80% of the customers come to shop, 15% come to eat, and 5% come to socialize (assume that each customer comes to do only one of these activities). Of those who come to shop, 95% are older than age 20 years; of those who come to eat, 65% are older than 20 years; and of those who come to socialize, 10% are older than 20. If one person is randomly selected, find the conditional probability that the person is there to shop given that the person is 18 years old.

5. Bag 1 contains 3 red and 7 green cubes, bag 2 contains 4 red and 6 green cubes, and bag 3 contains 5 red and 5 green cubes. The three bags are combined into one bag. If a red cube is randomly selected from this combination bag, find the conditional probability that the cube was originally in bag 1. (**Hint:** Let A_i be the event that the selected cube is from bag i. Because the bags are combined and they each contain the same number of cubes, $P(A_i) = \frac{1}{3}$ for each i.)

6. In Example 1.6.2, suppose the probability of locating the plane in region 2 given that it crashed in region 2 is p (that is, $P(B \mid A_2) = 1 - p$).

   a. Find $P(A_3 \mid B)$ in terms of p.

   b. Find the limit of the probability in part a as $p \to 0$. If p is "close" to 0 and we don't find the plane in region 2, what does this limit mean about the probability that the plane is in any one of the three regions? (That is, is the plane more likely to be in region 1, 2, or 3? Or is it equally likely to be in any one of the three regions?)

   c. Repeat part b, except replace the number 0 with 1.

7. Solve Example 1.6.3 using Bayes' theorem. Does this result agree with that found using natural frequencies?

1.6 Bayes' Theorem

8. Consider the first part of Example 1.6.1, where 0.5% of people actually use marijuana. Calculate $P(U \mid T^+)$ using natural frequencies. Round all the frequencies to the nearest whole number. Compare this result to the probability calculated in the example.

9. Suppose that among 50-year-old musicians, 2% smoke. Further suppose that 5% of smokers die in the next year and that 0.5% of nonsmokers die in the next year. If we randomly choose a 50-year-old musician and the person dies, use natural frequencies to find the probability that the person was a smoker. Round all the frequencies to the nearest whole number.

10. At a football game of the Bulldogs versus the Blue Jays, 30% of the spectators are wearing a Bulldogs T-shirt while 70% are wearing a Blue Jays T-shirt. A spectator is randomly selected, and a letter is randomly selected from the person's T-shirt. If the letter is a vowel, find the probability that the person is wearing a Bulldogs T-shirt (do not count y as a vowel). What if the letter is a consonant? (Assume the only words on the T-shirts are the team names.)

11. In Example 1.6.4, suppose that a red cube was selected rather than a blue one. If B is the event that we select a red cube, find $P(A_1 \mid B)$ if we believe that $P(A_1) = 0.9$. Would selecting a red cube make us more or less confident that there are 9 red cubes and 1 blue cube in the bag?

12. From past experience, a statistics professor knows that 85% of his students who do the test review problems pass the test, and of those who do not do the review problems, 90% fail. Before he starts to grade the latest test, he believes that 95% of his students did the review problems. The first randomly selected test he grades receives a failing grade. Find the conditional probability that 95% of the students did the review problems given that the first student failed the test. (**Hint:** Let A_1 be the event that the first student did the problems and A_2 be the event that the first student did not do the problems. Then we believe that $P(A_1) = 0.95$. Let B be the event that the first student fails the test. We want to find $P(A_1 \mid B)$.)

13. Consider the marijuana drug test described in Example 1.6.1, where $P(T^+ \mid NU) = 0.135$ and $P(T^- \mid U) = 0.025$. Suppose the principal of a high school with 1000 students believes there is a marijuana drug problem in the school and so decides to give *all* students this drug test. Further suppose that 50 students actually use marijuana and 950 students have never used marijuana.

  a. Approximately how many students will get a false-positive result? How many will get a false-negative result?

b. Approximately how many students will get "caught," and how many will "get away with it"? How many will be falsely accused of using marijuana?

c. Based on these calculations, would you recommend giving the test to every student? Explain why or why not.

1.7 Independent Events

A conditional probability $P(B\,|\,A)$ is the probability of event B occurring given that event A has already occurred. In many instances, the occurrence of one event does not affect the probability of the other event occurring. Such events are said to be *independent*. The next example illustrates this idea, shows what this means in terms of conditional probability, and motivates the formal definition of independence.

Example 1.7.1 **Rolling a Die and Selecting a Cube** A random experiment consists of rolling a fair six-sided die and then selecting a cube from a bag that contains one red, one blue, one green, and one yellow cube. The sample space is

$$S = \{1R, 1B, \ldots, 6Y\}$$

so that $n(S) = 24$, and we assume that each outcome is equally likely. Let A be the event that the roll of the die is a 4 and B be the event that the cube is blue. Then

$$A = \{4R, 4B, 4G, 4Y\} \quad \text{and} \quad B = \{1B, \ldots, 6B\}.$$

It seems that the outcome of the roll of the die should have no effect on the color of the cube. Thus it seems that A and B are independent. To understand what this means mathematically, note that

$$P(A) = \frac{n(A)}{n(S)} = \frac{4}{24} = \frac{1}{6}, \quad P(B) = \frac{n(B)}{n(S)} = \frac{6}{24} = \frac{1}{4},$$

and

$$A \cap B = \{4B\} \;\Rightarrow\; P(A \cap B) = \frac{1}{24}.$$

Next note that

$$P(A\,|\,B) = \frac{P(A \cap B)}{P(B)} = \frac{\frac{1}{24}}{\frac{1}{4}} = \frac{1}{6} = P(A),$$

$$P(B\mid A) = \frac{P(B\cap A)}{P(A)} = \frac{\frac{1}{24}}{\frac{1}{6}} = \frac{1}{4} = P(B),$$

and

$$P(A)P(B) = \frac{1}{6}\cdot\frac{1}{4} = \frac{1}{24} = A\cap B.$$

This last observation leads to the formal definition of independent events.

Definition 1.7.1 Two events A and B are said to be *independent* if

$$P(A\cap B) = P(A)P(B).$$

If they are not independent, they are said to be *dependent*.

To formally show that two events A and B are independent, we must verify that $P(A\cap B) = P(A)P(B)$. In many situations, it may be reasonable to *assume* that two events are independent if it appears that one has no effect on the other. The next example illustrates this.

Example 1.7.2 Redundancy Suppose a student has an 8:30 AM statistics test and is worried that her alarm clock will fail and not ring, so she decides to set two different battery-powered alarm clocks. If the probability that each clock will fail is 0.005, find the probability that at least one clock will ring.

Let F_1 and F_2 be the events that the first and second clocks fail, respectively. It is reasonable to assume that these events are independent because what happens to one clock has no effect on the other clock. If R is the event that at least one clock rings, then $\bar{R}$ is the event that both clocks fail ($F_1 \cap F_2$). Thus using properties of complementary events and the definition of independent events, we have

$$P(R) = 1 - P(\bar{R}) = 1 - P(F_1 \cap F_2) = 1 - P(F_1)P(F_2)$$
$$= 1 - (0.005)(0.005) = 0.999975.$$

So at least one clock will ring with probability 0.999975, which is higher than the probability that any one individual clock will ring.

The following theorem connects the ideas of independence and complements.

Theorem 1.7.1 If two events A and B are independent, then so are A and $\bar{B}$.

Proof. Note that $A = (A \cap B) \cup (A \cap \bar{B})$ and that $\varnothing = (A \cap B) \cap (A \cap \bar{B})$ so that

$$\begin{aligned} P(A) &= P[(A \cap B) \cup (A \cap \bar{B})] \\ &= P(A \cap B) + P(A \cap \bar{B}) \\ &= P(A)P(B) + P(A \cap \bar{B}) \end{aligned}$$

$$\begin{aligned} \Rightarrow P(A \cap \bar{B}) &= P(A) - P(A)P(B) \\ &= P(A)[1 - P(B)] \\ &= P(A)P(\bar{B}). \end{aligned}$$

Thus events A and $\bar{B}$ are independent by definition. Using similar arguments, we can show that $\bar{A}$ and B are independent as well as $\bar{A}$ and $\bar{B}$. $\square$

Example 1.7.3 Die and Cube In Example 1.7.1, we showed that A, the event that the roll of the die is a 4, and B, the event that the cube is blue, are independent. The set $\bar{A} \cap \bar{B}$ is the event that the roll of the die is not a 4 and the cube is not blue. By Theorem 1.7.1, $\bar{A}$ and $\bar{B}$ are independent so that

$$P(\bar{A} \cap \bar{B}) = P(\bar{A})P(\bar{B}) = \left(1 - \frac{1}{6}\right)\left(1 - \frac{1}{4}\right) = \frac{5}{6} \cdot \frac{3}{4} = \frac{5}{8}.$$
$\square$

Next we extend the definition of independence to three events.

Definition 1.7.2   Three events A, B, and C are said to be *pairwise independent* if

$$P(A \cap B) = P(A)P(B), \quad P(A \cap C) = P(A)P(C), \quad \text{and} \quad P(B \cap C) = P(B)P(C).$$

They are said to be *mutually independent* (or simply *independent*) if they are pairwise independent and

$$P(A \cap B \cap C) = P(A)P(B)P(C). \tag{1.3}$$

A set of n events $A_1, A_2, \ldots, A_n$ is mutually independent if any subset of these events satisfies a relationship analogous to Equation (1.3).

Example 1.7.4 Flipping a Coin Three Times Suppose we flip a coin three times and observe the sequence of heads and tails. The sample space is

$$S = \{HHH, HHT, HTH, THH, TTH, THT, HTT, TTT\}.$$

Let T_1 be the event that the first flip lands tails up, H_2 be the event that the second flip lands heads up, and T_3 be the event that the third flip lands tails up. Then

$$T_1 = \{THH, TTH, THT, TTT\}, \quad H_2 = \{HHH, HHT, THH, THT\},$$

and $T_3 = \{HHT, THT, HTT, TTT\}$

so that

$$P(T_1) = P(H_2) = P(T_3) = \frac{1}{2}.$$

These three events are pairwise independent because

$$T_1 \cap H_2 = \{THH, THT\} \;\Rightarrow\; P(T_1 \cap H_2) = \frac{2}{8} = \frac{1}{4} = P(T_1)P(H_2),$$

$$T_1 \cap T_3 = \{THT, TTT\} \;\Rightarrow\; P(T_1 \cap T_3) = \frac{2}{8} = \frac{1}{4} = P(T_1)P(T_3),$$

$$H_2 \cap T_3 = \{HHT, THT\} \;\Rightarrow\; P(H_2 \cap T_3) = \frac{2}{8} = \frac{1}{4} = P(H_2)P(T_3).$$

They are mutually independent because they are pairwise independent and

$$P(T_1 \cap H_2 \cap T_3) = P(\{THT\}) = \frac{1}{8} = P(T_1)P(H_2)P(T_3). \qquad \square$$

The random experiment in the previous example consisted of repeating an activity several times. We call each repetition a *trial*. There are three important observations we need to make regarding this example:

1. The event T_1 is characterized by what happens on only the first flip. The same is true for H_2 and T_3 in regard to the second and third flips, respectively.

2. The three events are mutually independent.

3. If we defined events H_1, T_2, and H_3 in the same fashion, point 1 is still true, and we could use a similar argument to that in the example to show that these events are also mutually independent. The same would be true for any set of three similar events.

Because of point 3, we say that the three trials are *independent*. In this case, we often abbreviate events such as $T_1 \cap H_2 \cap T_3$ by simply THT so that

$$P(THT) = P(T)P(H)P(T) = \frac{1}{2} \cdot \frac{1}{2} \cdot \frac{1}{2} = \frac{1}{8}.$$

When a random experiment consists of a sequence of trials and it appears that no trial has an effect on any other, it is often reasonable to assume the trials are independent. The next example illustrates this point.

Example 1.7.5 **Multiple-Choice Test** Suppose a student randomly guesses the answers on a multiple-choice test containing five questions. Each question has four possible answers, of which exactly one is correct. Find the probability that the first three questions are incorrect (I) and the last two are correct (C).

Because the guesses are random, the probability of a correct answer on any one question is $\frac{1}{4}$ and the probability of an incorrect answer is $\frac{3}{4}$. Assuming the five guesses are independent, we have

$$P(IIICC) = P(I)P(I)P(I)P(C)P(C) = \left(\frac{3}{4}\right)^3 \left(\frac{1}{4}\right)^2 \approx 0.0263. \qquad \square$$

We can extend this idea of independent trials to random experiments consisting of a sequence of activities that are different from one other.

Example 1.7.6 **Game of Chance** Suppose a simple game of chance consists of flipping a coin, rolling a six-sided die, and then drawing a card from a standard 52-card deck. The game is won if the player gets a tail, an even number on the die, and a diamond. Find the probability of winning.

Note that this experiment consists of a sequence of three different activities. Because no one activity appears to affect any other, it is reasonable to assume they are independent. There are three events in which the player wins, $T2D$, $T4D$, and $T6D$. Because no two of these events can occur at the same time (meaning $T2D \cap T4D = T2D \cap T6D = T4D \cap T6D = \emptyset$), they are disjoint. So we have

$$\begin{aligned} P(\text{winning}) &= P(T2D \cup T4D \cup T6D) \\ &= P(T2D) + P(T4D) + P(T6D) \\ &= \frac{1}{2} \cdot \frac{1}{6} \cdot \frac{13}{52} + \frac{1}{2} \cdot \frac{1}{6} \cdot \frac{13}{52} + \frac{1}{2} \cdot \frac{1}{6} \cdot \frac{13}{52} \\ &= \frac{1}{16}. \end{aligned} \qquad \square$$

Example 1.7.7 Opinion Poll Suppose that on a college campus of 1000 students (referred to as the *population*), 600 support the idea of building a new gym and 400 are opposed. The president of the college randomly selects five different students and talks with each about her or his opinion. Find the probability that they all oppose the idea.

Let A_1 be the event that the first student opposes the idea, A_2 be the event that the second student opposes it, etc. Because the president chooses five *different* students, the selections are made *without replacement*. Using the multiplication rule, we get

$$P(A_1 \cap A_2 \cap A_3 \cap A_4 \cap A_5) = \frac{400}{1000} \cdot \frac{399}{999} \cdot \frac{398}{998} \cdot \frac{397}{997} \cdot \frac{396}{996} \approx 0.0100867469.$$

Note that the occurrence of A_1 had to be taken into account when calculating the probabilities in this product. Thus these five events are dependent. Now suppose the president makes the selections *with replacement*. In this case, the selection of any student does not affect what happens on any other selection. Thus it is reasonable to assume these events are independent. In other words, the selections consist of a sequence of five independent trials. So we get

$$P(A_1 \cap A_2 \cap A_3 \cap A_4 \cap A_5) = \frac{400}{1000} \cdot \frac{400}{1000} \cdot \frac{400}{1000} \cdot \frac{400}{1000} \cdot \frac{400}{1000} = (0.4)^5 = 0.01024.$$

Notice that these two probabilities are, for all practical purposes, equal. This is so because the president is selecting such a small portion of the population that the fractions in the first calculation aren't that much different from the fractions in the second. However, the second calculation was a bit easier than the first. □

The previous example illustrates that it is often beneficial to treat events as being independent even when they are technically dependent. Treating them as independent simplifies the calculations and does not necessarily result in a significant error. However, there are limitations on this principle. If the population in Example 1.7.7 were only 15, for instance, then the two different calculations of probability would be significantly different. A general guideline for when selections such as these can be treated as independent is given below.

5% Guideline: If no more than 5% of the population is being selected, then the selections may be treated as independent, even though they are technically dependent.

The importance of this principle will be seen in Section 2.3.

60   CHAPTER 1   Basics of Probability

Exercises

1. In each part below, a random experiment and two events are described. Determine if it is reasonable to assume the two events are independent. Briefly explain your reasoning.

  a. A man asks two sisters for a date. Let A be the event that the first woman says no and B be the event that the second says no.

  b. A husband and wife are in two different cities. Let A be the event that the man has a sandwich for lunch and B be the event that the woman has pizza for lunch.

  c. A man and a woman eat lunch together in a shopping mall food court. Let A be the event that the man has a sandwich for lunch and B be the event that the woman has pizza for lunch.

  d. A mother takes her four-year-old son to the park. Let A be the event that the boy sits on a bench and plays video games the entire time and B be the event that he skins his knee.

  e. A man checks the weather forecasts for both New York City and Los Angeles. Let A be the event that the high temperature in New York City is above 65 degrees and B be the event that it is expected to rain in Los Angeles.

  f. A truck is randomly selected from an assembly line. Let A be the event that the truck has a dented rear fender and B be the event that the truck has a damaged taillight.

  g. A woman drives her car to a field to fly a radio-controlled airplane. Let A be the event that the car runs out of gas and B be the event that she forgets the gas for the airplane.

2. Suppose the probability that a single jet aircraft engine fails on a flight is 0.0001.

  a. Assuming independence, what is the probability that all three engines on a commercial airliner fail on the the same flight?

  b. Suppose a single mechanic changes the oil in all three engines before a flight. Is the assumption of independence reasonable in this case? Why or why not?

3. A dart player estimates that the probability of hitting the bull's-eye is 0.15.

a. Suppose he throws five different darts at the board. Assuming independent trials, what is the probability that all five miss the bull's-eye?

b. How many darts must be thrown so that the probability that at least one dart hits the bull's-eye is at least 0.75? (**Hint**: Let n be the number of darts thrown. The complement that at least one hits the bull's-eye is that they all miss.)

4. A fair four-sided die with sides labeled 1, 2, 3, and 4 is rolled, and the side that lands down is noted. Let $A = \{1, 2\}$, $B = \{1, 3\}$, and $C = \{1, 4\}$. Show that these three events are pairwise independent, but not mutually independent.

5. An introduction to statistics class contains six freshmen girls, six sophomore boys, nine sophomore girls, and n freshmen boys. Suppose one student is randomly chosen from the class and let A be the event that the student is a girl and B be the event that the student is a freshman. Find the value of n so that A and B are independent.

6. A bag contains 5 red and 10 blue cubes. A fair six-sided die is rolled, and that number of cubes is chosen from the bag without replacement. Find the probability that all of the cubes selected are blue. (**Hint**: This can happen in six disjoint ways: (1) a 1 is rolled, and the one cube selected is blue; (2) a 2 is rolled, and both cubes selected are blue; etc. Apply probability axiom 3 and the multiplication rule.)

7. In a certain small town, 25% of voters support a school bond issue. A newspaper reporter interviews five voters for an article on the issue.

    a. If $SSSNN$ denotes the event that the first three voters interviewed support the issue and the last two do not support it, find $P(SSSNN)$, treating the selections as independent.

    b. Find $P(SNNSS)$ and $P(NSSNS)$. Compare these probabilities to that found in part a.

    c. Count the number of ways in which the reporter could select three voters who support the issue and two who do not support it. (**Hint**: Think of it as choosing the three positions for the voters who support it.)

    d. Find the probability that the reporter selects three voters who support the issue and two who do not support it. (**Hint**: In part c, we found the number of ways this can happen. Generalize your results in part b to find the probabilities of these different ways.)

8. A state lottery sells a ticket that requires the player to scratch off an area that reveals four boxes. Each box contains a letter, and if at least one box contains the letter W, the player wins $25. The lottery claims that the probability of winning is 0.36. Assuming the letters in the four boxes are chosen independently, find the probability that a single box contains the letter W. (**Hint**: $0.36 = 1 - P(\text{losing})$. A player loses if all four boxes do not contain a W.)

9. When a certain factory that assembles bicycles receives a shipment of tires from a supplier, quality control engineers at the factory randomly select five different tires and inspect them. If the tires all pass inspection, the shipment is accepted. If even one fails inspection, the entire shipment is rejected (such a procedure is called *acceptance sampling*). Suppose that the factory receives a shipment of 500 tires and that 10 of the tires will not pass inspection.

   a. Let T_i be the event that the ith tire selected, $i = 1, \ldots, 5$, passes inspection. Are these events technically independent or dependent? Explain.

   b. Calculate the exact probability that the shipment is accepted.

   c. Suppose the selections are done with replacement so that the events T_i are independent. Calculate the probability that the shipment is accepted in this case.

   d. Is there much difference in your answers in parts b and c?

10. To save time and money, a laboratory that tests blood samples for the presence of marijuana combines samples from 10 different people into one sample and then tests the combined sample. If the combined sample tests negative, then it means all 10 individual samples do not contain the presence of marijuana and no further testing is required. If the combined sample tests positive, then all 10 individual samples are tested separately. If 1% of the individual samples contain the presence of marijuana, assuming independence, find the probability that a combined sample tests positive.

11. A fair six-sided die is rolled n times. Find the smallest value of n so that the probability of getting at least one 3 is greater than 0.5.

12. A spacecraft has four computer systems. The probability that each one works properly is 0.95. Assuming independence, find the probability that

    a. all four work properly,

    b. all four fail to work properly,

c. exactly one works properly (**Hint**: In how many ways can this happen?), and

d. at least two work properly (**Hint**: Use the results from parts b and c.).

13. A certain child's mood (happy or sad) is the same one minute as the previous minute with probability p and changes with probability $1 - p$. If the child is happy in minute 1, find the probability that the child is happy in minute 4 in terms of p. (**Hint**: List all the ways this can happen.)

14. Prove the following results:

a. If $P(A) \neq 0 \neq P(B)$ and A and B are disjoint, then A and B are dependent.

b. The empty set $\emptyset$ is independent with any event A.

15. If A and B are independent events, prove that the following pairs of events are also independent:

a. $\bar{A}$ and B

b. $\bar{A}$ and $\bar{B}$

16. The Monty Hall problem was first described in Example 1.2.7. Consider a generalization of this problem where there are N doors, exactly one of which contains the real prize. Without loss of generality, assume we play the game in the following way:

1. We always initially choose door 1.
2. If the door with the prize is door 1 or door N, then door 2 is opened, revealing a dummy prize. In this case, if the contestant switches, the door switched to is randomly selected between 3 and N, inclusive.
3. If the door with the prize is not door 1 or door N, then door N is opened, revealing a dummy prize. In this case, if the contestant switches, then the door switched to is randomly selected between 2 and $N - 1$, inclusive.

Show the following probabilities:

$$P(\text{winning if not switching}) = \frac{1}{N} \quad \text{and} \quad P(\text{winning if switching}) = \frac{N-1}{N(N-2)}.$$

(**Hint**: To calculate $P(\text{winning if switching})$, consider the following three disjoint events: (1) The real prize is behind door 1, (2) the real prize is behind door N, and (3) the real prize is behind one of doors 2 through $N-1$. Calculate the probability of each of these events. Then use the multiplication rule to calculate the probability of winning in each event.)

CHAPTER 2

Discrete Random Variables

Chapter Objectives

- Define random variables
- Define discrete random variables and the probability mass function
- Introduce the hypergometric, binomial, and Poisson distributions
- Define the mean, the variance, and functions of a random variable
- Introduce the moment-generating function

2.1 Introduction

Outcomes of random experiments take many different forms. They could be colors, such as when we select a cube from a bag and note its color. They could be yes-or-no answers to questions, such as when we ask people if they support a candidate running for office. Or they could be sequences of results from a set of trials, such as when we flip a coin three times.

It is often convenient to associate a number with each outcome of a random experiment. Such associations are called *random variables*. We illustrate the idea of a random variable with a few examples:

- A six-sided die is rolled once. Let the random variable

$$X = \text{the number of dots on the side that lands up.}$$

  Then X takes values between 1 and 6, inclusive.
- Fifty students are randomly selected. Let the random variable

$$Y = \text{the number of females in the sample.}$$

  Then Y takes values between 0 and 50, inclusive.
- A fish is randomly selected from a lake. Let the random variable

$$Z = \text{the length of the fish in inches.}$$

  Then Z takes values greater than 0.
- A voter is randomly selected and asked if he or she supports a candidate running for office. Let the random variable

$$W = \begin{cases} 1, & \text{if the person supports the candidate} \\ 0, & \text{otherwise.} \end{cases}$$

  Then W takes values of 0 and 1.

Each one of these random variables is a function because each one is a rule for assigning a real number to each outcome of a random experiment. This leads to the formal definition of a *random variable*.

Definition 2.1.1 Let S be the sample space of a random experiment. A *random variable* is a function

$$X : S \to \mathbb{R}.$$

The set of possible values of X is called the *range* of X (also called the *support* of X).

This definition is quite abstract, so a few practical comments are warranted:

1. A random variable is a *function*. It takes in an outcome from a random experiment and returns a real number.
2. A random variable is *not* a probability.
3. Random variables can be defined in practically any way.[1] Their values do not have to be positive or between 0 and 1 as with probabilities.
4. Random variables are typically named using capital, or uppercase, letters such as X, Y, or Z. Values of random variables are denoted with their respective lowercase letters. Thus, the expression

$$X = x$$

   means that the random variable X has the value x.

In this chapter, we study a particular type of random variable called *discrete*.

Definition 2.1.2 A random variable is *discrete* if its range is either finite or countable.

In Chapter 3 we study another type of random variable called *continuous*. Informally, a continuous random variable X has an uncountable range that typically consists of one or more intervals of real numbers and $P(X = a) = 0$ for any number a.

The random variables X, Y, and W defined earlier are all discrete while the variable Z is continuous. It is interesting to note that some random variables are neither discrete nor continuous (see Exercise 23 in Section 2.2 for an example). However, in this text, we will consider only random variables that are either discrete or continuous.

2.2 Probability Mass Functions

Because the values of a random variable are associated with outcomes of a random experiment, it seems natural to want to know the probabilities that the variable takes each value in its range. Such probabilities are described with a *probability mass function* (abbreviated pmf). We illustrate this idea with an example.

---

[1] Note that the only requirement for a function X to be a random variable is that for every $x \in \mathbb{R}$, the set $\{s \in S : X(s) \leq x\}$ is an event.

Example 2.2.1 Selecting Cubes A bag contains one red cube (R) and one blue cube (B). Consider the random experiment of selecting two cubes with replacement. The two cubes we select are called the *sample*. The same space is $S = \{RR, RB, BR, BB\}$, and we assume that each outcome is equally likely. Define the random variable

$$X = \text{the number of red cubes in the sample.}$$

Then as a function from S to the real numbers,

$$X(RR) = 2, \ X(RB) = 1, \ X(BR) = 1, \ \text{and} \ X(BB) = 0.$$

We can calculate the probabilities that X takes the values of 0, 1, and 2 in the following way:

$$P(X = 0) = P(\{BB\}) = \frac{1}{4}$$
$$P(X = 1) = P(\{RB\} \cup \{BR\}) = \frac{2}{4} = \frac{1}{2}$$
$$P(X = 2) = P(\{RR\}) = \frac{1}{4}.$$

We can also calculate the probability that X is greater than or equal to 1 in the following way:

$$\begin{aligned}P(X \geq 1) &= P(\{RB\} \cup \{BR\} \cup \{RR\}) \\ &= P(\{RB\}) + P(\{BR\}) + P(\{RR\}) \\ &= \frac{1}{4} + \frac{1}{4} + \frac{1}{4} \\ &= \frac{1}{2} + \frac{1}{4} \\ &= P(X = 1) + P(X = 2).\end{aligned}$$

□

We need to make several important observations about the previous example:

1. When we calculated $P(X = 0) = \frac{1}{4}$, $P(X = 1) = \frac{1}{2}$, and $P(X = 2) = \frac{1}{4}$, we associated each value of the random variable X with a probability (a number). In other words, we defined a function that maps values of the random variable to the real numbers. Such a function is called a *probability mass function*. Let f denote this function, and we define

$$f(x) = P(X = x).$$

In this case,

$$f(0) = P(X = 0) = \frac{1}{4}, \quad f(1) = P(X = 1) = \frac{1}{2}, \quad \text{and} \quad f(2) = P(X = 2) = \frac{1}{4}.$$

2. Note that $f(x) > 0$ for all values of x and

$$f(0) + f(1) + f(2) = \frac{1}{4} + \frac{1}{2} + \frac{1}{4} = 1.$$

3. If we let $A = \{1, 2\}$, then

$$\begin{aligned} P(X \in A) &= P(X = 1 \text{ or } X = 2) \\ &= P(X \geq 1) \\ &= P(X = 1) + P(X = 2) \\ &= f(1) + f(2) \\ &= \sum_{x \in A} f(x). \end{aligned}$$

These observations motivate our formal definition of a probability mass function for a discrete random variable.

Definition 2.2.1 Let X be a discrete random variable, and let R be the range of X. The *probability mass function* (abbreviated pmf) of X is a function $f: R \to \mathbb{R}$ that satisfies the following three properties:

1. $f(x) > 0$ for all $x \in R$
2. $\sum_{x \in R} f(x) = 1$
3. If $A \subset R$, then $P(X \in A) = \sum_{x \in A} f(x)$

We should note that some books and software call the probability mass function the *probability density function* (abbreviated pdf) or simply the *probability function*. However, in this text we reserve the term *pdf* for use with continuous random variables only (see Section 3.2 for the definition of a pdf).

For a given random variable, its pmf can be described in three different ways: table, graph, or formula. These descriptions are referred to as the *distribution* of the random variable.

Definition 2.2.2 The *distribution* of a random variable is a description of the probabilities of the values of the variable.

For the variable X in Example 2.2.1, its pmf is described in table and graph form in Figure 2.1. A graph such as that in Figure 2.1 is called a *probability histogram*.

x	0	1	2
$f(x)$	$\frac{1}{4}$	$\frac{1}{2}$	$\frac{1}{4}$

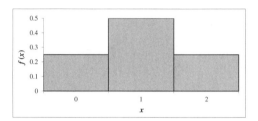

Figure 2.1

A formula for $f(x)$ is

$$f(x) = \frac{2 - |x - 1|}{4}, \quad \text{for } x = 0, 1, 2.$$

In high school algebra, certain types of functions are given special names. For instance, $f(x) = 2x^2 + 6x - 4$ is called a *quadratic* function, $g(x) = x^9$ is called a *power* function, and $h(x) = \tan 3x$ is called a *trigonometric* function. Likewise, certain types of pmfs (or distributions) are given special names. In the next example, we introduce the simplest type of distribution, the *uniform distribution*.

Example 2.2.2 **Rolling a Die** Consider the random experiment of rolling a fair six-sided die. The sample space is $S = \{1, 2, 3, 4, 5, 6\}$. Let the random variable X denote the number of dots on the side that lands up. Then

$$X(s) = s \quad \text{for each } s \in S.$$

Because the die is fair, the pmf of X is

$$f(x) = P(X = x) = \frac{1}{6} \quad \text{for each } x \in \{1, 2, 3, 4, 5, 6\}. \qquad \square$$

Notice that this pmf is a constant function. This is the essence of a uniform distribution.

Definition 2.2.3 Let X be a discrete random variable with k elements in its range R. The variable X has a *uniform distribution* (or is *uniformally distributed*) if its pmf is

$$f(x) = \frac{1}{k} \quad \text{for each } x \in R.$$

So far every random variable we have looked at has a finite set of possible values. The random variable in the next example has an infinite set of possible values.

Example 2.2.3 Drawing Cubes Consider a bag containing 10 cubes, of which 4 are blue (B), 3 are red (R), 2 are green (G), and 1 is yellow (Y). We randomly draw one cube at a time, with replacement, until we get a yellow cube. Define the random variable

$$X = \text{the number of draws until the first yellow cube.}$$

Then, for instance, $X(RBBY) = 4$ and the range of X is $\mathbb{Z}^+$ (the set of positive integers). To find the pmf, let $\bar{Y}$ denote the complement of getting a yellow cube, and note that

$$f(4) = P(X = 4) = P(\bar{Y}\bar{Y}\bar{Y}Y) = \frac{9}{10} \cdot \frac{9}{10} \cdot \frac{9}{10} \cdot \frac{1}{10} = \left(\frac{9}{10}\right)^3 \frac{1}{10},$$

and in general,

$$f(x) = \left(\frac{9}{10}\right)^{x-1} \frac{1}{10} \quad \text{for any } x \in \mathbb{Z}^+.$$

To verify that f satisfies the second property for a pmf, we use the formula for the sum of a convergent infinite geometric series:

$$\sum_{x \in R} f(x) = \sum_{x=1}^{\infty} f(x) = \sum_{x=1}^{\infty} \left(\frac{9}{10}\right)^{x-1} \frac{1}{10} = \frac{\frac{1}{10}}{1 - \frac{9}{10}} = 1 \qquad \square$$

Example 2.2.4 Rolling Dice Consider the random experiment of rolling two fair six-sided dice and calculating the sum. Let the random variable X denote this sum. Table 2.1 lists the outcomes, which are also the possible values of X.

	\multicolumn{6}{c}{Second Die}					
First Die	1	2	3	4	5	6
1	2	3	4	5	6	7
2	3	4	5	6	7	8
3	4	5	6	7	8	9
4	5	6	7	8	9	10
5	6	7	8	9	10	11
6	7	8	9	10	11	12

Table 2.1

Using the sample space and the fact that these 36 outcomes are all equally likely, we can easily calculate the values of the pmf as shown in Table 2.2.

x	2	3	4	5	6	7	8	9	10	11	12
$f(x)$	$\frac{1}{36}$	$\frac{2}{36}$	$\frac{3}{36}$	$\frac{4}{36}$	$\frac{5}{36}$	$\frac{6}{36}$	$\frac{5}{36}$	$\frac{4}{36}$	$\frac{3}{36}$	$\frac{2}{36}$	$\frac{1}{36}$

Table 2.2

Now suppose that in a simple casino game, we win if the sum of the dice is less than or equal to 4. Find the probability of winning.

In terms of the random variable X, we want to calculate

$$\begin{aligned} P(X \leq 4) &= P(X \in \{2, 3, 4\}) \\ &= f(2) + f(3) + f(4) \quad \text{by property 3 of the pmf} \\ &= \frac{1}{36} + \frac{2}{36} + \frac{3}{36} \\ &= \frac{1}{6}. \end{aligned}$$

Thus the desired probability is $\frac{1}{6}$. □

Probabilities such as $P(X \leq 4)$ calculated in the previous example occur frequently in theory and application. For this reason we define the *cumulative distribution function*.

Definition 2.2.4 The *cumulative distribution function* (abbreviated cdf) of a discrete random variable X is

$$F(b) = P(X \leq b) = \sum_{x \leq b} f(x) \quad \text{for all real numbers } b.$$

The cdf is also called the *distribution function*. Note that the domain of this function (the set of all inputs) is all real numbers, not just the values of X. For the random variable in

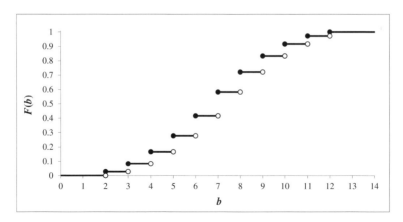

Figure 2.2

Example 2.2.4, some values of F include

$$F(-1) = 0$$
$$F(2) = f(2) = \frac{1}{36}$$
$$F(2.5) = F(2) = \frac{1}{36}$$
$$F(4) = f(2) + f(3) + f(4) = \frac{1}{36} + \frac{2}{36} + \frac{3}{36} = \frac{1}{6}$$
$$F(5) = f(2) + f(3) + f(4) + f(5) = F(4) + f(5) = \frac{1}{6} + \frac{4}{36} = \frac{5}{18}$$
$$F(20) = f(2) + \cdots + f(12) = 1$$

A graph of F over the interval $[0, 14]$ is shown in Figure 2.2. Note that $F(b) = 0$ for $b < 2$. This is so because X does not take values less than 2. Also note that $F(b) = 1$ for $b \geq 12$ because $X \leq 12$. In addition, F is an increasing function. This is true for all cdfs.

The next example illustrates how to find the pmf of a variable, given its cdf.

Example 2.2.5 Finding a PMF Suppose a discrete random variable X with range $\{1, 2, 3, 4\}$

has the cdf

$$F(b) = \begin{cases} 0, & b < 1 \\ \dfrac{1}{4}, & 1 \le b < 2 \\ \dfrac{3}{8}, & 2 \le b < 3 \\ \dfrac{7}{8}, & 3 \le b < 4 \\ 1, & 4 \le b. \end{cases}$$

To illustrate how the cdf can be used to calculate probabilities, consider the calculation of $P(2 < X \le 4)$:

$$P(2 < X \le 4) = P(X \le 4) - P(X \le 2) = F(4) - F(2) = 1 - \frac{3}{8} = \frac{5}{8}.$$

To find the pmf, we first calculate $P(X = 2)$:

$$P(X = 2) = P(X \le 2) - P(X \le 1) = F(2) - F(1) = \frac{3}{8} - \frac{1}{4} = \frac{1}{8}.$$

Using similar calculations, we see that the pmf is

$$f(1) = \frac{1}{4}, \ f(2) = \frac{1}{8}, \ f(3) = \frac{1}{2}, \text{ and } f(4) = \frac{1}{8}. \qquad \square$$

We end this section with the definition of the *mode* and the *median* of a random variable.

Definition 2.2.5 The *mode* of a discrete random variable X is a value of X at which the pmf f is maximized. The *median* of X is the smallest number m such that

$$P(X \le m) \ge 0.5 \quad \text{and} \quad P(X \ge m) \ge 0.5.$$

Informally, the mode of a random variable is a value that occurs with highest probability, or occurs "most often," while the median is the value "in the middle." From Figure 2.1, we see that the random variable in Example 2.2.1 has a mode of 1. The random variable in Example 2.2.4 has a mode of 7, and the variable in Example 2.2.2 has modes of 1, 2, 3, 4, 5, and 6. A random variable with two distinct modes is said to be *bimodal*. A variable with two or more modes is said to be *multimodal*.

The variable in Example 2.2.1 has a median of 1 because

$$P(X \leq 1) = \frac{1}{4} + \frac{1}{2} = \frac{3}{4} \quad \text{and} \quad P(X \geq 1) = \frac{1}{2} + \frac{1}{4} = \frac{3}{4},$$

and 1 is the smallest value of X for which both these probabilities are greater than 0.5.

Exercises

1. Consider the pmf described in Table 2.2. Calculate the following probabilities:

$$P(X \in \{2, 7, 11\}), \ P(X \text{ is even}), \ P(3 \leq X < 10), \text{ and } P(12 \leq X \leq 15).$$

2. The pmf for a random variable X is $f(x) = x/15$ for $x = 1, 2, 3, 4, 5$.

   a. Describe this pmf in the form of a table, and find the mode and median of X.
   b. Draw the probability histogram for this pmf.
   c. Graph the cdf for this variable over the interval $[0, 6]$.

3. In each of the following, find the value of a so that the function satisfies the first two properties of a pmf.

   a. $f(x) = x/a$ for $x = 1, 2, 3$
   b. $f(x) = a(\frac{1}{3})^{x-1}$ for $x = 1, 2, \ldots$
   c. $f(1) = 0.2$, $f(5) = 0.36$, and $f(10) = a$ for $x = 1, 5, 10$

4. Explain why properties 1 and 2 in the definition of a pmf $f(x)$ imply that $f(x) \leq 1$ for all $x \in R$.

5. Consider the function $f(x) = (\frac{1}{2})^x$ for $x = 1, 2, \ldots$. Determine if this function satisfies the first two properties of a pmf.

6. Consider the function $f(x) = a/x^2$ for $x = 1, 2, \ldots$. Show that for *some* value of a this function satisfies the first two properties of a pmf. (**Hint:** The sum $\sum_{x=1}^{\infty} (1/x^2)$ is a convergent p-series. Do not try to find the value of a.)

7. Let X be a uniformly distributed discrete random variable with range $R = \{1, 2, \ldots, k\}$. Find a formula for its cdf $F(x)$ in terms of x and k.

8. Consider the random experiment of rolling two fair six-sided dice. Define the random variable

$$X = \begin{cases} \text{the larger of the two rolls,} & \text{if they are different} \\ \text{the common value,} & \text{if they are equal.} \end{cases}$$

Describe all possible values of X in a table such as Table 2.1, and describe the distribution of X in the form of a table.

9. A random experiment consists of flipping a coin and then rolling a fair six-sided die. Let Y denote the result of the die. Define the random variable

$$X = \begin{cases} 1 + Y, & \text{if the coin lands heads up} \\ Y, & \text{if the coin lands tails up.} \end{cases}$$

Describe all possible values of X in a table such as Table 2.1, and describe the distribution of X in the form of a table.

10. A random variable X has the pmf described in the table below. Find $P(X \geq 3 \mid X \geq 1)$. (**Hint:** $P(X \geq 3 \cap X \geq 1) = P(X \geq 3)$.)

x	0	1	2	3	4
$f(x)$	0.1	0.1	0.3	0.4	0.1

11. A coin is tossed n times. Define the random variable

$$X = \text{total number of tails} - \text{total number of heads}.$$

List the numbers in the range of X, and find $P(X = n)$.

12. In a simple casino game, let X be the winnings (in dollars) from a single play of the game. Suppose that X has the pmf given in the table below, where a negative value of X means that the player loses money.

x	-5	-1	1	4
$f(x)$	0.4	0.3	0.2	0.1

In each of the following parts, an event is described in words. Describe the probability of the event in a form similar to $P(X < x)$, and calculate the probability.

a. The player wins money.

b. The player wins at least $3.

c. The player loses more than $3.

d. The player loses more than $8.

13. Consider the random variable X with pmf as described in Exercise 10. Define the random variable
$$Y = 3X - 1.$$
This means, for example, that when $X = 1$, $Y = 3(1) - 1 = 2$. Describe the pmf of Y in table form. (**Hint:** $P(Y = 2) = P(X = 1)$.)

14. Let the random variable X denote the number of students who visit a professor's office in a week. Suppose its pmf is
$$f(x) = \frac{1}{x+1} - \frac{1}{x+2}, \quad \text{for } x = 0, 1, \ldots.$$

a. Draw the probability histogram for X over the interval $[0, 6]$. What is the mode of X?

b. Graph the cdf $F(x)$ over the interval $[-1, 6]$.

c. Find $P(X \geq 5 \mid X \geq 2)$.

15. A fair coin is flipped repeatedly until the first tail appears. Let the random variable X denote the number of flips until the first tail appears. Find a formula for the pmf of X.

16. A bag contains four blue cubes, three red cubes, two green cubes, and one yellow cube. In a simple casino game, a player randomly chooses a cube and wins $2 if a green cube is selected and $3 if a yellow cube is selected. If a red cube is selected, the player loses $1, and if a blue cube is selected, the player neither wins nor loses any money. Let X denote the winnings from a single play of the game. Describe the pmf of X.

17. The cdf for a random variable X with range $R = \{-2, 0, 5, 15\}$ is given below. Describe the pmf of X in table form, and find its mode and median.

$$F(b) = \begin{cases} 0, & b < -2 \\ 0.2, & -2 \leq b < 0 \\ 0.6, & 0 \leq b < 5 \\ 0.99, & 5 \leq b < 15 \\ 1, & 15 \leq b. \end{cases}$$

18. Explain why a cdf must be an increasing function. That is, if $a \leq b$, explain why $F(a) \leq F(b)$. (**Hint**: See Theorem 1.4.3.)

19. A student is preparing for an exam and claims that the probability she will get 70% or below is 0.65, the probability she will get 80% or below is 0.95, the probability she will get 90% or below is 0.90, and the probability she will get 100% or below is 0.98. Give at least two mistakes in this reasoning. Assume the highest score on the exam is 100%.

20. A multiple-choice quiz consists of three questions. Each question has four possible choices, of which exactly one is correct. Suppose a student guesses the answers to all three questions. Let the random variable X denote the number of correct guesses. Assuming the guesses are independent, describe the distribution of X in the form of a table.

21. Suppose you have two six-sided dice with blank faces. Describe how you could label the 12 faces with numbers so that when the pair is rolled, the random variable X denoting the sum of the dice has a uniform distribution. (**Hint**: Consider putting 0 on at least one face.)

22. A random variable can have more than one mode. Can a random variable have more than one median? Explain why or why not.

23. Consider the following scenario: An auto insurance company will pay a claim for collision repair up to a maximum amount of $1000. Actual repair costs can take any value between $100 and $5000. Let the random variable X denote the actual amount paid by the insurer for a randomly selected claim. According to Definition 2.1.2 and the ensuing paragraph, is X discrete, continuous, or neither? Explain. (**Hint**: Can X take any value between 100 and 1000? Is $P(X = 1000)$ equal to 0?)

2.3 The Hypergeometric and Binomial Distributions

In this section, we introduce two different discrete distributions, the *hypergeometric* and the *binomial*. These distributions describe similar, but fundamentally different, types of random experiments. We introduce the first with an example.

Example 2.3.1 Choosing Cubes Consider a bag containing 10 red and 15 blue cubes (this collection of cubes is called the *population*). A sample of five cubes is randomly selected *without replacement*. Find the probability of obtaining exactly two red and three blue cubes.

We use the theoretical approach to find this probability. To find the size of the sample space $n(S)$, note that we are selecting 5 cubes from a group of 25 and the order does not matter, so

$$n(S) = \binom{25}{5}.$$

Now let A be the event that exactly 2 cubes are red and 3 cubes are blue. Because there are 10 red cubes and the order does not matter, there are $\binom{10}{2}$ ways of selecting 2 red cubes. Likewise, there are $\binom{15}{3}$ ways of selecting the 3 blue cubes. Thus by the fundamental counting principle,

$$n(A) = \binom{10}{2}\binom{15}{3}.$$

So

$$P(A) = \frac{n(A)}{n(S)} = \frac{\binom{10}{2}\binom{15}{3}}{\binom{25}{5}} \approx 0.385.$$

Now define the random variable

$$X = \text{the number of red cubes in the sample of 5,}$$

and note that if $X = 2$, then we have exactly 2 red cubes and 3 blue cubes. Thus

$$f(2) = P(X = 2) = P(A) = \frac{\binom{10}{2}\binom{15}{3}}{\binom{25}{5}} = \frac{\binom{10}{2}\binom{15}{5-2}}{\binom{25}{5}}.$$

In general, we have

$$f(x) = P(X = x) = \frac{\binom{10}{x}\binom{15}{5-x}}{\binom{25}{5}}, \quad \text{for } x = 0, 1, \ldots, 5.$$

We can use this formula to calculate the probability of getting any number of red cubes between 0 and 5. □

We generalize this example in the following definition.

Definition 2.3.1 Consider a random experiment that meets the following requirements:

1. From a population of N objects, n are selected *without replacement* $(n \leq N)$.

2. The N objects are of two types, call them type I and type II, where N_1 are of type I and N_2 are of type II $(N_1 + N_2 = N)$.

Define the random variable

$$X = \text{the number of type I objects in the sample of } n.$$

Then X is said to have a *hypergeometric distribution*, and its pmf is

$$f(x) = P(X = x) = \frac{\binom{N_1}{x}\binom{N_2}{n-x}}{\binom{N}{n}}, \qquad x = 0, 1, \ldots, n.$$

Example 2.3.2 Testing Transistors A manufacturer of radios receives a shipment of 200 transistors, 4 of which are defective. To determine if they will accept the shipment, they randomly select 10 transistors and test each. If there is more than one defective transistor in the sample, they reject the entire shipment. Find the probability that the shipment is *not* rejected.

Note that we are selecting 10 objects from a group of 200 without replacement, and 4 of the 200 are defective (type I) and 196 are not defective (type II). If we let

$$X = \text{the number of defective transistors in the sample of 10},$$

then X has a hypergeometric distribution with $N = 200$, $n = 10$, $N_1 = 4$, and $N_2 = 196$. So

$$\begin{aligned}
P(\text{shipment is not rejected}) &= P(X = 0 \cup X = 1) \\
&= P(X = 0) + P(X = 1) \\
&= f(0) + f(1) \\
&= \frac{\binom{4}{0}\binom{196}{10-0}}{\binom{200}{10}} + \frac{\binom{4}{1}\binom{196}{10-1}}{\binom{200}{10}} \\
&\approx 0.813 + 0.173 = 0.986.
\end{aligned}$$

Thus the probability that the shipment is not rejected is approximately 0.986. Likewise, the probability that the shipment *is* rejected is $1 - 0.986 = 0.014$. □

A graph of the pmf of the random variable in Example 2.3.2 is shown in Figure 2.3. The graph shows that the probability of getting 0 defective transistors in the sample of 10 ($f(0)$) is much higher than the probability of getting 1 or 2. The probabilities of getting 3 or 4 defective transistors ($f(3)$ and $f(4)$) are so small that we can consider them to be 0.

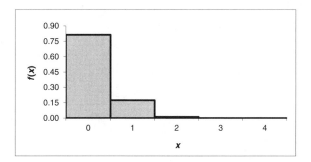

Figure 2.3

The numbers N, n, N_1, and N_2 used in the pmf are called the *parameters* of the random variable. We are often interested in analyzing what happens to the distribution when the values of these parameters change. For instance, suppose that there are more defective transistors in the shipment. In other words, suppose N_1 has increased. Graphs of the pmfs for the cases of $N_1 = 50$ and $N_2 = 125$ are shown in Figure 2.4.

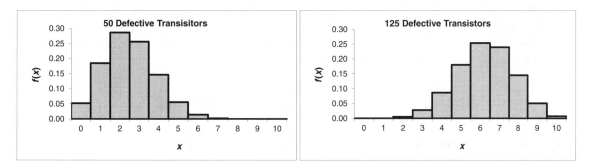

Figure 2.4

In the case of 50 defective transistors, we see that $f(0)$ and $f(1)$ are quite a bit smaller than in the original case. This indicates that the probability of not rejecting the shipment decreases. We also see that out of all the possibilities, getting exactly 2 defective transistors in the sample has the highest probability. In the case of 125 defective transistors, $f(0)$ and $f(1)$ are practically 0. This means that the shipment will almost certainly be rejected.

Binomial Distribution

To introduce the binomial distribution, we use an example similar to that used to introduce the hypergeometric distribution, but with a slight difference.

Example 2.3.3 **Choosing Cubes** Consider a bag containing 10 red and 15 blue cubes. Suppose that we randomly select five cubes *with replacement*. Find the probability that exactly two cubes are red.

Let A be the event that we get exactly two red cubes. One outcome in this event is $RRBBB$ and, because the selections of the cubes are independent,

$$P(RRBBB) = \frac{10}{25} \cdot \frac{10}{25} \cdot \frac{15}{25} \cdot \frac{15}{25} \cdot \frac{15}{25} = \left(\frac{10}{25}\right)^2 \left(\frac{15}{25}\right)^3.$$

Another outcome is $RBBBR$, and

$$P(RBBBR) = \frac{10}{25} \cdot \frac{15}{25} \cdot \frac{15}{25} \cdot \frac{15}{25} \cdot \frac{10}{25} = \left(\frac{10}{25}\right)^2 \left(\frac{15}{25}\right)^3.$$

We see that there are many different outcomes in this event and that each has the same probability. Thus to find $P(A)$, we need to count the number of outcomes and then multiply this number by the probability of the individual outcomes.

To count the number of outcomes, note that the only difference between the outcomes is the position of the red cubes (i.e., the first two cubes are red versus the first and fifth cubes). So we need to count the number of ways we can "choose" the positions of the two red cubes. Because there are five positions to choose from and the order of the selections does not matter, there are $\binom{5}{2}$ possibilities. Thus,

$$P(A) = \binom{5}{2}\left(\frac{10}{25}\right)^2 \left(\frac{15}{25}\right)^3 = 10(0.4)^2(0.6)^3 = 0.3456.$$

Now define the random variable

$$X = \text{ the number of red cubes selected.}$$

We have shown that

$$f(2) = P(X = 2) = \binom{5}{2}(0.4)^2(0.6)^3 = \binom{5}{2}(0.4)^2(0.6)^{5-2}.$$

2.3 The Hypergeometric and Binomial Distributions

In general, we have

$$f(x) = P(X = x) = \binom{5}{x}(0.4)^x(0.6)^{5-x}.$$

□

To generalize this example, first note that each selection of a cube is a random experiment and that each outcome falls into one of two categories, red or blue. Experiments such as this are called *Bernoulli experiments*.

Definition 2.3.2 A random experiment is called a *Bernoulli experiment* if each outcome is classified into exactly one of two distinct categories. These two categories are called *success* and *failure*. If a Bernoulli experiment is repeated several times in such a way that the probability of a success does not change from one iteration to the next, the experiments are said to be *independent*. A sequence of n *Bernoulli trials* is a sequence of n independent Bernoulli experiments.

A success does not necessarily mean that something good happened; nor does a failure mean that something bad happened. A *success* is a generic term meaning that we observed whatever it is we are interested in. In this example, we are interested in the number of red cubes, so a success is getting a red cube and a failure is getting a blue cube.

The random experiment in this example consists of a sequence of five Bernoulli trials, and the random variable X is the number of successes in these trials. This leads to our definition of a *binomial distribution*.

Definition 2.3.3 Consider a random experiment that meets the following requirements:

1. A sequence of n Bernoulli trials is performed.
2. The probability of a success in any one trial is p.

Define the random variable

$$X = \text{the number of successes in the } n \text{ trials.}$$

Then X is said to have a *binomial distribution*, and its pmf is

$$f(x) = P(X = x) = \binom{n}{x}p^x(1-p)^{n-x}, \quad x = 0, 1, \ldots, n.$$

The numbers n and p are the *parameters* of the distribution. The phrase "X is $b(n,p)$" means the variable X has a binomial distribution with parameters n and p.

Notice the similarities and differences between the hypergeometric and binomial distributions. Both involve a sequence of trials and counting the number of successes. But in a hypergeometric distribution, the trials are *not* independent. In the binomial distribution, they are independent.

Figure 2.5 shows the probability histograms for two different binomial distributions. In both cases $n = 20$, but they have different values of p. Note that for the larger value of p, $f(x)$ is large for large values of x and small for small values of x. This indicates that there is a high probability of many successes and a low probability of few successes. For the smaller value of p, the opposite is true.

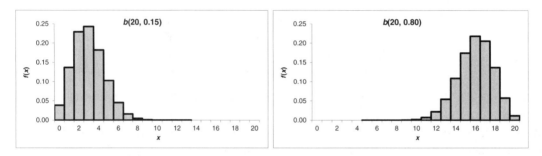

Figure 2.5

Example 2.3.4 Boys in a Family If a family with nine children is randomly chosen, find the probability of selecting a family with exactly four boys. Also find the probability of selecting a family with three or fewer boys.

Note that each child in a family is a Bernoulli experiment. We assume that the gender of one child does not affect the gender of any other child, so that each trial is independent. Because we are interested in the number of boys, a success is a boy and we further assume that the probability of a success p is 0.5. If X denotes the number of boys in the family, then X is $b(9, 0.5)$. So

$$P(X = 4) = f(4) = \binom{9}{4} 0.5^4 (1 - 0.5)^{9-4} \approx 0.246.$$

To find the probability of selecting a family with three or fewer boys in the family, we calculate

$$P(X \leq 3) = P(X = 0) + \cdots + P(X = 3)$$
$$= \sum_{x=0}^{3} \binom{9}{x} 0.5^x (1 - 0.5)^{9-x}$$
$$= \binom{9}{0} 0.5^0 (1 - 0.5)^{9-0} + \cdots + \binom{9}{3} 0.5^3 (1 - 0.5)^{9-3}$$
$$\approx 0.00195 + 0.0176 + 0.0703 + 0.164$$
$$\approx 0.254.$$

□

Example 2.3.4 illustrates a use of the cumulative distribution function (cdf) of a random variable. From the calculation we see that if X is $b(n,p)$, then its cdf is

$$F(b) = P(X \leq b) = \sum_{x=0}^{b} \binom{n}{x} p^x (1-p)^{n-x}, \quad \text{for integers } b \geq 0.$$

The next example illustrates the difference between the hypergeometric and binomial distributions and that we can use a binomial distribution to estimate a hypergeometric distribution.

Example 2.3.5 **Approximating Hypergeometric with Binomial** Suppose that in a population of 15,000 voters, 9000 support the passage of a water treatment initiative. If 50 voters are surveyed, find the probability that exactly 30 support it.

Because we are selecting 50 different voters, the selection is done *without* replacement. If X is the number of voters who support the initiative, then X is hypergeometric with $N = 15{,}000$, $N_1 = 9000$, $N_2 = 6000$, and $n = 50$. Thus,

$$P(X = 30) = \frac{\binom{9000}{30}\binom{6000}{50-30}}{\binom{15{,}000}{50}}.$$

The numbers involved with this calculation are rather large, and many calculators would not be able to handle it. Using more sophisticated software, this value comes out to be approximately 0.114750.

To simplify the calculations, note that the sample of 50 is only about 0.33% of the population. Thus using the 5% guideline, we can treat the selections as being independent, even though they are technically dependent. In this case, we have a sequence of Bernoulli trials, and so we can use the binomial distribution. It seems reasonable to assign the probability of a success in any one trial as the proportion of people who support the initiative. That is, $p = 9000/15{,}000 = 0.6$. So we can approximate X as $b(50, 0.6)$, and we have

$$P(X = 30) \approx \binom{50}{30} 0.6^{30}(1 - 0.6)^{50-30} \approx 0.114559.$$

This is indeed a reasonable approximation. □

Example 2.3.6 **Understanding Probabilities of a Random Variable** A bag has four blue cubes, three red cubes, two green cubes, and one yellow cube. Consider the random experiment of choosing three cubes without replacement, and let X denote the number of blue cubes in the sample of 3. Then X has a hypergeometric distribution with parameters $N = 10$, $N_1 = 4$, $N_2 = 6$, and $n = 3$. Its pmf and resulting probability histogram are shown in Figure 2.6.

x	0	1	2	3
$f(x)$	0.167	0.500	0.300	0.033

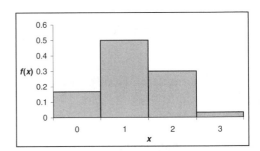

Figure 2.6

To better understand what these probabilities mean, a group of students does 240 trials of this random experiment. For each trial, they record the value of X as summarized in Figure 2.7 and then calculate the relative frequency of each value. In other words, they "observed" 240 values of the variable X and calculated the percentage of times they observed each value. These observations are examples of *data*. The graph of these relative frequencies in Figure 2.7 is called a *relative frequency histogram*.

Two important observations need to be made. First, the relative frequencies are very close in value to the theoretical probabilities. Second, the relative frequency histogram looks very similar to the probability histogram. This illustrates what the theoretical probabilities mean in a practical sense. For example, $P(X = 0) = f(0) = 0.167$ means that if we "observe" *many* values of X, then approximately 16.7% of them will equal 0. The law of large numbers says that the more values of X we observe, the closer the relative frequencies are to the theoretical probabilities. □

2.3 The Hypergeometric and Binomial Distributions

x	0	1	2	3
Freq	47	104	75	14
Rel Freq	0.196	0.433	0.313	0.058

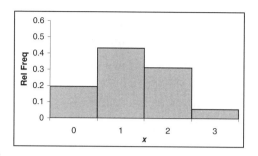

Figure 2.7

The last topic in this section is a fairly simple but important idea that forms the basis of much of what we do in Chapter 5. To illustrate the idea, suppose a gambler claims to have a fair coin, flips it 100 times, and obtains 85 tails. One would argue that if the coin were indeed fair, then getting 85 tails would be very unusual (meaning a very small probability), so his claim is probably incorrect. We generalize this idea in the *unusual-event principle*.

Unusual-Event Principle: If we make an assumption (or a claim) about a random experiment and then observe an event with a very small probability based on that assumption (called an *unusual event*), then we conclude that the assumption is most likely incorrect.

Defining an unusual event is somewhat subjective. When we are dealing with the binomial distribution, this guideline is often used:

Unusual-Event Guideline: If X is $b(n,p)$ and represents the number of successes in a random experiment, then an observed event consisting of exactly x successes is considered *unusual* if

$$P(X \leq x) \leq 0.05 \quad \text{or} \quad P(X \geq x) \leq 0.05.$$

There is nothing magical about the number 0.05, but it is often used. Other commonly used numbers include 0.10 and 0.01. We illustrate the use of these ideas in the next example.

Example 2.3.7   Selecting a Panel of Students Suppose a university student body consists of 70% females. To discuss ways of improving dormitory policies, the president selects a panel of 15 students. She claims to have randomly selected the students; however, one administrator questions this claim because there are only five females on the panel. Use the rare-event principle to test the claim of randomness.

Let X denote the number of females on the panel of 15. If the students were randomly selected as claimed, then X would be approximately $b(15, 0.7)$. The observed event consists of exactly five successes. Because this seems lower than expected, we calculate

$$P(X \leq 5) = P(X = 0) + \cdots + P(X = 5) \approx 0 + 0 + 0 + 0 + 0.0006 + 0.003 = 0.0036.$$

Because this probability is less than 0.05, the event of 5 females on the panel of 15 is considered unusual under the assumption of randomness. Therefore, we reject the claim of randomness.

Now suppose the panel had 13 females. This seems higher than expected, so we calculate

$$P(X \geq 13) = P(X = 13) + P(X = 14) + P(X = 15) \approx 0.092 + 0.031 + 0.005 = 0.128.$$

Because this probability is greater than 0.05, this event is not unusual under the assumption of randomness. Therefore, we do not reject the claim of randomness. □

Software Calculations

Minitab: To calculate values of the binomial pmf $f(x)$, begin by entering the desired values of x in a column. Then select **Calc → Probability Distributions → Binomial**. Select **Probability**, enter the number of trials n and the probability of a success p. Next to **Input column**, select the column with the desired values of x, and next to **Optional storage**, select the column where you want to store the values of $f(x)$. To calculate values of the cdf $F(x)$, follow the same steps, except select **Cumulative probability**. Hypergeometric probabilities are calculated in a similar fashion by selecting **Calc → Probability Distributions → Hypergeometric**

R: The syntax for calculating values of the binomial pmf $f(x)$ is dbinom(x, n, p). To calculate values of the binomial cdf $F(x)$, change "dbinom" to "pbinom."

Excel: The syntax for calculating values of the binomial pmf $f(x)$ is BINOM.DIST$(x, n, p, 0)$. To calculate values of the cdf $F(x)$, change the 0 to a 1. The syntax for calculating values of the hypergeometric pmf $f(x)$ is HYPGEOM.DIST$(x, n, N_1, N, 0)$.

> **TI-83/84 Calculators**: To calculate values of the binomial pmf $f(x)$, press **2nd** → **DISTR** → **A:binompdf**. The syntax for this function is binompdf(n, p, x). To calculate values of the cdf $F(x)$, follow the same steps except select **B:binomcdf**. There is no built-in command to calculate hypergeometric probabilities.

Exercises

Directions: In each problem asking for a probability, (1) define the random variable involved, (2) state what type of distribution the variable has, (3) give the variable's pmf, and (4) calculate the probability.

1. The dairy case at a local supermarket contains 25 gal of milk, of which exactly 6 gal have gone sour. If a person purchases 5 gal, find the probability that

    a. exactly 2 gal are sour,
    b. less than 3 gal are sour, and
    c. more than 3 gal are sour.

2. A forest contains a population of 500 birds, 50 of which are caught and tagged. Some time later, 10 birds are captured. Assuming the size of the population has not changed, find the probability that exactly 4 of the captured birds have been tagged.

3. A bag contains three red, eight blue, and two yellow cubes. If three cubes are randomly selected without replacement, find the probability that exactly two of them are not red.

4. A bag contains nine red and six blue cubes, and five cubes are drawn without replacement. Let X denote the number of red cubes in the sample. Find the conditional probability $P(X = 5 \,|\, X > 3)$.

5. Suppose the probability that a basketball team wins a game is 0.65. Assuming the results of each game are independent, find the probability that the team wins 8 of its next 10 games.

6. A family with seven children is randomly selected. Assuming the probability that any one child is a boy is 0.5 and that the genders of the children are independent, find the probability that there are

    a. exactly four boys,
    b. more boys than girls, and
    c. fewer than five boys.

7. Suppose that the probability that a "1-lb" package of shredded cheddar cheese weighs less than 1 lb is 0.02. Assuming that the weights of packages are independent, find the probability that at least one of eight randomly selected packages weighs less than 1 lb.

8. An airline has a policy of selling 25 tickets for a flight that can seat only 23. Records show that only 80% of ticket holders actually arrive for the flight. Assuming that the arrival of each passenger is independent of the arrivals of all other passengers, find the probability that if the airline sells 25 tickets, not enough seats will be available.

9. Suppose that the probability of suffering a headache after taking an experimental drug is 0.14. If the drug is given to 50 patients, find the probability that 10 or fewer suffer a headache.

10. If 10 points are randomly and independently selected from the rectangle $0 \leq x \leq 1$, $0 \leq y \leq 2$ on the x-y plane, find the probability that for eight or more of these points $x \leq y$. (**Hint:** Draw a picture of the rectangle. In what portion of this region is $x \leq y$?)

11. In 1941, Joe DiMaggio had a record 56-game hitting streak. A current baseball player has a batting average of 0.330, so that the probability of getting a hit at each at-bat is 0.330. Assume that the player gets four at-bats in each game and that the result of each at-bat is independent of all others.

   a. Find the probability of at least one hit in a game.
   b. Find the probability of getting at least one hit in each of 56 games.

12. A child's large bag of Halloween candy contains 20% chocolate candies and 80% hard candies. The child's father chose 10 pieces of candy to eat himself and claims that he randomly selected the pieces.

   a. Suppose six of the chosen pieces are chocolate. Find the probability of choosing this many or more chocolate candies, assuming the father's claim is true (also assume that the bag is large enough that the selections can be treated as independent).
   b. According to the unusual-event principle, does the father's claim appear to be valid? What does this say about the father's preferences in candy?
   c. Now suppose one of the pieces is chocolate. Find the probability of choosing this many or fewer chocolate candies, assuming the father's claim is true. Does the father's claim appear to be valid in this case?

13. The Nutty Goodness company claims that 15% of the nuts in its deluxe mix are peanuts. A random sample of 50 nuts from the deluxe mix contains 2 peanuts.

2.3 The Hypergeometric and Binomial Distributions

a. Is 2 peanuts in a sample of 50 more or less than we would expect if the claim were true? Explain.

b. Calculate the probability of getting 2 peanuts or fewer in a sample of 50, assuming the claim is true.

c. Repeat part b, but calculate the probability of getting 2 peanuts or more.

d. What does the unusual-event principle say about the validity of the claim? Which probability, that from part b or part c, leads you to this conclusion?

e. Generalize your observations. Suppose X denotes the number of successes and x successes are observed. If x is less than expected, do we need to calculate $P(X \leq x)$ or $P(X \geq x)$ to determine the validity of the claim? What if x is more than expected?

14. Suppose that 35% of the customers who enter a sporting goods store purchase shoes, 25% purchase other sporting equipment, and 40% are simply browsing. If five customers enter the store in a given hour, find the probability that exactly two buy shoes and one buys other sporting equipment. (**Hint**: Find the probability that the first two buy shoes, the third buys other equipment, and the last two browse. Find the total number of ways these customers could be arranged.)

15. In a factory that produces lead weights, the probability that any weight is too heavy or too light is $p_1 = 0.06$ or $p_2 = 0.08$, respectively, and the probability that a weight is within an acceptable range is $p_3 = 0.86$. Find the probability that in a batch of 50 weights, exactly 3 are too light, 5 are too heavy, and 45 are within an acceptable range. (**Hint**: Find the probability that the first 3 are too light, the next 2 are too heavy, and the last 45 are within an acceptable range. Then count the number of ways these weights can be arranged.)

16. Consider an infinite sequence of Bernoulli trials where the probability of a success in any one trial is p. Let r be a fixed positive integer, and define the random variable X to be the trial number on which the rth success occurs (i.e., if $r = 7$ and the 7th success occurs on the 15th trial, then $X = 15$). Explain why the pmf of X is

$$f(n) = P(X = n) = \binom{n-1}{r-1} p^r (1-p)^{n-r} \quad \text{for } n = r, r+1, \ldots .$$

A random variable with this pmf is said to have a *negative binomial distribution*. (**Hint**: $X = n$ means that the rth success occurs on the nth trial. This means there were $r - 1$ successes on the first $n - 1$ trials and a success on the nth trial. In how many ways can this occur?)

17. A bag contains three red and seven blue cubes. Consider the random experiment of repeatedly selecting a cube at random with replacement until a red cube is obtained.

a. Find the probability that the first red cube is obtained on the third selection. (**Hint**: The only event in which this occurs is *BBR*.)

b. Find the probability that the first red cube is obtained on the 15th selection.

c. Generalize your calculations in parts a and b. Consider a random experiment consisting of a sequence of Bernoulli trials where the probability of a success in any one trial is p. Let the random variable X denote the trial number on which the first success appears. Then X is discrete with range $\{1, 2, \dots\}$. Find a formula for the pmf of X, $f(x) = P(X = x)$. A random variable with this pmf is said to have a *geometric distribution*.

2.4 The Poisson Distribution

To illustrate the basic idea behind the Poisson distribution, consider a secretary who receives an "average" of two calls per hour over an 8-hr day. We want to find the probability of receiving exactly three calls in a given hour.

To find this probability, we divide 1 hr of time into n subintervals, each of length $1/n$, and make two assumptions about these subintervals:

1. The calls randomly arrive over the span of the 8-hr day so that the number of calls received in each subinterval is independent.

2. For n "large enough," the probability of receiving two or more calls in a subinterval is 0 and the probability of receiving exactly one call in a subinterval is $2(1/n) = 2/n$.

Based on these assumptions, we see that each subinterval of time is a Bernoulli experiment where a success means receiving a call and $p = 2/n$. The overall interval of 1 hr is a sequence of n Bernoulli trials. Thus if we let

$$X = \text{the number of calls received in 1 hr},$$

then X is (approximately) the number of successes in the n Bernoulli trials. In other words, X is approximately

$$b\left(n, \frac{2}{n}\right).$$

Thus

$$P(X = 3) \approx \binom{n}{3}\left(\frac{2}{n}\right)^3 \left(1 - \frac{2}{n}\right)^{n-3}$$

$$= \frac{n!}{3!(n-3)!} \left(\frac{2}{n}\right)^3 \left(1 - \frac{2}{n}\right)^{n-3}$$

$$= \frac{n!}{n^3(n-3)!} \left(\frac{2^3}{3!}\right) \left(1 - \frac{2}{n}\right)^n \left(1 - \frac{2}{n}\right)^{-3}$$

Now, for n large enough,

$$\frac{n!}{n^3(n-3)!} = \frac{n(n-1)(n-2)(n-3)!}{n^3(n-3)!} = \frac{n(n-1)(n-2)}{n^3} \approx 1,$$

and

$$\left(1 - \tfrac{2}{n}\right)^{-3} \approx (1-0)^{-3} = 1.$$

Using the fact that $\lim_{n \to \infty} \left(1 + \frac{x}{n}\right)^n = e^x$, we also have

$$\left(1 - \frac{2}{n}\right)^n \approx e^{-2},$$

for n large enough. Therefore,

$$P(X = 3) \approx \left(\frac{2^3}{3!}\right) e^{-2} \approx 0.180.$$

It seems reasonable to assign this value to the probability of receiving exactly three calls in an hour. These calculations motivate the definition of the *Poisson distribution*.

Definition 2.4.1 A random variable X has a *Poisson distribution* if its pmf is

$$f(x) = P(X = x) = \frac{\lambda^x}{x!} e^{-\lambda}, \quad x = 0, 1, \ldots$$

where $\lambda > 0$ is a constant.

To verify that this function satisfies the first two properties of a pmf, we use the Taylor series expansion of the exponential function, $e^y = \sum_{n=0}^{\infty} (y^n/n!)$:

$$\sum_{x=0}^{\infty} f(x) = \sum_{x=0}^{\infty} \frac{\lambda^x}{x!} e^{-\lambda} = e^{-\lambda} \sum_{x=0}^{\infty} \frac{\lambda^x}{x!} = e^{-\lambda} \cdot e^{\lambda} = 1.$$

The Poisson distribution is often used to describe the number of "occurrences" over a randomly selected "interval." The interval could be an interval of time, or it could be a two-dimensional area or a three-dimensional volume. The parameter λ is the "average" number of occurrences per interval. To use the Poisson distribution, assumptions such as those three we made regarding the number of received calls need to hold. Particularly, we need to assume that the occurrences are randomly scattered across the intervals so that the second assumption holds.

Example 2.4.1 Dandelions in a Lawn Suppose a 1000 ft² lawn contains 3000 dandelions. Find the probability that a randomly chosen 1 ft² section of lawn contains exactly 5 dandelions.

Let an occurrence be a dandelion and an interval be a 1 ft² section of lawn. Assuming the dandelions are randomly scattered across the lawn, it seems reasonable to assume that the random variable X denoting the number of dandelions in a 1 ft² section of lawn has a Poisson distribution. In this case,

$$\lambda = \frac{3000 \text{ dandelions}}{1000 \text{ ft}^2} = 3 \text{ dandelions/ft}^2$$

so that

$$P(X = 5) = \frac{3^5}{5!} e^{-3} \approx 0.101.$$

Now suppose we want to know the probability that a 1 yd² section of lawn contains exactly 5 dandelions. Because 1 yd² = 9 ft², we might be tempted to simply say the probability is $9(0.101) = 0.909$, or $\frac{1}{9}(0.101) = 0.0112$, but these are both incorrect. If we let Y denote the number of dandelions in a 1 yd² section of lawn, then Y is Poisson with

$$\lambda = \frac{3 \text{ dandelions}}{1 \text{ ft}^2} \cdot \frac{9 \text{ ft}^2}{1 \text{ yd}^2} = 27 \text{ dandelions/yd}^2$$

so that

$$P(Y = 5) = \frac{27^5}{5!} e^{-27} \approx 2.247 \times 10^{-7}.$$

Note that this probability is so small because 5 dandelions is much less than the average of 27 dandelions/yd². □

Example 2.4.2 Flaws in a Cable Suppose a manufacturer of cable observes an average of one flaw per 150 ft. Find the probability that a 600-ft section of cable has three or fewer flaws.

Let an occurrence be a flaw and an interval be a 600-ft section of cable. If X denotes the number of flaws per 600-ft section of cable, and assuming a Poisson distribution applies, then X has the parameter

$$\lambda = \frac{1 \text{ flaw}}{150 \text{ ft}} (600 \text{ ft}) = 4 \text{ flaws per section}.$$

Thus

$$P(X \leq 3) = \sum_{x=0}^{3} \frac{4^x}{x!} e^{-4}$$
$$\approx 0.0183 + 0.0733 + 0.1465 + 0.1954$$
$$= 0.4335.$$

Figure 2.8 shows the probability histogram of X for different values of λ. Notice that for larger values of λ (meaning more flaws per section) the probability that X has larger values increases. □

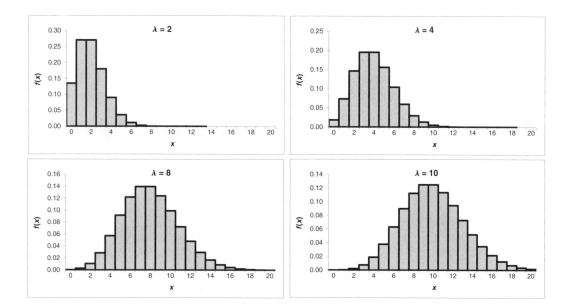

Figure 2.8

Example 2.4.3 Chocolate Chips When chocolate chip cookies are made, the chips are poured into the dough and mixed up, randomly scattered throughout the mixture. Let the random variable X denote the number of chocolate chips in a finished cookie. It seems reasonable to assume that X is described by a Poisson distribution. To test this assumption, a student counts the number of chips in 20 different cookies as recorded in Table 2.3.

| 2 | 3 | 0 | 1 | 4 | 1 | 2 | 0 | 5 | 2 | 2 | 3 | 1 | 4 | 3 | 1 | 1 | 2 | 3 | 0 |

Table 2.3

The average number of chocolate chips in a cookie is

$$\frac{2+3+\cdots+0}{20} = 2.$$

Thus if X is described by a Poisson distribution, then it should have parameter $\lambda = 2$ (at least approximately). Table 2.4 shows the number of chocolate chips x, the frequency of each number, the relative frequency, and the value of $f(x)$ calculated using a Poisson distribution with $\lambda = 2$.

x	0	1	2	3	4	5
Freq	3	5	5	4	2	1
Rel Freq	0.15	0.25	0.25	0.20	0.10	0.05
$f(x)$	0.135	0.271	0.271	0.180	0.090	0.036

Table 2.4

Observe that each relative frequency is very close to the corresponding value of $f(x)$. This indicates that it is indeed reasonable to assume that X has a Poisson distribution. □

From the introductory example, we see that if X has a Poisson distribution with parameter λ, then X is approximately

$$b\left(n, \frac{\lambda}{n}\right)$$

for n large enough and λ of moderate size. We can turn this approximation around and say that if Y is $b(n, p)$, then Y is approximately Poisson with parameter $\lambda = np$ as long as n is large enough and np is of moderate size. A rule of thumb for when these conditions are met is that

$$n \geq 100 \quad \text{and} \quad np \leq 10.$$

The next example illustrates the use of a Poisson distribution to approximate binomial probabilities.

Example 2.4.4 Approximating the Binomial with the Poisson Suppose that a manufacturer of watch batteries claims that 0.02% are defective. Assuming this is correct, find the probability that in a shipment of 150 batteries, exactly 4 are defective.

If Y is the number of defective batteries in the shipment, then Y is $b(150, 0.02)$. We can approximate $P(Y = 4)$ using a Poisson distribution with $\lambda = 150(0.02) = 3$ so that

$$P(Y = 4) \approx \frac{3^4}{4!} e^{-3} \approx 0.168.$$

□

Software Calculations

Minitab: To calculate values of the Poisson pmf $f(x)$, begin by entering the desired values of x in a column. Then select **Calc** → **Probability Distributions** → **Poisson**. Select **Probability** and enter the mean λ. Next to **Input column**: select the column with the desired values of x; next to **Optional storage**: select the column where you want to store the values of $f(x)$. To calculate values of the cdf $F(x)$, follow the same steps, except select **Cumulative probability**.

R: The syntax for calculating values of the Poisson pmf $f(x)$ is dpois(x, λ). To calculate values of the Poisson cdf $F(x)$, change "dpois" to "ppois."

Excel: The syntax for calculating values of the Poisson pmf $f(x)$ is POISSON($x, \lambda, 0$). To calculate values of the cdf $F(x)$, the syntax is POISSON($x, \lambda, 1$).

TI-83/84 Calculators: To calculate values of the Poisson pmf $f(x)$, press **2nd** → **DISTR** → **C:poissonpdf**. The syntax for this function is poissonpdf(λ, x). To calculate values of the cdf $F(x)$, follow the same steps except select **D:poissoncdf**.

Exercises

Directions: In each problem asking for a probability, (1) define the random variable involved, (2) state what type of distribution the variable has, (3) give the value(s) of the variable's parameter(s), and (4) calculate the probability.

1. In the scenario of the secretary receiving calls, we assumed the calls randomly arrive over the span of the 8-hr day. In general, to apply the Poisson distribution, we need to assume that all the occurrences are randomly scattered over the large interval from which the smaller intervals are selected. In each scenario below, determine if this assumption is reasonable. Explain your reasoning.

   a. A movie is scheduled to begin at 7:00 PM. Let X denote the number of people who arrive at the ticket counter in a 1-min interval selected between the times of 6:30 PM and 7:15 PM.
   b. Let X denote the number of calories consumed by a person in a 30-min interval selected between the times of 6:00 AM and 10:00 PM.
   c. A homeowner fertilizes her yard. Let X denote the number of fertilizer granules in a 1 ft^2 section of lawn.
   d. A student types a 25-page paper. Let X denote the number of typographical errors in a line.
   e. Let X denote the number of people who live in a randomly selected 1-acre section of land in the United States.

2. Customers arrive at a coffee shop at an average of 25 every hour. Assuming the Poisson distribution applies, find the probability that exactly 20 arrive in a randomly selected hour.

3. In a textile factory, 100-ft sections of cloth contain an average of five flaws. Assuming the Poisson distribution applies, find the probability that a randomly selected 100-ft section contains no flaws. Would it be considered unusual to receive a 100-ft section of cloth with no flaws from this factory? Explain.

4. A hospital records an average of 248 births per year. Find

   a. the average number of births per day,
   b. the probability of exactly 1 birth on a randomly selected day,
   c. the probability of fewer than 3 births on a randomly selected day, and
   d. the probability of 3 or more births on a randomly selected day.

5. Customers arrive at a supermarket checkout stand at an average of three per hour. Assuming the Poisson distribution applies, calculate the probabilities of the following events:

   a. Exactly two people arrive in a given hour.
   b. Exactly two people arrive between the hours of 9:00 and 11:00 AM.
   c. A total of exactly two people arrive between the hours of 9:00 and 10:00 AM or between 1:00 and 2:00 PM. (**Hint**: Use the fact that this could occur when two people arrive in one of the intervals and none in the other, or when one person arrives in each interval.) part b.
   d. Compare your answers to parts b and c. Does this make sense? Explain.

6. A random variable X has a Poisson distribution with the property that $P(X = 7) = P(X = 8)$. Find the values of λ and $P(X = 6)$.

7. A secretary receives an average of six calls per hour. If the secretary takes a 5-min break, find the probability that at least one call is received during that time.

8. A student counts the number of chocolate chips in cookies of two different brands, call them brands A and B, and records the data shown in the table below (data collected by Emily Brandt, 2010). Use an approach like that in Example 2.4.3 to determine if the number of chocolate chips in either brand is reasonably described by a Poisson distribution. Explain your reasoning.

Brand	Number of Chocolate Chips
A	28 19 23 19 17 23 24 25 23 27
	18 16 24 23 20 24 24 20 24 23
B	27 28 18 20 30 24 21 23 27 24
	30 23 22 25 17 18 22 25 29 30

9. A carpenter counts the number of knots, x, on 200 different 8-ft-long pine boards as summarized in the table below. Use an approach similar to that in Example 2.4.3 to determine if the number of knots on a board is reasonably described by a Poisson distribution. Explain your reasoning. (**Hint**: To find the value of λ, we need to find the average number of knots on a board. What is the total number of knots observed?)

x	0	1	2	3	4	5	6	7
Freq	28	56	51	35	19	7	3	1

10. The square in Figure 2.9 contains 30 randomly scattered dots. The square has been broken down into 100 smaller squares. Let X denote the number of dots in a randomly chosen small square. Use an approach such as that in Example 2.4.3 to verify that it is reasonable to describe X with a Poisson distribution. Explain your reasoning.

11. In a certain state lottery game, the probability that a single ticket is a winner is 0.001. If 2000 such tickets are sold, use the Poisson distribution to approximate the probability that

  a. exactly 1 ticket is a winner;

  b. between 4 and 6 tickets, inclusive, are winners; and

  c. fewer than 3 tickets are winners.

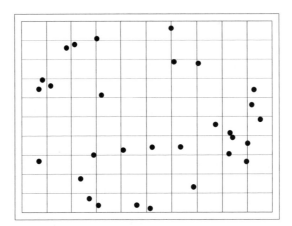

Figure 2.9

12. Suppose that during a typical flu season at a high school of 200 students, 4% get the flu. A flu vaccine program is started, and during the following flu season, only two students get the flu. Use the Poisson distribution to approximate the probability that two or fewer students get the flu, assuming that the rate of 4% getting the flu has not changed. Based on this result, does it appear that the vaccine program has been effective? Explain your reasoning.

13. Cars arrive at a tollbooth at an average of 4 per minute. Assuming that the number of cars that arrive in a 1-min interval is described with a Poisson distribution, use the Poisson distribution to approximate the probability that exactly 10 cars arrive in 8 out of 1000 different 1-min intervals of time. (**Hint**: What is the probability that exactly 10 cars arrive in a single 1-min interval?)

14. Consider a cow pasture that measures 1000 ft × 1000 ft. Suppose the pasture contains 20,000 "cow patties" that occupy an average of 0.5 ft^2 each. Assume the patties are randomly scattered around the pasture. If you take 400 randomly placed steps to walk across the pasture, use the Poisson distribution to estimate the probability that you step in at least one cow patty. (For simplicity, assume a step occupies a single point in the pasture.)

15. Let X have a Poisson distribution with parameter λ.

   a. Show that $\dfrac{P(X=x)}{P(X=x-1)} = \dfrac{\lambda}{x}$ for $x = 1, 2, \ldots$.

   b. Find the values of x such that $P(X = x) \geq P(X = x - 1)$.

   c. Use part b to explain why the mode of X is the greatest integer less than or equal to λ.

16. Assuming that a Poisson distribution applies, use the results of Exercise 15 above to answer these questions:

   a. If a secretary receives an average of 8.4 phone calls per hour, what is the most likely number of calls received in 1 hr?

   b. If the average is 8 calls per hour, show that receiving 8 calls in 1 hr is just as likely as receiving 7 calls in 1 hr.

2.5 Mean and Variance

Associated with every random variable are two numbers: the *mean* and the *variance*. We begin by defining the mean, which is also called the *expected value*.

Definition 2.5.1 The *mean* (or *expected value*) of a discrete random variable X with range R and pmf $f(x)$, denoted μ or $E(X)$, is

$$\mu = E(X) = \sum_{x \in R} x \cdot f(x)$$

provided this series converges absolutely.

The qualifier that the series must converge *absolutely* is necessary to avoid some complications that are beyond the scope of this text. For our purposes, we do not need to worry about this requirement. If the series does not converge to a finite number, then we say that $E(X)$ does not exist. In everyday language, the words *average* and *mean* are often used interchangeably. However, the word *average* is often used to describe quantities that do not fit the definition of a mean (for instance, the Dow Jones Industrial Average is simply the sum of the values of several different stocks). Therefore, we use the word *mean* to avoid any confusion.

Example 2.5.1 **Roulette** If we bet $5 on the number 7 in roulette, the probability that we win nothing is $\frac{37}{38}$ and the probability that we win the $180 prize is $\frac{1}{38}$. Define the random variable

$$X = \text{winnings} - \$5$$

to be the *profit* from one play of this game. Find the expected value of X.

To calculate the mean, or expected value, of a random variable, it is often helpful to describe its pmf in the form of a table. Note that X has only two values, –$5 and $175.

The associated probabilities are $\frac{37}{38}$ and $\frac{1}{38}$. The pmf is shown in Table 2.5. Therefore, the

x	$f(x)$
-5	$\frac{37}{38}$
175	$\frac{1}{38}$

Table 2.5

expected value of X is

$$E(X) = -5(\tfrac{37}{38}) + 175(\tfrac{1}{38}) \approx -0.26.$$

This number is called the *expected value of the game*. To understand what this expected value means, note that if we were to play this game *many times*, say, 3800 times, then $f(-5) = \frac{37}{38}$ means that we would expect to lose

$$\frac{37}{38}(3800) = 3700 \text{ times.}$$

Likewise, we would expect to win 100 times. Therefore, the total expected profit would be

$$-5(3700) + 175(100) = -1000$$

and the average expected profit would be

$$\frac{-1000}{3800} = \frac{-5(3700) + 175(100)}{3800} = -5\left(\frac{37}{38}\right) + 175\left(\frac{1}{38}\right) = E(X) \approx -0.26.$$

Stated another way, we expect to lose an average of $0.26 each time we play. The casino expects to make an average of $0.26 each time the game is played. In general, the mean of a random variable is the expected "average" of *many* values of the variable. □

Informally, the mean of a random variable measures how big its values are. The value of the mean is affected by two factors: (1) the size of the values of the variable and (2) the distribution. To better understand how these factors influence the mean, consider the four random variables W, X, Y, and Z whose distributions and means are shown in Figure 2.10.

First compare W and X. Note that both are uniformly distributed but that W has a much larger mean because it has values between 20 and 30 whereas W has values between 0 and 10. Next compare Y and Z. They both take on values between 0 and 10, but Y has higher probabilities of taking on the larger values. Informally, we might say that Y has "more" larger values and Z has "more" smaller values. Therefore, Y has a larger mean

2.5 Mean and Variance

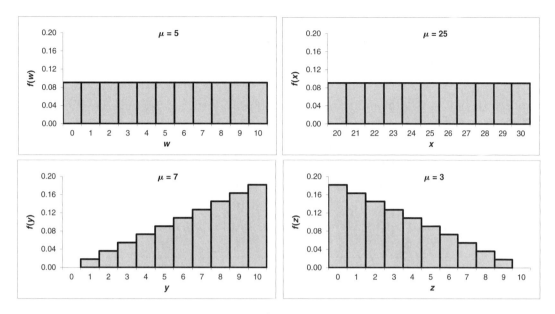

Figure 2.10

and Z has a smaller mean. Because the mean is affected by the probabilities, the mean is sometimes referred to as a *weighted mean* or *weighted average*.

Next we define the *variance* and the related *standard deviation*.

Definition 2.5.2 The *variance* of a discrete random variable X with range R and pmf $f(x)$, denoted σ^2 or $\text{Var}(X)$, is

$$\sigma^2 = \text{Var}(X) = \sum_{x \in R} (x - \mu)^2 f(x)$$

provided this series converges absolutely. The *standard deviation* of X, denoted σ, is the square root of the variance.

Example 2.5.2 Calculating Variance The pmf for a random variable X is given in Table 2.6. Find its mean, variance, and standard deviation.

The mean is relatively easy to calculate:

$$\mu = 0(0.19) + 1(0.35) + 2(0.33) + 3(0.13) = 1.4.$$

x	0	1	2	3
$f(x)$	0.19	0.35	0.33	0.13

Table 2.6

The standard deviation is a bit more tedious. It helps to form a table such as Table 2.7

x	$f(x)$	$(x-\mu)^2$	$(x-\mu)^2 f(x)$
0	0.19	$(0-1.4)^2 = 1.96$	$1.96(0.19) = 0.3724$
1	0.35	$(1-1.4)^2 = 0.16$	$0.16(0.35) = 0.0560$
2	0.33	$(2-1.4)^2 = 0.36$	$0.36(0.33) = 0.1188$
3	0.13	$(3-1.4)^2 = 2.56$	$2.56(0.13) = 0.3328$
			Sum $= 0.8800$

Table 2.7

Thus the variance is $\sigma^2 = 0.88$, and the standard deviation is $\sigma = \sqrt{0.88} \approx 0.938$. □

There are two informal ways to interpret the standard deviation and variance of a variable X: (1) They are measurements of how "spread out" the values of X are, and (2) they are measurements of the average distance of the values of X from the mean. To understand this first interpretation, consider the random variables W and X whose distributions, means, and standard deviations are shown in Figure 2.11.

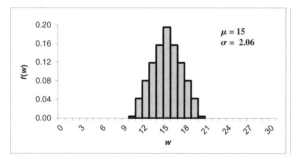
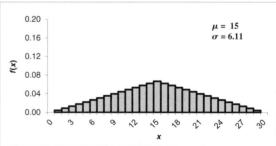

Figure 2.11

Notice that the mean of each W and X is 15 but that the standard deviation of X is much larger. This is so because W has values between 10 and 20 whereas X has values between 0 and 30. Thus the values of X are more "spread out," resulting in a larger standard deviation. To understand the second interpretation, consider the random variables Y and Z whose distributions, means, and standard deviations are shown in Figure 2.12.

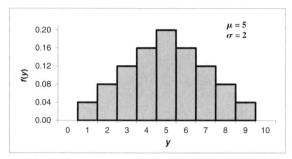 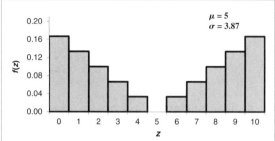

Figure 2.12

Notice that the mean of each Y and Z is 5 and that their ranges are the same but that the standard deviation of Y is much smaller. This is so because for values of Y close to the mean of 5 (for example, $y = 4, 5$, and 6), $f(y)$ is relatively large. For values of Y farther from the mean (for example, $y = 0, 1, 9$, and 10), $f(y)$ is relatively small. For Z, the opposite is true. Informally, we might say that Y has many values close to the mean and Z has many values far from the mean. This results in a smaller standard deviation for Y and a larger standard deviation for Z.

We can derive another formula for calculating the variance of a random variable by algebraically manipulating the definition:

$$\begin{aligned}
\sigma^2 &= \sum_{x \in R} (x - \mu)^2 f(x) = \sum_{x \in R} (x^2 - 2x\mu + \mu^2) f(x) \\
&= \sum_{x \in R} x^2 f(x) - \sum_{x \in R} 2\mu x f(x) + \sum_{x \in R} \mu^2 f(x) \\
&= \sum_{x \in R} x^2 f(x) - 2\mu \sum_{x \in R} x f(x) + \mu^2 \sum_{x \in R} f(x) \\
&= \sum_{x \in R} x^2 f(x) - 2\mu\mu + \mu^2 (1) \\
&= \sum_{x \in R} x^2 f(x) - \mu^2.
\end{aligned}$$

The next example illustrates a use of this alternate formula.

Example 2.5.3 Mean and Variance of a Uniform Distribution Consider a random variable X with range $R = \{1, 2, \ldots, k\}$ that has a uniform distribution. Its pmf is $f(x) = 1/k$. To find the mean of this random variable, we use the identity $\sum_{x=1}^{k} x = k(k+1)/2$:

$$\mu = E(X) = \sum_{x \in R} x \cdot f(x) = \sum_{x=1}^{k} x \cdot \frac{1}{k} = \frac{1}{k} \cdot \sum_{x=1}^{k} x = \frac{1}{k} \cdot \frac{k(k+1)}{2} = \frac{k+1}{2}.$$

To find the variance of X, we use the alternate formula. First we calculate $\sum_{x \in R} x^2 f(x)$ using the identity $\sum_{x=1}^{k} x^2 = k(k+1)(2k+1)/6$:

$$\sum_{x \in R} x^2 f(x) = \sum_{x=1}^{k} x^2 \cdot \frac{1}{k} = \frac{1}{k} \sum_{x=1}^{k} x^2 = \frac{1}{k} \cdot \frac{k(k+1)(2k+1)}{6} = \frac{(k+1)(2k+1)}{6}.$$

Thus

$$\sigma^2 = \sum_{x=1}^{k} x^2 f(x) - \mu^2 = \frac{(k+1)(2k+1)}{6} - \left(\frac{k+1}{2}\right)^2 = \frac{2(k+1)(2k+1) - 3(k+1)^2}{12}$$

$$= \frac{(k+1)\left[2(2k+1) - 3(k+1)\right]}{12} = \frac{(k+1)(k-1)}{12} = \frac{k^2 - 1}{12}.$$

The standard deviation of X is $\sigma = \sqrt{(k^2 - 1)/12}$. □

Example 2.5.4 Mean and Variance of a Poisson Distribution Consider a random variable X with range $R = \{0, 1, \ldots\}$ that has a Poisson distribution with parameter λ. The pmf of X is $f(x) = (\lambda^x/x!)e^{-\lambda}$. To begin to calculate $E(X)$, we use the identity $x/x! = 1/(x-1)!$:

$$\mu = E(X) = \sum_{x \in R} x \cdot f(x) = \sum_{x=0}^{\infty} x \cdot \frac{\lambda^x}{x!} e^{-\lambda} = e^{-\lambda} \sum_{x=1}^{\infty} \lambda^x \cdot \frac{x}{x!} = e^{-\lambda} \sum_{x=1}^{\infty} \lambda^x \cdot \frac{1}{(x-1)!}$$

(we drop the $x = 0$ term in the series because this term equals 0). Now, let $k = x - 1$ so that $x = k + 1$, and using the Taylor series expansion $e^\lambda = \sum_{k=0}^{\infty} (\lambda^k/k!)$, we have

$$\mu = E(X) = e^{-\lambda} \sum_{k=0}^{\infty} \lambda^{k+1} \cdot \frac{1}{k!} = \lambda e^{-\lambda} \sum_{k=0}^{\infty} \frac{\lambda^k}{k!} = \lambda e^{-\lambda} e^{\lambda} = \lambda.$$

Thus the mean of X is λ. This justifies our use of λ as the average number of occurrences per interval. To find the variance, we use the alternate formula and first evaluate $\sum_{x \in R} x^2 f(x)$:

$$\sum_{x \in R} x^2 f(x) = \sum_{x=0}^{\infty} x^2 \cdot \frac{\lambda^x}{x!} e^{-\lambda}$$

$$= \sum_{x=1}^{\infty} x \cdot \frac{\lambda^x}{(x-1)!} e^{-\lambda}$$

$$= \sum_{x=0}^{\infty} (x+1) \frac{\lambda^{x+1}}{x!} e^{-\lambda}$$

$$= \lambda \left(\sum_{x=0}^{\infty} x \cdot \frac{\lambda^x}{x!} e^{-\lambda} + \sum_{x=0}^{\infty} \frac{\lambda^x}{x!} e^{-\lambda} \right).$$

But the sum on the left within the parentheses is $E(X) = \lambda$, and the sum on the right is the sum of the pmf over the range of X, so this sum equals 1. Thus we have

$$\sum_{x \in R} x^2 f(x) = \lambda (\lambda + 1) = \lambda^2 + \lambda.$$

Therefore,

$$\sigma^2 = \sum_{x \in R} x^2 f(x) - \mu^2 = \lambda^2 + \lambda - \lambda^2 = \lambda.$$

Thus for a random variable with a Poisson distribution, $\mu = \sigma^2 = \lambda$. □

Exercises

1. Find the mean, variance, and standard deviation of each random variable with the given distribution.

a. The pmf is given in the table below.

x	0	1	2	3	4
$f(x)$	0.1	0.1	0.3	0.4	0.1

b. $f(x) = \frac{1}{6}$, $x = -2, 0, 3, 9, 11, 15$.

c. $f(x) = 1$, $x = 3$.

d. $f(x) = \dfrac{2 - |x - 1|}{4}$, $x = 0, 1, 2$.

2. Consider the random variable X with pmf as described in Exercise 1a. Define the random variable
$$Y = 3X - 1.$$
This means, for instance, that when $X = 3$, $Y = 3(3) - 1 = 8$ and $P(Y = 8) = P(X = 3) = 0.4$.

   a. Describe the pmf of Y in table form.
   b. Find the mean of Y.
   c. Find a relationship between the mean of X and the mean of Y.

3. A moving truck rental company sells three different sizes of boxes: 1 ft × 1.5 ft × 2 ft, 2 ft × 3 ft × 3.5 ft, and 3 ft × 4 ft × 5 ft. If 35% of the boxes sold are of the first size, 45% are of the second size, and 20% are of the third size, find the mean volume of a box sold.

4. A coffee shop sells nine different types of coffee. The first type sells for $1 a cup, the second for $2, the third for $3, and so on. Consider the random experiment of selecting a customer who purchases a cup of coffee. Let the random variable X denote the price the customer paid for the cup of coffee. If the coffee shop sells an equal number of cups of each type of coffee, find $E(X)$ and $\text{Var}(X)$.

5. Each child in a school is given a bag of mixed candies. Assume the number of yellow candies in a bag is described by a Poisson distribution. If 1.83% of the bags contain no yellow candies, find the mean number of yellow candies in a bag.

6. Suppose X is $b(5, 0.20)$.

   a. Describe the pmf of X in table form.
   b. Calculate the mean, variance, and standard deviation of X.
   c. Find a relationship between the mean, n, and p. Do the same for the variance.

7. Two cards are selected, with replacement, from a standard deck of 52 cards. Find the mean and variance of the number of hearts selected. (**Hint**: Describe the pmf of the number of hearts selected.)

8. A Bernoulli experiment, as described in Definition 2.3.2, is a random experiment in which each outcome is categorized as a success or a failure. Define the random variable

$$X = \begin{cases} 1, & \text{if the outcome is a success} \\ 0, & \text{if the outcome is a failure.} \end{cases}$$

Such a random variable is called a *Bernoulli random variable*. If the probability of a success is p and the probability of a failure is $1-p$, find $E(X)$ and $\text{Var}(X)$ in terms of p.

9. Figure 2.13 shows the probability histograms of three different random variables X, Z, and W, each of which has a mean of 5. Without doing any calculations, determine which variable has the largest variance and which has the smallest. Explain your reasoning. (**Suggestion**: Interpret variance as an average distance of the values of a variable from the mean.)

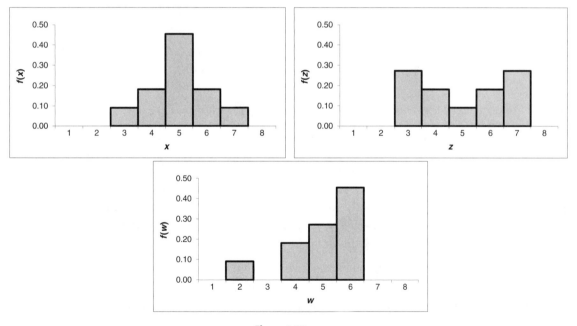

Figure 2.13

10. Suppose the probability that a randomly selected 25-year-old male lives one more year is 0.9985. A life insurance company charges \$155 for a one-year policy that will pay out a \$75,000 death benefit. If a 25-year-old male purchases this policy, let X denote his profit. Find the expected value of X. If the life insurance company sells many such policies, should it expect to make a profit?

11. In a simple casino game, a player spins a fair spinner with sections labeled 1, 2, 3, and 4. The player wins $1 if it lands on 1; $2 if it lands on 2; etc. If it costs $2.75 to play the game, find the expected value of the game (i.e., the expected value of the profit).

12. A gambler claims to have created a fair die with k sides such that the mean roll is 9.5. Find the value of k. (**Hint**: If X denotes the result of a single roll, what is the distribution of X?)

13. Suppose an office building contains 15 different offices. Ten have 20 employees each, four have 100 employees each, and one has 400 employees for a total of 1000 employees.

   a. Calculate the average number of employees per office.
   b. Consider the random experiment of selecting one employee from the 1000 employees. Let X denote the number of employees in the office in which this person works. Describe the pmf of X in the form of a table.
   c. Calculate $E(X)$.

14. A random variable X has the pmf $f(x) = 6/(\pi^2 x^2)$ for $x = 1, 2, \ldots$. Show that $E(X)$ does not exist. (**Hint**: A p-series $\sum_{n=1}^{\infty} 1/n^p$ diverges if $0 < p \leq 1$.)

15. A random variable T has the pmf $f(t) = \frac{1}{10}\left(\frac{9}{10}\right)^{t-1}$ for $t = 1, 2, \ldots$. Use the identity $\sum_{n=1}^{\infty} nx^{n-1} = x/(1-x)^2$ for $|x| < 1$ to find $E(T)$.

16. A simple casino game is played by rolling a fair six-sided die, of which three sides are labeled 1, two sides are labeled 2, and one side is labeled 3. If the die lands on a 1, the player wins $1; $2 if it lands on a 2; and $3 if it lands on 3. The player must pay a certain amount for each play of the game. The price is considered to be fair if the expected profit is 0. Find the fair price for this game.

17. Consider the following game of chance: A fair coin is flipped until the first tail appears. The player wins $1 if a tail appears on the first flip, $2 if the game lasts two flips, $4 if it lasts three flips, and so on. In general, the player wins 2^{n-1} if the game lasts n flips.

   a. Let the random variable X denote the winnings from a single play of the game. Describe the pmf of X in table form. (**Hint**: The range of X is infinite. The game lasts $n = 3$ flips and $X = 2^{3-1} = 4$ only when the sequence HHT is observed. What is the probability of this sequence?)
   b. Suppose it costs $25 to play this game. Find the probability that you win less than $25 (i.e., find $P(X \leq 25)$). Based on this result, would you want to play the game? Why or why not?

c. Show that $E(X)$ does not exist. What does this imply about the expected value of the game?

d. Based on your result in part c, would you want to play the game? Why or why not? Does this agree with your answer in part b? (This contradiction when comparing the expected value of a game to the probability of actually profiting is known as the *St. Petersburg paradox.*)

2.6 Functions of a Random Variable

Often, it is convenient to describe a random variable as a function of another random variable. We illustrate this idea with the following example.

Example 2.6.1 **Days in the Hospital** Suppose a man is injured in a car wreck. The doctor tells him that he needs to spend between 1 and 4 days in the hospital and that if X is the number of days, its distribution is given in Table 2.8.

x	1	2	3	4
$f(x)$	0.3	0.4	0.2	0.1

Table 2.8

The man's supplemental insurance policy will give him $500 plus $100 per day in the hospital to cover expenses. How much should the man expect to get from the insurance company?

If we let the random variable Y denote the amount the man receives from the insurance company, then

$$Y = 500 + 100X.$$

In other words, Y is a function of X. We want to find the expected value, or mean, of Y. The values of Y are shown in Table 2.9.

x	1	2	3	4
y	600	700	800	900

Table 2.9

Also note that $P(Y = 600) = P(X = 1) = f(1)$, and in general,

$$P(Y = 500 + 100x) = P(X = x) = f(x).$$

Therefore, the expected value of Y is

$$\mu = E(Y) = 600(0.3) + 700(0.4) + 800(0.2) + 900(0.1)$$
$$= [500 + 100(1)]f(1) + \cdots + [500 + 100(4)]f(4) = 710.$$

So the man can expect to receive \$710 from the insurance company. □

This example motivates our definition of the *mathematical expectation* or the *expected value* of a function of a random variable.

Definition 2.6.1   Let X be a discrete random variable with range R and pmf $f(x)$. Also let $u(X)$ be a function of X. The *mathematical expectation* or the *expected value* of $u(X)$, denoted $E[u(X)]$, is

$$E[u(X)] = \sum_{x \in R} u(x) f(x)$$

provided this series converges absolutely.

Example 2.6.2   Calculating Expected Values   Let X be a random variable whose distribution is shown in Table 2.10.

x	-2	1	5
$f(x)$	0.25	0.5	0.25

Table 2.10

Find (a) $E[X^2]$ and (b) $E[(X-1)^2]$.

a. $E[X^2] = \sum_{x \in R} x^2 f(x) = (-2)^2(0.25) + (1)^2(0.5) + (5)^2(0.25) = 7.75$

b. $E[(X-1)^2] = \sum_{x \in R} (x-1)^2 f(x)$
$= (-2-1)^2(0.25) + (1-1)^2(0.5) + (5-1)^2(0.25) = 6.25$ □

We have seen two formulas for calculating the variance of a random variable X:

$$\sigma^2 = \text{Var}(X) = \sum_{x \in R} (x - \mu)^2 f(x) = \sum_{x \in R} x^2 f(x) - \mu^2.$$

Remembering that μ is a number, note that the terms in the first summation look very similar to the terms in the summation in part b of the previous example. Also note that the terms in the second summation are identical to the terms in part a. Thus we see that

$$\sigma^2 = \text{Var}(X) = E[(X-\mu)^2] = E[X^2] - \mu^2.$$

This gives us two ways of describing the variance of X in terms of the expected values of functions of X.

Mathematical expectation E is a function. It takes in a function $u(X)$, does something to it (or "operates" on it), and returns a number. Functions that take in other functions are often called *operators*. In the next theorem, we list some properties of this operator E.

Theorem 2.6.1 Let X be a discrete random variable with range R and pmf $f(x)$. Whenever the expectations exist, the following three properties hold:

1. If a is a constant, then $E[a] = a$.

2. If a is a constant and $u(X)$ is a function of X, then
$$E[au(X)] = aE[u(X)].$$

3. If a_1 and a_2 are constants and $u_1(X)$ and $u_2(X)$ are functions of X, then
$$E[a_1 u_1(X) + a_2 u_2(X)] = a_1 E[u_1(X)] + a_2 E[u_2(X)].$$

Proof. In the first property, $E[a]$ really means $E[u(X)]$ where $u(X) = a$. Then by definition and the fact that $\sum_{x \in R} f(x) = 1$,

$$E[a] = E[u(X)] = \sum_{x \in R} af(x) = a \sum_{x \in R} f(x) = a(1) = a.$$

The proof of the second property is also very simple:

$$E[au(X)] = \sum_{x \in R} au(x)f(x) = a \sum_{x \in R} u(x)f(x) = aE[u(X)].$$

The proof of the third property uses similar arguments:

$$E[a_1 u_1(X) + a_2 u_2(X)] = \sum_{x \in R} [a_1 u_1(x) + a_2 u_2(x)] f(x)$$
$$= \sum_{x \in R} a_1 u_1(x) f(x) + \sum_{x \in R} a_2 u_2(x) f(x)$$
$$= a_1 E[u_1(X)] + a_2 E[u_2(X)]. \qquad \square$$

Using mathematical induction, we can extend the third property to any finite number of terms:

$$E\left[\sum_{i=1}^{n} a_i u_i(X)\right] = \sum_{i=1}^{n} a_i E\left[u_i(X)\right].$$

Because of this property, mathematical expectation E is called a *linear operator*. This property is very useful, as illustrated in the next examples.

Example 2.6.3 **Calculating Expected Values** Consider again the random variable X described in Example 2.6.2. Find (a) $E[X]$, (b) $E[3X + 4]$, and (c) $E[X(X + 2)]$.

a. $E[X] = (-2)(0.25) + (1)(0.5) + (5)(0.25) = 1.25$

b. $E[3X + 4] = E[3X] + E[4] = 3E[X] + 4 = 3(1.25) + 4 = 7.75$

c. $E[X(X + 2)] = E[X^2 + 2X] = E[X^2] + 2E[X] = 7.75 + 2.5 = 10.25$ □

Example 2.6.4 **Subtracting μ** Suppose a discrete random variable X has mean μ. Assuming $E[X - \mu]$ exists, find its value.

Using the linearity of E and remembering that $E[X] = \mu$, we have

$$E[X - \mu] = E[X] - E[\mu] = \mu - \mu = 0.$$

Thus, if we subtract the mean of a random variable from the variable, the resulting variable has mean 0.   □

Our last theorem is a simple but useful result whose proof we leave as an exercise.

Theorem 2.6.2   *If a is a constant, then*

$$Var[aX] = a^2 Var(X).$$

Exercises

1. Let X be a random variable whose distribution is shown in the table below. Calculate $E(X)$, $E[X^2]$, $\text{Var}(X)$, $E[X^2 + 3X + 1]$, and $E[\sqrt{X + 5}]$.

x	0	1	2	3
$f(x)$	$\frac{1}{4}$	$\frac{1}{8}$	$\frac{3}{8}$	$\frac{1}{4}$

2. A random variable X has the pmf

$$f(x) = \frac{2 - |x - 1|}{4}, \quad \text{for } x = 0, 1, 2.$$

Calculate $E(X)$, $E[X^2]$, $\text{Var}(X)$, and $E[4X^2 + 6X - 9]$.

3. In a simple casino game, a player spins a fair spinner with sections labeled 1, 2, 3, and 4. The player wins $1 if it lands on 1, $2 if it lands on 2, and so on.

   a. Let the random variable X denote the result of the spin. Find $E(X)$.
   b. If it costs $n to play the game, let $Y = X - n$ be the profit from a single play of the game. Find the expected value of the game in terms of n (i.e., find $E(Y)$).
   c. A game is considered to be fair if its expected value is 0. Find the value of n so that the game is fair.

4. Consider a factory that produces pencils. Suppose the daily production X is a random variable with mean $E(X) = \mu_X$ and variance $\text{Var}(X) = \sigma_X^2$. Management decides to double the number of machines and employees so that the new daily production Y is described by $Y = 2X$.

   a. How do you think the new mean daily production will be related to the original mean daily production?
   b. Use Theorem 2.6.1 to find $E(Y)$ in terms of μ_X. Does this result agree with your answer in part a?
   c. If you think of variance as a measure of how spread out the values of a random variable are, do you think the variance of the new daily production will be smaller or larger than the variance of the original daily production? Explain.
   d. Use Theorem 2.6.2 to find $\text{Var}(Y)$ in terms of σ_X^2. Does this result agree with your answer in part c?

5. Use the definition of variance and Theorem 2.6.1 to prove Theorem 2.6.2.

6. Let X be a discrete random variable with range $R = \{b\}$, where b is a constant. Show that $\text{Var}(X) = \text{Var}(b) = 0$.

7. Let X have a uniform distribution with range $\{1, \ldots, k\}$. Define the random variable

$$Y = X + a - 1$$

where a is some constant.

a. Describe the range and pmf of Y.

b. Show that $E(Y) = (a+b)/2$, where $b = k + a - 1$.

8. A home improvement contractor notes that customers who purchase windows always purchase between 5 and 11 windows, and if X denotes the number of windows purchased by a customer, the revenue from the sale is $Y = 120X - 200$. Assuming that each quantity of windows purchased (5, ..., 11) is equally likely, use the results of Exercise 7 above to find $E(Y)$.

9. Suppose X is a discrete random variable with mean μ_X and variance σ_X^2. Define the random variable

$$Y = \frac{X - \mu_X}{\sigma_X}.$$

Find $\mu_Y = E(Y)$ and $\sigma_Y^2 = \text{Var}(Y) = E\big[(Y - \mu_Y)^2\big]$.

10. Let X be a random variable such that $E[X - 5] = 5$ and $E\big[(X-5)^2\big] = 30$. Calculate $E(X)$, $E[X^2]$, and $\text{Var}(X)$.

11. Consider the following scenario: A girl stands on the middle step of a long staircase. She then walks up between 1 and 11 steps and then walks down 6 steps. Let X denote the number of steps she walks up, and assume X has a uniform distribution. Let the random variable Y denote the step on which she ends ($Y = 1$ if she ended 1 step above the middle step, $Y = -2$ if she ended 2 steps below, $Y = 0$ if she ended on the middle step, and so on). Find $E(Y)$.

12. At a factory that produces lightbulbs, the number of defective bulbs in a batch, call it X, is described by a Poisson distribution with mean $\lambda = 3$. Management figures that the profit Y (in dollars) from a batch of bulbs is $Y = 75 - 3X - X^2$. Find $E(Y)$. (**Hint:** See Example 2.5.4.)

13. Give a counterexample to the conjecture that $E[X^n] = [E(X)]^n$ for any real number n.

14. Show that the value of $E\big[(X - a)^2\big]$ is minimized when $a = E(X)$. (**Hint:** Expand this expected value using linearity properties, and take the derivative with respect to a.)

15. The manager of an electronics store is trying to figure out how many of a certain type of stereo to order for the holiday season. Let the random variable X denote the number of

such stereos sold in a holiday season. Based on records from previous years, the manager determines the distribution of X is as shown in the table below.

x	0	1	2	3	4
$f(x)$	0.15	0.35	0.25	0.15	0.1

The store makes a profit of $20 for each stereo sold and takes a loss of $6 for each unsold stereo. So if the manager orders s stereos, then the net profit is described with the random variable

$$P = \begin{cases} 20X - 6(s - X), & \text{if } X \leq s \\ 20s, & \text{if } X > s. \end{cases}$$

a. If the manager orders $s = 1$ stereo, describe the distribution of P and find $E(P)$.
b. Repeat part a for $s = 2, 3,$ and 4.
c. If the manager wants to maximize expected profit, how many stereos should be ordered?

2.7 The Moment-Generating Function

The *moment-generating function* of a random variable is a very important and useful theoretical tool. At first glance, its definition looks formidable.

Definition 2.7.1 Let X be a discrete random variable with pmf $f(x)$ and range R. The *moment-generating function* (mgf) of X, denoted $M(t)$, is

$$M(t) = E[e^{tX}] = \sum_{x \in R} e^{tx} f(x)$$

for all values of t for which this mathematical expectation exists.

Example 2.7.1 Finding the MGF Consider a random variable X with range $R = \{2, 5\}$ and pmf $f(2) = 0.25$ and $f(5) = 0.75$. Its mgf is

$$M(t) = e^{2t}(0.25) + e^{5t}(0.75).$$

Note that $M(t)$ is a function of t. □

There are two important observations we need to make about the mgf in the previous example:

1. The derivative of $M(t)$ with respect to t is

$$M'(t) = \frac{d}{dt}\left[e^{2t}(0.25) + e^{5t}(0.75)\right] = 2e^{2t}(0.25) + 5e^{5t}(0.75).$$

Evaluating this derivative at $t = 0$ gives

$$M'(0) = 2e^{2(0)}(0.25) + 5e^{5(0)}(0.75) = 2(0.25) + 5(0.75) = E(X).$$

The second derivative of $M(t)$ with respect to t is

$$M''(t) = \frac{d}{dt}\left[2e^{2t}(0.25) + 5e^{5t}(0.75)\right] = 2^2 e^{2t}(0.25) + 5^2 e^{5t}(0.75).$$

Evaluating $M''(0)$ yields

$$M''(0) = 2^2 e^{2(0)}(0.25) + 5^2 e^{5(0)}(0.75) = 2^2(0.25) + 5^2(0.75) = E[X^2].$$

2. The coefficients of e^{2t} and e^{5t} are the values of $f(2)$ and $f(5)$, respectively. Thus, if all we know about a random variable is its mgf, we know the range and its distribution.

These two observations illustrate the two very important properties of the mgf given in the next theorem.

Theorem 2.7.1 *If X is a random variable and its mgf $M(t)$ exists for all t in an open interval containing 0, then*

1. *$M(t)$ uniquely determines the distribution of X, and*
2. *$M'(0) = E[X]$ and $M''(0) = E[X^2]$.*

The proof of this theorem uses properties of the Laplace transform and is well beyond the scope of this book. The first part says that if we know the mgf of X, then we know its distribution (or at least are able to identify it). The second property can be extended to say that when the mgf exists, $M^{(r)}(0)$ exists for any integer $r > 0$ and $M^{(r)}(0) = E[X^r]$. The quantity $E[X^r]$ is called the *rth moment of the distribution about the origin*. This is where the mgf gets its name.

The mgf has at least two important uses:

1. If we know the mgf of a random variable, then we can use the first and second derivatives to find the mean and variance of the variable.
2. If we can show that two random variables have the same mgf, then we can conclude that they have the same distribution.

The next example illustrates a trivial application of the first use. In later chapters, we will illustrate the second use.

Example 2.7.2   **Finding the Mean and Variance**   Suppose the mgf of a random variable X is

$$M(t) = e^{-3t}(0.1) + 0.6 + e^{2t}(0.3).$$

Describe the distribution of X, and find its mean and variance.

The range of X is given in the exponents of the mgf. Because there is no exponent in the second term, we take it to be 0 so that the range is –3, 0, 2. The pmf is given by the coefficients. We see that $f(-3) = 0.1$, $f(0) = 0.6$, and $f(2) = 0.3$.

To find the mean and variance, we could use the distribution and the definition of the mean, but we will use derivatives instead. Note that

$$M'(t) = -3e^{-3t}(0.1) + 0 + 2e^{2t}(0.3) \quad \Rightarrow \quad E(X) = \mu = M'(0) = -3(0.1) + 2(0.3) = 0.3.$$

Now,

$$M''(t) = 9e^{-3t}(0.1) + 4e^{2t}(0.3) \quad \Rightarrow \quad E[X^2] = M''(0) = 9(0.1) + 4(0.3) = 2.1.$$

Thus,

$$\sigma^2 = E[X^2] - \mu^2 = M''(0) - [M'(0)]^2 = 2.1 - (0.3)^2 = 2.01. \quad \square$$

Using derivatives of the mgf to calculate the mean and variance of X in the previous example is more cumbersome than using the definitions directly. The real power of the mgf lies in theoretical calculations, as illustrated in the next example.

Example 2.7.3   **Mean and Variance of a Binomial Distribution**   Consider a random variable X with a binomial distribution. Its pmf is

$$f(x) = \binom{n}{x} p^x q^{n-x}, \quad x = 0, 1, \ldots, n$$

where $q = 1 - p$. First we find the mgf by using the binomial theorem $(a + b)^n = \sum_{k=0}^{n} \binom{n}{k} a^k b^{n-k}$:

$$M(t) = E[e^{tX}] = \sum_{x=0}^{n} e^{tx} \binom{n}{x} p^x q^{n-x} = \sum_{x=0}^{n} \binom{n}{x} (pe^t)^x q^{n-x}$$
$$= (q + pe^t)^n.$$

Note that the summation in this derivation has a finite number of terms, so it converges, and hence $M(t)$ exists for all values of t. Now we find $M'(t)$ and $M''(t)$, remembering that n, p, and q are constants:

$$M'(t) = n(q + pe^t)^{n-1}(pe^t)$$

$$M''(t) = n(n-1)(q + pe^t)^{n-2}(pe^t)^2 + n(q + pe^t)^{n-1}(pe^t).$$

Thus, noting that $p + q = 1$,

$$\mu = E(X) = M'(0) = n(q + pe^0)^{n-1}(pe^0) = n(q + p)^{n-1}(p) = np$$

and

$$\sigma^2 = E[X^2] - \mu^2 = M''(0) - \mu^2$$
$$= n(n-1)(q+p)^{n-2}p^2 + n(q+p)^{n-1}p - (np)^2$$
$$= n(n-1)p^2 + np - (np)^2$$
$$= np\,[(n-1)p + 1 - np]$$
$$= np\,(1-p) = npq.$$

So a random variable with a binomial distribution has mean $\mu = np$ and variance $\sigma^2 = npq$. The mean, or expected value, of a binomial distribution has a special meaning. It is the *expected* number of successes in the n trials. □

Not every random variable has an mgf defined on an open interval containing 0, as in the next example.

Example 2.7.4   A Random Variable with No MGF   Consider the random variable X with pmf $f(x) = a/x^2$ for $x = 1, 2, \ldots$ where $a = \sum_{x=1}^{\infty} (1/x^2)$ as considered in Exercise 6 of Section 2.2. To try to find its mgf, we calculate

$$M(t) = E(e^{Xt}) = \sum_{x=1}^{\infty} e^{xt} \frac{a}{x^2} = a \sum_{x=1}^{\infty} \frac{e^{xt}}{x^2}.$$

When $t = 0$, this is a convergent p-series. For this to converge for other values of t, by the ratio test we need the following limit to be less than 1:

$$\lim_{x \to \infty} \frac{e^{(x+1)t}/(x+1)^2}{e^{xt}/x^2} = \lim_{x \to \infty} e^t \frac{x^2}{(x+1)^2} = e^t.$$

So we need $e^t < 1$, which means that $t < 0$. Thus $M(t)$ is defined only for $t \leq 0$. Because $M(t)$ is not defined on an *open* interval containing 0, we say that X has no mgf. □

Exercises

1. In each part below, the mgf for a discrete random variable X is given. For each, describe the pmf of X in table form, find $M'(t)$ and $M''(t)$, and use these derivatives to calculate μ and σ^2.

   a. $M(t) = \frac{1}{5}e^{-t} + \frac{2}{5}e^{3t} + \frac{2}{5}e^{4t}$

   b. $M(t) = \sum_{x=1}^{5} 0.2 e^{tx}$

2. If a random variable X has the mgf $M(t) = (0.25 + 0.75 e^t)^{25}$, determine what type of distribution X has. (**Hint**: Compare this mgf to that in Example 2.7.3 and use the first part of Theorem 2.7.1.)

3. Show that for any mgf $M(t)$, $M(0) = 1$.

4. A random variable X is said to have a *Bernoulli distribution* if its pmf is

$$f(x) = p^x (1-p)^{1-x}, \quad x = 0, 1$$

where $0 \leq p \leq 1$. Find the mgf of X and use it to find $E(X)$ and $\text{Var}(X)$ in terms of p.

5. Let $M(t)$ be the mgf of a random variable X with mean μ and variance σ^2. Define $R(t) = \ln[M(t)]$. Show that $R'(0) = \mu$ and $R''(0) = \sigma^2$.

6. Find the expected value of the random variable X in Example 2.7.4 (if it exists). Explain why it is that if the expected value of a random variable does not exist, then the variable cannot have an mgf. (**Hint**: Consider the second part of Theorem 2.7.1.)

7. Suppose a random variable X has the mgf $M_X(t)$. Define the random variable $Y = aX + b$ where a and b are any real numbers. Show that the mgf of Y is $M_Y(t) = e^{bt} M_X(at)$. (**Hint**: By definition, $M_Y(t) = E[e^{Yt}] = E[e^{(aX+b)t}]$. Rewrite this, using properties of the exponential and the linearity properties of mathematical expectation.)

8. The variable X has the mgf $M_X(t) = 0.5 + 0.3e^t + 0.2e^{3t}$. Define the variable $Y = 2X+4$. Use the results of Exercise 7 to find the mgf of Y.

9. Use Exercise 7 to show that if $Y = aX + b$, then $E(Y) = aE(X) + b$ and $\text{Var}(Y) = a^2 \text{Var}(X)$ where a and b are any real numbers.

10. Suppose X is $b(n,p)$. Define the random variable $F = n - X$ to be the number of failures in the n trials. Use Exercise 7 and Example 2.7.3 to show that the mgf of F is $M(t) = (p + qe^t)^n$. Based on this result, what type of distribution does F have?

11. Consider the scenario of a laboratory that tests blood samples for the presence of marijuana by combining individual samples and then testing the combined sample as first described in Exercise 10 of Section 1.7. If 1% of the individual samples contain the presence of marijuana and 10 individual samples are combined, we calculated that the probability of a combined sample testing positive is 0.0956.

a. Suppose 500 individual samples are collected. This means that 50 combined samples are tested. Let X denote the number of combined samples that test positive. What type of distribution does X have? Find $E(X)$.

b. If a combined sample tests positive, then each of the 10 individual samples is tested separately. Let T denote the total number of samples tested (combined and individual). Find an expression for T in terms of X, and calculate $E(T)$.

c. Based on this result, is the laboratory really saving any work by testing combined samples rather than testing each individual sample? Explain why or why not.

d. Let p denote the proportion of individual samples that contain the presence of marijuana. For which values of p does the method of testing combined samples save work? That is, for which values of p is $E(T) < 500$?

12. A random variable X is said to have a *geometric distribution* if its pmf is $f(x) = q^{x-1}p$ for $x = 1, 2, \ldots$ where $0 \leq p \leq 1$ and $q = 1 - p$. Show that the mgf of X is

$$M(t) = \frac{pe^t}{1 - qe^t}, \quad \text{for} \quad t < \ln \frac{1}{q}.$$

Use the mgf to find $E(X)$ and $\text{Var}(X)$. (**Hint**: Use the formula for the sum of a convergent geometric sequence: $\sum_{n=1}^{\infty} ar^{n-1} = a/(1-r)$ when $|r| < 1$.)

13. It can be shown that the mgf for a random variable X with a negative binomial distribution as defined in Exercise 16 of Section 2.3 is

$$M(t) = \left(\frac{pe^t}{1 - qe^t}\right)^r$$

where $q = 1 - p$. Use this to calculate $E(X)$.

14. Let X have a Poisson distribution with parameter λ. Show that the mgf of X is

$$M(t) = e^{\lambda(e^t - 1)}.$$

(**Hint**: Use the Taylor series $e^{\lambda e^t} = \sum_{x=0}^{\infty} [(\lambda e^t)^x / x!]$.)

15. Use Exercise 14 to show that if X has a Poisson distribution, then $E(X) = \lambda$, $E[X^2] = \lambda^2 + \lambda$, and $\text{Var}(X) = \lambda$. Compare these calculations to those done in Example 2.5.4. Which do you think were easier?

16. Another way to identify unusual events as discussed in Section 2.3 is with the *range rule of thumb*, which states that usual values of a random variable X lie within 2 standard deviations of its mean. In other words, usual values occur in the interval $[\mu - 2\sigma, \mu + 2\sigma]$. Any value of X outside this interval is considered unusual. The quantity $\mu - 2\sigma$ is called the *usual minimum value* of X whereas $\mu + 2\sigma$ is called the *usual maximum value*.

   a. A candy manufacturer claims that 20% of its plain chocolate candies are blue. Suppose that a random sample of 100 such candies is selected. Let the random variable X denote the number of blue candies in the sample. Assuming the claim is true, find the mean and standard deviation of X. (**Hint**: What type of distribution does X have?)

   b. Find the usual minimum and maximum values of X.

   c. Suppose that a sample of 100 candies contains 10 blue candies. Is this considered unusual according to the range rule of thumb? What does this say about the validity of the claim?

Directions for Exercises 17 to 19: Let X be a discrete random variable with range $0, 1, \ldots$ and pmf $f(x)$. Its *probability-generating function* is

$$P(t) = E[t^X] = \sum_{x=0}^{\infty} f(x) t^x$$

for all values of t for which this series converges. It can be shown that $E(X) = P'(1)$. Use this information in the following exercises.

17. Show that the probability-generating function of a random variable X with a Poisson distribution with parameter λ is $P(t) = e^{\lambda(t-1)}$ for all t. Use this to find $E(X)$. (**Hint**: Use the Taylor series $e^{\lambda t} = \sum_{x=0}^{\infty} [(\lambda t)^x / x!]$.)

18. Show that the probability-generating function of a random variable X that is $b(n, p)$ is $P(t) = (q + pt)^n$ for all t where $q = 1 - p$. Use this to find $E(X)$. Note that by definition $0 \leq X \leq n$, so take $f(x) = 0$ for $x > n$. (**Hint**: Use the binomial theorem.)

19. Show that the probability-generating function of a random variable X with a geometric distribution as defined in Exercise 12 of this section is

$$P(t) = \frac{pt}{1 - qt}, \quad \text{if } |t| < \frac{1}{q}.$$

Use this to find $E(X)$. Note that by definition $X > 0$, so take $f(0) = 0$. (**Hint**: Use the formula for the sum of a convergent geometric sequence: $\sum_{n=1}^{\infty} ar^{n-1} = a/(1-r)$ when $|r| < 1$.)

CHAPTER 3

Continuous Random Variables

Chapter Objectives

- Define continuous random variables and the probability density function
- Discuss similarities between discrete and continuous random variables
- Introduce the uniform, exponential, normal, and other common continuous distributions
- Introduce joint distributions
- Present the central limit theorem
- Discuss approximation techniques

3.1 Introduction

Chapter 2 dealt with discrete random variables, which have a finite or countable range. This chapter deals with continuous random variables, which do not have a countable range. Discrete and continuous random variables share many of the same concepts. Both are functions that associate a real number with each outcome of a random experiment. With both we calculate probabilities, means, variances, and so on. The biggest difference is that with continuous variables we use integration instead of summation for these calculations.

To motivate some of the concepts related to continuous random variables, consider the random experiment of selecting a plain M&M candy and measuring its mass. Let the random variable X denote the mass of the candy in grams. Because X could, in principle, take on *any* value between 0 and some maximum value, X is not discrete. Instead, X is a continuous random variable. To better understand this random variable, a student observes 30 values of X by measuring the masses of 30 such candies on an electronic scale, as recorded in Table 3.1 (data collected by Frank Ohlinger, 2009). These observed values are called *data*.

0.76	0.82	0.85	0.86	0.88	0.90
0.80	0.84	0.86	0.86	0.88	0.91
0.81	0.84	0.86	0.86	0.88	0.91
0.82	0.84	0.86	0.87	0.88	0.92
0.82	0.85	0.86	0.88	0.90	0.94

Table 3.1

The first step in analyzing data such as these is to construct a relative frequency histogram. Because the variable X does not have a discrete set of values, it does not make sense to calculate the relative frequency of each value of X in Table 3.1 as we do with discrete random variables. The basic idea behind a relative frequency histogram of continuous data is that we divide the data into several *subintervals* (also called *bins* or *classes*) and then calculate the relative frequency of each bin.

To construct the histogram, first we select the number of bins. There is no one correct way to select this number. Here we choose to use 10 bins. Next we calculate the *bin width*

$$\text{Bin width} \approx \frac{(\text{maximum value}) - (\text{minimum value})}{\text{number of bins}}.$$

The quantity (maximum value) − (minimum value) is called the *range* of the data. For these data, bin width $\approx (0.94 - 0.76)/10 = 0.018$. We round this number to two decimal places and use a bin width of 0.02.

Next we construct the bins as in the first column of Table 3.2. The left endpoint of the first interval needs to be smaller than the minimum data value whereas the right endpoint of the last interval needs to be at least as large as the maximum data value. Then we count the number of data values in each bin and calculate the relative frequencies as recorded in the rightmost two columns of the table. The resulting table is called the *relative frequency distribution* of the data, or the *sample distribution*.

Bin	Freq	Rel Freq
$(0.75, 0.77]$	1	0.033
$(0.77, 0.79]$	0	0.000
$(0.79, 0.81]$	2	0.067
$(0.81, 0.83]$	3	0.100
$(0.83, 0.85]$	5	0.167
$(0.85, 0.87]$	8	0.267
$(0.87, 0.89]$	5	0.167
$(0.89, 0.91]$	2	0.133
$(0.91, 0.93]$	1	0.033
$(0.93, 0.95]$	1	0.033

Table 3.2

Next we draw a *bar graph* of the relative frequencies as in Figure 3.1. Each bin has a bar that is centered at the midpoint of the bin and whose height equals the relative frequency. This graph is the relative frequency histogram of the data.

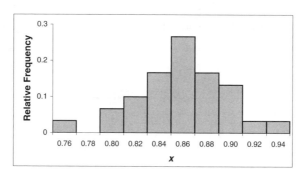

Figure 3.1

Each relative frequency is an *approximation* of the probability that X takes on a value in the respective interval. From the histogram, we see that X has a relatively large probability of taking a value in the interval $(0.85, 0.87]$ and a smaller probability of taking a value in the interval $(0.93, 0.95]$. In mathematical notation, we write

$$P(0.85 < X \leq 0.87) > P(0.93 < X \leq 0.95). \tag{3.1}$$

One problem with the histogram is that it makes the variable appear to be discrete. If one were to only look at the histogram, it would appear that X is discrete with range $\{0.76, 0.78, \ldots, 0.94\}$. One way to correct this problem is to draw a smooth curve that captures the "shape" of the histogram as in Figure 3.2. Such a curve is called a *density curve*.

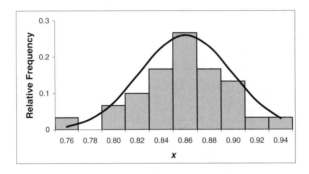

Figure 3.2

The density curve is the graph of the *probability density function* (pdf) of the random variable, which is the continuous analogy of the discrete probability mass function. Notice that over the interval $(0.85, 0.87]$ the area under the density curve is much larger than it is over the interval $(0.93, 0.95]$. This mirrors the relationship in Equation (3.1). Thus we use areas under the density curve to find probabilities of X (i.e., we integrate the pdf).

3.2 Definitions

We begin with a formal definition of a *continuous random variable* and its associated *probability density function*.

Definition 3.2.1 A random variable X is said to be *continuous* if there exists a function f called the *probability density function* (pdf) that is continuous at all but a finite number of points and satisfies the following three properties:

1. $f(x) \geq 0$ for all x,
2. $\int_{-\infty}^{\infty} f(x)\,dx = 1$, and
3. $P(a < X \leq b) = \int_{a}^{b} f(x)\,dx$.

These three properties are very similar to the properties for the pmf of a discrete random variable. The first property says that $f(x)$ cannot be negative. This is so because the pdf is used to find probabilities. The second property says that the total area under the curve $y = f(x)$ must equal 1. The third property describes how the pdf is used to find probabilities. Informally, it says to find the probability that X takes on a value in some interval, we "add up" all values of f over the interval by integrating f over the interval. The graph of a pdf can take many different shapes. Figure 3.3 shows the graph of a certain pdf and illustrates these three properties.

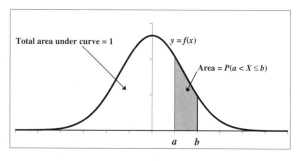

Figure 3.3

Note the differences between the pmf of a discrete random variable and the pdf of a continuous variable. For a discrete variable X, its pdf is $f(x) = P(X = x)$. That is, $f(x)$ gives the probability that X takes on a certain value x. This is not the case for continuous variables. *Areas* under the curve $y = f(x)$ give probabilities. The pdf in and of itself does not give probabilities. Note that if X is continuous, then by property 3 of the pdf and properties of the definite integral,

$$P(X = a) = P(a < X \leq a) = \int_{a}^{a} f(x)\,dx = 0.$$

That is, the probability that X takes on any certain value is 0.

The *cumulative distribution function* (cdf), or simply the *distribution function* $F(x)$ of a continuous random variable X is

$$F(x) = P(X \leq x) = \int_{-\infty}^{x} f(t)\, dt.$$

In terms of the pdf, $F(x)$ is the area under the density curve $y = f(t)$ to the left of $t = x$. This is illustrated in Figure 3.4.

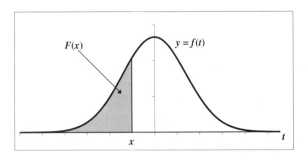

Figure 3.4

One important relationship between F and f is that

$$F'(x) = \frac{d}{dx} \int_{-\infty}^{x} f(t)\, dt = f(x).$$

Example 3.2.1 A Continuous Random Variable Suppose that X is a continuous random variable with range $[1, 2]$ and pdf

$$f(x) = \begin{cases} 2x - 2, & 1 \leq x \leq 2 \\ 0, & \text{elsewhere.} \end{cases}$$

The graph of f is shown in Figure 3.5. Notice that because the curve $y = f(x)$ is "taller" near 2 than it is near 1, X has higher probabilities of taking on values near 2 than near 1.

First we verify that this function satisfies property 2 of a pdf:

$$\int_{-\infty}^{\infty} f(x)\, dx = \int_{1}^{2} 2x - 2\, dx = x^2 - 2x \Big|_{1}^{2} = (4 - 4) - (1 - 2) = 1.$$

Geometrically, we see that the region under the curve $y = f(x)$ is a triangle with base 1 and height 2, which has area 1.

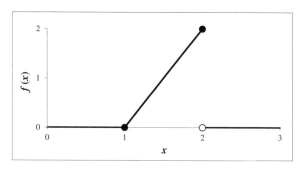

Figure 3.5

To illustrate how the pdf is used to find probabilities, consider the calculation of $P(1 < X \leq 1.25)$. This is done by integrating the pdf over the interval $(1, 1.25]$:

$$P(1 < X \leq 1.25) = \int_1^{1.25} 2x - 2 \, dx = x^2 - 2x \Big|_1^{1.25} = [1.25^2 - 2(1.25)] - (1 - 2) = 0.0625.$$

The cdf is

$$F(x) = P(X \leq x) = \int_{-\infty}^x f(t) \, dt = \begin{cases} 0, & x \leq 1 \\ \int_1^x 2t - 2 \, dt = x^2 - 2x + 1, & 1 < x \leq 2 \\ 1, & x > 2 \end{cases}$$

The graph of $F(x)$ is shown in Figure 3.6. □

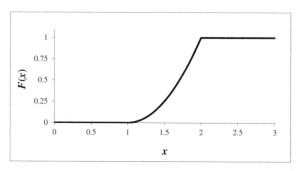

Figure 3.6

The definitions of the expected value, variance, standard deviation, and moment-generating function of a continuous random variable are all similar to those for a discrete random variable.

Definition 3.2.2 Let X be a continuous random variable with pdf $f(x)$. Its *expected value*, or *mean*, is

$$\mu = E(X) = \int_{-\infty}^{\infty} x f(x)\,dx.$$

Its *variance* is

$$\sigma^2 = \text{Var}(X) = E\big[(X-\mu)^2\big] = \int_{-\infty}^{\infty} (x-\mu)^2 f(x)\,dx.$$

Its *standard deviation* is

$$\sigma = \sqrt{\text{Var}(X)}.$$

Its *moment-generating function* is

$$M(t) = E\big[e^{tX}\big] = \int_{-\infty}^{\infty} e^{tx} f(x)\,dx$$

for all values of t for which this integral exists. Lastly, if $u(X)$ is a function of X, then the *expected value of $u(X)$* (or *mathematical expectation* of $u(X)$) is

$$E[u(X)] = \int_{-\infty}^{\infty} u(x) f(x)\,dx.$$

All of these terms are defined only if the corresponding integrals exist.

The interpretations and properties of these terms are the same for continuous variables as they are for discrete variables.

Example 3.2.2 Finding Mean and Variance To illustrate how to calculate the mean and variance of a continuous random variable, consider the random variable X in Example 3.2.1. The mean of X is

$$\mu = \int_{1}^{2} x(2x-2)\,dx = \frac{2}{3}x^3 - x^2 \Big|_{1}^{2} = \frac{5}{3},$$

and its variance is

$$\sigma^2 = \int_1^2 \left(x - \frac{5}{3}\right)^2 (2x - 2)\,dx = \frac{1}{18}$$

so that its standard deviation is $\sigma = \sqrt{\frac{1}{18}} \approx 0.236$. Using integration by parts, we find that the moment-generating function is

$$M(t) = \int_1^2 e^{tx}(2x-2)\,dx = 2e^{tx}\left(\frac{x-1}{t} - \frac{1}{t^2}\right)\bigg|_{x=1}^{x=2} = 2e^{2t}\left(\frac{1}{t} - \frac{1}{t^2}\right) - 2e^t\frac{1}{t^2}, \qquad \text{for } t \neq 0$$

and

$$M(0) = \int_1^2 e^0 (2x-2)\,dx = 1. \qquad \square$$

In the previous examples, the pdf was nonzero on only a finite interval. In the next example, the pdf is nonzero on an unbounded interval.

Example 3.2.3 **Milk Production** Suppose that a local dairy always produces at least 2000 gallons of milk a day. Let the random variable X denote the output on a day, in thousands of gallons, and further suppose that the pdf of X is

$$f(x) = \begin{cases} \dfrac{8}{x^3}, & \text{for } x \geq 2 \\ 0, & \text{otherwise.} \end{cases}$$

A portion of the graph of this pdf is shown in Figure 3.7. Notice that the pdf is decreasing on the interval $[2, \infty)$; thus there is a relatively large probability that X takes values close to 2 and a small probability that X takes larger values.

First we verify that the total area under this curve is 1:

$$\int_{-\infty}^{\infty} f(x)\,dx = \int_2^{\infty} \frac{8}{x^3}\,dx = \lim_{b \to \infty} \left[\frac{-4}{x^2}\right]_2^b = -4 \lim_{b \to \infty} \left(\frac{1}{b^2} - \frac{1}{4}\right) = 1.$$

Next we find the probability that there is more than 3000 gallons of milk produced on a randomly selected day:

$$P(X > 3) = 1 - P(X \leq 3) = 1 - \int_2^3 \frac{8}{x^3}\,dx = 1 - \left[\frac{-4}{x^2}\right]_2^3 = \frac{1}{2}.$$

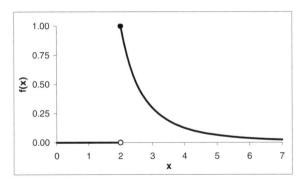

Figure 3.7

The mean amount of milk produced is

$$E(X) = \int_2^\infty x \cdot \frac{8}{x^3}\, dx = \lim_{b\to\infty} \left[\frac{-8}{x}\right]_2^b = -8 \lim_{b\to\infty} \left(\frac{1}{b} - \frac{1}{2}\right) = 4.$$

Thus there is a mean of 4000 gallons of milk produced per day. □

We end this section with the definition of *percentiles*, *quantiles*, and *quartiles*.

Definition 3.2.3 Let X be a continuous random variable with pdf f and cdf F, and let p be a number between 0 and 1. The $(100p)$th *percentile* is a value of X, denoted π_p, such that

$$p = \int_{-\infty}^{\pi_p} f(x)\, dx = F(\pi_p).$$

The $(100p)$th percentile is also called the *quantile of order p*. The 25th, 50th, and 75th percentiles are called the *first*, *second*, and *third quartiles* and are denoted by $p_1 = \pi_{0.25}$, $p_2 = \pi_{0.50}$, and $p_3 = \pi_{0.75}$, respectively. The 50th percentile is also called the *median* and is denoted $m = p_2$. The *mode* of X is the value x for which f is maximum.

From the definition of π_p, we see that the probability that the value of X is less than π_p is p. The 25th, 50th, and 75th percentiles are called *quartiles* because they divide the range of X into four subintervals. The probability that X is less than p_1 is 0.25, the probability that X is between p_1 and p_2 is 0.25, and so on. The 50th percentile is given the special name of *median* because it is the value of X that is "in the middle," meaning that the probability that X is less than the median is 0.50. Likewise, the probability that X is greater than the median is also 0.50.

Example 3.2.4 Finding Quartiles The random variable X in Example 3.2.1 has a cdf $F(x) = x^2 - 2x + 1$ for $1 < x \leq 2$. To find the first quartile, $p_1 = \pi_{0.25}$, we need to solve

$$0.25 = F(\pi_{0.25}) = (\pi_{0.25})^2 - 2(\pi_{0.25}) + 1.$$

Solving this quadratic yields two solutions, $\pi_{0.25} = 1.5$ or $\pi_{0.25} = 0.5$. Because X only has values between 1 and 2, the only solution that makes sense is $p_1 = \pi_{0.25} = 1.5$. Similar calculations show that $p_2 = \pi_{0.50} \approx 1.707$ and $p_3 = \pi_{0.75} \approx 1.867$.

To find the mode of X, we examine Figure 3.5 and observe that the maximum value of the pdf $f(x)$ occurs at $x = 2$. Therefore, the mode of X is 2. □

Exercises

1. A random variable X has the pdf

$$f(x) = \begin{cases} 5(1-x)^4, & 0 < x \leq 1 \\ 0, & \text{otherwise.} \end{cases}$$

   a. Verify that this function meets the first two requirements for a pdf.
   b. Draw a graph of this pdf over the interval $0 < x \leq 1$.
   c. Find the cdf and draw its graph.
   d. Calculate $P(0 < X \leq 0.5)$, $P(X \leq 0)$, and $P(X > 0.25)$.

2. Repeat Exercise 1 above for the pdf

$$f(x) = \begin{cases} 0.5, & 0 < x \leq 0.25 \\ 2, & 0.25 < x \leq 0.4 \\ 1, & 0.4 < x \leq 0.75 \\ 0.9, & 0.75 < x \leq 1 \\ 0, & \text{otherwise.} \end{cases}$$

3. Repeat Exercise 1 above for the pdf

$$f(x) = \begin{cases} 0.25, & 0 < x \leq 0.5 \\ -x + 2.5, & 0.5 < x \leq 1 \\ 0, & \text{otherwise.} \end{cases}$$

4. For each function below, find the value of a so that the function satisfies the requirements for a pdf. In each case, $f(x) = 0$ for x outside the given interval.

a. $f(x) = (\frac{5}{6})x^3$ for $0 < x < a$
b. $f(x) = 0.16x^2$ for $-a < x < a$
c. $f(x) = a(1 - x^4)$ for $-1 < x < 1$
d. $f(x) = a\sqrt{3x}$ for $0 < x < 2$
e. $f(x) = a$ for $1 < x < 5$
f. $f(x) = ae^{-2x}$ for $0 < x$
g. $f(x) = a/x^2$ for $10 < x$

5. For each pdf below, calculate the mean μ and variance σ^2. In each case, $f(x) = 0$ for x outside the given interval.

a. $f(x) = \frac{1}{10}$ for $0 < x < 10$
b. $f(x) = \frac{3}{8}(4x - 2x^2)$ for $0 < x < 2$
c. $f(x) = \frac{1}{5}e^{-x/5}$ for $0 < x$
d. $f(x) = \frac{3}{x^4}$ for $1 < x$

6. A random variable X has the pdf $f(x) = \frac{3}{4}(1 - x^2)$ for $-1 < x < 1$.

a. Find the cdf.
b. Find the first, second, and third quartiles.

7. A random variable X has the pdf $f(x) = 5/x^2$ for $5 < x$.

a. Find the cdf.
b. Find $\pi_{0.10}$ and $\pi_{0.99}$.

8. Lightbulbs of a certain brand last between 1000 and 1200 hr. Let the random variable X denote the life span of a randomly selected bulb, and suppose the pdf of X is

$$f(x) = \begin{cases} \frac{1}{200}, & 1000 < x < 1200 \\ 0, & \text{otherwise.} \end{cases}$$

a. Find the mean life span.

b. Find the probability that a randomly selected bulb will last longer than 1150 hr.

c. Find the conditional probability $P(X > 1150 \,|\, X > 1100)$. (**Hint:** Use the definition of conditional probability. The numerator is the probability of the event $(X > 1150) \cap (X > 1100)$, which means the bulb lasts longer than 1150 *and* 1100 hr. How long must the bulb last in this event?)

9. Let the random variable X denote the weekly demand for gasoline (in thousands of gallons) at a local filling station. Suppose that the pdf of X is

$$f(x) = \begin{cases} \frac{3}{4}(1 - (x-2)^2), & 1 < x < 3 \\ 0, & \text{otherwise.} \end{cases}$$

a. Sketch a graph of this pdf. Based on this graph, is the weekly demand more likely to be closer to 1000, 2000, or 3000? Explain.

b. Find the mean and mode of the weekly demand.

c. If the station has 2700 gal of gasoline in stock at the beginning of a week, find the probability that there is not enough to meet the demand.

d. Find how much gasoline needs to be in stock at the beginning of a week so that the probability of there being enough to meet demand is 0.90. (**Hint:** If a is the amount in stock, then we need $P(X < a) = 0.90$.)

10. If the cdf of a random variable X is $F(x) = 1 - e^{-x^2}$ for $x \geq 0$, find the pdf.

11. The cdf of a random variable X is

$$F(x) = \begin{cases} 0, & x \leq 0 \\ x/4, & 0 < x \leq 1 \\ x^2/4, & 1 < x \leq 2 \\ 1, & x > 2. \end{cases}$$

Calculate the following probabilities:

a. $P(X \leq 1.5)$,
b. $P(X \geq 0.5)$,
c. $P(0.5 < X \leq 1.75)$, and
d. $P(X \geq 0.5 \,|\, X \leq 1.5)$.
e. Find the pdf of X.

12. At his one-year-old check-up, a boy weighs 25 lb. The doctor explains that this weight is the 90th percentile of the weights of all one-year-olds. Explain what this means in a practical sense.

13. A random variable X has the pdf $f(x) = \frac{1}{3}$, for $1 < x < 4$. Find the mgf $M(t)$.

14. A random variable X has the pdf $f(x) = e^{-x}$, for $0 \leq x < \infty$. Find the mgf $M(t)$.

15. Explain why $M(0) = 1$ for every mgf $M(t)$.

16. Explain why $\sigma^2 = \text{Var}(X)$, when it exists, must be greater than or equal to 0 for every continuous random variable X.

17. Let $\epsilon > 0$ be a small number, and let X be a random variable with range $[a, b]$ and pdf $f(x)$. Further assume that f is continuous on $[a, b]$. Explain why $P(x_0 - \epsilon/2 < X < x_0 + \epsilon/2) \approx \epsilon f(x_0)$ for any $x_0 \in (a + \epsilon/2, b - \epsilon/2)$. (**Hint**: Think of this probability as an area under the density curve, and use the mean value theorem for integrals.)

18. Two high school students, Tyson and Kylie, each drive to school every day. Tyson tries to park as close to the front door of the school as possible, so he always looks for a parking spot close to the door. Sometimes he finds a spot very close to the door; however, he does not always find such a parking spot and must park farther from the door. Sometimes he must park very far from the door. Kylie, on the other hand, does not even try to find a parking spot close to the door. Instead, she always drives directly to a parking lot far from the door that always has at least one open spot, parks there, and then walks to the front door.

   a. Let X_T and X_K be the random variables denoting the distance that Tyson and Kylie, respectively, park from the front door of the school. Sketch possible graphs of the pdfs of X_T and X_K.

   b. Which is probably larger, the mean of X_T or the mean of X_K? Likewise, compare the variance of X_T to the variance of X_K. Explain your reasoning.

   c. Suppose Tyson and Kylie live next door to each other. Let Y_T and Y_K be the random variables representing the amount of time it takes each to drive from home and reach the front door of the school. Compare the variance of Y_T to Y_K.

   d. If Tyson were to leave for school at the same time every day, would he always have the same amount of free time before classes start? What about Kylie? Explain your reasoning.

3.3 The Uniform and Exponential Distributions

As with discrete random variables, different types of pdfs (or distributions) of continuous variables are given special names. In this section, we introduce two of the simplest types, the *uniform* and *exponential* distributions.

> **Definition 3.3.1** A continuous random variable X has a *uniform distribution* if its pdf is
>
> $$f(x) = \begin{cases} \dfrac{1}{b-a}, & a \leq x \leq b \\ 0, & \text{otherwise} \end{cases}$$
>
> where $a < b$ are any real numbers. The phrases "X is $U(a,b)$" and "X is uniformly distributed over $[a,b]$" mean that X has this type of distribution.

A graph of the uniform pdf is shown in Figure 3.8. Note the rectangular shape of the graph. For this reason, a uniform distribution is also referred to as a *rectangular distribution*. In the discrete case, a uniform distribution means that all values of X are equally likely. Informally, we can think of a continuous uniform distribution in the same way.

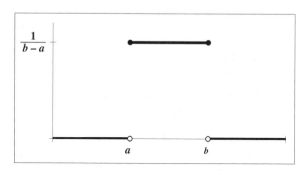

Figure 3.8

Observe that

$$P(X < a) = \int_{-\infty}^{a} 0\, dx = 0 \quad \text{and} \quad P(X > b) = \int_{b}^{\infty} 0\, dx = 0,$$

thus X takes on values only between a and b. The mean of a uniform distribution is

$$\mu = \int_a^b x \frac{1}{b-a} dx = \frac{1}{b-a} \left(\frac{x^2}{2}\right)\bigg|_a^b = \frac{1}{b-a}\left(\frac{b^2}{2} - \frac{a^2}{2}\right) = \frac{(b-a)(b+a)}{2}\frac{1}{b-a} = \frac{a+b}{2}.$$

We leave it as an exercise to verify that the variance and mgf of X are

$$\sigma^2 = \frac{(b-a)^2}{12} \quad \text{and} \quad M(t) = \frac{e^{tb} - e^{ta}}{t(b-a)}, \quad t \neq 0.$$

Example 3.3.1 Dropping a Needle To illustrate how these formulas are used to calculate the mean and variance of a continuous random variable with a uniform distribution, and how to calculate probabilities of such a variable, consider the following scenario: Suppose we randomly drop a needle on a wood floor where the joints between the planks are 1 in apart. Let the random variable X denote the distance from the middle point of the needle to the nearest joint between the planks. Because the drop is random, it seems reasonable to assume that X is $U(0, 0.5)$. Its mean and variance are

$$\mu = \frac{0.5 + 0}{2} = 0.25 \quad \text{and} \quad \sigma^2 = \frac{(0.5-0)^2}{12} \approx 0.0208.$$

The probability that the middle point of the needle lands between 0.2 and 0.4 in from the nearest joint, $P(0.2 < X < 0.4)$, is

$$P(0.2 < X < 0.4) = \int_{0.2}^{0.4} \frac{1}{0.5 - 0} dx = \frac{0.2}{0.5} = 0.40.$$ □

Example 3.3.2 Purchasing Gasoline The owner of a local gas station observes that customers typically purchase between 5 and 20 gallons of gasoline at each fill-up and that no amount is more frequent than any other. If gas costs $2.50 per gallon, find the probability that a randomly selected customer spends more than $30 on gas.

If X denotes the number of gallons of gas purchased by a customer, we assume that X is $U(5, 20)$. Thus its pdf is $f(x) = \frac{1}{15}$ for $5 \leq x \leq 20$. Let Y be the amount of money spent by a customer on gas. In terms of X,

$$Y = 2.5X.$$

We want to find

$$P(Y > 30) = P(2.5X > 30) = P(X > 12) = \int_{12}^{20} \frac{1}{20-5} dx = \frac{8}{15} \approx 0.533.$$

Thus the probability is approximately 0.533. This means that about 53% of customers will spend more than $30 on gas, assuming that the number of gallons they purchase is $U(5, 20)$. □

The next example motivates our next distribution.

Example 3.3.3 Wait Time Consider again the scenario of a secretary who receives a mean of two telephone calls per hour. Let X denote the number of calls received in an hour of time. X has a Poisson distribution with pdf

$$f(x) = P(X = x) = \left(\frac{2^x}{x!}\right)e^{-2}.$$

Now suppose the secretary has just received a call, and consider the time until the *next* call arrives. Let W denote the value of this continuous random variable. To find the pdf of W, we first consider its cdf,

$$F(w) = P(W \leq w) = 1 - P(W > w) = 1 - P(\text{no calls in } w \text{ hr}).$$

Now, to find $P(\text{no calls in } w \text{ hr})$, let Y denote the number of calls received in w hr. Because the mean number of calls received in an hr is 2, the mean of Y is $2w$, and Y has a Poisson distribution. Thus

$$f(y) = P(Y = y) = \left(\frac{2w^y}{y!}\right)e^{-2w}$$

and

$$P(\text{no calls in } w \text{ hr}) = P(Y = 0) = \left(\frac{2w^0}{0!}\right)e^{-2w} = e^{-2w}$$

$$\Rightarrow F(w) = 1 - e^{-2w}$$

$$\Rightarrow f(w) = F'(x) = 2e^{-2w}. \qquad \square$$

We generalize this example in the following definition.

Definition 3.3.2 A random variable X has an *exponential* distribution if its pdf is

$$f(x) = \begin{cases} \lambda e^{-\lambda x}, & x \geq 0 \\ 0, & \text{otherwise} \end{cases}$$

where $\lambda > 0$.

Example 3.3.4 Calculating a Probability of an Exponential Distribution Suppose X has an exponential distribution with parameter $\lambda = 5$. Find $P(1 < X < 3)$.

The pdf of X is $f(x) = 5e^{-5x}$ for $x > 0$, so

$$P(1 < X < 3) = \int_1^3 5e^{-5x}\,dx = -e^{-5x}\Big|_1^3 = e^{-15} + e^{-5} \approx 0.00674. \qquad \square$$

Example 3.3.5 Mean and Variance of an Exponential Distribution To find the mean and variance of an exponential distribution, first we find its mgf:

$$M(t) = E\bigl[e^{tX}\bigr] = \int_0^\infty e^{tx}\left(\lambda e^{-\lambda x}\right)dx = \lim_{b \to \infty} \int_0^b \lambda e^{(t-\lambda)x}\,dx$$

$$= \lim_{b \to \infty} \left[\frac{\lambda}{t-\lambda} e^{(t-\lambda)x}\right]_0^b = \lim_{b \to \infty} \frac{\lambda}{t-\lambda}\left(e^{(t-\lambda)b} - e^0\right)$$

$$= \frac{\lambda}{t-\lambda}(-1), \quad \text{for } t < \lambda$$

$$= \frac{\lambda}{\lambda - t}.$$

Thus,

$$M'(t) = \frac{\lambda}{(\lambda - t)^2} \quad \Rightarrow \quad \mu = M'(0) = \frac{1}{\lambda}$$

and

$$M''(t) = \frac{2\lambda}{(\lambda - t)^3} \quad \Rightarrow \quad M''(0) = \frac{2}{\lambda^2}$$

$$\Rightarrow \quad \sigma^2 = M''(0) - [M'(0)]^2 = \frac{2}{\lambda^2} - \left(\frac{1}{\lambda}\right)^2 = \frac{1}{\lambda^2}.$$

Therefore, a random variable with an exponential distribution has a mean and standard deviation both equal to $1/\lambda$. For this reason, we often let $\theta = 1/\lambda$ so that the pdf is

$$f(x) = \frac{1}{\theta} e^{-x/\theta}$$

and the mean and standard deviation both equal θ. $\qquad \square$

A graph of a typical exponential pdf is shown in Figure 3.9. Notice that $f(x)$ is much larger for values of x near 0 than for larger values of x. This indicates that X has higher a probability of taking values near 0 than larger values.

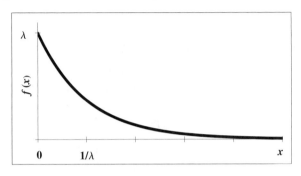

Figure 3.9

As illustrated in Example 3.3.3, exponential distributions are often used to describe the waiting time *between* occurrences that are described by a Poisson distribution. The parameter λ in both distributions has the same meaning—it is the mean number of occurrences in an interval.

The cdf of a variable with an exponential distribution is

$$F(x) = P(X \leq x) = F(x) = \int_{-\infty}^{x} f(t)\, dt = \int_{0}^{x} \lambda e^{-\lambda t}\, dt = 1 - e^{-\lambda x}, \quad \text{for } x > 0. \quad (3.2)$$

From the cdf, we also get

$$P(X > x) = 1 - P(X \leq x) = 1 - \left(1 - e^{-\lambda x}\right) = e^{-\lambda x}, \quad \text{for } x > 0. \quad (3.3)$$

Example 3.3.6 **Defects in Wire** Suppose a manufacturer of electrical wire observes a mean of 0.5 defect per 100 ft of wire. Find the mean distance between defects (in feet) and find the probability that the distance between one defect and the next is less than 125 ft.

It seems reasonable to assume that the number of defects in a foot of wire has a Poisson distribution with parameter $\lambda = (0.5 \text{ defect})/(100 \text{ ft}) = 0.005$ defect/ft. If we let X denote the distance in feet between defects, then we could think of X as the "wait time" between events. Thus it seems reasonable to assume that X is exponential with $\lambda = 0.005$. The mean of X is

$$\mu = \frac{1}{0.005} = 200,$$

and from Equation (3.2) we have

$$P(X < 125) = F(125) = 1 - e^{-0.005(125)} \approx 0.465. \qquad \square$$

Example 3.3.7 **Verifying Assumptions** Many customers at a local grocery store have been complaining about the amount of time they wait in line at the checkout stand. So the manager decides to offer a $0.50 coupon to anyone who waits longer than 2.5 min. Before starting the program, she wants to find the probability that a randomly selected customer will receive a coupon. If the random variable X denotes the time that a customer spends waiting in line, then she wants to know $P(X > 2.5)$. She assumes that because the exponential distribution is used to describe the wait time between events, then X should also be exponential.

To test this assumption, she observes 30 values of X by measuring the time, in minutes, that 30 different customers spend waiting in line. These values (or data) are shown in Table 3.3.

0.03	0.21	0.33	0.66	0.91	1.33	1.56	1.84	2.49	3.46
0.04	0.21	0.36	0.87	0.92	1.40	1.60	1.87	2.79	3.80
0.10	0.22	0.47	0.90	0.96	1.51	1.76	2.36	3.24	4.10

Table 3.3

We begin by constructing a relative frequency distribution of the data as in the first three rows of Table 3.4 and the relative frequency histogram as in Figure 3.10. Notice that the shape of the histogram resembles the shape of the graph of an exponential pdf. This is evidence that X does indeed have an exponential distribution.

Interval	(0, 1]	(1, 2]	(2, 3]	(3, 4]	(4, 5]
Freq	15	8	3	3	1
Rel Freq	0.50	0.27	0.10	0.10	0.03
Theor Prob	0.51	0.25	0.12	0.06	0.03

Table 3.4

To further verify the assumption, note that the sum of these 30 wait times is 42.30. Therefore, the average wait time is

$$\text{Average} = \frac{42.30}{30} = 1.41 \text{ min.}$$

Thus the mean of X is approximately 1.41, and the parameter λ is approximately

$$\lambda = \frac{1}{1.41} \approx 0.71.$$

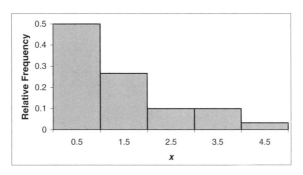

Figure 3.10

So if X is exponential with $\lambda = 0.71$, then, for instance,

$$P(0 < X \leq 1) = \int_0^1 0.71 e^{-0.71x} dx \approx 0.51.$$

This means that if X is indeed exponential with $\lambda = 0.71$, then about 51% of observed values of X will be between 0 and 1. Note that the relative frequency of the bin $(0, 1]$ is 0.5, which is very close to the theoretical probability.

The relative frequencies and theoretical probabilities for other intervals are shown in Table 3.4. Note that these values are very close in each case. This leads us to conclude that the assumption that X is exponential with parameter $\lambda = 0.71$ is reasonable.

Thus the probability that a randomly selected customer will receive a coupon is, using Equation (3.3),

$$P(X > 2.5) = e^{-0.71(2.5)} \approx 0.169.$$

This means that about 17% of all customers will receive a coupon. □

Software Calculations

Minitab: To calculate values of the exponential pdf $f(x)$, begin by entering the desired values of x in a column. Then select **Calc** → **Probability Distributions** → **Exponential**. Select **Probability density** and enter the scale (the mean $1/\lambda$). Set the threshold to 0. Next to **Input column**: select the column with the desired values of x; next to **Optional storage**: select the column where you want to store the values of $f(x)$. To calculate values of the cdf $F(x)$, follow the same steps, except select **Cumulative probability**.

R: The syntax for calculating values of the exponential pdf $f(x)$ is dexp(x, λ). To calculate values of the binomial cdf $F(x)$, change "dexp" to "pexp."

Excel: The syntax for calculating values of the exponential pdf $f(x)$ is EXPONDIST($x, \lambda, 0$). To calculate values of the cdf $F(x)$, the syntax is EXPONDIST($x, \lambda, 1$).

TI-83/84 Calculators: There is no built-in command to evaluate the exponential pdf or cdf. However, the formulas $f(x) = \lambda e^{-\lambda x}$ and $F(x) = 1 - e^{-\lambda x}$ can easily be evaluated by entering either one as **Y1** in the **Y=** menu. Then press **VARS** → **Y-VARS** → **1:FUNCTION** → **1:Y1** from the home screen. Then press (, the desired value of x, and).

Exercises

1. In each scenario, determine if it is reasonable to assume that the given random variable has a uniform distribution. If so, give the values of a and b. Explain your reasoning.

   a. A dart is randomly thrown at a dartboard. Let X denote the distance the dart lands from the center.

   b. A teacher gives a student a 12-in piece of string and tells her to cut it at a random point. Let X denote the length of the shorter piece.

   c. A student arrives for a class. Let X denote the number of minutes before or after the class start time that the student arrives.

   d. A man and a woman agree to meet at a restaurant sometime between 6:00 and 6:15 PM. Let X denote the number of minutes past 6:00 that the man arrives.

   e. A man and a woman agree to meet at a restaurant at 6:00 PM. Let X denote the number of minutes before or after 6:00 the woman arrives (where a negative value means she arrives before 6:00 and a positive value means she arrives afer 6:00).

2. If X is $U(a,b)$, use the definitions to verify that $\text{Var}(X) = (b-a)^2/12$ and that the mgf of X is $(e^{tb} - e^{ta})/[t(b-a)]$ for $t \neq 0$.

3. Suppose X, the amount of time a child spends playing with a new toy, is $U(15, 25)$. Find

   a. $E(X)$,
   b. $\text{Var}(X)$,
   c. $P(18 < X < 23)$, and
   d. $P(12 < X < 18)$.

4. If X is $U(0,1)$, find the following probabilities by rewriting them in the form $P(a < X < b)$.

   a. $P(4X + 2 < 4)$
   b. $P(3 < 4X + 2 < 5)$

5. If X is $U(0,1)$, and d and c are numbers such that $d - c > 1.5$, show the following is true:
$$P[c + 0.5 < X \cdot (d - c) + c < c + 1.5] = 2P[c < X \cdot (d - c) + c < c + 0.5]$$

(**Hint**: Rewrite both probabilities in the form $P(a < X < b)$ and use the fact that because X is $U(0,1)$, $P(a < X < b) = b - a$ when $0 \leq a < b \leq 1$.)

6. If X is $U(-2, 2)$, find $P(|X| > 1)$.

7. If X is $U(a, b)$, find values of c and d in terms of a and b such that

   a. $P(X < d) = 0.75$ where $a < d < b$,
   b. $P(c < X) = 0.90$ where $a < c < b$, and
   c. $P\left(\frac{a+b}{2} - c < X < \frac{a+b}{2} + c\right) = 0.60$ where $0 < c < \frac{b-a}{2}$.
   d. Explain why there is not a unique solution to the problem of finding c and d in the interval (a, b) such that $P(c < X < d) = 0.30$.

8. An engineer designing a certain miniature lightbulb believes that the life span of the bulb, in hours, is $U(15, 19)$. A customer asks the question, "What is the probability a bulb will last between 14 and 16 hr?" The engineer responds with the answer of $\frac{1}{2}$. Is this answer correct based on the assumption that the life span is $U(15, 19)$? If not, find the correct answer and explain what mistake the engineer made.

9. Commuter trains arrive at a certain train station every 20 min, starting at 6:00 AM. If a passenger arrives at a time that is uniformally distributed between 6:00 and 6:40 AM, find the probability the passenger has to wait

   a. less than 5 min for a train, and
   b. more than 10 min for a train.

(**Hint**: If the passenger waits the specified amount of time, what are the possible times the passenger could have arrived?)

10. Suppose the number of calls a secretary receives over a 1-hr period is described by a Poisson distribution. If it is known that the secretary received exactly 1 call in that 1-hr period, it can be shown that the time the call was received is uniformly distributed over that 1-hr period. If the secretary receives exactly 1 call between 9:00 and 10:00 AM, find the probability that the call was received between 9:20 and 9:45 AM.

11. Calls arrive at a tech support center with a mean of 4 per hour. Assuming the number of calls received in 1 hr is described by a Poisson distribution, find the probability that there is more than 20 min between the arrivals of randomly selected successive calls.

12. Suppose an expert dart player throws a dart at a dartboard such that X, the distance the dart lands from the center, is exponentially distributed with mean of 0.5 in. If the bull's-eye has a diameter of $\frac{5}{8}$ in, find the probability of hitting the bull's-eye.

13. Let X be exponentially distributed with parameter $\lambda = 2$.

   a. Show that $P(X > 2+1 \mid X > 1) = P(X > 2)$.

   b. Generalize your calculations in part a. Suppose X is exponentially distributed with parameter λ. Show that $P(X > a+b \mid X > b) = P(X > a)$ for all $a, b \geq 0$. A random variable X with this property is said to be *memoryless*.

14. An engineer assumes that the random variable X denoting the life span (in hours) of a particular type of lightbulb is exponentially distributed with parameter $\lambda = 0.001$.

   a. Find the probability that a randomly selected lightbulb lasts at least 200 hr. That is, find $P(X > 200)$.

   b. Suppose a lightbulb has lasted at least 900 hr. Use the results from Exercise 13 above to find the probability that it lasts at least 200 hr more. (**Hint:** Evaluate $P(X > 1100 \mid X > 900)$.)

   c. Do the results from parts a and b cause you to question the assumption that X has an exponential distribution? Why or why not?

15. Let X have an exponential distribution. In parts a and b, find the percentiles for the given value of λ.

   a. $\pi_{0.25}$, $\lambda = \frac{1}{4}$
   b. $\pi_{0.95}$, $\lambda = 0.1$
   c. Generalize your results. Find a formula for π_p in terms of p and λ.

16. Let X be $U(a,b)$. In parts a and b, find the percentiles for the given values of a and b.

   a. $\pi_{0.12}$, $a = 0$, $b = 1$
   b. $\pi_{0.99}$, $a = -8$, $b = 15.5$
   c. Generalize your results. Find a formula for π_p in terms of p, a, and b.

17. Find the median of an exponential distribution in terms of λ. Do the same for a uniform distribution in terms of a and b.

18. The table below shows the number of gallons of gas purchased by 30 different customers at a local gas station. Use an approach similar to that in Example 3.3.7 to determine if it is reasonable to assume that X, the number of gallons of gas purchased by each customer, is $U(5, 20)$. (**Suggestion**: Use bins of width 3.)

5.08	6.99	9.12	9.97	10.72	11.49	13.22	14.78	16.15	19.02
5.28	7.72	9.38	10.40	11.04	12.12	14.72	15.83	16.53	19.25
6.48	7.92	9.79	10.55	11.34	12.13	14.73	15.86	18.33	19.96

19. The table below shows the number of minutes 30 different cars spent waiting for food at a drive-in diner. Use an approach similar to that in Example 3.3.7 to determine if it is reasonable to assume that X, the number of minutes spent waiting at this diner, is exponentially distributed. (**Suggestion**: Use bins of width 2.)

0.87	3.39	3.74	4.04	4.34	4.99	5.71	5.92	7.14	8.07
2.42	3.61	3.74	4.06	4.36	5.38	5.85	6.63	7.19	8.20
2.86	3.61	3.91	4.18	4.79	5.43	5.91	6.71	7.21	10.24

20. The *skewness* of a random variable X is defined as

$$\gamma_1 = E\left[\left(\frac{X-\mu}{\sigma}\right)^3\right] = \frac{E[X^3] - 3\mu\sigma^2 - \mu^3}{\sigma^3}.$$

The skewness is a measure of the asymmetry of the density curve of X. If the density curve is symmetric, then the skewness is 0.

   a. Use the results of Example 3.3.5 and the fact that $E[X^3] = M'''(0)$ to show that if X has an exponential distribution, then $\gamma_1 = 2$.

b. Find the skewness of a random variable that is $U(a,b)$.

21. The *excess kurtosis*, or simply *kurtosis*, of a random variable X is

$$\gamma_2 = \frac{E[(X-\mu)^4]}{\sigma^4} = \frac{E[X^4] - 4\mu E[X^3] + 2\mu^2 E[X^2] + 5\mu^4}{\sigma^4} - 3.$$

The kurtosis measures the "peakedness" of the density curve of X. A large kurtosis means the density curve has a sharper peak while a lower kurtosis means the density curve has a more rounded peak. Show that for a random variable X with an exponential distribution, $\gamma_2 = 6$.

3.4 The Normal Distribution

In this section, we introduce one of the most important distributions, the *normal* distribution. We begin by defining a simple case of the more general normal distribution, the *standard normal* distribution.

Definition 3.4.1 A continuous random variable Z has a *standard normal* distribution if its pdf is

$$f(z) = \frac{1}{\sqrt{2\pi}} e^{-z^2/2}, \quad \text{for } -\infty < z < \infty.$$

The graph of the standard normal pdf is shown in Figure 3.11. Notice the distinctive bell curve shape. Note that the mode is 0 (which, as we will see, is also the mean) and that the graph is symmetric with respect to the y axis. Because $f(z) > 0$ for all z, the variable Z could take on *any* value. However, note that for values of z less than -3 or greater than $+3$, $f(z)$ is practically 0.

To show that this function satisfies the second property of a pdf, we need to show that

$$\int_{-\infty}^{\infty} \frac{1}{\sqrt{2\pi}} e^{-z^2/2} \, dz = 1.$$

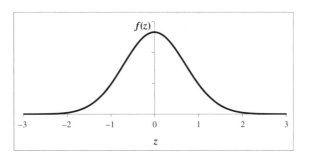

Figure 3.11

To this end, let $A = \int_{-\infty}^{\infty} (1/\sqrt{2\pi}) e^{-z^2/2} \, dz$. Then

$$A^2 = \left(\int_{-\infty}^{\infty} \frac{1}{\sqrt{2\pi}} e^{-z^2/2} \, dz \right) \left(\int_{-\infty}^{\infty} \frac{1}{\sqrt{2\pi}} e^{-y^2/2} \, dy \right)$$
$$= \frac{1}{2\pi} \int_{-\infty}^{\infty} \int_{-\infty}^{\infty} e^{-(z^2+y^2)/2} \, dz \, dy.$$

Using polar coordinates with $z = r \cos \theta$ and $y = r \sin \theta$ so that $z^2 + y^2 = r^2$ and $dz \, dy = r \, dr \, d\theta$, we have

$$A^2 = \frac{1}{2\pi} \int_0^{2\pi} \left(\int_0^{\infty} e^{-r^2/2} r \, dr \right) d\theta.$$

Now,

$$\int_0^{\infty} e^{-r^2/2} r \, dr = \lim_{b \to \infty} -e^{-r^2/2} \Big|_0^b = \lim_{b \to \infty} -e^{-b^2/2} + e^0 = 0 + 1 = 1.$$

Thus

$$A^2 = \frac{1}{2\pi} \int_0^{2\pi} 1 \, d\theta = \frac{1}{2\pi} (2\pi) = 1.$$

Because $A > 0$, we conclude that $A = 1$ as required. To find the mean and variance of this distribution, we first find the mgf:

$$M(t) = E[e^{Zt}] = \int_{-\infty}^{\infty} e^{zt} \frac{1}{\sqrt{2\pi}} e^{-z^2/2} \, dz = \int_{-\infty}^{\infty} \frac{1}{\sqrt{2\pi}} e^{-(z^2-2zt)/2} dz$$

$$= \int_{-\infty}^{\infty} \frac{1}{\sqrt{2\pi}} e^{-(z^2-2zt+t^2-t^2)/2} \, dz$$

$$= \int_{-\infty}^{\infty} \frac{1}{\sqrt{2\pi}} e^{-[(z-t)^2-t^2]/2} \, dz$$

$$= e^{t^2/2} \int_{-\infty}^{\infty} \frac{1}{\sqrt{2\pi}} e^{-(z-t)^2/2} \, dz$$

Notice that this integrand is simply the pdf of a standard normal distribution with z replaced by $(z - t)$. As shown above, this integral equals 1 so that

$$M(t) = e^{t^2/2}.$$

Now,

$$M'(t) = te^{t^2/2} \quad \Rightarrow \quad M'(0) = \mu = 0,$$

and

$$M''(t) = e^{t^2/2} + t^2 e^{t^2/2} \quad \Rightarrow \quad M''(0) = e^0 + 0 = 1$$

$$\Rightarrow \quad \sigma^2 = M''(0) - [M'(0)]^2 = 1 - 0^2 = 1.$$

Thus the mean of a standard normal distribution is 0, and its variance and standard deviation are 1.

The cdf of a standard normal distribution is given the name Φ instead of F and is described by

$$\Phi(z) = P(Z \leq z) = \int_{-\infty}^{z} \frac{1}{\sqrt{2\pi}} e^{-t^2/2} \, dt.$$

Evaluating this integral is not a trivial matter. Values of Φ are often given in tables or found with software. Table C.1 in Appendix C gives the values of Φ to four decimal places for values of z between -3.50 and 3.50. Using this table to find probabilities other than those of the form $P(Z \leq z)$ requires the use of two important properties:

1. $P(Z > z) = 1 - P(Z \leq z) = 1 - \Phi(z)$, and
2. $P(a < Z \leq b) = \Phi(b) - \Phi(a)$.

Example 3.4.1 Finding Probabilities To illustrate the use of Table C.1, consider the calculations of the following probabilities where Z has a standard normal distribution:

$$P(Z \leq 2.05) = 0.9798,$$
$$P(Z > -1.23) = 1 - \Phi(-1.23) = 1 - 0.1093 = 0.8907,$$

and

$$P(-2.65 < Z \leq 0.23) = \Phi(0.23) - \Phi(-2.65) = 0.5910 - 0.0040 = 0.5870. \quad \square$$

Next we define the more general *normal* distribution.

Definition 3.4.2 A continuous random variable X has a *normal distribution* if its pdf is

$$f(x) = \frac{1}{\sigma\sqrt{2\pi}} e^{-(x-\mu)^2/(2\sigma^2)}, \quad \text{for } -\infty < x < \infty$$

where μ is any real number and $\sigma > 0$.

The phrase "X is $N(\mu, \sigma^2)$" means that X has a normal distribution with parameters μ and σ. Notice that when $\mu = 0$ and $\sigma = 1$, this pdf is the same as that for the standard normal distribution. Thus if Z has a standard normal distribution, it is $N(0, 1)$.

Using techniques similar to that for the standard normal distribution, we can show that the mgf of a normal distribution is

$$M(t) = \exp\left\{\mu t + \frac{\sigma^2 t^2}{2}\right\}$$

so that

$$M'(t) = (\mu + \sigma^2 t) \exp\left\{\mu t + \frac{\sigma^2 t^2}{2}\right\} \Rightarrow M'(0) = \mu$$

and

$$M''(t) = \sigma^2 \exp\left\{\mu t + \frac{\sigma^2 t^2}{2}\right\} + (\mu + \sigma^2 t)^2 \exp\left\{\mu t + \frac{\sigma^2 t^2}{2}\right\} \Rightarrow M''(0) = \sigma^2 + \mu^2.$$

Thus the mean of a random variable x that is $N(\mu, \sigma^2)$ is μ and the variance of x is $M''(0) - M'(0)^2 = \sigma^2 + \mu^2 - \mu^2 = \sigma^2$, thus justifying the use of the symbols μ and σ as the parameters.

Figure 3.12 shows the graphs of the normal pdf for several different values of μ and σ. Note that the larger σ is, the wider and shorter the bell curve is, and that in all cases, the graph has a bell curve shape, the mode is μ, the graph is symmetric about the line $x = \mu$, and $f(x)$ is practically 0 for x outside of the interval $(\mu - 3\sigma, \mu + 3\sigma)$.

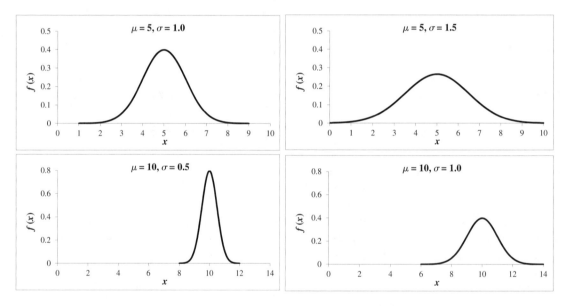

Figure 3.12

Normally distributed random variables satisfy the *empirical rule* (also called the 68-95-99.7 rule) illustrated in Figure 3.13. This rule states that if X is $N(\mu, \sigma^2)$, then

$$P(\mu - \sigma < X \leq \mu + \sigma) \approx 0.68, \quad P(\mu - 2\sigma < X \leq \mu + 2\sigma) \approx 0.95,$$

$$\text{and} \quad P(\mu - 3\sigma < X \leq \mu + 3\sigma) \approx 0.997$$

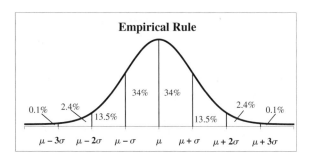

Figure 3.13

(see Exercise 4 in this section). In practical terms, this rule means that if we have *many* observed values of X, then

- about 68% of the values lie within 1 standard deviation of the mean,
- about 95% of the values lie within 2 standard deviations of the mean, and
- about 99.7% of the values lie within 3 standard deviations of the mean.

If X is $N(\mu, \sigma^2)$ where $\mu \neq 0$ or $\sigma \neq 1$, we say that X has a *nonstandard* normal distribution. The next theorem tells us how we can use Table C.1 to find probabilities of such a variable.

Theorem 3.4.1 *If X is $N(\mu, \sigma^2)$, then the random variable $Z = (X - \mu)/\sigma$ is $N(0,1)$.*

Proof. We proceed by finding the cdf of Z:

$$F(z) = P(Z \leq z) = P\left(\frac{X - \mu}{\sigma} \leq z\right) = P(X \leq z\sigma + \mu)$$

$$= \int_{-\infty}^{z\sigma+\mu} \frac{1}{\sigma\sqrt{2\pi}} e^{-(x-\mu)^2/(2\sigma^2)} \, dx$$

Now we make a change of variables by letting $y = (x - \mu)/\sigma$ so that $dy = (1/\sigma)\, dx$,

$$x \to -\infty \;\Rightarrow\; y \to -\infty,$$

and
$$x = z\sigma + \mu \;\Rightarrow\; y = \frac{(z\sigma+\mu)-\mu}{\sigma} = z.$$

Thus

$$F(z) = \int_{-\infty}^{z} \frac{1}{\sqrt{2\pi}} e^{-y^2/2} \, dy = \Phi(z),$$

so Z has the same cdf as that of a standard normal random variable. Therefore, Z must have the same pdf, and we conclude that Z is $N(0,1)$ as desired. □

To use Theorem 3.4.1 to find probabilities of a random variable with a nonstandard normal distribution, observe that if X is $N(\mu, \sigma^2)$, then

$$P(X \leq x) = P\left(\frac{X-\mu}{\sigma} \leq \frac{x-\mu}{\sigma}\right),$$

but the random variable $(X-\mu)/\sigma$ is $N(0,1)$. The quantity $(x-\mu)/\sigma$ is called the *z-score* of the value. Thus we define

$$z = \frac{x-\mu}{\sigma},$$

so that

$$P(X \leq x) = P(Z \leq z) = \Phi(z),$$

which can be found with the use of Table C.1.

Example 3.4.2 Heights of Women Consider the random experiment of choosing a woman at random and measuring her height. Let the random variable X denote the height in inches. Assuming that X is $N(63.6, 2.5^2)$, find the probability of selecting a woman who is between 60 and 62 in tall.

In mathematical notation, we want to find $P(60 < X \leq 62)$. We apply Theorem 3.4.1 by calculating the z-scores of both 60 and 62:

$$z_1 = \frac{60 - 63.6}{2.5} = -1.44 \quad \text{and} \quad z_2 = \frac{62 - 63.6}{2.5} = -0.64.$$

Thus,

$$P(60 < X \leq 62) = P(-1.44 < Z \leq -0.64)$$
$$= \Phi(-0.64) - \Phi(-1.44) = 0.2611 - 0.0749 = 0.1862.$$

In practical terms, this probability means that approximately 18.62% of women are between 60 and 62 in tall, assuming that X is $N(63.6, 2.5^2)$. □

Example 3.4.3 Finding Percentiles Consider again the random variable X denoting the height of a woman in inches, which we assume is $N(63.6, 2.5^2)$. Find the 33rd percentile, $\pi_{0.33}$, of this random variable.

The 33rd percentile is a height that is larger than 33% of all other heights of women. In mathematical notation,

$$0.33 = P(X \leq \pi_{0.33}).$$

To find the value of $\pi_{0.33}$, we first find the 33rd percentile of the standard normal distribution. By examining Table C.1, we note that $\Phi(-0.44) = 0.3300$. This means that $P(Z \leq -0.44) = 0.3300$. Thus -0.44 is the 33rd percentile of the standard normal distribution by definition. (Note that we are really evaluating $\Phi^{-1}(0.33)$.)

Next, note that

$$0.3300 = P(Z \leq -0.44) = P\left(\frac{X - \mu}{\sigma} \leq -0.44\right) = P\left(\frac{X - 63.6}{2.5} \leq -0.44\right) = P(X \leq 62.5).$$

Thus the 33rd percentile of X is 62.5. This means that about 33% of women are shorter than 62.5 in, assuming that X really is $N(63.6, 2.5^2)$. □

Software Calculations

Minitab: To calculate $P(X \leq b)$ where X is $N(\mu, \sigma^2)$, select **Calc → Probability Distributions → Normal**. Select **Cumulative probability** and enter the mean and standard deviation. Next to **Input constant**: enter the value of b. To find a percentile π_p, select **Inverse cumulative probability** and enter p next to **Input constant**.

R: The syntax for calculating $P(X \leq b)$ where X is $N(\mu, \sigma^2)$ is dnorm(b, μ, σ). To find a percentile π_p, the syntax is qnorm(p, μ, σ).

Excel: The syntax for calculating $P(X \leq b)$ where X is $N(\mu, \sigma^2)$ is NORM.DIST$(b, \mu, \sigma, 1)$. To find a percentile π_p, the syntax is NORMINV(p, μ, σ).

TI-83/84 Calculators: To calculate $P(a < X \leq b)$ where X is $N(\mu, \sigma^2)$, press **2nd → DISTR → 2:normalcdf**. The syntax for this function is normalcdf(a, b, μ, σ). To evaluate $P(X \leq b)$, follow the same steps and let a be a large negative number. To find a percentile π_p, select **3:invNorm**, and the syntax is invNorm(p, μ, σ).

Exercises

1. If Z is $N(0,1)$, use Table C.1 to find these probabilities:

   a. $P(Z > 1.78)$ b. $P(-1.54 < Z \leq 0.98)$
   c. $P(|Z| > 2.5)$ d. $P(Z^2 + 1 < 2.6)$

2. If X is $N(12, 1.4^2)$, find these probabilities:

   a. $P(X > 13.4)$ b. $P(10.5 < X \leq 14.5)$
   c. $P(|X - 12| > 0.9)$ d. $P(X^2 + 10 < 110)$
   e. $P(X > 14 \,|\, X > 13.4)$

3. If X, the weight of a randomly selected person in lb, is $N(172, 29^2)$, find

   a. the percentage of people who weigh between 150 and 190 lb,

   b. the 45th percentile, and

   c. the 95th percentile.

4. Verify the probabilities in the empirical rule. That is, if X is $N(\mu, \sigma^2)$, show that

   a. $P(\mu - \sigma < X \leq \mu + \sigma) \approx 0.68$ b. $P(\mu - 2\sigma < X \leq \mu + 2\sigma) \approx 0.95$
   c. $P(\mu - 3\sigma < X \leq \mu + 3\sigma) \approx 0.997$

5. Let X denote the weight, in ounces, of a "2-lb" package of carrots, and assume that X is $N(32.5, 1.1^2)$. Use the empirical rule to find the percentages of these packages that weigh between

   a. 31.4 and 33.6 oz   b. 30.3 and 34.7 oz
   c. 29.2 and 35.8 oz

6. If X is $N(\mu, \sigma^2)$, use the empirical rule to estimate the following percentiles of X in terms of μ and σ:

   a. $\pi_{0.84}$ b. $\pi_{0.975}$ c. $\pi_{0.999}$

7. If Z is $N(0,1)$, use Table C.1 to approximate the values of $c > 0$ such that

   a. $P(-c < Z \leq c) = 0.90$ b. $P(-c < Z \leq c) = 0.95$
   c. $P(-c < Z \leq c) = 0.99$

8. When planning a family trip, a father budgets $200 for gasoline. If the actual amount spent on gasoline has a normal distribution with mean $185 and standard deviation $10, find the probability that the amount exceeds the budgeted amount. Does this budgeted amount appear to be reasonable, or should it be increased?

9. A machine is designed to produce bolts with a 3-in diameter. The actual diameter of the bolts has a normal distribution with mean of 3.002 in and standard deviation of 0.002 in. Each bolt is measured and accepted if the length is within 0.005 in of 3 in; otherwise the bolt is scrapped. Find the percentage of bolts that are scrapped.

10. A corn syrup bottling machine fills bottles that are labeled as containing 12 oz of syrup. The actual amount dispensed into a bottle has a normal distribution with mean μ and standard deviation of 0.1 oz. The machine can be adjusted so that μ is any desired value. Find the value of μ so that the bottles are underfilled only 1.5% of the time.

11. Define the random variable $Y = e^X$ where X is $N(5, 1^2)$ (the variable Y is said to have a *lognormal distribution*). Find $P(125 < Y \leq 175)$.

12. The table below shows the heights of 30 different randomly selected women. Use an approach similar to that in Example 3.3.7 to determine if it is reasonable to assume that X, the height of a woman, is $N(63.6, 2.5^2)$. You do not need to try to calculate the mean or variance from the data. (**Suggestion**: Use five intervals of width 2, the first of which is $[59, 61)$.)

59.38	61.62	62.11	63.02	63.48	64.33	64.60	65.32	65.91	66.81
59.80	62.02	62.16	63.16	63.96	64.34	64.69	65.34	66.08	67.64
61.00	62.05	62.82	63.45	64.06	64.59	65.03	65.45	66.29	68.31

13. Three different random variables X, Y, and Z are each normally distributed. The graphs of their pdfs are shown in Figure 3.14. Without doing any calculations, determine which has the largest mean and which has the smallest mean. Which has the largest variance and which has the smallest variance? Explain your reasoning.

14. A student breaks a piece of 10-in spaghetti into two pieces. Let X denote the length of the left piece, in inches, and assume that X is $N(6, 1^2)$. The student then constructs a rectangle using the long piece as the length and the short piece as the height. Find the expected value of the area of the rectangle. (**Hint**: Use the fact that $E(X^2) = M''(0)$.)

15. Solve these problems pertaining to the standard normal and normal distributions.

   a. Show that the standard normal pdf is an even function. That is, show that $f(-z) = f(z)$.

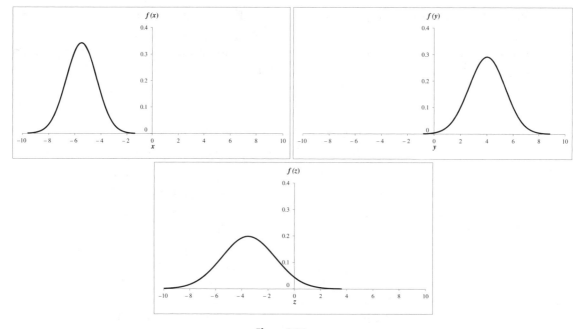

Figure 3.14

b. Show that for the normal pdf $f(x)$, if $h > 0$ is any number, then $f(\mu - h) = f(\mu + h)$.

c. Find the median of the normal distribution.

d. Show that μ is the mode of the normal distribution. (**Hint**: Solve $f'(x) = 0$ and argue that the solution is the location of a local maximum.)

e. Show that the maximum value of $f(x)$, the normal pdf, is $1/(\sigma\sqrt{2\pi})$.

16. In the text, we proved Theorem 3.4.1 by finding the cdf of Z. Prove this theorem in another way by using moment-generating functions. Specifically, use the fact that if a random variable X has the mgf $M_X(t)$, then the random variable $Y = aX + b$, where a and b are any real numbers, has the mgf $M_Y(t) = e^{bt} M_X(at)$.

17. Show that the graph of the normal pdf $y = f(x)$ has points of inflection at $x = \mu \pm \sigma$. (An inflection point is where the second derivative changes signs.)

18. Use the definitions of skewness and kurtosis in Exercises 20 and 21 of Section 3.3 to show that the skewness and kurtosis of a variable X with a standard normal distribution are both 0.

19. Calculating the values in Table C.1 is not a trivial matter. One way to calculate them is to use numeric integration techniques such as the trapezoidal rule. Another way is to use the following formula for $z > 0$:

$$P(Z \leq z) \approx \frac{1}{2} + \frac{1}{2}\left\{1 - \frac{1}{30}\left[7e^{-z^2/2} + 16e^{-z^2(2-\sqrt{2})} + \left(7 + \frac{\pi}{4}z^2\right)e^{-z^2}\right]\right\}^{1/2}.$$

(See Richard J. Bagby, "Calculating Normal Probabilities," *The American Mathematical Monthly*, Vol. 102, No. 1 (Jan. 1995), pp. 46–49 for a detailed derivation of this formula.) Use this formula to find $P(Z \leq 0.1236)$.

20. It can be shown that

$$\lim_{z \to \infty} \frac{1 - \Phi(z)}{[1/(z\sqrt{2\pi})]\,e^{-z^2/2}} = 1.$$

Informally, this means that for large z, $1 - \Phi(z) \approx [1/(z\sqrt{2\pi})]\,e^{-z^2/2}$ and that as z gets larger, this approximation gets better. Use this to show that for any $a > 0$,

$$\lim_{z \to \infty} \frac{P(Z > z + \frac{a}{z})}{P(Z > z)} = e^{-a}$$

where Z is $N(0,1)$.

3.5 Functions of Continuous Random Variables

With continuous random variables, as with discrete random variables, it is often necessary to work with functions of a random variable. In this section, we describe some basic techniques for finding the distributions of functions of a random variable. These techniques are based on this relationship between the cdf and the pdf:

$$\frac{d}{dx}F(x) = f(x).$$

Example 3.5.1 **A Function of a Uniform Distribution** Let X be $U(0,1)$ and define the variable $Y = X^2$. Find the pdf of Y.

First note that Y has the range $(0,1)$. We begin by finding the cdf of Y:

$$F_Y(y) = P(Y \leq y) = P(X^2 \leq y) = P(X \leq \sqrt{y}).$$

But X is $U(0,1)$ so $P(X \le x) = x$ for $0 < x < 1$ so that

$$F_Y(y) = P(X \le \sqrt{y}) = \sqrt{y}.$$

Therefore,

$$f_Y(y) = \frac{d}{dy}F_Y(y) = \frac{d}{dy}\sqrt{y} = \frac{1}{2\sqrt{y}}, \quad 0 < y < 1. \quad \square$$

Calculating $d/dy F_Y(y)$ sometimes requires the use of the second fundamental theorem of calculus

$$\frac{d}{dx}\int_a^{u(x)} f(t)\,dt = f(u(x))\frac{du}{dx}$$

where $u(x)$ is any differentiable function of x.

Example 3.5.2 Using the Fundamental Theorem of Calculus Consider the random variable X with range $0 < x < 1$ and pdf $f_X(x) = 5x^4$. Define the variable $Y = X^2$. Find the pdf of Y.

As in the previous example, Y has the range $(0,1)$. The cdf of Y is

$$F_Y(y) = P(Y \le y) = P(X^2 \le y) = P(X \le \sqrt{y}) = \int_0^{\sqrt{y}} 5x^4\,dx.$$

Now, using the fundamental theorem of calculus, we get

$$f_Y(y) = \frac{d}{dy}\int_0^{\sqrt{y}} 5x^4\,dx = 5\left(\sqrt{y}\right)^4 \frac{1}{2\sqrt{y}} = \frac{5}{2}y^{3/2}, \quad 0 < y < 1. \quad \square$$

Example 3.5.3 A Piecewise PDF Consider a random variable X that is $U(-1,2)$. Define the variable $Y = X^2$. Find the pdf of Y.

Because X has the range $-1 < x < 2$, the range of Y is $0 < y < 4$. The cdf of Y is

$$F_Y(y) = P(Y \le y) = P(X^2 \le y) = P(-\sqrt{y} \le X \le \sqrt{y}).$$

If $y \le 1$, then it is possible for X to be between $-\sqrt{y}$ and $\sqrt{y}$ so that

$$F_Y(y) = P(-\sqrt{y} \le X \le \sqrt{y}) = \frac{1}{3}(2\sqrt{y}) = \frac{2}{3}(\sqrt{y}).$$

However, if $1 < y < 4$, then X cannot be as small as $-\sqrt{y}$ because the smallest value of X is -1. In this case,

$$F_Y(y) = P(-1 \leq X \leq \sqrt{y}) = \frac{1}{3}(\sqrt{y} + 1).$$

Putting it all together, we get

$$F_Y(y) = \begin{cases} \frac{2}{3}\sqrt{y}, & 0 < y \leq 1 \\ \frac{1}{3}(1 + \sqrt{y}), & 1 < y < 4. \end{cases}$$

Differentiating, we get

$$f_Y(y) = \begin{cases} \dfrac{1}{3\sqrt{y}}, & 0 < y \leq 1 \\ \dfrac{1}{6\sqrt{y}}, & 1 < y < 4. \end{cases} \qquad \square$$

Simulations

Suppose we would like to create a computer program to replicate the roll of a fair six-sided die. Such a program is called a *simulation*. In this program, we would need to generate rolls of this die. If the random variable X denotes the value of a single roll, then we need to generate values of this random variable.

Many programming languages have built-in algorithms for generating values of a random variable that is $U(0, 1)$ (see Exercise 16 in this section for a description of one such algorithm). The next theorem describes how we can use such an algorithm to generate values of any other random variable by using the cdf of the variable.

Theorem 3.5.1 Let Y be a random variable that is $U(0, 1)$, and let $F(x)$ satisfy the requirements of a cdf of a continuous random variable with the following properties:

1. $F(a) = 0$ and $F(b) = 1$ where a and b could possibly be $-\infty$ and ∞, respectively, and

2. $F(x)$ is a strictly increasing function on the domain $a < x < b$.

Let F^{-1} denote the inverse of the function F. Define the random variable $X = F^{-1}(Y)$. Then X has the cdf $F(x)$.

Proof. Our goal is to show that $P(X \le x) = F(x)$ for any x between a and b. By definition of X,

$$P(X \le x) = P\left[F^{-1}(Y) \le x\right], \quad \text{for } a < x < b.$$

But because $F(x)$ is a strictly increasing function on the domain $a < x < b$, the set $\{y : F^{-1}(y) \le x\}$ equals the set $\{y : y \le F(x)\}$ so that

$$P(X \le x) = P[Y \le F(x)], \quad \text{for } a < x < b.$$

Now because Y is $U(0,1)$, $P(Y \le y) = y$ for $0 < y < 1$, and therefore

$$P(X \le x) = P[Y \le F(x)] = F(x), \quad \text{for } a < x < b$$

as desired. □

This theorem suggests a relatively simple algorithm for generating values of a random variable with a desired distribution by using the uniform distribution:

1. Find the cdf $F(x) = \int_{-\infty}^{x} f(t)\,dt$ of the desired distribution.

2. Find the inverse cdf $F^{-1}(y)$.

3. Generate values $y_1, y_2, \ldots, y_n$ of a random variable Y that is $U(0,1)$ and calculate $x_i = F^{-1}(y_i)$. The resulting values $x_1, x_2, \ldots, x_n$ are observed values of a random variable with the desired distribution.

Example 3.5.4 Generating Values of an Exponential Random Variable We use the above algorithm to generate values of a random variable with an exponential distribution:

1. From Equation (3.2), the cdf is $F(x) = 1 - e^{-\lambda x}$.
2. To find $F^{-1}(y)$, we set $F(x) = y$ and solve for x:

$$y = 1 - e^{-\lambda x} \quad \Rightarrow \quad x = -\frac{1}{\lambda}\ln(1-y) \quad \Rightarrow \quad F^{-1}(y) = -\frac{1}{\lambda}\ln(1-y)$$

3. Using software, we generated the eight values of a random variable that is $U(0,1)$ shown in the first row of Table 3.5. Letting $\lambda = 2$, the corresponding values x_i are shown in the second row. □

y_i	0.858	0.421	0.472	0.591	0.623	0.854	0.248	0.639
$x_i = -0.5\ln(1-y_i)$	0.977	0.273	0.319	0.447	0.488	0.963	0.142	0.509

Table 3.5

Software Calculations

Minitab: Minitab has built-in functions for generating values of many different types of distributions. Select **Calc** → **Random Data** and then choose the desired type of distribution. Enter the number of values you would like to generate, the column in which they are to be stored, and the appropriate parameters.

R: The following table gives a sample of functions for generating values of a random variable with a certain distribution. Each syntax generates n values and stores the results in the vector x.

Distribution	Syntax
$U(a,b)$	x <- runif(n, a, b)
$N(\mu, \sigma^2)$	x <- rnorm(n, μ, σ)
Exponential	x <- rexp(n, λ)
Poisson	x <- dpois(n, λ)

Excel: The syntax for generating values of a random variable that is $U(0,1)$ is RAND(). To generate values of random variables with other types of distributions, select **Data** → **Data Analysis** → **Random Number Generation**. Enter the number of columns of numbers you would like to generate next to **Number of Variables** and the number of rows next to **Number of Random Numbers**. Choose the desired type of distribution, enter the appropriate parameters, and select an **Output Range**.

TI-83/84 Calculators: To generate values of a random variable that is $U(0,1)$, press **MATH** → **PRB** → **1:rand** and press **ENTER**. Each time you press **ENTER**, a new value will be generated. To generate values of a random variable that is $N(\mu, \sigma^2)$, select **6:randNorm**. The syntax for this function is randNorm(μ, σ).

Exercises

1. Let $Y = (b-a)X + a$ where $a < b$ and X is $U(0,1)$. Find the pdf of Y. What type of distribution does Y have?

2. Let $Y = X^2$ where X has the pdf $f_X(x) = 2x$, $0 < x < 1$. Show that Y is $U(0,1)$.

3. Let $Y = X^2$ where X has the pdf $f_X(x) = (x/4)e^{-x/2}$, $0 < x$. Find the pdf of Y.

4. Let $Y = X^2$ where X is $U(a,b)$, $0 \leq a < b$. Show that the pdf of Y is $f(y) = 1/[2\sqrt{y}(b-a)]$ for $a^2 \leq y \leq b^2$.

5. A square is constructed such that the length of a side, X, is $U(2,3)$. Use the results of Exercise 4 to find the probability that the area of the square is between 5 and 6.

6. Let $Y = (1 - 2X)^3$ where X has the pdf $f_X(x) = 3(1 - 2x)^2$, $0 < x < 1$. Show that Y is $U(-1,1)$.

7. Let $Y = X^2$ where X is $U(-4,5)$. Find the pdf of Y.

8. Let $Y = X^2$ where X has the pdf $f_X(x) = x^2/42$, $-1 < x < 5$. Find the pdf of Y.

9. Let $Y = X^2$ where X is $N(0,1)$. Show that the pdf of Y is $f(y) = (1/\sqrt{2\pi y})e^{-y/2}$.

10. Let Y be $U(0,1)$.

    a. Show that the variable $1 - Y$ is also $U(0,1)$.

    b. In Example 3.5.4, we showed the formula $x = -(1/\lambda)\ln(1 - y)$ where y is a value of a random variable that is $U(0,1)$ and that can be used to generate values of an exponentially distributed random variable. Use the result in part a to explain why an equally valid formula is $x = -(1/\lambda)\ln y$.

11. Let $Y = e^X$ where X is $N(\mu, \sigma^2)$ (the variable Y is said to have a *lognormal distribution*). Find the pdf of Y.

12. Consider a random variable X with the pdf

$$f_X(x) = \frac{e^{-x}}{(1 + e^{-x})^2}, \quad -\infty < x < \infty$$

(this pdf is a special case of the *logistic distribution*). Define the variable

$$Y = \frac{1}{1 + e^{-x}}.$$

Show that Y is $U(0,1)$.

13. Let $f_X(x)$ be the pdf of a variable X. Define the variable $Y = aX + b$ where $a \neq 0$ and b are constants. Show that the pdf of Y is

$$f_Y(y) = \frac{1}{|a|} f_X\left(\frac{y - b}{a}\right).$$

(**Hint:** Consider two cases, $a > 0$ and $a < 0$.)

14. Use Exercise 13 above to show that if X is $N(\mu, \sigma^2)$ then the variable $Y = aX + b$ is $N(a\mu + b, (a\sigma)^2)$.

15. Let X be a continuous random variable with range $a < x < b$ and cdf $F_X(x)$ that is continuous and strictly increasing on $a < x < b$. Define the random variable $Y = F_X(X)$. Show that Y is $U(0,1)$. (**Hint:** Show that $P(Y \leq y) = y$ by using the fact that $P(X \leq x) = F_X(x)$.)

16. A computer generates a list of "random" numbers by using an iterative function where one output becomes the next input, $x_{n+1} = f(x_n)$. Each output is a number in the list. The initial input x_0 is called the *seed*. Because the list is generated using a deterministic algorithm, there *will* be a pattern to the list so the list is not truly random. Therefore, the numbers are called *pseudorandom*.

One such *pseudorandom generator* that gives values of a random variable that is $U(0,1)$ is given by the following algorithm:

1. $x_0 =$ an arbitrary number between 0 and 1,
2. $x_{n+1} =$ fractional part of $(9821 * x_n + 0.211327)$.

Use this algorithm and any available software to generate a list of 100 pseudorandom numbers. Create a relative frequency histogram of these numbers. Do they indeed appear to have a uniform distribution? Explain.

17. A random variable X has the density function

$$f(x) = \begin{cases} 1/(2\sqrt{x}), & 0 < x < 1 \\ 0, & \text{elsewhere.} \end{cases}$$

a. Find the cdf $F(x)$.

b. Find the inverse cdf $F^{-1}(y)$.

c. Use $F^{-1}(y)$ and any available software to generate 100 values of X.

18. Repeat Exercise 17 above with the density function

$$f(x) = \begin{cases} 2(1-x), & 0 < x < 1 \\ 0, & \text{elsewhere.} \end{cases}$$

3.6 Joint Distributions

Often in a random experiment there is more than one random variable involved. For example, if we choose a person at random, one random variable could be the person's weight and another could be the height. Both of these variables have their own distributions, but we may want to consider probabilities involving some combinations of the variables. In such cases, we need to consider a pdf (or pmf in the discrete case) that combines these variables. Such a function is called a *joint pdf*.

First consider two discrete random variables X and Y. Let the notation $P(X = x, Y = y)$ denote the probability that $X = x$ and $Y = y$. The joint pmf is described by

$$f(x, y) = P(X = x, Y = y).$$

We illustrate how to calculate values of a joint pmf with an example.

Example 3.6.1 **Choosing Marbles** A bag contains three marbles that weigh 1, 2, and 3 oz, respectively. Suppose we choose two marbles *with replacement* and let the random variable X denote the weight of the heavier marble and Y denote the weight of the lighter marble. Then both X and Y are discrete variables that take values of 1, 2, and 3.

To find the joint pdf of X and Y, first note that

$$f(1, 1) = \frac{1}{9}$$

because there are 9 different outcomes and $X = Y = 1$ only when both marbles weigh 1 oz. Similarly,

$$f(2, 1) = \frac{2}{9}$$

because there are 2 ways that $X = 2$ and $Y = 1$, when the first marble weighs 1 oz and the second weighs 2 oz or vice versa. Note that

$$f(1, 2) = 0$$

because X must be larger than Y by definition. The other values of f are shown in the center of Table 3.6.

We might ask, What is the probability that the heavier marble weighs 3 oz? or What is the probability that the lighter marble weighs 1 oz? In terms of the random variables, these questions translate to, What are the values of $P(X = 3)$ and $P(Y = 1)$? To answer this, we define the *marginal pmfs* of X and Y:

$$f_X(x) = P(X = x) \quad \text{and} \quad f_Y(y) = P(Y = y).$$

$f(x,y)$	x = 1	2	3	$P(Y=y)$
y = 1	$\frac{1}{9}$	$\frac{2}{9}$	$\frac{2}{9}$	$\frac{5}{9}$
2	0	$\frac{1}{9}$	$\frac{2}{9}$	$\frac{1}{3}$
3	0	0	$\frac{1}{9}$	$\frac{1}{9}$
$P(X=x)$	$\frac{1}{9}$	$\frac{1}{3}$	$\frac{5}{9}$	

Table 3.6

Then, for instance,

$$\begin{aligned}f_Y(1) = P(Y=1) &= P(\{X=1,Y=1\} \cup \{X=2,Y=1\} \cup \{X=3,Y=1\}) \\ &= P(\{X=1,Y=1\}) + P(\{X=2,Y=1\}) + P(\{X=3,Y=1\}) \\ &= f(1,1) + f(2,1) + f(3,1) \\ &= 1/9 + 2/9 + 2/9 = 5/9.\end{aligned}$$

Notice that this quantity is simply the sum of the entries of the first row of Table 3.6. By a similar argument, $f_Y(2)$ and $f_Y(3)$ are the sums of the second and third rows, respectively. Likewise, the values of $f_X(x)$ are given by the sums of the columns of the table. All the values of $f_X(x)$ are shown in the bottom row of Table 3.6 while the values of $f_Y(y)$ are shown in the right column. □

We formalize the definitions of joint and marginal pmfs below.

Definition 3.6.1 Let X and Y be two discrete random variables with ranges R_X and R_Y, respectively, and let $R = R_X \times R_Y = \{(x,y) | x \in R_X, y \in R_Y\}$ be the set of all possible pairs of values of X and Y. The *joint probability mass function*, or joint pmf, is a function $f(x,y)$ described by

$$f(x,y) = P(X=x, Y=y)$$

that satisfies the following three properties:
1. $0 \leq f(x,y) \leq 1$ for all $x \in R_X$ and $y \in R_Y$,
2. $\sum\sum_{(x,y) \in R} f(x,y) = 1$, and
3. if $S \subset R$, then $P[(X,Y) \in S] = \sum\sum_{(x,y) \in S} f(x,y)$.

The *marginal pmf* of X is
$$f_X(x) = P(X = x) = \sum_{y \in R_Y} f(x,y)$$
and the *marginal pmf* of Y is
$$f_Y(y) = P(Y = y) = \sum_{x \in R_X} f(x,y).$$

A marginal pmf is given this name because its values appear along the edges, or margins, of a table such as Table 3.6. The marginal pmf of a variable X is the same as the pmf of X as defined in Definition 2.2.1. The qualifier *marginal* simply indicates that the variable is being used in the context of another variable. The definitions of a joint pmf and marginal pmfs can be extended to more than two random variables.

The third property of a joint pmf says that to find the probability that X and Y take on some set of values, we simply sum the joint pmf over all the values in the set. The next example illustrates this as well as the other properties.

Example 3.6.2 **A Joint PMF** To illustrate the calculation of marginal pmfs, consider the variables X and Y with the joint pmf

$$f(x,y) = \frac{x^3 y}{54} \quad \text{for } x = 1, 2, \, y = 1, 2, 3.$$

First note that this function satisfies the first two properties of a joint pmf because $0 \leq f(x,y) \leq 1$ for all $x = 1, 2$ and $y = 1, 2, 3$ and

$$\sum\sum_{(x,y) \in R} f(x,y) = \sum_{x=1}^{2}\sum_{y=1}^{3} \frac{x^3 y}{54} = \frac{1^3(1)}{54} + \frac{1^3(2)}{54} + \frac{1^3(3)}{54} + \frac{2^3(1)}{54} + \frac{2^3(2)}{54} + \frac{2^3(3)}{54} = 1.$$

The marginal pmfs are

$$f_X(x) = \sum_{y=1}^{3} \frac{x^3 y}{54} = \frac{x^3(1)}{54} + \frac{x^3(2)}{54} + \frac{x^3(3)}{54} = \frac{6x^3}{54} = \frac{x^3}{9}$$

and

$$f_Y(y) = \sum_{x=1}^{2} \frac{x^3 y}{54} = \frac{(1^3) y}{54} + \frac{(2^3) y}{54} = \frac{9y}{54} = \frac{y}{6}.$$

To calculate a probability involving both X and Y, we sum the joint pmf over all values of interest. For instance, to calculate the probability that the sum $X + Y$ is greater than or equal to 4, denoted $P(X + Y \geq 4)$, we sum the joint pmf over all values x and y such that $x + y \geq 4$:

$$P(X + Y \geq 4) = \sum_{x=1}^{2} \sum_{y=4-x}^{3} \frac{x^3 y}{54} = \frac{1^3(3)}{54} + \frac{2^3(2)}{54} + \frac{2^3(3)}{54} = \frac{43}{54}. \qquad \square$$

Next we extend the definition of the joint pmf, and marginal pmfs, to continuous random variables.

Definition 3.6.2 Let X and Y be two continuous random variables. The *joint probability density function*, or joint pdf, is an integrable function $f(x, y)$ that satisfies the following three properties:

1. $0 \leq f(x, y)$ for all $-\infty < x < \infty$ and $-\infty < y < \infty$,
2. $\int_{-\infty}^{\infty} \int_{-\infty}^{\infty} f(x, y) \, dy \, dx = 1$, and
3. if S is a subset of the two-dimensional plane, then

$$P[(X, Y) \in S] = \iint_S f(x, y) \, dy \, dx.$$

The *marginal probability density functions*, or marginal pdfs, of X and Y are

$$f_X(x) = \int_{-\infty}^{\infty} f(x, y) \, dy \quad \text{and} \quad f_Y(y) = \int_{-\infty}^{\infty} f(x, y) \, dx,$$

respectively.

Example 3.6.3 **A Joint PDF** Suppose the joint pdf of X and Y is

$$f(x, y) = \begin{cases} \dfrac{1}{2}, & 0 < x < 1, \ 0 < y < 2 \\ 0, & \text{otherwise.} \end{cases}$$

This function satisfies the first two properties of a joint pdf because $0 \leq f(x,y)$ for all x and y and

$$\int_{-\infty}^{\infty}\int_{-\infty}^{\infty} f(x,y)\,dy\,dx = \int_0^1 \int_0^2 \frac{1}{2}\,dy\,dx = \frac{1}{2}\int_0^1 2\,dx = \frac{1}{2}(2) = 1.$$

The marginal pdfs are

$$f_X(x) = \int_0^2 \frac{1}{2}\,dy = 1, \quad \text{for } 0 < x < 1$$

and

$$f_Y(y) = \int_0^1 \frac{1}{2}\,dx = \frac{1}{2}, \quad \text{for } 0 < y < 2.$$

These marginal pdfs show that X and Y are uniformly distributed over $[0,1]$ and $[0,2]$, respectively.

To illustrate how the joint pdf is used to find probabilities, consider the calculation of the probability that X is less than Y, denoted $P(X < Y)$. To do this, we graph all possible values of X and Y as illustrated by the rectangle in Figure 3.15. Next we determine the region in which $x < y$. We see that this is the region inside the rectangle, but above the line $y = x$. To calculate $P(X < Y)$, we integrate over this region with a double integral:

$$P(X < Y) = \int_0^1 \int_x^2 \frac{1}{2}\,dy\,dx = \int_0^1 \frac{1}{2}(2-x)\,dx = \frac{1}{2}\left(2x - \frac{x^2}{2}\right)\Big|_0^1 = \frac{3}{4}. \qquad \square$$

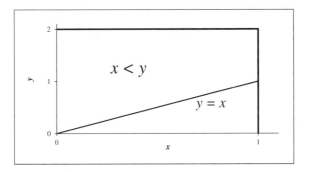

Figure 3.15

In the next definition, we extend the idea of independent events to random variables. Two events A and B are independent if

$$P(A \cap B) = P(A)P(B).$$

We define two independent *random variables* in a similar way.

Definition 3.6.3 Two discrete or continuous random variables X and Y are said to be *independent* if

$$f(x,y) = f_X(x) f_Y(y).$$

That is, they are independent if their joint pmf (or pdf) is the product of their marginal pmfs (or pdfs). If X and Y are not independent, they are said to be *dependent*. A set of n variables $X_1, X_2, \ldots, X_n$ is said to be *mutually independent* if their joint pdf is

$$f(x_1, x_2, \ldots, x_n) = f_{X_1}(x_1) \cdot f_{X_2}(x_2) \cdots f_{X_n}(x_n).$$

Informally, two events are independent if the occurrence of one does not affect the probability of the other. Likewise, two random variables are independent if the value of one does not affect the probability of the other. In Example 3.6.2, note that

$$f_X(x) f_Y(y) = \frac{x^3}{9} \cdot \frac{y}{6} = \frac{x^3 y}{54} = f(x,y)$$

so that X and Y are independent by definition. Also, in Example 3.6.3,

$$f_X(x) f_Y(y) = 1 \cdot \frac{1}{2} = \frac{1}{2} = f(x,y)$$

so that these two variables are also independent.

As with discrete variables, the marginal pdf of a continuous random variable X is no different than the pdf of X as defined in Definition 3.2.1. Again, the qualifier *marginal* simply indicates that the variable is being used in the context of another. If we know a variable's pdf, then we also know its marginal pdf. Also, if we assume that two variables are independent, then we can find their joint pdf by finding the marginal pdf of each one while ignoring the other, and then multiplying them. The next two examples illustrate these points.

Example 3.6.4 **Waiting at a Restaurant** Suppose a man and a woman agree to meet at a restaurant sometime between 6:00 and 6:15 PM. Find the probability that the man has to wait longer than 5 minutes for the woman to arrive.

Let the random variables X and Y denote the number of minutes past 6:00 PM that the man and the woman arrive, respectively, and we make the following two assumptions about these random variables:

1. Both X and Y are $U(0,15)$ so that $f_X(x) = f_Y(y) = \frac{1}{15}$ for $0 \le x \le 15$, $0 \le y \le 15$, and

2. X and Y are independent so that $f(x,y) = f_X(x) f_Y(y) = (\frac{1}{15})^2$.

174 CHAPTER 3 Continuous Random Variables

In mathematical notation, we want to find

$$P(Y > X + 5) = \int_0^{10} \int_{x+5}^{15} \left(\frac{1}{15}\right)^2 dy\, dx = \left(\frac{1}{15}\right)^2 \int_0^{10} [15 - (x+5)]\, dx$$

$$= \left(\frac{1}{15}\right)^2 \left(10x - \frac{x^2}{2}\right)\bigg|_0^{10} = \frac{2}{9}.$$

Thus the desired probability is $\frac{2}{9}$. □

Example 3.6.5 Buffon's Needle Problem A version of this problem was first solved by the French naturalist and mathematician, the Comte de Buffon (1707–1788). Suppose we randomly drop a needle of length $L \leq 1$ on a wood floor in which the joints between the planks are 1 unit apart. Find the probability that the needle "hits," or intersects, one of the joints.

Let the random variables X be the distance from the midpoint of the needle to the nearest joint between the planks and Θ be the angle as illustrated in Figure 3.16.

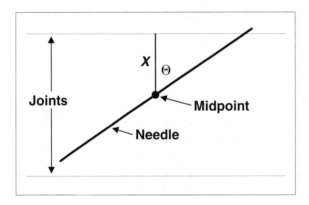

Figure 3.16

Because the drop is random, it seems reasonable to assume that X is $U(0, \frac{1}{2})$ and that Θ is $U(0, \pi/2)$ so that

$$f_X(x) = 2 \quad \text{and} \quad f_\Theta(\theta) = \frac{2}{\pi}.$$

Furthermore, it seems reasonable to assume that X and Θ are independent so that

$$f(x, \theta) = f_X(x) f_\Theta(\theta) = \frac{4}{\pi}, \quad 0 < x < \frac{1}{2},\ 0 < \theta < \frac{\pi}{2}.$$

Now, the needle will intersect a joint if the hypotenuse of the right triangle in Figure 3.16 is less than $L/2$ (if the needle does not intersect a joint, then the needle must be "extended" to form the triangle, making the hypotenuse greater than $L/2$). If h denotes the length of the hypotenuse, then we need to find

$$P\left(h < \frac{L}{2}\right).$$

Using trigonometry, we have

$$\cos\Theta = \frac{X}{h} \quad\Rightarrow\quad h = \frac{X}{\cos\Theta}.$$

So

$$P\left(h < \frac{L}{2}\right) = P\left(\frac{X}{\cos\Theta} < \frac{L}{2}\right) = P\left(X < \frac{L}{2}\cos\Theta\right)$$

$$= \int_0^{\pi/2} \int_0^{\frac{L}{2}\cos\theta} \frac{4}{\pi}\, dx\, d\theta$$

$$= \frac{4}{\pi} \int_0^{\pi/2} \frac{L}{2}\cos\theta\, d\theta$$

$$= \frac{2L}{\pi} \sin\theta\Big|_0^{\pi/2}$$

$$= \frac{2L}{\pi}(1 - 0) = \frac{2L}{\pi}.$$

Thus the desired probability is $2L/\pi$. $\square$

Exercises

1. Consider the random experiment of selecting two cubes at random without replacement from a bag that contains three red and two blue cubes.

   a. If the cubes are named R_1, R_2, R_3, B_1, and B_2, list all 20 equally likely outcomes.
   b. Let X denote the number of red cubes and Y denote the number of blue cubes in the sample of 2. Describe the joint pmf $f(x,y)$ as in Table 3.6. Give the values of the marginal pmfs $f_X(x)$ and $f_Y(y)$ in the margins of the table.
   c. Find $P(X + Y \geq 2)$ and $P(2X - Y \leq 3)$.

2. Two fair four-sided dice are rolled; one is colored red and the other is blue. Let X denote the result of the red die and Y denote the larger of the two results.

   a. Describe the joint pmf $f(x, y)$ as in Table 3.6. Give the values of the marginal pmfs $f_X(x)$ and $f_Y(y)$ in the margins of the table.
   b. Find $P(X + Y = 4)$ and $P(X \geq Y)$.

3. The joint pmf of two variables X and Y is

$$f(x, y) = \frac{3x + y}{39}, \quad x = 1, 2, \quad y = 1, 2, 3.$$

   a. Verify that this function satisfies the first two properties of a joint pmf.
   b. Find the marginal pmfs. Are X and Y independent?
   c. Calculate $P(Y \geq X)$, $P(X + 1 \leq Y)$, and $P(X^2 \leq Y^2)$.

4. Suppose we randomly choose a point on the xy plane within the square $-1 \leq x \leq 1$, $-1 \leq y \leq 1$. Let X and Y be the x and y coordinates of the point, respectively.

   a. Assuming that both X and Y are $U(-1, 1)$ and independent, give the joint pdf $f(x, y)$.
   b. Find $P(Y \leq X)$, $P(Y \leq |X|)$, and $P(X^2 + Y^2 \leq 1)$.

5. Consider the scenario of a man and woman who agree to meet at a restaurant as described in Example 3.6.4.

   a. Find the probability that the first to arrive has to wait longer than 5 min.
   b. Find the probability that the man has to wait longer than 10 min.
   c. Find the probability that the woman arrives before the man.
   d. Suppose they agree to arrive sometime between 6:00 and 6:30 PM, and that both X and Y are $U(0, 30)$ and independent. Find the probability that the man has to wait longer than 5 min.

6. Consider again the scenario of a man and woman who agree to meet at a restaurant as described in Example 3.6.4. However, suppose the marginal pdfs are

$$f_X(x) = \frac{2}{225} x, \quad f_Y(y) = -\frac{2}{225} y + \frac{2}{15}, \quad 0 \leq x \leq 15, 0 \leq y \leq 15.$$

a. Sketch the graphs of these marginal pdfs, and briefly explain what they tell about how close to 6:00 PM each person expects to arrive.

b. Assuming that X and Y are independent, find the probability that the man has to wait more than 5 min for the woman to arrive.

7. The variables X and Y have the joint pdf $f(x, y) = (x + 0.5)(y + 0.5)$ for $0 \leq x \leq 1$, $0 \leq y \leq 1$ and $f(x, y) = 0$ otherwise. Find the marginal pdfs and determine if X and Y are independent.

8. The joint pdf of two variables X and Y is

$$f(x, y) = \frac{24}{11}\left(x^2 + \frac{xy}{2}\right), \quad 0 \leq x \leq 1, 0 \leq y \leq 1.$$

a. Verify that this function satisfies the first two properties of a joint pdf.
b. Find the marginal pdfs.
c. Calculate $P(Y > X)$.

9. The joint pdf of two variables X and Y is

$$f(x, y) = 6xy^2, \quad 0 \leq x \leq 1, 0 \leq y \leq 1.$$

a. Find the marginal pdfs.
b. Verify that X and Y are independent.
c. Calculate $P(X + Y < 1)$.

10. The joint pdf of two variables X and Y is

$$f(x, y) = \frac{1}{2}, \quad 0 \leq y \leq x \leq 2.$$

a. Find the marginal pdfs.
b. Verify that X and Y are dependent.
c. Calculate $P(Y < X/2)$.

11. Consider a fair six-sided die with two sides labeled 1, two sides labeled 2, and two sides labeled 3. The die is rolled once, and let the random variable X denote the outcome. Also let the random variable Y denote the outcome.

a. Describe the marginal distributions of X and Y, $f_X(x)$ and $f_Y(y)$, in the form of tables.
b. Describe the joint distribution of X and X, $f(x,y)$, in the form of a table.
c. Use part b to show that X and Y are not independent. That is, show that $f(x,y) \neq f_X(x) f_Y(y)$ for some values of x and y.
d. Notice that X and Y are the same random variable with different names. In part c we showed that X and Y are not independent. In other words, we showed that this random variable is not independent with itself. Generalize this result. Thinking of independence as meaning that the value of one variable does not affect the value of the other, explain why no random variable is independent with itself.

12. Let X, Y, and Z each be uniformly distributed over $[0,1]$ and mutually independent. Find $P(X \geq YZ)$.

13. Let $f(x,y)$ be the joint pdf of variables X and Y, and let $\epsilon > 0$ be a small number. Use geometric reasoning to argue that

$$P\left(x_0 - \frac{\epsilon}{2} < X < x_0 + \frac{\epsilon}{2}, y_0 - \frac{\epsilon}{2} < Y < y_0 + \frac{\epsilon}{2}\right) \approx \epsilon^2 f(x_0, y_0)$$

for any x_0 in the domain of X and y_0 in the domain of Y. (**Hint**: This probability is the double integral of $f(x,y)$ over the square $x_0 - \epsilon/2 < y < x_0 + \epsilon/2$, $y_0 - \epsilon/2 < y < y_0 + \epsilon/2$.)

14. Suppose two random variables X and Y are independent and both $U(0,1)$.

a. Find $P(Y/X < 0.5)$.
b. Find $P(1.5 \leq Y/X < 2.5)$. (**Hint**: Draw the region inside the square $0 \leq x \leq 1$, $0 \leq y \leq 1$ such that $1.5x \leq y < 2.5x$ and observe that this region is a triangle. Find the area of this triangle by taking the leg of the triangle on the top of the square as the base. The height of the triangle is 1.)

15. The number $\phi = 1.618$ is called the *golden ratio*. Many people claim that this number appears frequently as the ratio of different measurements in famous works of architecture, art, and even literature. In this exercise, we will test this claim, to some extent.

a. Suppose two random variables X and Y are independent and both $U(0,1)$. Find the probability that the ratio Y/X or X/Y is within 0.05 of 1.618. (**Hint**: To find the probability that Y/X is within 0.05 of 1.618, we need to find $P(1.618 - 0.05 < Y/X < 1.618 + 0.05)$. Thus we need to integrate over the region $1.568x < y < 1.668x$ within the square $0 < x < 1$, $0 < y < 1$, which forms a triangle. The probability that X/Y is within 0.05 of 1.618 has the same value.)

b. If we have 100 independent pairs of random variables as those in part a, find the probability that in at least one pair, the ratio Y/X or X/Y is within 0.05 of 1.618. (**Hint**: Use complementary events.)

c. Suppose a researcher measures several different dimensions of a building and finds that the ratio of the overall height to the width is 1.62. She then says that this is close to the golden ratio and therefore the building was designed according to the golden ratio. What does your result in part b say about the validity of her claim, if anything? Explain.

3.7 Functions of Independent Random Variables

In this section, we give several important practical and theoretical results concerning independent random variables. These results are given in terms of two continuous variables; however, they can all be extended to more than two variables and to discrete variables as well. We begin with an example to motivate our first definition.

Example 3.7.1 **Functions of Random Variables** Suppose a golf cart that seats two people can carry a total of 500 lb (people and equipment), and we want to determine if the cart has enough capacity. If we randomly choose two different people, then the weight of each person with equipment is a random variable. Let X denote the weight of the first person and Y denote the weight of the second person. Of interest here is the total weight of these two people, which is a random variable. Let W denote this total weight. Then

$$W = X + Y.$$

To help answer the question, we may want to know the mean of W. Consider another scenario where we cut two sticks to the same random length, next do the same to another two sticks, and then form a rectangle with the four cut sticks. Suppose we want to know the mean of the area of the rectangle. The lengths of the two pairs of sticks are random variables, call them X and Y. The area of the rectangle is another random variable, call it A. Then

$$A = XY$$

and we want to find the mean of A. □

In both scenarios in this example, we want to find the mean of a random variable that is determined by two other random variables X and Y. In other words, these random variables are *functions* of X and Y. This motivates the need for the definition of the mean of a function of two random variables.

Definition 3.7.1 The *mean* or *expected value* of a function of two random variables X and Y, $u(X,Y)$, is denoted $E[u(X,Y)]$ and defined by

$$E[u(X,Y)] = \int_{-\infty}^{\infty} \int_{-\infty}^{\infty} u(x,y) f(x,y) \, dy \, dx$$

where $f(x,y)$ is the joint pdf of X and Y (assuming this integral exists).

In the notation for the expected value, we often replace $u(X,Y)$ with the description of the function. For instance, if $u(X,Y) = X+Y$, then the notation $E(X+Y)$ denotes the expected value $E[u(X,Y)]$.

Example 3.7.2 Calculating an Expected Value Suppose the joint pdf of X and Y is

$$f(x,y) = \begin{cases} \dfrac{1}{2}, & 0 < x < 1, \ 0 < y < 2 \\ 0, & \text{otherwise.} \end{cases}$$

Find $E[XY^2 + 3X + 2Y]$.

By definition,

$$\begin{aligned}
E[XY^2 + 3X + 2Y] &= \int_{-\infty}^{\infty} \int_{-\infty}^{\infty} (xy^2 + 3x + 2y) f(x,y) \, dy \, dx \\
&= \int_0^1 \int_0^2 \frac{1}{2}(xy^2 + 3x + 2y) \, dy \, dx \\
&= \int_0^1 \frac{4x}{3} + 3x + 2 \, dx \\
&= \frac{25}{6}.
\end{aligned}$$

□

The next theorem explains how to calculate the means of the variables in Example 3.7.1 without using Definition 3.7.1. Hopefully this theorem agrees with our intuition.

Theorem 3.7.1 *The mean of the sum of two discrete or continuous random variables X and Y is the sum of the means. That is,*

$$E(X+Y) = E(X) + E(Y).$$

If X and Y are independent, then the mean of their product is the product of their means. That is,

$$E(XY) = E(X)E(Y).$$

Proof. Using properties of the double integral, the definition of the marginal pdf, and the definition of expected values, we have

$$E(X+Y) = \iint (x+y)f(x,y)\,dy\,dx$$
$$= \iint xf(x,y)\,dy\,dx + \iint yf(x,y)\,dy\,dx$$
$$= \int x \left[\int f(x,y)\,dy\right] dx + \int y \left[\int f(x,y)\,dx\right] dy$$
$$= \int x f_X(x)\,dx + \int y f_Y(y)\,dy$$
$$= E(X) + E(Y)$$

where all integrals are from $-\infty$ to ∞. Now if X and Y are independent, then $f(x,y) = f_X(x)f_Y(y)$ and we have

$$E(XY) = \iint xy f(x,y)\,dy\,dx$$
$$= \iint xy f_X(x) f_Y(y)\,dy\,dx$$
$$= \int x f_X(x)\,dx \int y f_Y(y)\,dy$$
$$= E(X)E(Y). \qquad \square$$

Note that in Theorem 3.7.1, no particular distribution of X or Y is assumed (i.e., they could be normal, exponential, uniform, etc.). Also note that independence is not necessary for $E(X+Y) = E(X) + E(Y)$ to be true.

Example 3.7.3 **Expected Total Weight of Two People** In the golf cart scenario in Example 3.7.1, if we assume that $E(X) = E(Y) = 220$, then

$$E(W) = E(X+Y) = 220 + 220 = 440.$$

Thus the expected weight of two randomly selected people with equipment is less than the capacity of the cart. This indicates that the cart does have enough capacity. It does *not* mean that we are guaranteed that the total weight is less than 500 lb, however.

In the cutting sticks scenario, if we assume that X and Y are independent and that $E(X) = 6$ and $E(Y) = 8$, then

$$E(A) = E(XY) = 6 \times 8 = 48.$$ □

It is often convenient to describe one random variable as the sum of several others. This can simplify the process of calculating the expected value, as illustrated in the next example.

Example 3.7.4 Sum of 15 Rolls of a Die A fair six-sided die is rolled 15 times. Let X denote the sum of the 15 rolls. Find $E(X)$.

To calculate this expected value, we could, in principle, describe the pmf of X and then calculate $E(X)$ using the definition. However, the range of X is $\{15, \ldots, 90\}$, and there are 6^{15} different outcomes to this random experiment, so this approach is not reasonable. Instead, consider the random variables

$$X_i = \text{value of the } i\text{th roll}, \quad i = 1, 2, \ldots, 15.$$

Then the sum of all 15 rolls is

$$X = X_1 + X_2 + \cdots + X_{15}.$$

For each i, X_i has a range of $\{1, \ldots, 6\}$ and is uniformly distributed. Thus

$$E[X_i] = \frac{(6+1)}{2} = \frac{7}{2}$$

and

$$E(X) = E[X_1 + \cdots + X_{15}] = E[X_1] + \cdots + E[X_{15}] = 15 \cdot \frac{7}{2} = \frac{105}{2} = 52.5.$$ □

Consider again the random variable $W = X + Y$ in Example 3.7.1. We know that its mean is less than 500, but we might want to know the probability that its value is greater than 500, $P(W > 500)$. To calculate this, we also need to know its variance. By definition, its variance is

$$\sigma^2 = \text{Var}(W) = E\big[(W - \mu)^2\big] = E(W^2) - [E(W)]^2. \tag{3.4}$$

To simplify this formula, note that by Theorem 3.7.1,

$$[E(W)]^2 = [E(X+Y)]^2 = [E(X) + E(Y)]^2 = [E(X)]^2 + 2E(X)E(Y) + [E(Y)]^2, \quad (3.5)$$

and using the fact that expectation is a linear operator,

$$E(W^2) = E[(X+Y)^2] = E(X^2 + 2XY + Y^2) = E(X^2) + 2E(XY) + E(Y^2). \quad (3.6)$$

Now, putting Equations (3.4), (3.5), and (3.6) together, we have

$$\begin{aligned}\sigma^2 &= E(W^2) - [E(W)]^2 \\ &= E(X^2) + 2E(XY) + E(Y^2) - [E(X)]^2 - 2E(X)E(Y) - [E(Y)]^2 \\ &= \{E(X^2) - [E(X)]^2\} + \{E(Y^2) - [E(Y)]^2\} + 2[E(XY) - E(X)E(Y)]\end{aligned}$$

On the final line, the quantity in the first set of curly braces is simply Var(X), which we denote σ_X^2. Likewise, the quantity in the second set is Var(Y), denoted σ_Y^2. The quantity in the square brackets is called the *covariance* of X and Y and is denoted

$$\sigma_{XY} = E(XY) - E(X)E(Y).$$

(The covariance is discussed further in Section 6.2.) Thus our formula for σ^2 becomes

$$\sigma^2 = \sigma_X^2 + \sigma_Y^2 + 2\sigma_{XY}.$$

If X and Y are independent, then by Theorem 3.7.1, $E(XY) = E(X)E(Y)$ so that $\sigma_{XY} = 0$ and our formula simplifies to

$$\sigma^2 = \sigma_X^2 + \sigma_Y^2.$$

This result establishes the next theorem.

Theorem 3.7.2 *If X and Y are independent random variables, then the variance of their sum is the sum of their variances. That is,*

$$Var(X+Y) = Var(X) + Var(Y). \qquad \square$$

So for the random variable $W = X + Y$ in Example 3.7.1, if we assume that both X and Y have variances of 29^2, then

$$Var(W) = 29^2 + 29^2 = 1682.$$

The next theorem is an important theoretical result.

Theorem 3.7.3 *If X and Y are independent random variables with respective moment-generating functions $M_X(t)$ and $M_Y(t)$, and a and b are constants, then the mgf of $W = aX + bY$ is*

$$M_W(t) = M_X(at)M_Y(bt).$$

Proof. By the definition of a mgf,

$$\begin{aligned}
M_W(t) = E\left[e^{Wt}\right] &= E\left[e^{(aX+bY)t}\right] \\
&= \iint e^{(ax+by)t} f(x,y)\,dy\,dx \\
&= \iint e^{axt} e^{byt} f_X(x) f_Y(y)\,dy\,dx \\
&= \int e^{axt} f_X(x)\,dx \int e^{byt} f_Y(y)\,dy \\
&= M_X(at) M_Y(bt),
\end{aligned}$$

which establishes the theorem. $\square$

This theorem leads to the following important practical and theoretical result.

Theorem 3.7.4 *If X is $N(\mu_X, \sigma_X^2)$, Y is $N(\mu_Y, \sigma_Y^2)$, X and Y are independent, and a and b are constants, then the variable*

$$W = aX + bY$$

is $N\left(a\mu_X + b\mu_Y, a^2\sigma_X^2 + b^2\sigma_Y^2\right)$.

Proof. The mgfs of X and Y are

$$M_X(t) = \exp\left\{\mu_X t + \frac{\sigma_X^2 t^2}{2}\right\} \quad \text{and} \quad M_Y(t) = \exp\left\{\mu_Y t + \frac{\sigma_Y^2 t^2}{2}\right\}.$$

By Theorem 3.7.3, the mgf of $W = aX + bY$ is

$$\begin{aligned}
M_W(t) &= M_X(at) M_Y(bt) \\
&= \exp\left\{\mu_X at + \frac{\sigma_X^2 (at)^2}{2}\right\} \exp\left\{\mu_Y bt + \frac{\sigma_Y^2 (bt)^2}{2}\right\} \\
&= \exp\left\{(a\mu_X + b\mu_Y)t + \frac{(a^2\sigma_X^2 + b^2\sigma_Y^2)t^2}{2}\right\}
\end{aligned}$$

which is the mgf of a normally distributed random variable with mean $a\mu_X + b\mu_Y$ and variance $a^2\sigma_X^2 + b^2\sigma_Y^2$. This establishes the result. □

Example 3.7.5 Weight of Two People Again consider the random variable $W = X + Y$ in Example 3.7.1. Assuming that both X and Y are $N(220, 29^2)$ and are independent, find the probability that the cart will be overloaded.

In mathematical notation, we want to find $P(W > 500)$. Theorem 3.7.4 tells us that W is $N(220 + 220, 29^2 + 29^2) = N(440, 1682)$. The z-score of 500 is

$$z = \frac{500 - 440}{\sqrt{1682}} \approx 1.46.$$

Thus using Table C.1,

$$P(W > 500) = P(Z > 1.46) = 0.0721.$$

In practical terms, this means that the cart will be overloaded about 7.2% of the time, assuming that the weight of a person with equipment is $N(220, 29^2)$ and that the two people on the cart are randomly chosen so that their weights are independent. □

Example 3.7.6 Packages of Cheese Let X denote the weight, in pounds, of a randomly selected "1-lb" package of shredded cheddar cheese, and let Y be the weight of a randomly selected "3-lb" package. Assume that X is $N(1.05, 0.05^2)$ and that Y is $N(3.11, 0.09^2)$. Find the probability that the total weight of three randomly chosen 1-lb packages exceeds the weight of one randomly chosen 3-lb package.

Let X_i denote the weight of the ith 1-lb package selected, $i = 1, 2, 3$. Then each X_i has the same distribution as X. In mathematical notation, we want to find

$$P(X_1 + X_2 + X_3 > Y) = P(X_1 + X_2 + X_3 - Y > 0).$$

Let $W = X_1 + X_2 + X_3 - Y$. Assuming that X_1, X_2, X_3, and Y are mutually independent, by Theorem 3.7.4 the random variable W is

$$N\left[3(1.05) + (-1)(3.11), 3\left(0.05^2\right) + (-1)^2\left(0.09^2\right)\right] = N(0.04, 0.0156).$$

Then to find $P(W > 0)$, the z-score of 0 is

$$z = \frac{0 - 0.04}{\sqrt{0.0156}} \approx -0.32.$$

Thus,
$$P(W > 0) \approx P(Z > -0.32) = 0.6255.$$

This means there is about a 63% chance that three 1-lb packages will weigh more than one 3-lb package (whether this difference is practically significant is another issue). □

Our last theorem is an important theoretical tool.

Theorem 3.7.5 *Let X and Y be independent discrete or continuous random variables, and let $g(x)$ and $h(y)$ be any functions. Then the random variables $Z = g(X)$ and $W = h(Y)$ are independent.* □

Informally, this theorem says that functions of independent variables are independent. Intuitively, this result should make sense. The following example contains a proof of a very special case of this theorem. A proof of the theorem in general is given in Appendix A.1.

Example 3.7.7 Squares of Independent Random Variables Let X and Y be independent discrete random variables, both with nonnegative ranges. Define the random variables $Z = X^2$ and $W = Y^2$. Let $f_{ZW}(z,w)$ denote the joint pdf of Z and W and $f_Z(z)$ and $f_W(w)$ denote the marginal distributions. We will prove that Z and W are independent by showing that $f_{ZW}(z,w) = f_Z(z) \cdot f_W(w)$.

Let z and w be any elements in the ranges of Z and W, respectively. Then

$$\begin{aligned}
f_{ZW}(z,w) &= P(Z = z, W = w) \quad \text{by definition} \\
&= P(X^2 = z, Y^2 = w) \\
&= P(X = \sqrt{z}, Y = \sqrt{w}) \\
&= P(X = \sqrt{z}) \cdot P(Y = \sqrt{w}) \quad \text{because } X \text{ and } Y \text{ are independent} \\
&= P(X^2 = z) \cdot P(Y^2 = w) \\
&= P(Z = z) \cdot P(W = w) \\
&= f_Z(z) \cdot f_W(w)
\end{aligned}$$

as desired. □

Exercises

1. The joint pdf of two variables X and Y is
$$f(x,y) = 6xy^2, \quad 0 \le x \le 1, 0 \le y \le 1.$$

3.7 Functions of Independent Random Variables

Use the definition of the expected value of a function of random variables to find the following expected values:

a. $E[X+Y]$ b. $E[X]$ c. $E[Y]$ d. $E[X^2+Y]$

2. The joint pdf of two variables X and Y is

$$f(x,y) = \frac{1}{2}, \quad 0 \leq y \leq x \leq 2.$$

Use the definition of the expected value of a function of random variables to find the following expected values:

a. $E[2X+Y]$ b. $E[3X^2+Y^3]$ c. $E[|Y-X|]$

3. Consider the scenario of a man and woman who agree to meet at a restaurant as described in Example 3.6.4. If the man arrives first, he waits $X-Y$ min. If the woman arrives first, she waits $Y-X$ min. In general, the wait time experienced by the couple is $|X-Y|$. Find the expected value of the wait time experienced by the couple. (**Hint:** Divide the region $0 \leq x \leq 15$, $0 \leq y \leq 15$ into two subregions, one where $y > x$ and one where $y < x$. Integrate over these subregions and add the results.)

4. If X is $U(2,10)$, Y is $U(10,20)$, and X and Y are independent, find $E[X^2Y+3]$.

5. Consider Buffon's needle problem as solved in Example 3.6.5. The length of the hypotenuse of the right triangle in Figure 3.16 is

$$h = \frac{X}{\cos \Theta}.$$

Find $E(h)$, if it exists.

6. In each of the following problems, define the random variables involved, describe the desired expected value or variance in terms of these variables, and calculate its value.

a. A boy always gets money from his grandparents for his birthday. If he expects to get $12 from his maternal grandparents and $15 from his paternal grandparents, find the expected total amount.

b. A produce department at a local supermarket sells three sizes of fruit baskets, small, medium, and large. This holiday season, the manager expects to sell 25 small, 18 medium, and 9 large baskets. If the department makes a profit of $5 for each small basket, $8 for each medium basket, and $10 for each large basket, find the expected total profit.

c. Bowls of punch are made by mixing ginger ale and fruit juice. The amount of ginger ale poured into the bowl has variance of 0.25^2 cups^2, and the amount of fruit juice has variance of 0.12^2 cups^2. Assuming the amounts of ginger ale and fruit juice are independent, find the variance of the volume of punch in the bowls.

d. A girl randomly cuts rectangles out of a piece of paper in such a way that the length of the long side has a variance of 2^2 in^2, the length of the short side has a variance of 0.5^2 in^2, and the lengths of the sides are independent. Find the variance of the perimeter of the resulting rectangles. (**Hint**: Use Theorem 2.6.2.)

7. Consider a random variable X with mean $E(X) = 3$ and variance $\text{Var}(X) = 5$. A student applies the first part of Theorem 3.7.1 to calculate $E[2X]$ in the following way:

$$E[2X] = E[X + X] = E(X) + E(X) = 3 + 3 = 6.$$

She then applies Theorem 3.7.2 to calculate $\text{Var}[2X]$ in the following way:

$$\text{Var}[2X] = \text{Var}[X + X] = \text{Var}(X) + \text{Var}(X) = 5 + 5 = 10.$$

Explain why this first calculation is valid but the second is not valid. What is the correct value of $\text{Var}[2X]$?

8. Consider a random experiment consisting of a sequence of n Bernoulli trials where the probability of a success in any one trial is p. Let X denote the total number of successes in the n trials. Then X is $b(n, p)$. In this exercise, we derive a formula for $E(X)$ and $\text{Var}(X)$ using a sum of random variables.

a. For $i = 1, 2, \ldots, n$, define the random variable

$$X_i = \begin{cases} 1, & \text{if the } i\text{th trial is a success} \\ 0, & \text{otherwise.} \end{cases}$$

Describe the pmf of X_i in table form and find $E(X_i)$ and $\text{Var}(X_i)$.

b. The sum $X_1 + X_2 + \cdots + X_n$ is the total number of successes in the n trials. That is,

$$X = X_1 + X_2 + \cdots + X_n.$$

Use this representation of X to find $E(X)$ and $\text{Var}(X)$.

9. In this exercise, we will attempt to derive a formula for the expected value of a hypergeometric distribution. Consider the random experiment of drawing two cubes without replacement from a bag containing three red and two blue cubes. Let X denote the number of red cubes in the sample of 2. Then X has a hypergeometric distribution. We begin by finding $E(X)$.

a. Call the five cubes R_1, R_2, R_3, B_1, and B_2. List all $5 \times 4 = 20$ equally likely outcomes of this random experiment.

b. For $i = 1, 2$, define the random variable

$$X_i = \begin{cases} 1, & \text{if the } i\text{th ball is red} \\ 0, & \text{otherwise.} \end{cases}$$

Describe the pmfs of X_1 and X_2 in table form, and find the expected value of each variable.

c. The sum $X_1 + X_2$ is the total number of red cubes in the sample of 2. That is,

$$X = X_1 + X_2.$$

Use this representation of X to find $E(X)$.

d. Generalize your results in part c. Suppose a bag contains N_1 red and N_2 blue cubes and n cubes are drawn without replacement. Let X denote the number of red cubes in the sample of n. Use your result in part c to conjecture a formula for $E(X)$.

10. Let X be the life span (in hours) of one particular brand of lightbulb and Y be the life span of another brand. Suppose that X is $N(965, 19^2)$ and that Y is $N(995, 31^2)$. If one of each brand of bulb is randomly selected, find $P(X > Y)$.

11. Consider the packages of cheese described in Example 3.7.6.

a. Find the probability that the total weight of six randomly chosen 1-lb packages is greater than the weight of two randomly chosen 3-lb packages.

b. Generalize the calculations in part a. Suppose n packages of 3 lb are randomly chosen and $3n$ packages of 1 lb are randomly chosen. Let T_1 denote the total weight of the 1-lb packages and T_2 denote the total weight of the 3-lb packages. Show $P(T_1 > T_2) \approx P(Z > -0.32\sqrt{n})$.

c. Find the limit of the probability in part b as $n \to \infty$. Based on this result, if we wanted a large quantity of cheese, would we actually get more cheese if we purchased several 3-lb packages or three times as many 1-lb packages?

d. Generalize the results in part b. Suppose that X, the weight of a randomly selected 1-lb package, is $N(\mu_X, \sigma_X^2)$ and that Y, the weight of a randomly selected 3-lb package, is $N(\mu_Y, \sigma_Y^2)$. Let $L = \lim_{n \to \infty} P(T_1 > T_2)$. Show that $L = 1$ if $3\mu_X > \mu_Y$, $L = 0$ if $3\mu_X < \mu_Y$, and $L = 0.5$ if $3\mu_X = \mu_Y$.

12. Consider the packages of cheese described in Example 3.7.6. Suppose a customer buys a 3-lb package and divides it into three bags of equal weight. Find the probability that the weight of one of these bags is greater than the weight of a randomly chosen 1-lb package.

13. Consider the calculations in Example 3.7.6.

    a. We might be tempted to simplify the notation by describing the total weight of the three 1-lb packages as $3X$ rather than $X_1 + X_2 + X_3$. Explain why this is *not* appropriate. (**Hint**: Do all three packages have to weigh the same?)
    b. Compare the variance of $3X$ to the variance of $X_1 + X_2 + X_3$.
    c. Do the variables $3X$ and $X_1 + X_2 + X_3$ have the same distribution? Explain. (Remember, to have the same distribution, they must have the same mean and variance.)

14. A standardized calculus test consists of two parts, one where calculators are allowed and another where they are not. Suppose that X, the score on the part with calculators, is $N(85, 4^2)$ and that Y, the score on the part with no calculators, is $N(80, 5^2)$. Define the variable $W = (X + Y)/2$ to be the *average* score. If one student's test is randomly selected, find $P(W > 90)$.

15. Over fall break, a student has an 8-hour stretch of free time and decides to spend it watching movies. He goes to the movie rental store and randomly picks out n movies. If X, the length of a movie in minutes, is $N(110, 15^2)$, find the smallest value of n so that the total length of all the movies is at least 8 hr with probability of at least 0.95. Completely justify your solution.

16. Let X and Y be independent random variables such that $\text{Var}[X^2] = 2$ and $\text{Var}[Y^2] = 3$. Use Theorems 3.7.2 and 3.7.5 to find $\text{Var}[X^2 + Y^2]$. Explain how each of these theorems is used in your calculation.

3.8 The Central Limit Theorem

Many random experiments involve randomly choosing a set of objects from a larger set of objects. Consider the following examples:

1. Five cards are drawn from a deck of 52.
2. Two bags of shredded cheese are selected from a packaging line.
3. Two people are seated in a golf cart.

Situations such as these lead to the following definitions:

> **Definition 3.8.1** When a set of objects is selected from a larger set of objects, the larger set is called the *population* and the smaller set is called the *sample*. The number of objects in the sample is called the *sample size*.

The ideas of population and sample will be explored more in the next chapter. In the first example above, the population is the deck of 52 cards, and the sample is the 5 cards drawn. In the second example, the population is all bags of cheese that come off the packaging line, and the sample is 2 bags. In the third example, the population is all people, and the sample is 2 people. Note that a population does not have to be a group of people.

Associated with a sample is the *sample mean*, which we motivate with the next example.

Example 3.8.1 Sample Mean Suppose we randomly choose a 1-lb package of shredded cheddar cheese as in Example 3.7.6. Let the random variable X denote its weight, in pounds, and assume that X is $N(1.05, 0.05^2)$. This mean of 1.05 is called the *population mean*, and the variance of 0.05^2 is called the *population variance*. The distribution of X is called the *distribution of the population*.

Now suppose we choose two such packages and let X_i be the weight of the ith package, $i = 1, 2$. Each one of these two random variables has the same distribution as X. Assuming the sample is random, these two variables are independent. We calculate the *average* weight of these two packages by adding the weights and dividing by 2. That is,

$$\text{Average weight} = \frac{X_1 + X_2}{2}.$$

Notice that this average weight is a function of the variables X_1 and X_2, so it is a random variable itself, call it

$$\bar{X}_2 = \frac{X_1 + X_2}{2} = \frac{1}{2}X_1 + \frac{1}{2}X_2.$$

By Theorem 3.7.4, $\bar{X}_2$ is

$$N\left(\frac{1}{2}(1.05) + \frac{1}{2}(1.05), \left(\frac{1}{2}\right)^2 0.05^2 + \left(\frac{1}{2}\right)^2 0.05^2\right) = N\left(1.05, \frac{0.05^2}{2}\right).$$

Thus we see $\bar{X}_2$ is normally distributed with mean equal to the population mean and variance equal to the population variance divided by the sample size. □

We generalize this example in the following theorem.

Theorem 3.8.1 If $X_1, X_2, \ldots, X_n$ are mutually independent random variables and each is $N(\mu, \sigma^2)$, then the sample mean

$$\bar{X}_n = \frac{X_1 + X_2 + \cdots + X_n}{n}$$

is normally distributed with mean

$$\mu_{\bar{X}} = \mu$$

and variance

$$\sigma_{\bar{X}}^2 = \frac{\sigma^2}{n}.$$ □

Example 3.8.2 Calculating a Sample Mean Suppose we select a sample of five packages of shredded cheddar cheese as in Example 3.8.1 and record the weight of each as shown below:

$$0.955, 1.033, 1.046, 1.085, 1.135.$$

The value of the sample mean $\bar{X}_5$, denoted $\bar{x}$, for this sample is

$$\bar{x} = \frac{0.955 + 1.033 + 1.046 + 1.085 + 1.135}{5} = 1.0508.$$

This value, $\bar{x} = 1.0508$, is called an *observed value* of $\bar{X}_5$. Other observed values of $\bar{X}_5$ could be obtained by selecting other samples of five packages and performing similar calculations. □

Example 3.8.3 Passenger Ferry Suppose that a passenger ferry has a safe operating capacity of 3500 lb and that the captain allows at most 20 people onboard. Let X denote the weight of a randomly selected person, and assume that X is $N(172, 29^2)$. Find the probability that the boat will be overloaded with 20 randomly selected people.

First note that

$$\frac{3500 \text{ lb}}{20 \text{ people}} = 175 \text{ lb/person}.$$

Thus if the sample mean $\bar{X}_{20}$ is greater than 175, then the boat will be overloaded. So we want to know

$$P(\bar{X}_{20} > 175).$$

By Theorem 3.8.1, $\bar{X}_{20}$ is normally distributed with mean $\mu_{\bar{X}} = 172$ and variance

$$\sigma_{\bar{X}}^2 = \frac{29^2}{20} = 42.05,$$

so that the z-score of 175 is

$$z = \frac{175 - 172}{\sqrt{42.05}} \approx 0.46.$$

Thus

$$P(\bar{X}_{20} > 175) \approx P(Z > 0.46) = 0.3228.$$

So there is approximately a 32% chance that the boat will be overloaded with 20 randomly selected people, assuming the weight of the population of people is $N(172, 29^2)$. Most people would argue that this probability is too high, and thus the number of people allowed on the boat should be decreased. The question of how much it should be decreased is analyzed in Exercise 6 of this section. □

Not all populations have a normal distribution. The next example illustrates that even in these cases the sample mean has an approximately normal distribution.

Example 3.8.4 Mean Waiting Times Let X denote the amount of time a person spends waiting in line at a grocery store checkout line. Assume that X has an exponential distribution with parameter $\lambda = \frac{1}{2}$. Then the population mean is $\mu = 1/\lambda = 2$.

Suppose we take random samples of $n = 30$ people and calculate the sample mean $\bar{X}_{30}$ of each sample. Because X does not have a normal distribution, Theorem 3.8.1 does not tell us anything about the distribution of $\bar{X}_{30}$.

To estimate the distribution of $\bar{X}_{30}$, we simulated selecting 1000 different samples of 30 people and calculating the mean of each sample as in Example 3.8.2. Figure 3.17 shows a relative frequency histogram of these 1000 values of $\bar{X}_{30}$. Notice that the histogram has a bell curve shape centered near 2. This indicates that $\bar{X}_{30}$ has an approximately normal distribution with mean 2, which is the same as the population mean. □

We summarize the results of the previous example in the very important *central limit theorem*, which is a generalization of Theorem 3.8.1. A proof of the central limit theorem is given in Appendix A.2.

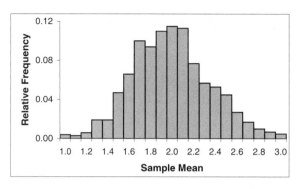

Figure 3.17

Theorem 3.8.2 Central Limit Theorem *If $X_1, X_2, \ldots, X_n$ are mutually independent random variables with a common distribution, mean μ, and variance σ^2, then as $n \to \infty$, the distribution of the sample mean*

$$\bar{X}_n = \frac{X_1 + \cdots + X_n}{n} \qquad approaches \qquad N\left(\mu, \frac{\sigma^2}{n}\right).$$ □

Informally, the central limit theorem says that for n "large" enough, the sample mean $\bar{X}_n$ is approximately normally distributed with mean equal to the population mean and variance equal to the population variance divided by the sample size *regardless of the distribution of the population.* The population can be continuous or discrete.

Just how large n needs to be for this approximation to be reasonable depends on the distribution of the population (i.e., the common distribution). If the distribution is "close" to normal (meaning the graph of the pdf resembles a bell curve), then n can be rather small. In the discrete case, *close* to normal means that the pattern of the tops of the bars in the probability histogram graph of the pmf resembles a bell curve. If the graph of the pdf (or pmf) is far from normal, then n must be larger. A general rule of thumb says that in either case, this approximation is reasonable for $n > 30$. If the population is indeed normal, then n can be any size. In the next chapter, we will look at techniques for determining if a population is indeed normal.

Example 3.8.5 Using the Central Limit Theorem Suppose people at a certain movie theater wait a mean of 30 sec with a variance of 5^2 sec^2 to buy tickets. Find the probability that the mean wait time of a group of 50 moviegoers is less than 28 sec.

We want to find $P(\bar{X}_{50} < 28)$. According to the central limit theorem, $\bar{X}_{50}$ is approximately normal with mean $\mu_{\bar{X}} = 30$ and variance

$$\sigma_{\bar{X}}^2 = \frac{5^2}{50} = 0.5.$$

(Note that wait times are often described with an exponential distribution, so the population is probably not normally distributed. However, because $n > 30$, this does not matter.) The z-score of 28 is

$$z = \frac{28 - 30}{\sqrt{0.5}} \approx -2.83.$$

So

$$P(\bar{X}_{50} \leq 28) \approx P(Z \leq -2.83) = 0.0023.$$

Thus there is less than a 1% chance that the mean wait time for a group of 50 is less than 28 sec. □

Exercises

1. Let X be the mass, in grams, of a randomly chosen chocolate candy. Assume that X is $N(0.920, 0.003)$. A random sample of 12 such candies is taken, and the sample mean is calculated. Approximate the probability that

   a. the sample mean is less than 0.915, and

   b. the sample mean is between 0.915 and 0.922.

   c. If the sample mean is less than 0.920, does this mean that every candy in the sample has a mass less than 0.920 g? Why or why not?

2. IQ scores have a normal distribution with mean $\mu = 100$ and variance $\sigma^2 = 15^2$.

   a. If one person is randomly chosen, find the probability that her or his IQ is greater than 110.

   b. If a sample of 25 people is randomly chosen, find the probability that the sample mean is greater than 110.

3. Consider bottles of corn syrup labeled as containing 16 oz each. The contents of 45 such bottles are measured and found to have a sample mean of $\bar{x} = 16.04$ oz. Assuming that the population of such bottles has a mean of $\mu = 16.00$ and standard deviation $\sigma = 0.08$, calculate the probability that the sample mean of 45 bottles is greater than 16.04 oz. Based on these results, does it appear that the bottles are filled with an amount greater than 16.00 oz? Explain.

4. Let the random variable X denote the number of Sunday newspapers sold by a paperboy in a large city. Assume that X is $b(500, \frac{1}{5})$. If 50 such paperboys are randomly selected, approximate the probability that the sample mean is less than 99.

5. Scores on a certain standardized test have a normal distribution with mean of $\mu = 1515$ and variance of $\sigma^2 = 300^2$. A random sample of n scores is selected.

   a. Calculate $P(1510 < \bar{X}_n < 1520)$ for $n = 5, 50, 250, 500,$ and 5000.

   b. Based on your results in part a, as n gets larger, what happens to the probability that the sample mean $\bar{X}_n$ is close to the population mean μ?

   c. Let $\varepsilon > 0$. Use your results to estimate $\lim\limits_{n \to \infty} P(\mu - \varepsilon < \bar{X}_n < \mu + \varepsilon)$.

6. Refer to Example 3.8.3. If n is the number of people onboard, find the largest value of n so that the boat is overloaded with probability less than 0.01. (**Hint**: The boat is overloaded only if $\bar{X}_n > 3500/n$.)

7. A random sample of 109 healthy adults has a mean body temperature of 98.3°F.

   a. Assuming that the mean body temperature of the population of healthy adults is 98.6°F (as often believed) with a standard deviation of $\sigma = 0.6°F$, approximate the probability that a random sample of 109 healthy adults has a mean body temperature of 98.3°F or less.

   b. What do your results in part a suggest about the validity of the assumption that the mean body temperature of the population of healthy adults is 98.6°F?

8. Consider a population with mean μ and variance σ^2.

   a. Define the random variable $Y = \bar{X}_n - \mu$. Show that Y is approximately $N(0, \sigma^2/n)$. (**Hint**: Use the fact that if a random variable X has the mgf $M_X(t)$, then the random variable $W = aX + b$, where a and b are any real numbers, has the mgf $M_W(t) = e^{bt} M_X(at)$.)

   b. Use part a to confirm your answer to part c of Exercise 5 above.

9. Let μ denote the mean SAT score of the population of all students who attend private universities. The variance of this population is assumed to be $\sigma^2 = 325^2$, but the value of μ is unknown. To estimate the value of μ, a researcher plans to collect the SAT scores of a random sample of 100 students and calculate the sample mean $\bar{X}_{100}$. Use Exercise 8 above to calculate the probability that the difference between this sample mean and μ is no more than 10 units.

10. Suppose the researcher in Exercise 9 above wants the difference between the sample mean and μ to be no more than 50 units with probability of at least 0.95. Find the minimum sample size required.

11. Suppose that the random variable X_1 is exponentially distributed with $\lambda = 1$.

   a. Find the mean and variance of X_1, and sketch a graph of its pdf. Does this distribution appear to be close to normal?

   b. Suppose that X_2 is independent of X_1 but has the same distribution. Find the joint pdf of X_1 and X_2, $f(x_1, x_2)$.

   c. Let $\bar{X}_2 = (X_1 + X_2)/2$ be the sample mean of these variables. Find the exact value of $P(\bar{X}_2 < \frac{1}{2})$.

   d. Use the central limit theorem to try to estimate $P(\bar{X}_2 < \frac{1}{2})$. Compare this estimation to the exact value. How good is the estimate? Explain why this estimate is so poor.

12. Suppose a sample of size n is taken from one population with mean μ_X and variance σ_X^2 and that a sample also of size n is taken from a different population with mean μ_Y and variance σ_Y^2. Let $\bar{X}_n$ and $\bar{Y}_n$ denote the respective sample means. Define the random variable

$$D_n = \bar{X}_n - \bar{Y}_n$$

to be the difference between the sample means. Assuming all samples are mutually independent, use the central limit theorem and Theorem 3.7.4 to show that D_n is approximately normally distributed with mean $\mu_X - \mu_Y$ and variance $(\sigma_X^2 + \sigma_Y^2)/n$.

13. Men's heights have a mean of 69.0 in and standard deviation of 2.8 in while women's heights have a mean of 63.6 in and standard deviation of 2.5 in. If we measure the heights of a sample of 30 men and a sample of 30 women, use Exercise 12 above to approximate the probability that the difference between the respective sample means is between 5.25 and 5.75 in.

14. To apply the central limit theorem, we need our sample values $X_1, \ldots, X_n$ to be mutually independent. If we have an infinitely large population, then the values are indeed independent. However, if the population is finite, then these values are *not* independent.

If the population contains N members, then it can be shown that the mean and variance of $\bar{X}_n$ are

$$E(\bar{X}_n) = \mu \quad \text{and} \quad \text{Var}(\bar{X}_n) = \frac{\sigma^2}{n}\left(1 - \frac{n-1}{N-1}\right)$$

where μ and σ^2 are the population mean and variance, respectively. The factor

$$\left(1 - \frac{n-1}{N-1}\right)$$

is called the *finite population correction*.

a. Calculate $\lim_{N \to \infty} \left(1 - \frac{n-1}{N-1}\right)$ where n is some fixed value.

b. What does the value of this limit mean if the sample size is very small in comparison to the population size? Is the finite population correction really necessary in this case?

3.9   The Gamma and Related Distributions

In this section, we introduce four very important continuous distributions: the gamma, chi-square, Student-t, and F-distributions. The last three will be very important starting in the next chapter. To motivate the first distribution, consider a secretary who receives an average of two telephone calls per hour. If the random variable X denotes the number of calls received in 1 hr, then X has a Poisson distribution with pmf

$$f(x) = P(X = x) = \frac{2^x}{x!} e^{-2}.$$

In Section 3.3, we showed that the wait time between calls has an exponential distribution. In this section, we consider the random variable W denoting the time until the rth call arrives.

First we derive the cdf of W:

$$F(W) = P(W_n \le w) = 1 - P(W > w) = 1 - P(\text{fewer than } r \text{ calls in } [0, w]).$$

Now, to find $P(\text{fewer than } r \text{ calls in } [0, w])$, let Y be the number of calls received in w hr. Then Y has a mean of $2w$ and a Poisson distribution. Thus

$$f(y) = P(Y = y) = \frac{(2w)^y}{y!} e^{-2w}$$

and

$$P(\text{fewer than } r \text{ calls in } [0, w]) = P(Y \leq r-1) = \sum_{k=0}^{r-1} \frac{(2w)^k}{k!} e^{-2w}$$

$$\Rightarrow \quad F(W) = 1 - \sum_{k=0}^{r-1} \frac{(2w)^k}{k!} e^{-2w} = 1 - e^{-2w} - \sum_{k=1}^{r-1} \frac{(2w)^k}{k!} e^{-2w}.$$

Now, using properties of telescoping series, we have

$$\begin{aligned}
f(w) = \frac{d}{dw} F(w) &= 2e^{-2w} - \sum_{k=1}^{r-1} \left[\frac{k(2w)^{k-1}(2)}{k!} e^{-2w} + \frac{(2w)^k}{k!}(-2) e^{-2w} \right] \\
&= 2e^{-2w} - 2e^{-2w} \sum_{k=1}^{r-1} \left[\frac{(2w)^{k-1}}{(k-1)!} - \frac{(2w)^k}{k!} \right] \\
&= 2e^{-2w} - 2e^{-2w} \left[1 - \frac{(2w)^{r-1}}{(r-1)!} \right] \\
&= \frac{2(2w)^{r-1}}{(r-1)!} e^{-2w} \\
&= \frac{2^r}{(r-1)!} w^{r-1} e^{-2w}.
\end{aligned}$$

We generalize this derivation in the following definition:

Definition 3.9.1 A random variable W has a *gamma distribution with parameters r and λ*, where r is a positive integer and $\lambda > 0$, if its pdf is

$$f(w) = \frac{\lambda^r}{(r-1)!} w^{r-1} e^{-\lambda w}, \quad \text{for } w > 0.$$

We see from the above derivation that if the number of occurrences in an interval is described by a Poisson distribution with parameter λ, then W denoting the time until the rth occurrence has a gamma distribution with parameters r and λ. Note that if $r = 1$, meaning W is the wait time until the first occurrence, the pdf is

$$f(w) = \lambda e^{-\lambda w},$$

which is the pdf of an exponential distribution, as expected.

Example 3.9.1 Flaws in a Wire A manufacturer of telephone wire experiences an average of 1 flaw per 1000 ft of wire. Suppose that a customer receives a shipment of wire, randomly selects a 150-ft section of wire, inspects it, and rejects the shipment if there are 2 or more flaws in the 150-ft section. Assuming that X, the number of flaws in a 1-ft section of wire, has a Poisson distribution, find the probability that the shipment is rejected.

The shipment is rejected if the distance to the second flaw is less than 150 ft. So define the random variable W to be the distance to the second flaw. Because X has a Poisson distribution with parameter $\lambda = 1/1000 = 0.001$, W has a gamma distribution with parameters $r = 2$ and $\lambda = 0.001$, so its pdf is

$$f(w) = \frac{0.001^2}{(2-1)!} w^{2-1} e^{-0.001w} = 0.001^2 w e^{-0.001w}.$$

Now, using integration by parts or a computer algebra system, we get

$$P(\text{reject the shipment}) = P(W < 150) = \int_0^{150} 0.001^2 w e^{-0.001w}\, dw \approx 0.01.$$

Thus about 1% of such shipments will be rejected. □

Example 3.9.2 Mean and Variance of a Gamma Distribution Consider a random variable W with a gamma distribution with parameters r and λ. We can think of this representing the wait time until the rth occurrence where the number of occurrences is described by a Poisson distribution with parameter λ.

Define the random variable W_1 to be the wait time until the first occurrence, W_2 to be the wait time between the first and second occurrences, ..., W_r to be the wait time between the $(r-1)$st and rth occurrence. Then we have

$$W = W_1 + W_2 + \cdots + W_r.$$

Now each W_i, $i = 1, \ldots, r$, has an exponential distribution, and as shown in Example 3.3.5, each has mean $1/\lambda$ and variance $1/\lambda^2$. Furthermore, these variables are independent.

Thus, using Theorem 3.7.1, we have

$$E(W) = E(W_1 + \cdots + W_r) = E(W_1) + \cdots + E(W_r) = \frac{r}{\lambda},$$

and using Theorem 3.7.2, we have

$$\text{Var}(W) = \text{Var}(W_1 + \cdots + W_r) = \text{Var}(W_1) + \cdots + \text{Var}(W_r) = \frac{r}{\lambda^2}.$$

Thus the expected value of a random variable with a gamma distribution is r/λ and the variance is r/λ^2. □

To extend the gamma distribution to allow for values of r other than integers, we define the *gamma function*:

$$\Gamma(t) = \int_0^\infty y^{t-1} e^{-y}\, dy, \quad \text{for} \quad 0 < t.$$

Example 3.9.3 **Evaluating the Gamma Function** To calculate, for instance, $\Gamma(2.3)$, we need to evaluate the integral

$$\Gamma(2.3) = \int_0^\infty y^{2.3-1} e^{-y}\, dy.$$

Evaluating this integral is not a trivial matter. Using numeric approximation techniques on a calculator yields the approximate value of 1.167. When $t = 1$, the calculations are much easier:

$$\Gamma(1) = \int_0^\infty e^{-y}\, dy = -e^{-y}\Big|_0^\infty = 0 + 1 = 1.$$

We get an interesting recursion formula for $\Gamma(t)$ by using integration by parts where $u = y^{t-1}$ and $dv = e^{-y}\, dy$:

$$\begin{aligned}
\Gamma(t) &= \int_0^\infty y^{t-1} e^{-y}\, dy \\
&= -y^{t-1} e^{-y}\Big|_0^\infty + \int_0^\infty (t-1) y^{t-2} e^{-y}\, dy \\
&= 0 + (t-1) \int_0^\infty y^{t-2} e^{-y}\, dy \\
&= (t-1)\Gamma(t-1).
\end{aligned}$$

Using this formula, we have, for instance,

$$\Gamma(2) = (1)\Gamma(1) = 1 \cdot 1 = 1! \quad \text{and} \quad \Gamma(3) = (2)\Gamma(2) = 2 \cdot 1! = 2!,$$

and in general, when n is an integer,

$$\Gamma(n) = (n-1)!$$

For this reason, the gamma function is called the *generalized factorial*. □

Now we use the gamma function to define the general gamma distribution.

Definition 3.9.2 A random variable X has a *gamma distribution with parameters r and λ* where $r > 0$ (not necessarily an integer) and $\lambda > 0$ if its pdf is

$$f(x) = \frac{\lambda^r}{\Gamma(r)} x^{r-1} e^{-\lambda x} \quad \text{for } x > 0.$$

The parameter r is called the *shape* parameter, λ is called the *rate* parameter, and the number $1/\lambda$ is called the *scale* parameter.

It can be shown that the mgf of a gamma distribution is

$$M(t) = \left(1 - \frac{t}{\lambda}\right)^{-r}, \quad \text{for } t < \lambda$$

and that its mean and variance are

$$\mu = \frac{r}{\lambda} \quad \text{and} \quad \sigma^2 = \frac{r}{\lambda^2}.$$

The next distribution is motivated by the following theorem.

Theorem 3.9.1 Let $Y_1, Y_2, \ldots, Y_n$ be independent random variables each with a standard normal distribution. Define the random variable

$$X = Y_1^2 + Y_2^2 + \cdots + Y_n^2.$$

The pdf of X is

$$f(x) = \frac{1}{\Gamma(n/2)\, 2^{n/2}} x^{n/2-1} e^{-x/2} \quad \text{for } x > 0.$$

A proof of this theorem is outlined in Exercise 19 of this section. A random variable with the pdf given in this theorem is said to have a *chi-square distribution*.

Definition 3.9.3 A random variable X with the pdf given in Theorem 3.9.1, where $n > 0$ is an integer, is said to have a *chi-square distribution with n degrees of freedom*.

Comparing the chi-square pdf to the gamma pdf, we see that the chi-square distribution is a special case of the gamma distribution with $\lambda = \frac{1}{2}$ and $r = n/2$. The phrase "X is $\chi^2(n)$" is used to denote that X has a chi-square distribution with n degrees of freedom. Chi-square density curves for several different values of n are shown in Figure 3.18.

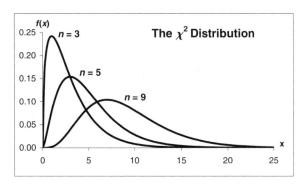

Figure 3.18

Using the formulas for the mean and variance of a gamma distribution, we see that the mean and variance of a chi-square distribution are

$$\mu = \frac{n/2}{1/2} = n \quad \text{and} \quad \sigma^2 = \frac{n/2}{(1/2)^2} = 2n.$$

If X is $\chi^2(n)$, then evaluating a probability such as $P(X < a)$ would require evaluating

$$P(X < a) = \int_0^a \frac{1}{\Gamma(n/2)2^{n/2}} x^{n/2-1} e^{-x/2} \, dx.$$

This integral looks rather formidable. So we might consider forming a table of values of this integral for different values of a as we did with the standard normal distribution. However, the value is also determined by the value of n, which could be any positive integer. Thus we would need one table for each value of n where each table contained different values of a. This is totally impractical. So instead, we define *critical values*.

Definition 3.9.4 Let p be a number between 0 and 1, and let X be $\chi^2(n)$. A *critical value* $\chi_p^2(n)$ is a positive number such that

$$P(X \leq \chi_p^2(n)) = 1 - p.$$

Figure 3.19 illustrates the definition of $\chi_p^2(n)$. Values of $\chi_p^2(n)$ for various values of p are given in Table C.3 (in the table, the degrees of freedom is denoted r). The next two examples illustrate how these critical values can be used to find probabilities.

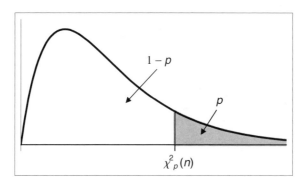

Figure 3.19

Example 3.9.4 Finding Chi-Square Probabilities Suppose X is $\chi^2(13)$. To find $P(X \leq 5.892)$, we inspect Table C.4 in Appendix C and see that $\chi^2_{0.95}(13) = 5.892$. By definition, this means that

$$P(X \leq 5.892) = 1 - 0.95 = 0.05.$$

Likewise, to find $P(4.107 < X \leq 22.36)$, we note that $\chi^2_{0.99}(13) = 4.107$ and $\chi^2_{0.05}(13) = 22.36$. This means that

$$P(X \leq 4.107) = 1 - 0.99 = 0.01 \quad \text{and} \quad P(X \leq 22.36) = 1 - 0.05 = 0.95.$$

Thus we have

$$P(4.107 < X \leq 22.36) = P(X \leq 22.36) - P(X \leq 4.107) = 0.95 - 0.01 = 0.94. \quad \square$$

Example 3.9.5 Coffee Shop Suppose customers arrive at a coffee shop at an average rate of 1 every 2 min. Assuming that the number of customers who arrive in 1 min is described by a Poisson distribution, approximate the probability that the 20th customer arrives within 1 hr after the shop opens.

Let W denote the wait time (in minutes) until the 20th customer arrives. We need to find $P(W < 60)$. Because customers arrive according to a Poisson distribution with parameter $\lambda = \frac{1}{2}$, W has a gamma distribution with parameters $r = 20$ and $\lambda = \frac{1}{2}$. However, this distribution is a special case of the chi-square distribution where

$$20 = \frac{n}{2} \quad \Rightarrow \quad n = 40.$$

Thus W is $\chi^2(40)$. Examining Table C.4, we see that $\chi^2_{0.05}(40) = 55.76$ and $\chi^2_{0.01}(40) = 63.69$. This means that

$$P(W < 55.76) = 0.95 \quad \text{and} \quad P(W < 63.69) = 0.99.$$

Taking the average of these two probabilities, we get

$$P(W < 60) \approx 0.97.$$

□

The next type of random variable and its associated distribution are named after W. S. Gosset, who did pioneering work in statistics during the early 1900s while working for an Irish brewery. He published under the pseudonym *Student*.

Definition 3.9.5 Let Z and C be independent random variables where Z is $N(0,1)$ and C is $\chi^2(n)$. The random variable

$$T = \frac{Z}{\sqrt{C/n}} \tag{3.7}$$

is called the *Student-t ratio with n degrees of freedom*.

The following theorem, which we present without proof, describes the pdf of T.

Theorem 3.9.2 *The pdf for the Student-t ratio with n degrees of freedom is*

$$f(t) = \frac{\Gamma[(n+1)/2]}{\sqrt{n\pi}\,\Gamma(n/2)\,(1+t^2/n)^{(n+1)/2}}, \quad -\infty < t < \infty.$$

A random variable with this pdf is said to have a *Student-t distribution* with n degrees of freedom. Student-t density curves for different values of n along with the standard normal density curve are shown in Figure 3.20. A Student-t density curve resembles a "flattened" standard normal bell curve. As in a bell curve, the Student-t pdf is symmetric with respect

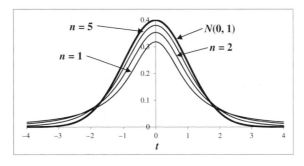

Figure 3.20

to the y axis; that is, $f(-t) = f(t)$. Also note that for larger values of n, the Student-t density curve gets closer to the standard normal bell curve.

As with the chi-square distribution, we define critical values of the Student-t distribution.

Definition 3.9.6 Let p be a number between 0 and 0.5, and let T have a Student-t distribution with n degrees of freedom. A *critical t-value* is a number $t_p(n)$ such that

$$P(T \leq t_p(n)) = 1 - p.$$

Figure 3.21 illustrates this definition. The number p is often referred to as the *area to the right* or the *area in one tail*. Table C.3 lists critical t-values for various values of p and n.

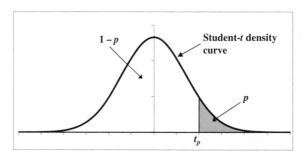

Figure 3.21

Example 3.9.6 **Student-t Probabilities** Let T have a Student-t distribution with 10 degrees of freedom. To find $P(T \leq 1.812)$, we examine Table C.3 and find that $t_{0.05}(10) = 1.812$. This means that

$$P(T \leq 1.812) = 1 - 0.05 = 0.95.$$

To find $P(-2.764 < T \leq 2.764)$, we note that $t_{0.01}(10) = 2.764$. This means that

$$P(T \leq 2.764) = 1 - 0.01 = 0.99,$$

and by symmetry of the Student-t pdf,

$$P(-2.764 < T) = 0.01.$$

Thus we have

$$P(-2.764 < T \leq 2.764) = P(T \leq 2.764) - P(-2.764 < T) = 0.99 - 0.01 = 0.98. \quad \square$$

3.9 The Gamma and Related Distributions

We end this section with the F-distribution that is named after Sir Roland A. Fisher, who was a contemporary of Gosset. This distribution is motivated by the following theorem, which we present without proof.

Theorem 3.9.3 *Let U and V be two independent chi-square random variables with n and d degrees of freedom, respectively. Define the random variable*

$$F = \frac{U/n}{V/d}. \tag{3.8}$$

Then the pdf of F is

$$f(x) = \frac{\Gamma\left[(n+d)/2\right] n^{n/2} d^{d/2} x^{(n/2)-1}}{\Gamma(n/2)\,\Gamma(d/2)\,(d+nx)^{(n+d)/2}}, \quad x > 0.$$

This theorem leads to the following definition.

Definition 3.9.7 A random variable with pdf given in Theorem 3.9.3 is said to have an F-distribution with n and d degrees of freedom. The number n is called the *numerator degrees of freedom* while d is called the *denominator degrees of freedom*.

Graphs of the F-distribution pdf for several different degrees of freedom are shown in Figure 3.22. Note the similarities with the χ^2 distribution.

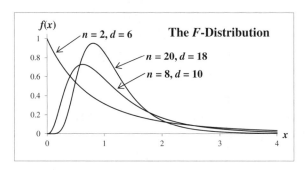

Figure 3.22

As with the chi-square and Student-t distributions, we define critical values of the F-distribution. Table C.5 in Appendix C lists critical F values for different values of n, d, and p.

Definition 3.9.8 Let p be a number between 0 and 1, and let F have an F-distribution with n and d degrees of freedom. A *critical F-value* is a number $f_p(n, d)$ such that

$$P(F \leq f_p(n, d)) = 1 - p.$$

Example 3.9.7 *F-Distribution Probabilities* Suppose F has an F-distribution with 15 and 7 degrees of freedom (d.f.). To find $P(F \leq 3.51)$, we examine Table C.5 and see that $f_{0.05}(15, 7) = 3.51$. This means that

$$P(F \leq 3.51) = 1 - 0.05 = 0.95.$$

To find $P(4.57 < F \leq 6.31)$, we see that $f_{0.025}(15, 7) = 4.57$ and $f_{0.01}(15, 7) = 6.31$. This means that

$$P(4.57 < F \leq 6.31) = 0.99 - 0.975 = 0.015. \qquad \square$$

Software Calculations

Minitab: Minitab has built-in functions for calculating probabilities of the form $P(X \leq b)$ where X has a gamma, χ^2, Student-t, or F-distribution. Select **Calc** → **Probability Distributions** and then choose the desired type of distribution. Select **Cumulative probability** and enter the appropriate parameters (for the gamma distribution, the shape parameter is r and the scale parameter is $1/\lambda$). Next to **Input constant**: enter the value of b. To calculate critical values, select **Inverse cumulative probability**.

R: R has functions for calculating probabilities and critical values for the distribution in this section. The syntax for these functions is summarized below:

Distribution	$P(X \leq b)$	Critical Value
Gamma	pgamma(b, r, λ)	
$\chi^2(n)$	pchisq(b, n)	qchisq(p, n)
Student-t with n d.f.	pt(b, n)	qt$(1 - p, n)$
F with n and d d.f.	pf(b, n, d)	qf$(1 - p, n, d)$

Excel: Excel has functions for calculating probabilities and critical values for the distribution in this section. The syntax for these functions is summarized below:

Distribution	$P(X \leq b), b > 0$	Critical Value
Gamma	GAMMADIST$(b, r, 1/\lambda, 1)$	
$\chi^2(n)$	1-CHIDIST(b, n)	CHIINV(p, n)
Student-t with n d.f.	1-TDIST$(b, n, 1)$	TINV$(2^*p, n)$
F with n and d d.f.	1-FDIST(b, n, d)	FINV(p, n, d)

TI-83/84 Calculators: To calculate probabilities for the χ^2, Student-t, and F-distributions, press **2ND** → **DISTR** and select the appropriate function. The syntax for these functions is summarized below:

Distribution	$P(a < X \leq b)$	Critical Value
$\chi^2(n)$	χ^2cdf(a, b, n)	
Student-t with n d.f.	tcdf(a, b, n)	invT$(1 - p, n)$
F with n and d d.f.	Fcdf(a, b, n, d)	

Exercises

1. Calls arrive at a 911 call center an average of 10 per hour. Assume that the number of calls in a 1-hr interval of time is described by a Poisson distribution.

   a. Let W denote the wait time (in minutes) until the second call arrives. Find the pdf, mean, and variance of W.

   b. Suppose an operator begins work at 8:00 AM. Find the probability that the second call arrives before 8:15 AM.

   c. Use a calculator or other software to approximate the probability that the eighth call arrives before 8:45 AM.

2. A family is planning an 8-hr road trip and is going to take along a portable video game system for the kids to play. The father is worried that the batteries will go dead before the end of the trip, so he takes along an extra set. From experience, he knows the batteries typically need to be replaced once every 10 hr. Assuming that the number of times the batteries need to be replaced in 1 hr has a Poisson distribution and that the kids play the videogame for all 8 hr, find the probability that the two sets of batteries last the entire trip. (**Hint**: The two sets of batteries last the entire trip if the time until the second replacement is greater than 8 hr.)

3. Use a calculator or other software to approximate the value of $\Gamma(0.75)$. Then use this value and the formula $\Gamma(t) = (t-1)\Gamma(t-1)$ to approximate the values of $\Gamma(1.75)$ and $\Gamma(2.75)$.

4. Let X be $\chi^2(15)$. Use Table C.4 in Appendix C to find the following probabilities:

   a. $P(X \leq 25)$ b. $P(X > 7.261)$ c. $P(5.229 < X \leq 8.547)$

5. Moviegoers arrive at a ticket booth an average of 2 every 4 min. Assume that the number who arrive in 1 min has a Poisson distribution. Find the probability that the ticket seller has to wait less than 34.17 min for the 10th customer to arrive.

6. Let T have a Student-t distribution with 6 degrees of freedom. Use Table C.3 to find the following probabilities:

   a. $P(T \leq 2.447)$ b. $P(T > 3.143)$
   c. $P(-1.943 < T \leq 1.943)$ d. $P(-3.707 < T \leq 1.943)$

7. Prove that the Student-t density curve is symmetric with respect to the y axis for any number of degrees of freedom n. That is, show that $f(t) = f(-t)$ where $f(t)$ is the Student-t pdf.

8. According to Theorem 3.9.1, if Z is $N(0,1)$, then the variable Z^2 is $\chi^2(1)$. Use this information to show that if T has a Student-t distribution with n degrees of freedom, then T^2 has an F-distribution with 1 and n degrees of freedom. (**Hint**: Square Equation (3.7) and compare it to Equation (3.8).)

9. Another distribution involving the gamma function is the *beta distribution*, whose pdf is

$$f(x) = \frac{\Gamma(a+b)}{\Gamma(a)\Gamma(b)} x^{a-1}(1-x)^{b-1}, \quad \text{for } 0 \leq x \leq 1$$

where $a, b > 0$ are the parameters.

   a. If $a = b = 1$, sketch a graph of the beta pdf. What type of distribution is this?
   b. Sketch the graph of the beta pdf for these different combinations of a and b, (a,b): $(1,3)$, $(2,2)$, and $(2,5)$.
   c. If X has a beta distribution with $a = 5$ and $b = 1$, find $P(X > 0.5)$.

10. A random variable X is said to have a *Weibull distribution* if its pdf is

$$\frac{k}{\lambda} \left(\frac{x}{\lambda}\right)^{k-1} e^{-(x/\lambda)^k}, \quad x \geq 0$$

where $k > 0$, $\lambda > 0$. This distribution is often used to describe the lifetime of objects. Show that $E(X) = \lambda \Gamma(1 + 1/k)$. (**Hint**: Use the definition of $E(X)$. To evaluate the integral, use the substitution $y = (x/\lambda)^k$.)

11. If X is $\chi^2(n)$, show that $E(X) = n$. (**Hint**: Use the fact that the χ^2 distribution is a special case of the gamma distribution where $\lambda = \frac{1}{2}$ and $r = n/2$.)

12. Suppose that X is $\chi^2(n)$, that Y is $\chi^2(m)$, and that X and Y are independent. Show that their sum $W = X + Y$ is $\chi^2(n+m)$. (**Hint**: The mgf of X is $M_X(t) = (1 - 2t)^{-n/2}$, and the mgf of Y is $M_Y(t) = (1 - 2t)^{-m/2}$. By Theorem 3.7.3, the mgf of W is $M_W(t) = M_X(t)M_Y(t)$. Show that $M_W(t)$ has the form of an mgf of a variable that is $\chi^2(n+m)$.)

13. Show that the moment-generating function of a random variable with a gamma distribution and parameters λ and r where $r = n$ is an integer is

$$M(t) = \left(\frac{\lambda}{\lambda - t}\right)^n \quad \text{for } t < \lambda$$

by letting $W = W_1 + W_2 + \cdots + W_n$ as in Example 3.9.2, finding the moment-generating function of each W_i using the results of Example 3.3.5, and then combining the results using Theorem 3.7.3.

14. Generalize the results of Exercise 13 above, and show that the moment-generating function of a random variable W with a gamma distribution and parameters λ and r (not necessarily an integer) is

$$M(t) = \left(\frac{\lambda}{\lambda - t}\right)^r \quad \text{for } t < \lambda$$

by using the definition of the mgf. (**Hint**: When evaluating the integral in the expected value, make the substitution $y = (\lambda - t)w$.)

15. If F has an F-distribution with 24 and 60 degrees of freedom, use Table C.5 in Appendix C to find the following probabilities:

  a. $P(F \leq 1.88)$ b. $P(F > 2.12)$ c. $P(1.70 \leq F \leq 2.12)$

16. Let U and V be independent χ^2 random variables with 6 and 5 degrees of freedom, respectively. Define $F = (U/6)/(V/5)$. Use Table C.5 to find

  a. $P(F > 4.95)$ b. $P(F \leq 10.67)$ c. $P(F > 10.67 \cup F \leq 4.95)$

17. If F has an F-distribution with 20 and 18 degrees of freedom, use Figure 3.22 to determine which of these probabilities is larger: $P(0.5 < F \leq 1.5)$ or $P(2 < F \leq 4)$. Explain your reasoning.

18. If a random variable $F = (U/n)/(V/d)$ has an F-distribution with n and d degrees of freedom, show that $1/F$ has an F-distribution with d and n degrees of freedom.

19. In this exercise we outline a proof of Theorem 3.9.1.

   a. Show that $\Gamma\left(\frac{1}{2}\right) = \int_0^\infty y^{-1/2} e^{-y}\, dy = \sqrt{\pi}$ by using the substitution $u = \sqrt{2y}$ and the fact that $\int_{-\infty}^\infty (1/\sqrt{2\pi}) e^{-x^2/2}\, dx = 1$.

   b. Use the results of Exercise 9 in Section 3.5 and part a of this exercise to show that if Y is $N(0,1)$, then $X = Y^2$ has a gamma distribution with $\lambda = r = \frac{1}{2}$ and that its mgf is $M(t) = (1 - 2t)^{-1/2}$.

   c. Let $Y_1, Y_2, \ldots, Y_n$ be independent random variables, each with a standard normal distribution. Define the random variable

   $$X = Y_1^2 + Y_2^2 + \cdots + Y_n^2.$$

   Use the results of part b and Theorems 3.7.3 and 3.7.5 to show that the mgf of X is $M_X(t) = (1 - 2t)^{-n/2}$.

   d. Conclude from part c that X has a gamma distribution with $\lambda = \frac{1}{2}$ and $r = n/2$ so that its pdf is as given in Theorem 3.9.1.

3.10 Approximating the Binomial Distribution

In this section, we discuss a technique for using the normal distribution to approximate the binomial distribution. This technique has useful practical applications as well as important theoretical uses.

The left half of Figure 3.23 shows the probability histogram of a random variable X that is $b(17, 0.5)$. Note the shape of the tops of the bars; it resembles a bell curve. The mean of X is $np = 8.5$, and its variance is $\sigma^2 = np(1-p) = 4.25$. The right half of Figure 3.23 shows the pdf of a normally distributed random variable with this mean and variance graphed on top of the histogram. Note that the graph closely follows the tops of the bars. Graphically this indicates that X is approximately normally distributed.

Next we make three observations about the bar centered at $x = 11$:

1. Its height is approximately 0.10, meaning that $P(X = 11) \approx 0.10$.
2. The width of this bar is 1 so that the area of the bar is also approximately 0.10.
3. The area under the bell curve between 10.5 and 11.5 is approximately the area of this bar.

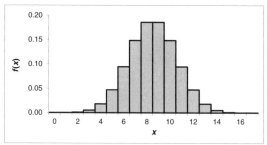 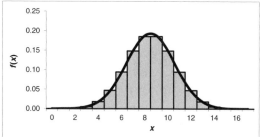

Figure 3.23

Putting these observations together, we have

$$P(X = 11) \approx P(10.5 < Y \leq 11.5)$$

where Y is $N(8.5, 4.24)$. Using a similar argument, we also have, for instance

$$P(5 \leq X \leq 12) \approx P(4.5 < Y \leq 12.5).$$

These results illustrate the following theorem, whose proof is given in Appendix A.3.

Theorem 3.10.1 Limit Theorem of De Moivre and Laplace Let X be $b(n,p)$. Then as $n \to \infty$, the distribution of X approaches

$$N(np, np(1-p)).$$ □

Informally, this theorem says that for n "large enough," X is approximately $N(np, np(1-p))$ and that as n gets larger, this approximation gets better. A rule of thumb for how large n needs to be for this approximation to be reasonable is that $np \geq 5$ and $n(1-p) \geq 5$.

We apply this theorem to approximate probabilities of X in the following way: Let a and b be integers such that $0 \leq a \leq b \leq n$, and let Y be $N(np, np(1-p))$. Then

$$P(a \leq X \leq b) \approx P(a - 0.5 < Y \leq b + 0.5). \tag{3.9}$$

Note that X is discrete and Y is continuous. The addition and subtraction of 0.5 in the expression $P(a - 0.5 \leq Y < b + 0.5)$ are used to correct for this change in continuity (called a *continuity correction*).

Example 3.10.1 Overbooking An airline has a policy of selling 375 tickets for a plane that seats only 365 people (called overbooking). Records indicate that about 95% of people who buy tickets for this flight actually show up. Find the probability that there are enough seats.

Let X denote the number of ticket holders who show up. Assuming the 95% rate is correct and that the arrivals of the 375 ticket holders are mutually independent, then X is $b(375, 0.95)$. We need to find $P(X \leq 365)$. According to Equation (3.9),

$$P(X \leq 365) \approx P(Y < 365.5)$$

where Y is $N(375 \cdot 0.95, 375 \cdot 0.95 \cdot 0.05) \approx N(356.25, 17.81)$ (note that $np = 356.25$ and $n(1-p) = 18.75$ are both greater than 5). The z-score of 365.5 is

$$z = \frac{365.5 - 356.25}{\sqrt{17.81}} \approx 2.19.$$

So, using Table C.1, we get

$$P(X \leq 365) \approx P(Y \leq 365.5) \approx P(Z \leq 2.19) = 0.9857.$$

Thus there is approximately a 98.6% chance that there will be enough seats. This approximation agrees with the exact value of approximately 0.9911 found with software. □

We end this section with a formal statement of the law of large numbers and its proof. There are many different forms of the law of large numbers; we present here one of the simplest forms.

Theorem 3.10.2   Bernoulli's Law of Large Numbers *Let the random variable X be the number of times a specified event A is observed in n trials of a random experiment. Let $p = P(A)$. Then for any small number $\epsilon > 0$,*

$$P\left(\left|\frac{X}{n} - p\right| \leq \epsilon\right) \to 1 \quad \text{as } n \to \infty.$$

Before we begin the proof, note that the variable X/n is simply the relative frequency approximation of $P(A)$. The expression $|X/n - p|$ is the error in this approximation. The law of large numbers says that the probability that the error is less than any specified number ϵ approaches 1 as n approaches infinity. In simpler terms, the more trials, the better the approximation.

Proof. The inequality $|X/n - p| \leq \epsilon$ is equivalent to

$$-\epsilon \leq \frac{X}{n} - p \leq \epsilon.$$

Rewriting this expression, we get

$$n(p - \epsilon) \leq X \leq n(p + \epsilon)$$

so that

$$P\left(\left|\frac{X}{n} - p\right| \leq \epsilon\right) = P[n(p - \epsilon) \leq X \leq n(p + \epsilon)].$$

Now, X is $b(n, p)$, so by Theorem 3.10.1 as $n \to \infty$,

$$P[n(p - \epsilon) \leq X \leq n(p + \epsilon)] \to P[n(p - \epsilon) \leq Y \leq n(p + \epsilon)]$$

where Y is $N(np, npq)$ and $q = 1 - p$. The z-scores of $n(p - \epsilon)$ and $n(p + \epsilon)$ are

$$\frac{n(p - \epsilon) - np}{\sqrt{npq}} = -\epsilon\sqrt{\frac{n}{pq}} \quad \text{and} \quad \frac{n(p + \epsilon) - np}{\sqrt{npq}} = \epsilon\sqrt{\frac{n}{pq}},$$

respectively. Now, note that

$$\lim_{n \to \infty} -\epsilon\sqrt{\frac{n}{pq}} = -\infty \quad \text{and} \quad \lim_{n \to \infty} \epsilon\sqrt{\frac{n}{pq}} = +\infty.$$

Thus, as $n \to \infty$,

$$P[n(p - \epsilon) \leq X \leq n(p + \epsilon)] \to P\left[-\epsilon\sqrt{\frac{n}{pq}} \leq Z \leq \epsilon\sqrt{\frac{n}{pq}}\right]$$
$$\to P(-\infty < Z < +\infty) = 1.$$

This proves the theorem. □

Exercises

1. In Example 3.10.1, we assumed that the arrivals of the 375 ticket holders are mutually independent. How reasonable an assumption do you think this is? Explain.

2. If X is $b(950, 0.35)$, use Equation (3.9) to approximate the following probabilities.

   a. $P(310 \leq X \leq 350)$
   b. $P(X \leq 315)$
   c. $P(340 \leq X)$

3. Suppose X is $b(75, 0.02)$.

   a. Calculate np and $n(1-p)$. According to the rule of thumb, can Equation (3.9) be used to approximate values of the probability of X?

   b. Find the exact value of $P(X \leq 1)$.

   c. Use Equation (3.9) to approximate the value of $P(X \leq 1)$. How good an approximation is this?

4. If a bet is placed on the number 7 in the game of roulette, the probability of winning is $\frac{1}{38}$. If 5000 different gamblers bet on the number 7, estimate the probability that more than 140 of them win.

5. In a recent election, 40% of voters voted for a clean-air proposition. In a survey of 825 voters after the election, 429 said they voted for it.

   a. Given that 40% of voters did vote for it, estimate the probability that in a survey of 825 voters, at least 429 of them actually did vote for it.

   b. What do the results in part a suggest about the honesty of the people in the survey?

6. If 500 points are randomly and independently selected from the rectangle $0 \leq x \leq 1$, $0 \leq y \leq 2$ on the xy plane, approximate the probability that for 365 or more of these points, $x \leq y$. (**Hint**: Draw a picture of the rectangle. In what portion of this region is $x \leq y$?)

7. A researcher wants to study the attitudes of college students toward credit cards. She plans to randomly select a sample of n students and give each a survey. She would like to have at least 50 students in the sample who regularly use credit cards. If 25% of all college students regularly use credit cards, estimate the minimum size of n so that the sample will contain 50 students who regularly use credit cards with a probability of at least 0.95.

8. Let X be any random variable with mean μ and variance σ^2. It can be shown that for any $\alpha > 0$,

$$P(|X - \mu| < \alpha) > 1 - \frac{\sigma^2}{\alpha^2}.$$

This result is known as *Chevyshev's inequality*. Consider the random experiment of flipping a fair coin n times, and let X denote the number of tails obtained.

   a. Find the mean and variance of X. (**Hint**: X has a binomial distribution.)

b. Use Chevyshev's inequality to show that

$$P\left(\left|\frac{X}{n} - \frac{1}{2}\right| < \epsilon\right) > 1 - \frac{1}{4n\epsilon^2}$$

for any $\epsilon > 0$. (**Hint**: Write $P(|X/n - \frac{1}{2}| < \epsilon) = P((1/n)|X - n/2| < \epsilon) = P(|X - n/2| < n\epsilon)$ and then apply Chevyshev's inequality.)

c. Find

$$\lim_{n \to \infty} \left(1 - \frac{1}{4n\epsilon^2}\right)$$

for any $\epsilon > 0$. What does this imply about the relative frequency of tails as the number of flips gets larger?

CHAPTER 4

Statistics

Chapter Objectives

- Define basic terms in statistics
- Stress the connections among statistics, probability, and random variables
- Introduce estimators
- Introduce sampling distributions
- Introduce confidence intervals
- Discuss ways of determining the necessary sample size
- Discuss techniques of assessing normality

4.1 What Is Statistics?

Suppose a grocery store manager is studying ways to improve customer satisfaction. One item of interest may be the amount of time customers spend waiting in line at the checkout stand. Some questions the manager might ask include these:

1. What is the expected waiting time?
2. What is the probability that a randomly selected customer waits longer than 10 min?
3. How much variation is there in the amount of time customers wait (i.e., do some customers wait very little time while others wait quite a bit)?

Let X denote the waiting time of a customer. Then X is a continuous random variable, and the answers to these questions are, in terms of X,

1. $E(X)$,
2. $P(X > 10)$, and
3. $\text{Var}(X)$.

If we know the distribution and the values of the appropriate parameters of X, then these values are easy to calculate. To determine an appropriate distribution and approximate the parameters of X, we need to observe values of X by measuring the amounts of time many different customers wait. These observed values are examples of *data*. This leads us to the following definitions of *data* and *statistics*:

> **Definition 4.1.1** *Data* are observed values of random variables. The field of *statistics* is a collection of methods for estimating distributions and parameters of random variables through the collection and analysis of data.

We must admit that these definitions are a bit narrow. Data do not have to be numbers, and statistics includes topics not directly related to estimating distributions and parameters of random variables. However, any set of data or statistical concept can be related to one or more random variables of some type.

A *statistical study* is a plan for collecting and analyzing data. The first step in a statistical study is to identify the *population* and select a *sample*.

> **Definition 4.1.2** The *population* is the set of all objects of interest in a statistical study. A *sample* is a subset of the population.

In the example of the grocery store, the population is the set of all customers. A sample is a set of customers whose waiting times are measured. In this statistical study, we want to know something about all the customers of the grocery store. Stated another way, we want to know something about the population. To know this precisely, we would need to *survey* (i.e., get information from) every customer. This is virtually impossible to do. However, surveying a sample of customers is relatively easy. Our goal is to use this information to draw conclusions about the population. This leads to the following, more practical definitions of data and statistics.

> **Definition 4.1.3**   *Data* are information that has been collected. The field of *statistics* is a collection of methods for drawing conclusions about a population by collecting and anlyzing data from a sample.

Note that the word *data* is plural. The singular form of the word is *datum*.

Types of Data

Data are divided into many categories and subcategories. First we define the difference between a *parameter* and a *statistic*.

> **Definition 4.1.4**   A *parameter* is a number calculated using information from *every* member of a population. A *statistic* is calculated using information from a sample.

Ideally, we want to know the value of a parameter. However, calculating it requires us to obtain information from every member of the population. In most cases, this is virtually impossible, so we try to estimate it with a statistic.

Example 4.1.1   Parameter Versus a Statistic   In a *census* of 6487 residents of a small town (in a *census*, every member of the population is surveyed), it was found that 79% use credit cards to pay for their groceries. Because every member of the population was surveyed, the number 79% is a *parameter*.

In a survey of 738 college students from across the United States, it was found that 52% of them regularly use credit cards. Because the population of interest is all college students in the United States, the data were collected from a sample, so the number 52% is a *statistic*.   □

Data are divided into two general types: *quantitative* and *qualitative*.

Definition 4.1.5 *Quantitative data* are numbers denoting measurements or counts. *Qualitative data* are information that can be separated into different categories.

Example 4.1.2 Quantitative versus Qualitative A student measures the masses of several pennies. These masses are numbers denoting measurements; therefore, the data collected are quantitative.

Suppose three candidates are running for state senate, Marten, Jones, and Smith. A pollster asks 735 randomly selected voters from across the state whom they support. The results are summarized in Table 4.1. At first glance, these data appear to be quantitative because there are numbers in the table. However, these numbers are not the data. These numbers *summarize* the data. The data collected are a list of names. The first name in the list is the candidate supported by the first voter surveyed, the second name is the person whom the second voter supports, etc. Thus the data are *qualitative*. The number 383, for example, simply states that the name *Marten* appears 383 times in the list.

Candidate	Marten	Jones	Smith
Freq	383	277	75

Table 4.1

Quantitative data are divided into two subcategories: *discrete* and *continuous*.

Definition 4.1.6 *Discrete data* are observed values of a discrete random variable. They are numbers that have a finite or countable set of values. *Continuous data* are observed values of a continuous random variable. They are numbers that can take any value within some range.

Example 4.1.3 Discrete Versus Continuous A market researcher surveys several households and asks how many cars each owns. Because the number of cars must be a whole number, these data are *discrete*.

Now suppose this researcher asks each household how many miles its members typically drive in a year. This number does not have to be a whole number, and it can be any nonnegative number; so these data are *continuous*. Note that the number of miles is typically reported as an integer, so it appears to be discrete. But the number of miles does not *have* to be an integer, so it is technically continuous. Also, because there are so many different possible values of the number of miles, it will be more convenient to treat it as continuous than discrete. □

Data are also divided into four *levels of measurement*.

> **Definition 4.1.7**   Levels of Measurement
> - Data are at the *nominal level of measurement* if they consist of only names, labels, or categories. They cannot be ordered (such as smallest to largest) in a meaningful way.
> - Data are at the *ordinal level of measurement* if they can be ordered in a meaningful way, but differences between data values cannot be calculated or are meaningless.
> - Data are at the *interval level of measurement* if they can be ordered in a meaningful way and differences between data values are meaningful.
> - Data are at the *ratio level of measurement* if they are at the interval level, ratios of data values are meaningful, and there is meaningful zero starting point.

Example 4.1.4   **Levels of Measurement**   Identify the level of measurement of each set of data.

a. A student observes 100 cars in the parking lot outside the student union and records what type of car each is (sedan, van, truck, etc.).

b. A questionnaire asks respondents to respond to a statement with $1 =$ strongly disagree, $2 =$ disagree, $3 =$ neutral, $4 =$ agree, or $5 =$ strongly agree. The responses of 100 people are recorded.

c. A political pollster records the years of birth of 500 registered voters.

d. A political pollster records the ages of 500 registered voters.

Solution

a. Each item in these data is a category. So these data are at the nominal level of measurement.

b. These data can be ordered in a meaningful way, but differences between values are meaningless. For instance, the difference between agree and disagree is $4 - 2 = 2$, but this difference is meaningless. So these data are at the ordinal level of measurement.

c. These data can be ordered in a meaningful way, and the differences are meaningful. For instance, the difference $1980 - 1975$ means that one voter is 5 years older than the other. However, the ratio $1980/1975 \approx 1.003$ is meaningless. So these data are at the interval level of measurement.

d. These data can be ordered in a meaningful way, the differences are meaningful, and ratios are meaningful. For instance, the ratio $50/25 = 2$ means that one voter is twice as old as another. Also, the age of 0 is a natural starting point (even though there is no voter of age 0). So these data are at the ratio level of measurement. □

Types of Studies

As defined earlier, a *statistical study* is a plan for collecting and analyzing data. Statistical studies are divided into two broad categories: *observational studies* and *experiments*.

Definition 4.1.8   Types of Studies

- In an *observational study*, data are obtained in a way such that the members of the sample are not changed, modified, or altered in any way.
- In an *experiment*, something is done to the members of the sample, and the resulting effects are recorded. The "something" that is done is called a *treatment*.

Do not confuse an *experiment* as defined here with a *random experiment*. A *random experiment* is an activity where the outcome is unknown until the activity is performed whereas an *experiment* is a plan for conducting a statistical study.

Example 4.1.5   Observational Study Versus Experiment

- A political poll is an example of an observational study. A reputable pollster will ask questions in such a way as to not influence the person's response. The pollster simply wants to "observe" what the person thinks.
- Tests of new medications are examples of experiments. In these tests, subjects are given a medication, or treatment, and the effects are recorded. □

Observational studies are subdivided into three categories according to when the data are collected.

Definition 4.1.9   Types of Observational Studies

- In a *cross-sectional study*, data are collected at one specific point in time.
- In a *retrospective study*, data are collected from studies done in the past.
- In a *prospective study*, data are collected by observing a sample for some time into the future.

When conducting an experiment, a researcher is usually looking for the effects of a single variable, or *factor*. For instance, a medical researcher may want to know how the dosage of a new allergy medication is related to the incidence of fatigue. However, factors such as gender and age may also influence fatigue. Such factors may not be of interest to the researcher and are called *nuisance factors* or *lurking variables*. If a study is not designed properly, the researcher may not be able to differentiate the effects of different factors. This is called *confounding*.

One way to help control confounding is through a *randomized block design*.

> **Definition 4.1.10** A *block* is a subset of the population with a similar characteristic. Different blocks of a population have different characteristics that may affect the variable of interest differently. A *randomized block design* is a type of experiment where
> 1. the population is divided into blocks, and
> 2. members from each block are randomly chosen to receive the treatment.

Example 4.1.6 Randomized Block Design An agronomist is testing the effectiveness of a new herbicide in field corn. Suppose the agronomist has 50 fields of corn, 25 of which have sandy soil, and the other 25 fields have clay soil. Because the type of soil could affect the performance of the herbicide, it makes sense to define two blocks: those fields with sandy soil and those with clay soil. To perform a randomized block design study, the agronomist would randomly choose several fields from each block, treat those with the new herbicide, leave the remaining fields untreated, and measure the results of the treated and untreated fields. Statistical procedures can then be used to compare the results.

In a poorly designed study, only fields with one type of soil would be treated. This design would not allow the researcher to distinguish between effects of the soil type and effects of the herbicide. Techniques for analyzing the results of a randomized block design are discussed in Section 5.10. □

Sampling Techniques

Once the study has been designed, the next step is to select the sample. This step is extremely important, and doing it properly is extremely difficult. The goal of choosing a sample is this:

The sample should accurately represent the population.

Just exactly what *accurately represent* means is subject to interpretation. However, everyone agrees that the sample should be chosen in a way that does not favor or exclude any portion of the population over any other.

There are many different types of sampling techniques, some of which we define below.

> **Definition 4.1.11 Sampling Techniques**
> - A *convenience sample* is a sample that is very easy to get.
> - A *voluntary response sample* is obtained when members of the sample decide whether to participate or not.
> - A *systematic sample* is obtained by arranging the population in some order, next selecting a starting point, and then selecting every kth member (such as every 20th).
> - A *cluster sample* is obtained by dividing the population into subsets (or *clusters*) where the members of each cluster have a common characteristic, then randomly choosing some of the clusters, and surveying *every* member of the chosen clusters.
> - A *stratified sample* is obtained by dividing the population into subsets and then randomly choosing some members from *each* of the subsets.
> - A *multistage sample* is obtained by successively applying a variety of sampling techniques. At each stage, the sample becomes smaller, and at the last stage, a cluster sample is chosen.

Example 4.1.7 Sampling Techniques
- A statistics student is interested in the mean amount of money spent by all students at his university on textbooks, so he surveys his classmates. This is an example of a *convenience sample*. Obviously this sample is not representative of the population of all students, so this is not a good sample.
- Restaurants often provide survey cards for customers to rate food quality and service. Because people decide whether to fill out a card or not, the resulting data come from a *voluntary response sample*. People typically fill out such a card only if they had a very bad or very good experience, so the sample is not representative of the population of all customers at the restaurant.
- Consider again the statistics student who wants to know the mean amount of money spent by college students on textbooks. To get a better sample, he might obtain a list of all students at his university, select the 14th name on the list, and then select every 30th name after that. This is an example of a *systematic sample*, which is a much better sample than the convenience sample.
- A city is divided into several voting precincts. To gauge the attitudes of voters toward the construction of a new middle school, the school board might randomly select a few of these precincts and then survey every voter in each of the chosen precincts. This is an example of a *cluster sample*.

- Another approach to gauging attitudes toward the new middle school is to divide voters into two subsets—those who live near the proposed location of the new school and those who live far away—and then choose several members from each subset. This is an example of a *stratified sample*.
- Yet another approach is to divide the precincts into two subsets—those close to the proposed location and those far away—then randomly choose a few precincts from each subset, divide each chosen precinct into neighborhoods, select a few neighborhoods from each chosen precinct, and survey each voter in the chosen neighborhoods. This approach combines stratified and cluster sampling, so the resulting sample is an example of a *multistage sample*. □

When calculating probabilities, we dealt with situations where items were "randomly" chosen. Randomly choosing items is the opposite of choosing items in any systematic way. Randomness is important so that the choices are independent. For this same reason, we need samples to be randomly chosen.

> **Definition 4.1.12   Random and Simple Random Samples**
> - A *random sample* is chosen in such a way that every *individual* member of the population has the same probability of being chosen.
> - A *simple random sample* of size n is chosen in such a way that every *group* of size n has the same probability of being chosen.

Example 4.1.8   Random Versus Simple Random Lightbulbs at a certain factory are produced in batches of 50, and 10 batches are produced each day. A quality control engineer at the factory needs to select 100 bulbs from a day's production for testing. One approach is to randomly select two batches and test each bulb. This is a *random sample* because each bulb has a $\frac{2}{10} = 0.2$ probability of being selected. However, this is not a *simple random sample* because not every group of size 100 has the same probability of being selected. For instance, the group consisting of the bulbs in the first batch has a 0.2 probability of being selected whereas the group consisting of the first 40 bulbs from the first batch, the first 20 bulbs from the fifth batch, and the first 40 bulbs from the sixth batch has a 0 probability of being selected.

To obtain a simple random sample, the engineer could assign a number between 1 and 500 to each bulb, use a computer to randomly choose 100 different numbers between 1 and 500, and select the corresponding bulbs. This is a random sample because each bulb has a $\frac{1}{500}$ probability of being selected. But it is also a *simple random sample* because each group of 100 bulbs has the same probability of being selected, $1/\binom{500}{100}$. □

A simple random sample is the "gold standard" in samples. Nearly every statistical formula and technique we discuss in this text has the requirement that the data come from a simple random sample. When selecting a sample, we should keep this quote in mind (see M. F. Triola, *Elementary Statistics*, 10th ed., Pearson Education, Inc., Boston, 2006, p. 31).

> The method used to collect data is absolutely and critically important, and we should know that *randomness* is particularly important. If sample data are not collected in an appropriate way, the data may be so completely useless that no amount of statistical torturing can salvage them.

Once the data have been collected, the "number crunching" begins. In the following sections and chapters, we discuss many basic techniques for doing just this. We stress how probability forms the foundation for everything we do in statistics. The study of this underlying mathematical theory is called *mathematical statistics*.

Exercises

1. In each of the following scenarios, describe the population(s) and the sample(s). Determine if the sample is biased. Explain your reasoning.

  a. A pollster wants to know the mean income of all voters in a city. She surveys 50 voters from a neighborhood where each house has a three-car garage.

  b. A math professor wants to study the attitudes toward calculators of students at her university. She surveys each student in her four classes.

  c. A math professor wants to study the attitudes toward calculators of students at his university. He obtains the student directory, assigns each student a number, and uses a computer to randomly choose 50 numbers. He surveys the corresponding students.

  d. A student wants to compare the weights of three varieties of potatoes. She purchases one 10-lb bag of each variety and weighs each potato.

  e. A student wants to study the ability of fellow students to estimate distances. He asks 15 of his friends to draw a line 14 cm long, and then he measures the line and calculates $|\text{length} - 14|$.

  f. To monitor the quality of fiber-optic cable being manufactured, a quality control technician selects pieces of cable, each 3.5 m long, randomly throughout the day and counts the number of defects in each piece.

2. Identify whether the given percentages are parameters or statistics. Explain your reasoning.

  a. A biologist walks along a beach and notices that 60% of observed ghost crabs have one claw that is larger than the other.

  b. The phonebook for a small town contains 1236 names, 178 of which begin with the letter S. The percentage of names in this phonebook that begin with the letter S is 14.4%.

  c. A dentist looks over the records of all her patients and finds that 25% have not had a checkup in the past year.

  d. When walking through a mall, a mother observes that 65% of teenagers are wearing hats.

  e. After a winter storm, a highway patrol officer observes that highways are 45% snow-packed.

3. In each of the following scenarios, determine if the data collected are quantitative or qualitative. If they are quantitative, determine if they are discrete or continuous.

  a. A pollster asks several voters their annual income.

  b. A math professor asks each student in a class if he or she has a calculator.

  c. A math professor counts the number of students in five different classes who have calculators.

  d. A student weighs 14 potatoes.

  e. A student asks fellow students to draw a line 14 cm long, measures the length of the line, and calculates $|\text{length} - 14|$.

  f. A quality control technician selects 19 different pieces of fiber-optic cable, each 3.5 m long, and counts the number of defects in each piece.

4. Identify whether the given data values belong to continuous or discrete sets of data.

  a. A first-grader counts the number of empty hangers in his closet.

b. An online homework system records the amount of time a student spends completing an assignment.

c. A brochure for a new car gives the weight of the car.

d. A student counts the number of chocolate candies in a small bag of candies.

e. A chemist measures the percentage of alcohol in a solution of water and alcohol.

5. Determine which level of measurement (nominal, ordinal, interval, ratio) best categorizes the data in each scenario.

a. A racing fan records the number on each car that competes in the Indy 500.

b. A climatologist records the daily high temperature in the month of June.

c. A sociologist records the annual incomes of 50 migrant workers.

d. A student records the weights of 140 potatoes.

e. A professor rates each student in a research class on her or his potential for success in graduate school as 1 = low, 2 = neutral, 3 = high.

6. Determine which level of measurement (nominal, ordinal, interval, ratio) best categorizes the data in each scenario.

a. The registrar of a university records the number of students enrolled in each class offered by the college of education.

b. A movie fan makes a list of the movies he has watched in the past year.

c. A sleep specialist records the time her patients go to bed.

d. A fisherman records the species of fish he catches on a fishing trip.

e. A nurse records the body temperatures of a group of sick children.

7. Identify each statistical study as an observational study or an experiment.

a. A police officer records the colors of cars that are speeding.

b. A statistics professor uses a new textbook in a class and compares the class test scores to those from the previous semester.

c. A group of overweight people is put on a new type of diet, and their weight losses are recorded.

d. A pharmacist selects a bottle of aspirin and measures the amount of aspirin in each tablet.

e. A corn researcher applies different fertilizers to different plots of corn.

8. Identify the type of each of these samples (convenience, voluntary response, systematic, stratified, or cluster).

a. To determine what type of foods kids like to eat, a mother asks several of her friends.

b. To determine the level of support for the construction of a new athletics facility, a university calls every 10th person on its alumni/alumnae list.

c. A company asks customers to make comments about a new product on its Facebook page.

d. To assess the rate of job growth, a staffing company randomly selects 15 different Zip codes and surveys every business with an address in those Zip codes.

e. A high school principle surveys a few randomly selected students from each grade (9th through 12th) about their summer plans.

9. Describe how confounding could occur in each of these studies.

a. When testing the effectiveness of a new cancer treatment, researchers give the treatment to many cancer patients with different types of cancer.

b. The statistics department chair wants to compare three different online homework systems. She assigns one system to each of three different professors teaching an introduction to statistics class and compares homework scores from each class.

c. For his science fair experiment, a fifth-grade student wants to compare parachutes of different shapes. He constructs parachutes in different shapes, attaches a weight to each, drops each from a balcony, and records the amount of time each takes to reach the ground.

d. To determine if red wine "causes" good health, a researcher interviews several hundred people, asks each about any diseases he or she has, and asks how much red wine each drinks.

e. To compare sizes of different varieties of tomatoes, a horticulturist collects samples of different varieties grown by different gardners.

10. Determine if each of the following samples is a random sample, a simple random sample, or neither. Explain your reasoning.

   a. A quality control engineer at a candy factory collects all the candies produced in 1 hr, thoroughly mixes them, and scoops a sample of 50 candies.

   b. The students in a classroom are arranged in five rows of four students each. To select a sample of four students for a study, the teacher randomly selects one row, and the four students in that row form the group.

   c. A sample of 100 people is chosen by randomly selecting 50 males and 50 females from a population that is half male and half female. selected so the sample is random. However, we could not, for instance, get a sample of 25 males and 75 females, so the sample is not a simple random sample.

   d. Starting with the first person who enters a movie theater, every fifth person is surveyed.

11. One day in February the temperature was 20°F, and the next it was 40°F. Would it be appropriate to say, "It was twice as hot the second day as the first"? Explain.

12. Consider a school consisting of 6 classrooms. Each classroom contains 6 rows of desks, and each row contains 6 students. Explain how you could use a set of 3 fair 6-sided dice to select a simple random sample of 20 students from this school.

4.2 Summarizing Data

After data have been collected, the first step in analysis is to create a graph. There are many different types of statistical graphs including pie charts, bar graphs, and relative frequency histograms.

Example 4.2.1 **Polling Results** Suppose there are three candidates running for state senate, Marten, Jones, and Smith. A pollster asks 735 randomly selected voters from across the state whom they support. The results are summarized in Table 4.2.

As stated in the previous section, these data are qualitative. Two appropriate types of graphs of qualitative data such as these are pie charts and bar graphs, as shown in Figure 4.1. □

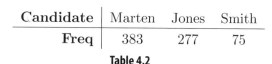

Table 4.2

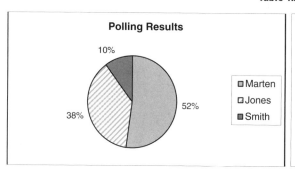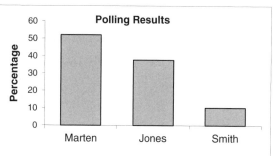

Figure 4.1

We first constructed relative frequency histograms in Section 2.3. In the next example, we construct histograms of discrete data and show how they can be used to compare sets of data.

Example 4.2.2 **Credit Hours** Two students are interested in comparing the number of credit hours taken by freshmen males and females at a certain university. They ask 17 males and 22 females how many hours they take. The data are shown in Table 4.3 (data collected by Megan Damron and Spencer Solomon, 2009).

Males	15	16	16	16	18	14	17	17	16	15	14
	15	16	15	14	15	14					
Females	14	16	16	14	14	14	14	16	14	16	16
	16	16	14	16	16	14	16	16	16	15	15

Table 4.3

Note that these data are discrete because they are all integers. Relative frequency histograms of each set of data are shown in Figure 4.2. Comparing them, we see that females typically take either 14 or 16 hr and that males have a wider range of values. Note that it is much easier to analyze the graphs than it is to analyze the raw data. This is one of the most important uses of graphs. □

The details for constructing a relative frequency histogram of continuous data are given in Section 3.1. In the next example, we construct such a histogram as well as a *cumulative relative frequency histogram*.

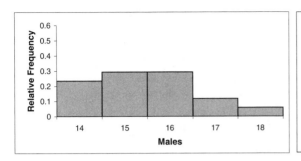
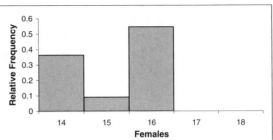

Figure 4.2

Example 4.2.3 Checkout Stand Wait Times Consider the example of grocery store wait times from Section 4.1. Suppose we measure the wait times, in minutes, of 30 randomly selected customers, as shown in Table 4.4.

0.0	0.2	0.5	0.8	1.5	1.9	2.2	2.9	3.3	4.3
0.0	0.2	0.5	1.1	1.7	1.9	2.5	2.9	3.4	5.1
0.0	0.3	0.7	1.2	1.7	2.1	2.6	3.1	4.1	7.3

Table 4.4

Note that the wait time could be any number greater than 0. It does not have to be an integer or be rounded off to one decimal place (as the data are). Thus these data are continuous. Because the data range from 0 to 7.3, we choose to use eight bins with a bin width of 1. The relative frequency distribution of the data is shown in the first three rows of Table 4.5.

Bin	$[0,1]$	$(1,2]$	$(2,3]$	$(3,4]$	$(4,5]$	$(5,6]$	$(6,7]$	$(7,8]$
Freq	10	7	6	3	2	1	0	1
Rel Freq	0.333	0.233	0.200	0.100	0.067	0.030	0.000	0.030
Cumul Rel Freq	0.333	0.566	0.766	0.866	0.933	0.963	0.963	1.000

Table 4.5

The histogram is shown in the left half of Figure 4.3. Notice that the bars close to 0 are relatively tall whereas the bars farther from 0 are short. This shows that a large proportion of the customers wait relatively little time. Only a very few wait a long time. This would be good news to the manager. The shape of the histogram indicates what type of distribution the underlying random variable has. This is one of the most important uses of a histogram. The tops of the bars form a pattern that resembles the graph of an exponential pdf. This is an indication that the random variable representing customer wait time has an exponential distribution.

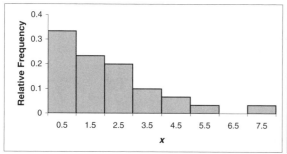 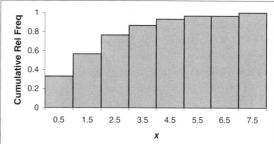

Figure 4.3

The fourth row of Table 4.5 contains the cumulative relative frequency of each interval, which is simply the sum of the relative frequencies of the interval and all previous intervals. The cumulative relative frequency of the bin $(5, 6]$, 0.963, tells us that 96.3% of all the data values are less than or equal to 6. The right half of Figure 4.3 shows the graph of these values, called the *cumulative relative frequency histogram* of the data. The shape of this graph is an approximation of the shape of the graph of the cumulative distribution function of the underlying random variable. □

One important point to remember about histograms is the following:

> The *shape* of a relative frequency histogram is an approximation of the graph of the pdf (or pmf) of the underlying random variable.

For this reason, a relative frequency histogram is also called the *sample distribution*.

Another way to summarize data is through the calculation of *sample statistics* (or simply *statistics*). Below we define several commonly used statistics.

Definition 4.2.1 Let $\{x_1, x_2, \ldots, x_n\}$ be a set of quantitative data collected from a sample of the population.

1. The *mean* of the data is

$$\bar{x} = \frac{1}{n} \sum_{i=1}^{n} x_i. \tag{4.1}$$

2. The *variance* of the data is

$$s^2 = \frac{1}{n-1} \sum_{i=1}^{n} (x_i - \bar{x})^2. \tag{4.2}$$

3. The *standard deviation* of the data is

$$s = \sqrt{s^2} = \sqrt{\frac{1}{n-1} \sum_{i=1}^{n} (x_i - \bar{x})^2}. \tag{4.3}$$

4. The *range* of the data is (max value) − (min value).

These statistics are also called the *sample mean, sample variance, sample standard deviation*, and *sample range*, respectively. Collectively, these statistics are referred to as *summary statistics*.

The sample mean, variance, and standard deviation are *estimates* of the mean, variance, and standard deviation of the underlying random variable, respectively. That is, if the data are observed values of the random variable X, then

$$\bar{x} \approx E(X) = \mu, \quad s^2 \approx \mathrm{Var}(X) = \sigma^2, \quad \text{and} \quad s \approx \sigma.$$

These claims will be justified and made more formal over the next few sections.

The mean of a set of data can be thought of as measuring how big the data values are. The variance and standard deviation are called *measures of variation* and can be thought of as measuring the average distance of the values from the mean. Informally, they measure how spread out the data are.

Example 4.2.4 Calculating Statistics Consider the data in Table 4.4. The mean of the data is

$$\bar{x} = \frac{1}{30}(0.0 + 0.0 + \cdots + 5.1 + 7.3) = 2 \min.$$

The variance of the data is

$$s^2 = \frac{1}{30-1}\left[(0.0-2)^2 + (0.0-2)^2 + \cdots + (5.1-2)^2 + (7.3-2)^2\right] \approx 2.946 \min^2,$$

the standard deviation of the data is

$$s = \sqrt{2.96} \approx 1.72 \,\text{min},$$

and the range of the data is $7.3 - 0 = 7.3$ min. □

Note the units on these different statistics. The units of the mean and standard deviation are the same as the units of the data. The units of variance are the square of the units of the data.

The $(100p)$th percentile of a continuous random variable X is a value of X, denoted π_p, such that

$$p = F(\pi_p) = P(X \leq \pi_p)$$

where p is between 0 and 1. We define percentiles of a set of data in a similar way.

Definition 4.2.2   Let p be a number between 0 and 1. The $(100p)$th *percentile* of a set of quantitative data is a number, denoted π_p, that is greater than $(100p)\%$ of the data values.

The 25th, 50th, and 75th percentiles are called the *first, second,* and *third quartiles* and are denoted $p_1 = \pi_{0.25}$, $p_2 = \pi_{0.50}$, and $p_3 = \pi_{0.75}$, respectively. The 50th percentile is also called the *median* of the data and is denoted by $m = p_2$. The *mode* of the data is the data value that occurs most frequently.

The *five-number summary* of a set of data consists of the minimum value, p_1, p_2, p_3, and the maximum value.

Percentiles of a set of data can be thought of as estimates of the percentiles of the underlying random variable. There is no universally agreed-upon method for calculating percentiles. Different statistical software use different methods, and some are more complex than others. Here we present a rather simple method for calculating the $(100p)$th percentile:

1. Arrange the data in increasing order: $x_1 \leq x_2 \leq \cdots \leq x_n$.
2. Calculate $L = np$.
3. If L is *not* an integer, then round it up to the next-larger integer and $\pi_p = x_L$.
4. If L is an integer, then $\pi_p = \frac{1}{2}(x_L + x_{L+1})$.

Example 4.2.5   Calculating the Five-Number Summary   Consider again the data in Table 4.4. To calculate the first quartile, $p_1 = \pi_{0.25}$, first note that the data are already arranged in

increasing order. Next,
$$L = 0.25(30) = 7.5.$$
Because L is not an integer, $p_1 = x_8 = 0.5$. By similar calculations, $p_3 = \pi_{0.75} = 2.9$. To find the median, or $p_2 = \pi_{0.5}$, $L = 0.5(30) = 15$. So
$$p_2 = \frac{1}{2}(x_{15} + x_{16}) = \frac{1}{2}(1.7 + 1.9) = 1.8.$$
The min and max of the data are 0 and 7.3, respectively. Thus the five-number summary is 0, 0.5, 1.8, 2.9, 7.3. The *box plot* in Figure 4.4 is a graphical description of the five-number summary. A box plot can be thought of as a simple histogram in the sense that the graph shows the distribution of the data. By the definition of p_1, the smallest 25% of the data values are between the min and p_1. We see that this is a very narrow range. Also, by the definition of p_3, the largest 25% of the data values are between p_3 and the max. We see that this is a much wider range. This illustrates that small data values are bunched together and larger values are more spread out. Box plots are especially useful for comparing sets of data (see Exercise 6 in this section).

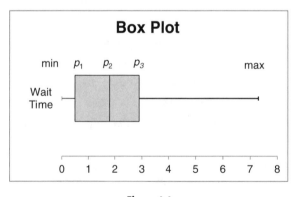

Figure 4.4

□

In the previous example, we saw that $\pi_{0.75} = 2.9$. Stated another way, the 75th percentile of the data is 2.9. This means that about 75% of the data values are less than 2.9. The number 75% is called the *percentile rank* of the data value 2.9. We can easily calculate the percentile rank of a data value x with the formula

$$\text{Percentile rank of } x = \frac{\text{number of data values less than } x}{\text{total number of data values}} \times 100\%.$$

Example 4.2.6 Calculating a Percentile Rank For the data in Table 4.4, find the percentile rank of the data value 2.5.

Note that 2.5 is greater than 19 other data values, so

$$\text{Percentile rank of } 2.5 = \frac{19}{30} \times 100\% \approx 63\%.$$

Thus 2.5 is the 63rd percentile of the data. □

Software Calculations

Minitab: Descriptive statistics can be calculated by first entering the data in a column. Then select **Stat → Basic Statistics → Display Descriptive Statistics**. Under **Variables**, select the column containing the data. The types of statistics calculated can be chosen by selecting **Statistics**. Graphs can be drawn by selecting **Graphs** and then choosing the desired type of graph.

Excel: To calculate many of the statistics discussed in this section, enter the data in a worksheet and then select **Data → Data Analysis → Descriptive Statistics**. The **Input Range** is the range containing the data. Select a blank cell for the **Output Range** and check **Summary statistics**. The mean, standard deviation, and median can be individually calculated with the functions AVERAGE, STDEV, and MEDIAN, respectively. The input for each of these functions is the range containing the data. The $(100p)$th percentile can be calculated with the syntax PERCENTILE(range, p). (Note that Excel uses a slightly different algorithm than given in the text for calculating percentiles.) Various types of graphs can be drawn by first selecting the data and then selecting the **Inset** tab. Select the type of graph you want. Histograms can be drawn by first entering the left endpoints of the bins in a column. Then select **Data → Data Analysis → Histogram**. The **Bin Range** is the range containing the left endpoints of the bins. Select a blank cell for the **Output Range** and check **Chart Output**.

R: Enter the data $\{x_1, x_2, \ldots, x_n\}$ as a vector with the syntax x<−c($x_1, x_2, \ldots, x_n$). The following syntax can be used to calculate sample statistics:

Syntax	Description of Output
summary(x)	Five-number summary and the sample mean
sd(x)	Standard deviation
quantile(x, p)	$(100p)$th percentile
hist(x, breaks=n, freq=FALSE)	Relative frequency histogram with n bins

> **TI-83/84 Calculators**: To calculate many of the statistics discussed in this section, first enter sample data by pressing **STAT** → **EDIT** → **1:Edit**. Enter the data in one of the lists. Then press **STAT** → **CALC** → **1:1-Var Stats**. Enter the name of the list containing the data by pressing **2ND** → **1** for list L1, **2** for list L2, etc. The output displays the sample mean, denoted $\bar{x}$; the sample standard deviation, denoted Sx; and the five-number summary, among other results. To generate a histogram of the data, press **2ND** → **STAT PLOT** and select one of the plots. Select **ON**; next to **Type:**, select the third type of graph; and next to **Xlist**, enter the name of the list containing the data. Then press **ZOOM** → **9:ZoomStat**.

Exercises

1. A statistics professor asks members of her class to give their home states. The results are summarized in the table below.

State	Nebraska	Kansas	Missouri	Colorado
Freq	11	6	4	2

   a. Are these data quantitative or qualitative? If they are quantitative, are they discrete or continuous? Explain.

   b. Draw a pie chart and a bar chart of this data.

   c. Would it be appropriate to calculate the mean, standard deviation, or the 32nd percentile of these data? Why or why not?

   d. What is the mode of these data?

2. The table below lists the weights of 20 randomly selected adult catfish from a lake.

6.1	9.4	10.1	10.9	11.2	11.9	12.1	13.4	14.9	15.8
8.8	9.6	10.4	11.0	11.6	12.0	13.2	13.6	15.5	17.0

   a. Construct a relative frequency histogram and a cumulative relative frequency histogram of these data, using six bins. If the random variable X denotes the weight of a randomly selected adult catfish in this lake, what type of distribution does X appear to have?

   b. Calculate the mean, variance, and standard deviation of these data.

c. Calculate the first, second, and third quartiles of the data.

d. Find the percentile rank of the weight 13.2.

3. The table below lists the masses (in grams) of 30 randomly selected regular M&Ms (data collected by Brian Maxson, 2011).

0.783	0.897	0.813	0.900	0.823	0.900	0.837	0.905	0.854	0.909
0.855	0.911	0.858	0.912	0.865	0.912	0.869	0.919	0.874	0.930
0.875	0.935	0.875	0.944	0.876	0.967	0.878	0.974	0.885	0.976

a. Calculate the mean, variance, and standard deviation of these data.

b. Construct a relative frequency histogram of the data, using eight bins over the interval $(0.765, 1.005]$. What type of distribution does the mass appear to have?

4. In the manufacture of polyethylene water tanks, resin is melted and poured into a mold. As the resin cools, it shrinks a bit. The table below lists the percentage of shrink of 30 randomly selected tanks (data collected by Russell D. Bartling, 2010).

2.08	2.37	2.37	2.37	2.67	2.67	2.67	2.67	2.97	2.97
2.08	2.37	2.37	2.67	2.67	2.67	2.67	2.67	2.97	2.97
2.08	2.37	2.37	2.67	2.67	2.67	2.67	2.97	2.97	3.26

a. Calculate the mean, variance, and standard deviation of these data.

b. Construct a relative frequency histogram of the data, using six bins over the interval $(2, 3.5]$. What type of distribution does the percentage of shrink appear to have?

c. Let the random variable X denote the percentage of shrink of a randomly selected tank. According to the empirical rule, if X has a normal distribution, then

$$P(\mu - 3\sigma < X \leq \mu + 3\sigma) \approx 0.997$$

where μ and σ are the mean and standard deviation of X, respectively. In practical terms, this means that about 99.7% of all tanks will shrink between $(\mu - 3\sigma)\%$ and $(\mu + 3\sigma)\%$. Use the results of part a to estimate this range.

d. The manufacturer of the resin claims that the normal range for shrink is 1% to 4%. Based on your results in part c, do the data appear to support this claim? Explain.

5. A critic of the safety of pickup trucks selects a sample of cars and a sample of trucks, records the braking distance of each, and finds that the cars have a mean braking distance of 150 ft and the trucks have a mean of 200 ft. To illustrate this difference, he constructs the graph in Figure 4.5.

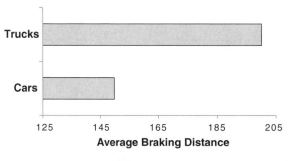

Figure 4.5

a. If a person were to simply look at the bars (and ignore the numbers), explain why he or she might conclude that the average stopping distance for trucks is more than twice as long as that for cars. Is this an accurate conclusion?

b. Redraw the graph so that it gives a more accurate comparison of the average stopping distances. Explain how your graph is different from the original.

6. The table below gives the lengths (in minutes) of samples of movies rated G, PG, PG-13, and R (data collected by Meredith Hein and Rachel Dahlke, 2011).

G			PG			PG-13			R		
25	75	88	87	97	104	95	116	130	93	98	123
63	76	97	94	99	114	97	117	130	95	100	125
68	82	98	95	100	120	99	117	153	95	103	129
74	83	106	96	103	143	105	122	157	96	109	138
74	83	117	96	104	152	111	124	178	97	119	148

a. Calculate the five-number summary of each of the four samples.

b. On the same xy-plane, draw a box plot of each of the four samples. Use the box plots to compare the four samples. That is, use the box plots to describe differences between the lengths of the four types of movies.

7. The table below gives the heights (in centimeters) of World Cup soccer players from Europe and the Americas (data collected by Santiago Keinbaum, 2011). For each sample of data, construct a histogram over the interval (165, 190] with a bin width of 5. Compare the two histograms. Does there appear to be much difference between the distributions of the heights of European World Cup players and those from the Americas?

Europe
167 169 170 170 170 171 172 174 175 175 175 175 175 177 178 178 178 180 180 180 181
181 181 181 181 182 182 182 182 182 182 183 183 183 183 183 184 184 184 184 184 185
185 185 185 185 185 185 187 187 187 187 187 187 188 188 188 188 188 189 190
Americas
166 166 167 168 168 168 168 169 170 170 170 173 173 173 175 175 175 176 176 176 176
176 177 177 178 178 178 178 178 178 180 180 180 180 180 180 180 181 181 181 182 182
182 182 183 183 183 183 184 185 185 185 185 185 185 186 187 187 187 187 188

8. A set of data is said to be *bimodal* if it has two modes. Graphically this means that a relative frequency histogram of the data has two peaks. The table below lists the scores of 17 different students on a certain calculus test. Find the mode(s) of the data and draw a relative frequency histogram, using five bins over the interval (50, 100]. Are these data bimodal? What does this say about the students in this class?

82	55	66	72	73	77	96	98	99
60	63	67	73	74	80	97	99	

9. Explain your answers to these questions about statistics.

   a. Can the mean of a set of data ever be negative?

   b. Can the variance of a set of data ever be negative?

   c. Can the variance of a set of data ever be larger than the mean? If so, give an example.

   d. For a given set of data, can $\pi_{0.25}$ ever be greater than $\pi_{0.75}$?

   e. Is it true that the mean of a set of data is always larger than one-half of the data values and less than one-half of the other data values? If not, give an example.

   f. Is it true that the mean of a set of data is always larger than at least one of the data values? If not, give an example.

10. A summary statistic similar to the median is the *midrange*:

$$\text{Midrange} = \frac{\text{max data value} - \text{min data value}}{2}.$$

  a. Calculate the midrange of the data in Table 4.4.
  b. Compare the midrange of these data to the median found in Example 4.2.5. Are they the same?
  c. A newspaper article refers to a certain statistic as being "in the middle, halfway between the min and max." Is the reporter referring to the median or the midrange? Explain your reasoning.

11. If x is a value of a normally distributed random variable with mean μ and standard deviation σ, the z-score of x is defined to be $z = (x - \mu)/\sigma$. Now, if x is a data value from a set of data with mean $\bar{x}$ and standard deviation $s \neq 0$, we define the *z-score* of x to be

$$z = \frac{x - \bar{x}}{s}.$$

The z-score is a measure of how far x is from the mean.

  a. Show that if $x > \bar{x}$, then the z-score of x is positive, and if $x < \bar{x}$, then z is negative.
  b. Show that if $x = \bar{x} + n \cdot s$, where n is any number, then the z-score of x is $z = n$. If x_1 is "farther" from the mean than x_2, how does the the magnitude of the z-score of x_1 compare to that of x_2?
  c. Bob got a score of 80 on a psychology test where the scores have a mean of 90 and a standard deviation of 5. Joe got a score of 50 on an economics test where the scores have a mean of 55 and a standard deviation of 10. Calculate the z-score of each test score. Who did relatively better, Bob or Joe? Explain your reasoning.

12. When working with the chi-square distribution, we use the term *degrees of freedom*. To illustrate one meaning of this term, consider the following problem: Find a set of 5 data values whose mean is $\bar{x} = 3.2$. How many of these values could you freely choose? Once these values are chosen, how many of the values are determined by these values? Generalize your results. If we want a sample of n data values with a specified mean, how many values can we freely choose? The number of freely chosen values is called the *degrees of freedom of the sample mean*.

13. Three different statistics classes each took the same 8-point quiz. The scores of 11 students from each class were randomly selected and are shown in the table below. The mean of each sample is 5.

Class 1:	3	4	4	5	5	5	5	5	6	6	7
Class 2:	3	3	3	4	4	5	6	6	7	7	7
Class 3:	2	4	4	5	5	5	6	6	6	6	6

a. Sketch a relative frequency histogram of the scores of each class.

b. Without doing any calculations, which class do you think has the largest sample standard deviation? Which has the smallest? Explain your answers. (**Hint:** Think of a sample standard deviation as the average distance of the data values from the mean.)

c. Calculate the sample standard deviation of each class. Do your answers to part b agree with these calculations?

14. The shape of a relative frequency histogram of a set of data is classified into one of three categories:

- *symmetric* if the left half of the histogram is (roughly) a mirror image of the right half and the mean, median, and mode of the data are (roughly) equal;
- *skewed left*, or *negatively skewed*, if the left tail of the histogram is longer than the right tail and the mean and median of the data are less than the mode; and
- *skewed right*, or *positively skewed*, if the right tail of the histogram is longer than the left tail and the mean and median of the data are greater than the mode.

The *skewness* of a set of data can be measured with the following statistic:

$$\text{Sample skewness} = \frac{\sqrt{n(n-1)}}{n-2} \cdot \frac{(1/n)\sum (x_i - \bar{x})^3}{\left[(1/n)\sum (x_i - \bar{x})^2\right]^{3/2}}$$

where n is the number of data values; x_i, $i = 1, \ldots, n$ are the data values; $\bar{x}$ is the sample mean; and all sums are taken from $i = 1$ to n. The sample skewness is an estimate of the skewness of the underlying random variable. For each set of data in the table below, calculate the median, mean, mode, and skewness. Also sketch the relative frequency histogram and classify each as symmetric, skewed left, or skewed right. How is the sign of the skewness related to the classification of the histogram?

Set 1	2	2	2	2	2	3	3	3	4	4	5
Set 2	3	4	4	5	5	5	5	5	6	6	7
Set 3	2	4	4	5	5	5	6	6	6	6	6

15. The kurtosis of a set of data is measured with the following statistic:

$$\text{Sample kurtosis} = \frac{(n+1)n(n-1)}{(n-2)(n-3)} \cdot \frac{\sum (x_i - \bar{x})^4}{\left[\sum (x_i - \bar{x})^2\right]^2} - \frac{3(n-1)^2}{(n-2)(n-3)}$$

where $n > 3$ is the number of data values; x_i, $i = 1, \ldots, n$ are the data values; $\bar{x}$ is the sample mean; and all sums are taken from $i = 1$ to n. The sample kurtosis is an estimate of the kurtosis of the underlying random variable.

a. For each set of data in the table below, calculate the sample kurtosis. Also sketch the relative frequency histogram.

Set 1	4	5	5	5	5	5	5	5	5	5	6
Set 2	2	3	3	4	4	4	4	4	5	5	6
Set 3	0	1	2	2	3	3	3	4	4	5	6

b. How is the value of the sample kurtosis related to the "peakedness" of the histogram?

c. Show that the sample kurtosis must be greater than $-3(n-1)^2/[(n-2)(n-3)]$. Also show that as $n \to \infty$, this lower bound approaches -3.

16. An *outlier* of a set of data is a data value that is much larger or smaller than other data values. In this exercise, we explore the effects of outliers on the sample mean and standard deviation. Suppose a class has 10 students, one-half of whom are 60 in tall and one-half of whom are 66 in tall.

a. Calculate the mean and standard deviation of the heights of the students in this class.

b. Suppose another class has 20 students, one-half of whom are 60 in tall and one-half of whom are 66 in tall. Calculate the mean and standard deviation of the heights of the students in this class.

c. Suppose the class with 10 students gets a new student who is 84 in tall. This height is considered to be an *outlier* because it is much different from the others. Calculate the mean and standard deviation of the heights including this outlier. How much does it affect the mean and standard deviation?

d. Suppose the class with 20 students also gets a new student who is 84 in tall. Calculate the mean and standard deviation of the heights including this outlier. Is the effect on the mean and standard deviation as much as with the class of only 10 students?

e. Based on your observations in parts c and d, is the effect of an outlier on the mean and standard deviation of a large set of data more or less significant than on a small set of data?

17. The mean and median of a set of data are called *measures of center*. As shown in Exercise 16 above, the mean is sensitive to the presence of outliers. In this exercise we examine the sensitivity of the median to outliers and the relation of the mean to the median. Consider the three sets of data in the table below.

Set 1	2.1	2.3	2.5	2.5	2.8
Set 2	1	2.1	2.5	2.5	2.8
Set 3	1	2.1	2.5	2.8	4

a. Calculate the mean and median of each set of data, and identify any outlier(s) in each set.
b. Consider the following four statements:

1. The median is not as sensitive to the presence of outliers as the mean.
2. If there are no outliers, then the median will be close to the mean.
3. If there is an outlier, then the median will not be close to the mean.
4. If the mean and median are close to each other, then there are no outliers.

Based on your calculations, which of these statements are true and which are false? Explain your answers.

18. As shown in Exercise 16 above, the mean of a set of data is sensitive to the presence of outliers. One way to deal with this problem is to use a *trimmed mean* rather than a sample mean. In a trimmed mean, a certain percentage of the largest and smallest values is discarded, and a sample mean is calculated using the remaining values. For instance, if a set of data has 100 values, to calculate a 10% trimmed mean we would discard the largest 5 values and the smallest 5 values and calculate the mean of the remaining 90 values.

10.5	10.7	11.5	11.9	12.8	12.9	13.3	13.5	13.8	14.2

a. Calculate a 20% trimmed mean of the values in the table above.
b. Now suppose the data value 14.2 is replaced with the outlier 25.8. Calculate the resulting 20% trimmed mean. Does this outlier affect the trimmed mean at all?

19. Another measure of variation of a set of data is called the *coefficient of variation*, denoted CV, and is

$$CV = \frac{s}{\bar{x}}.$$

This is usually expressed as a percentage. There are two benefits of using the coefficient of variation. First, it is unitless because both s and $\bar{x}$ have the same units. This allows us to compare variation between sets of data with different units. Second, it puts the standard deviation s in the context of the size of the data values. If the data values are large (meaning $\bar{x}$ is large), then a large standard deviation may not indicate as much variation as a small standard deviation in a set of small data values (meaning $\bar{x}$ is small). So the coefficient of variation allows us to compare variation between sets of data with numbers of different sizes.

   a. A sample of 50 students is selected. The height of each is measured, and each is asked how much money is in his or her wallet. The mean and standard deviation of the height and money data are shown in the table below. Calculate the coefficient of variation for the height data and for the money data.

	$\bar{x}$	s
Height	67.54 in	3.19 in
Money	$9.35	$2.56

   b. Which set of data has more variation, the height or money? Does this surprise you? Why or why not?

20. The table below gives the frequency distribution of the number of piglets in 15 different litters of pigs (data collected by Alexa Hopping and Brett Troyer, 2011). The table tells us 1 litter had 11 piglets, 1 litter had 12 piglets, 5 litters had 13 piglets, and so on. The mean and standard deviation of a set of discrete data described with a distribution like this can be calculated using the formulas

$$\bar{x} = \frac{1}{n}\sum(f \cdot x) \quad \text{and} \quad s = \sqrt{\frac{n\sum(f \cdot x^2) - [\sum(f \cdot x)]^2}{n(n-1)}} \qquad (4.4)$$

where n is the sample size. Use these formulas to calculate the mean and standard deviation of the litter sizes in this sample.

Litter size x	11	12	13	14	15
Frequency f	1	1	5	7	1

21. The formulas in Equation (4.4) can be used to estimate the mean and standard deviation of a set of continuous data that have been summarized with a frequency distribution. In these formulas,

- x is the midpoint of a bin in the distribution, and
- f is the corresponding frequency of the bin.

Use these formulas to estimate the mean and standard deviation of the data in Table 4.4, using the frequency distribution given in Table 4.5. Compare these estimates to the exact values given in Example 4.2.4. How accurate are these estimates?

22. Show that for a set of data $\{x_1, x_2, \ldots, x_n\}$ where $x_1 = x_2 = \cdots = x_n$, the standard deviation is $s = 0$.

23. Another formula for the variance of a set of data $\{x_1, x_2, \ldots, x_n\}$ is

$$s^2 = \frac{n \sum x_i^2 - (\sum x_i)^2}{n(n-1)}$$

where all sums are taken from $i = 1$ to n. Derive this formula from Equations (4.1) and (4.2).

4.3 Maximum Likelihood Estimates

As mentioned in Section 4.2, the shape of a histogram of observed values of a random variable indicates the type of distribution the variable has. For instance, in Example 4.2.3, the histogram suggested that the random variable X denoting wait time has an exponential distribution. This means that the pdf of X has the form

$$f(x) = \lambda e^{-\lambda x}$$

where $\lambda > 0$ is some unknown parameter. In order to use this pdf, we need to estimate the value of λ using the given data. We know that for an exponential distribution,

$$\mu = \frac{1}{\lambda} \quad \Rightarrow \quad \lambda = \frac{1}{\mu}.$$

As claimed in Section 4.2, a sample mean $\bar{x}$ is an estimate of μ. Thus it seems reasonable to estimate λ as

$$\lambda \approx \frac{1}{\bar{x}}.$$

In this section, we justify such an estimate and demonstrate where the formulas for the sample mean and sample variance come from. We introduce the basic ideas behind estimating a parameter with a relatively simple example.

Example 4.3.1 **Biased Coin** Suppose weight has been added to one side of a coin in an attempt to increase the probability of getting tails. Five flips of the coin result in the sequence TTHTT. Use these data to estimate p, the probability of getting tails.

If p is the probability of tails, then $(1-p)$ is the probability of heads. Because there are four tails and one head observed in this event, the probability of this event in terms of p is

$$P(\text{TTHTT}) = p^4(1-p).$$

This probability depends on the value of p; thus it is a function of p. Call this function $L(p)$. Now the goal is to find a reasonable approximation of the value of p. Because the event TTHTT has actually been observed, it seems reasonable to stipulate that our estimate of p make the probability of this event as large as possible. That is, our estimate of p should be the value of the input that maximizes the function $L(p)$.

To find the maximum value of $L(p)$, we take the derivative with respect to p and set it equal to 0:

$$\frac{d}{dp}[L(p)] = \frac{d}{dp}\left[p^4(1-p)\right] = 4p^3(1-p) - p^4 \stackrel{\text{SET}}{=} 0$$

$$\Rightarrow 4(1-p) - p = 0$$

$$\Rightarrow p = \frac{4}{5}.$$

Since $0 \leq p \leq 1$, $L(0) = L(1) = 0$, and $L(\frac{4}{5}) \approx 0.08$, L attains its unique global maximum at $\frac{4}{5}$. Thus, according to our criteria, our estimate of p is $\frac{4}{5}$. We denote this estimate $\hat{p} = \frac{4}{5}$. This estimate is called the *maximum likelihood estimate* of p based on the data. Notice that this value is simply the relative frequency of tails, as we might expect. □

To generalize the calculations in the previous example, define the random variable

$$X = \begin{cases} 1, & \text{if the coin lands tails up} \\ 0, & \text{if the coin lands heads up.} \end{cases}$$

Then the pmf of X is

$$f(x) = P(X = x) = \begin{cases} p, & \text{for } x = 1 \\ 1-p, & \text{for } x = 0 \end{cases}$$

or, alternately,
$$f(x) = p^x(1-p)^{1-x} \quad \text{for } x = 0, 1.$$
The event TTHTT constitutes five observed values of X, $x_1 = x_2 = 1$, $x_3 = 0$, and $x_4 = x_5 = 1$. Next note that
$$f(x_1) \cdot f(x_2) \cdot f(x_3) \cdot f(x_4) \cdot f(x_5) = p \cdot p \cdot (1-p) \cdot p \cdot p = p^4(1-p) = L(p).$$

Thus $L(p)$ is simply the product of the pmf of X evaluated at each observed value of X.

This leads to the formal definition of the *likelihood function* and the *maximum likelihood estimate*.

Definition 4.3.1 Let X be a random variable with pdf (or pmf) $f(x;\theta)$ where θ is some unknown parameter that we want to estimate, and let $x_1, \ldots, x_n$ be n random and independent observed values of X. The *likelihood function* is the product of $f(x;\theta)$ evaluated at each observed value,

$$L(\theta) = \prod_{i=1}^{n} f(x_i;\theta).$$

The *maximum likelihood estimate* (MLE) of θ, denoted $\hat{\theta}$, is the value of the input at which $L(\theta)$ attains its maximum. That is,

$$L(\hat{\theta}) \geq L(\theta)$$

for all possible values of θ.

Note that θ is a generic symbol that denotes a parameter. The parameter may be p, as in the first example, or other commonly used parameters such as λ, μ, or σ. The pdf of X is denoted $f(x;\theta)$ to indicate that the pdf depends on the value of θ. This pdf is in no way a different type of pdf than we have seen before.

To simplify calculations, it is often convenient to maximize the natural logarithm of $L(\theta)$, $\ln[L(\theta)]$, rather than $L(\theta)$ itself. Because ln is an increasing function, $\ln[L(\theta)]$ is maximized if and only if $L(\theta)$ is maximized. Thus we get the same value of $\hat{\theta}$.

For example, consider Example 4.3.1. The natural logarithm of $L(p)$ is

$$\ln[L(p)] = \ln\left[p^4(1-p)\right] = \ln p^4 + \ln(1-p) = 4\ln p + \ln(1-p).$$

To find the maximum value of $\ln(L(p))$, we take the derivative with respect to p:

$$\frac{d}{dp}[\ln(L(p))] = \frac{4}{p} + \frac{-1}{1-p}.$$

Setting this derivative equal to 0 and solving yields $\hat{p} = \frac{4}{5}$, as before.

Example 4.3.2 **Exponential Distribution** Consider the random variable X denoting checkout stand wait time as described in Example 4.2.3. The histogram of the data in Table 4.4 suggests that X has an exponential distribution. This means that the pdf of X has the form

$$f(x;\lambda) = \lambda e^{-\lambda x}$$

for some $\lambda > 0$. Use the data in this table to find the MLE of λ.

To simplify the calculations and help illustrate the basic idea, we begin by considering only three observed values of X from Table 4.4, $x_1 = 0.2$, $x_2 = 1.1$, and $x_3 = 1.9$. The likelihood function using these three values is

$$\begin{aligned} L(\lambda) = f(x_1) \cdot f(x_2) \cdot f(x_3) &= \left(\lambda e^{-0.2\lambda}\right) \cdot \left(\lambda e^{-1.1\lambda}\right) \cdot \left(\lambda e^{-1.9\lambda}\right) \\ &= \lambda \cdot \lambda \cdot \lambda \cdot e^{-0.2\lambda} \cdot e^{-1.1\lambda} \cdot e^{-1.9\lambda} \\ &= \lambda^3 e^{(-0.2\lambda - 1.1\lambda - 1.9\lambda)} \\ &= \lambda^3 e^{(-0.2 - 1.1 - 1.9)\lambda} \\ &= \lambda^3 e^{-3.2\lambda}. \end{aligned}$$

Taking the natural logarithm of this yields

$$\ln[L(\lambda)] = \ln\left(\lambda^3 e^{-3.2\lambda}\right) = \ln \lambda^3 + \ln\left(e^{-3.2\lambda}\right) = 3\ln\lambda - 3.2\lambda.$$

Now,

$$\frac{d}{d\lambda}[\ln(L(\lambda))] = \frac{d}{d\lambda}[3\ln\lambda - 3.2\lambda] = \frac{3}{\lambda} - 3.2.$$

Setting this derivative equal to 0 and solving yields the MLE

$$\hat{\lambda} = \frac{3}{3.2} = 0.9375.$$

(Note that the second derivative of $\ln(L(\lambda))$ is $-3/\lambda^2$, which is negative for all $\lambda > 0$, so 0.9375 is indeed the location of the global maximum.)

Observe that the number 3.2 is simply the sum of the three observed values. Generalizing these calculations (see Exercise 7 of this section), we see that for n observed values of X, $x_1, \ldots, x_n$, the MLE of λ is

$$\hat{\lambda} = \frac{n}{\sum_{i=1}^{n} x_i} = \frac{1}{\bar{x}}.$$

The sample mean of all 30 values in Table 4.4 is $\bar{x} = 2$. So these values give the MLE $\hat{\lambda} = \frac{1}{2}$. □

Example 4.3.3 Poisson Distribution A random variable X with a Poisson distribution has a pmf of the form

$$f(x; \lambda) = \frac{\lambda^x e^{-\lambda}}{x!}$$

where $\lambda > 0$ is the parameter. If n values of X, $x_1, x_2, \ldots, x_n$, have been observed, show that the MLE of λ is $\bar{x}$, the sample mean.

The likelihood function is

$$L(\lambda) = \prod_{i=1}^{n} \frac{\lambda^{x_i} e^{-\lambda}}{x_i!} = \left(\frac{1}{\prod x_i!} \right) \lambda^{\sum x_i} e^{-n\lambda}$$

where the sums and products are taken from $i = 1$ to n. Taking the natural logarithm of this yields

$$\ln[L(\lambda)] = \ln\left(\frac{1}{\prod x_i!} \right) + \ln\left(\lambda^{\sum x_i} \right) + \ln\left(e^{-n\lambda} \right)$$
$$= -\ln\left(\prod x_i! \right) + \left(\sum x_i \right) \ln \lambda - n\lambda.$$

The derivative of this function with respect to λ is

$$\frac{d}{d\lambda}[\ln(L(\lambda))] = 0 + \frac{\sum x_i}{\lambda} - n.$$

Setting this derivative equal to 0 and solving yields the MLE

$$\hat{\lambda} = \frac{\sum x_i}{n} = \bar{x},$$

as desired. □

Some random variables have two or more parameters that need to be estimated. We can find MLEs of each by taking the partial derivatives of the likelihood function, setting them equal to 0, and then solving for the desired parameters. We illustrate this idea in the next example and justify the claim that a sample variance is an estimate of the variance of a random variable.

Example 4.3.4 Normal Distribution A random variable X with a normal distribution has a pdf of the form

$$f(x; \mu, \sigma^2) = \frac{1}{\sqrt{2\pi\sigma^2}} e^{-(x-\mu)^2/(2\sigma^2)}$$

where $-\infty < \mu < \infty$ and $\sigma^2 > 0$ are the parameters. If n values of X, $x_1, x_2, \ldots, x_n$, have been observed, find the MLE of the parameter σ^2.

To simplify the notation, let $\theta = \sigma^2$ so that the pdf is

$$f(x; \mu, \theta) = \frac{1}{\sqrt{2\pi\theta}} e^{-(x-\mu)^2/(2\theta)}.$$

The likelihood function is then

$$L(\mu, \theta) = \prod_{i=1}^{n} \frac{1}{\sqrt{2\pi\theta}} e^{-(x_i-\mu)^2/(2\theta)}$$

$$= \frac{1}{(2\pi\theta)^{n/2}} e^{(-1/2\theta)\sum(x_i-\mu)^2}$$

$$= (2\pi\theta)^{-n/2} e^{k/\theta}$$

where $k = -\frac{1}{2}\sum(x_i - \mu)^2$ and the sum is taken from $i = 1$ to n. The natural logarithm of L is

$$\ln[L(\mu, \theta)] = -\frac{n}{2}\ln(2\pi\theta) + \frac{k}{\theta}.$$

Now, because k is constant with respect to θ, we get

$$\frac{\partial}{\partial \theta}[\ln(L(\mu, \theta))] = -\frac{n}{2\theta} - \frac{k}{\theta^2}.$$

Setting this derivative equal to 0 and solving for θ yields

$$\theta = -\frac{2}{n}k = -\frac{2}{n}\left[-\frac{1}{2}\sum_{i=1}^{n}(x_i - \mu)^2\right] = \frac{1}{n}\sum_{i=1}^{n}(x_i - \mu)^2.$$

It can be shown that the MLE of μ is the sample mean $\bar{x}$ (see Exercise 8 in this section). Thus, the MLE estimate of σ^2 is

$$\hat{\sigma}^2 = \frac{1}{n}\sum_{i=1}^{n}(x_i - \bar{x})^2.$$

Note that this formula for $\hat{\sigma}^2$ is very similar to the formula for the sample variance s^2 given in Equation (4.2). The reason for the slight difference is explained in Example 4.3.6. □

As shown in Example 4.3.3, the MLE of the parameter λ of Poisson distribution based on a set of data is the sample mean $\bar{x}$. Also, as shown in Example 3.8.2, the sample mean $\bar{x}$ is an observed value of the random variable $\bar{X}_n$ as defined in the central limit theorem (Theorem 3.8.2). Thus we can say that the random variable $\bar{X}_n$ is used to estimate λ. This leads to the following definition.

Definition 4.3.2 A random variable $\hat{\Theta}$ whose values are used to estimate the value of a parameter θ is called an *estimator* of θ. A value of $\hat{\Theta}$, $\hat{\theta}$, is called an *estimate* of θ. An estimator $\hat{\Theta}$ is called an *unbiased estimator* of θ if

$$E(\hat{\Theta}) = \theta.$$

If this equation is not true, then $\hat{\Theta}$ is called a *biased estimator* of θ.

Example 4.3.5 Unbiased Estimator If X has any distribution, show that the sample mean $\bar{X}_n$ is an unbiased estimator of $\mu = E(X)$.

By definition,

$$\bar{X}_n = \frac{1}{n}\sum_{i=1}^{n} X_i$$

where $X_1, X_2, \ldots, X_n$ are independent random variables each having the same distribution as X. Then because $E(X_i) = \mu$ for each $i = 1, 2, \ldots, n$ and by properties of the expected value,

$$E(\bar{X}_n) = E\left[\frac{X_1 + \cdots + X_n}{n}\right] = \frac{1}{n}[E(X_1) + \cdots + E(X_n)] = \frac{1}{n}(n\mu) = \mu$$

as desired. □

Example 4.3.6 Biased Estimator In Example 4.3.4, we showed that if X is $N(\mu, \sigma^2)$, then the MLE of the parameter σ^2 is

$$\hat{\sigma}^2 = \frac{1}{n}\sum_{i=1}^{n}(x_i - \bar{x})^2.$$

This value is an observed value of the random variable

$$\hat{\Sigma}^2 = \frac{1}{n}\sum_{i=1}^{n}(X_i - \bar{X})^2$$

where $X_1, X_2, \ldots, X_n$ are independent random variables each having the same distribution as X. Show that $\hat{\Sigma}^2$ is a biased estimator of σ^2.

First note that $\hat{\Sigma}^2$ may be written as

$$\begin{aligned}
\hat{\Sigma}^2 &= \frac{1}{n}\sum_{i=1}^{n}(X_i - \bar{X})^2 = \frac{1}{n}\sum_{i=1}^{n}(X_i^2 - 2X_i\bar{X} + \bar{X}^2) \\
&= \frac{1}{n}\left(\sum_{i=1}^{n}X_i^2 - 2\bar{X}\sum_{i=1}^{n}X_i + \sum_{i=1}^{n}\bar{X}^2\right) \\
&= \frac{1}{n}\left[\sum_{i=1}^{n}X_i^2 - 2\bar{X}(n\bar{X}) + n\bar{X}^2\right] \\
&= \frac{1}{n}\left(\sum_{i=1}^{n}X_i^2 - n\bar{X}^2\right)
\end{aligned}$$

Now, to find $E\left(\hat{\Sigma}^2\right)$, we use one property of the variance that applies to any random variable Y and two results from Theorem 3.8.1 that apply to any variable X that is $N(\mu, \sigma^2)$:

1. $\text{Var}(Y) = E(Y^2) - [E(Y)]^2 \Rightarrow E(Y^2) = \text{Var}(Y) + [E(Y)]^2$
2. $E(\bar{X}) = \mu$
3. $\text{Var}(\bar{X}) = \dfrac{\sigma^2}{n}$

Using these three properties and the linearity properties of expected value, we get

$$E\left(\hat{\Sigma}^2\right) = \frac{1}{n}\left[\sum_{i=1}^{n} E(X_i^2) - nE(\bar{X}^2)\right]$$

$$= \frac{1}{n}\left[\sum_{i=1}^{n}(\sigma^2 + \mu^2) - n\left(\frac{\sigma^2}{n} + \mu^2\right)\right]$$

$$= \frac{1}{n}\left(n\sigma^2 + n\mu^2 - \sigma^2 - n\right)$$

$$= \frac{n-1}{n}\sigma^2.$$

Thus $E\left(\hat{\Sigma}^2\right) \neq \sigma^2$ so $\hat{\Sigma}^2$ is a biased estimator of σ^2 by definition. Note that

$$\lim_{n\to\infty} E\left(\hat{\Sigma}^2\right) = \lim_{n\to\infty} \frac{n-1}{n}\sigma^2 = 1 \cdot \sigma^2 = \sigma^2,$$

so as the sample size n gets larger, $\hat{\Sigma}^2$ get "less biased." To make $\hat{\Sigma}^2$ unbiased, we can simply multiply it by $n/(n-1)$. This yields the random variable

$$S^2 = \frac{n}{n-1} \cdot \frac{1}{n}\sum_{i=1}^{n}(X_i - \bar{X})^2 = \frac{1}{n-1}\sum_{i=1}^{n}(X_i - \bar{X})^2.$$

This random variable, called the *sample variance*, is an unbiased estimator of σ^2. The sample variance s^2 defined in Equation (4.2) is simply an observed value of this variable. □

Exercises

1. If X has a Poisson distribution with parameter λ and if we observe the values $x_1 = 3$, $x_2 = 6$, and $x_3 = 5$, find the MLE of λ.

2. A discrete random variable X is said to have a *geometric distribution* if its pmf is

$$f(x;p) = p(1-p)^{x-1}, \quad x = 1, 2, \ldots$$

where $0 < p < 1$ is the parameter. If we observe the values $x_1 = 2$, $x_2 = 5$, and $x_3 = 6$, find the MLE of p.

3. Suppose that a random variable X has the pdf

$$f(x;\theta) = \frac{3x^2}{\theta^3} e^{-(x/\theta)^3}, \quad 0 < x < \infty$$

where $\theta > 0$ is some unknown parameter. If we observe the values $x_1 = 0.8$, $x_2 = 1.5$, and $x_3 = 2.3$, find the MLE of θ.

4. Suppose that a random variable X is uniformly distributed over the interval $[0, \theta]$ where $\theta > 0$ is some unknown parameter. The pdf of X is

$$f(x; \theta) = \frac{1}{\theta}, \quad 0 \leq x \leq \theta.$$

If we observe the values $x_1 = 1.8$, $x_2 = 5.2$, and $x_3 = 9.6$, find the MLE of θ. (**Hint:** No derivative is needed. Remember that θ must be at least as large as the largest observed value of X.)

5. Here we consider a generalization of Example 4.3.1. Suppose a random experiment has two distinct outcomes, called *success* and *failure*, where the probability of a success is p and the probability of a failure is $(1 - p)$. Define the random variable

$$Y = \begin{cases} 1, & \text{if experiment is a success} \\ 0, & \text{if experiment is a failure.} \end{cases}$$

Then the pmf of Y is

$$f(y) = p^y (1-p)^{1-y} \quad \text{for } y = 0, 1.$$

If n trials of the experiment are performed, yielding n observed values of Y, $y_1, y_2, \ldots, y_n$, show that the MLE of the parameter p is

$$\hat{p} = \frac{k}{n}$$

where $k = \sum_{i=1}^{n} y_i$ is the total number of successes. This MLE is called the *sample proportion*.

6. Consider the random experiment described in Exercise 5 above. Suppose the experiment is repeated n times, and define the random variable

$$X = \text{the total number of successes in } n \text{ trials.}$$

Then X is $b(n, p)$. Suppose n is some fixed number and p is to be estimated. Define the random variable

$$\hat{P} = \frac{X}{n}.$$

Show that $\hat{P}$, called the *sample proportion*, is an unbiased estimator of p. (Note that the sample proportion $\hat{p}$ from Exercise 5 above is an observed value of $\hat{P}$.)

7. Generalize Example 4.3.2 by showing that if we observe n values of a random variable X with an exponential distribution, $x_1, x_2, \ldots, x_n$, then the MLE of the parameter λ is

$$\hat{\lambda} = \frac{1}{\bar{x}}.$$

8. If X is $N(\mu, \sigma^2)$, show that the MLE of μ is the sample mean $\bar{x}$.

9. If X is a random variable and $\mu = E(X)$ is *known*, show that the random variable

$$\hat{\Sigma}^2 = \frac{1}{n} \sum_{i=1}^{n} (X_i - \mu)^2,$$

where each X_i has the same distribution as X, is an unbiased estimator of $\sigma^2 = \text{Var}(X)$.

4.4 Sampling Distributions

In many circumstances, we would like to know something about a large population. For instance, we might want to know the proportion of the population of voters in the United States who support a certain candidate for President. Or we might want to know the mean weight of all fish in a certain lake. To find the exact answers to these questions, we would need to get information from every single member of these populations. These populations contain several thousand or million members, so this is practically impossible.

Instead, it is more reasonable to try to *estimate* the answers by using information from samples. Using the terminology from Section 4.1, we would like to know the value of a population *parameter*, but we estimate it using a sample *statistic*. In this section, we begin to discuss how random variables are used to analyze such estimates; how estimators, as described in Section 4.3, are involved; and what the distributions of these estimators are.

The first example illustrates the need to estimate a population parameter and introduces some important definitions.

Example 4.4.1 Polling Results Consider the scenario of Marten, Jones, and Smith, who are all running for state senate as first described in Example 4.2.1. Suppose Marten would like to know how many voters support him. If the percentage is greater than 50% (or the *proportion* is greater than 0.50), then he figures he will win. To determine the exact value of this proportion, he would need to ask every single voter whom he or she supports. The proportion who support him, p, called the *population proportion*, is then

$$p = \frac{\text{the number who support him}}{\text{the total number of voters}}.$$

This number is an example of a population parameter. The total number of voters is denoted N and called the *population size*. This may sound easy in principle, but the population of voters is possibly several hundred thousand or million. It would be virtually impossible to survey every single voter and calculate the exact value of p. Instead, it would be more reasonable to *estimate* p by using sample data.

Referring to Table 4.2, we see that in the sample of 735 voters, 383 of them supported Marten. It seems reasonable that p should be approximately equal to the relative frequency of voters who support Marten. This relative frequency, called the *sample proportion* and denoted $\hat{p}$, is

$$\hat{p} = \frac{383}{735} \approx 0.521.$$

This sample proportion is an estimate of p and is an example of a sample statistic. □

To begin to analyze how close the estimate $\hat{p}$ is to the true value of p, suppose we randomly choose one voter and ask whom he or she supports. Then it seems reasonable to assign the following probability:

$$P(\text{selecting a person who supports Marten}) = p.$$

If we choose a random sample of n voters and let the random variable X denote the number who support Marten, then X is $b(n,p)$. Our goal is to estimate the value of this parameter p. The sample proportion who support Marten is then described by the random variable

$$\hat{P} = \frac{X}{n},$$

called the *sample proportion*, as first defined in Exercise 6 of Section 4.3. A sample proportion $\hat{p}$ calculated using sample data as in the previous example is an observed value of this random variable. As described in Definition 4.3.2, the variable $\hat{P}$ is an *estimator* of p while $\hat{p}$ is an *estimate* of p.

The next example shows results from a simulation designed to explore the distribution of the random variable $\hat{P}$.

Example 4.4.2 **Simulating Sample Proportions** Suppose that exactly 53% of voters support Marten so that $p = 0.53$. We simulated selecting 1000 different samples of $n = 735$ voters and calculating the sample proportion that support Marten, $\hat{p}$, for each. This results in 1000 different observed values of the random variable $\hat{P}$. Figure 4.6 shows the histogram of these values. Notice that the histogram has a bell curve shape, suggesting that $\hat{P}$ has a normal distribution.

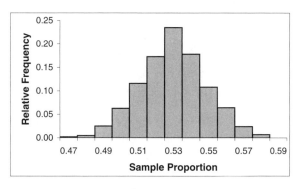

Figure 4.6

The mean of these 1000 values is $\bar{x} = 0.53$, and the variance is $s^2 = 0.000344$. Thus $E(\hat{P}) \approx 0.53$ and $\text{Var}(\hat{P}) \approx 0.000344$. However, note that $0.53 = p$ and

$$\frac{p(1-p)}{n} = \frac{0.53(0.47)}{735} \approx 0.000339.$$

Thus the simulation suggests that $\hat{P}$ is approximately $N(p, p(1-p)/n)$. □

This example illustrates the validity of the following theorem, whose proof is given in Appendix A.3.

Theorem 4.4.1 *Let X be $b(n,p)$. Then as $n \to \infty$, the distribution of*

$$\hat{P} = \frac{X}{n} \quad \text{approaches} \quad N\left(p, \frac{p(1-p)}{n}\right).$$
□

Informally this theorem says that $\hat{P}$ is approximately $N(p, p(1-p)/n)$ and as $n \to \infty$, this approximation gets better. A rule of thumb for how large n needs to be for this approximation to be reasonable is that $np \geq 5$ and $n(1-p) \geq 5$ (some sources use the number 10 rather than 5). This distribution is called the *sampling distribution of the proportion*. We make several observations about the results of the simulation.

1. Roughly one-half of the sample proportions are greater than p and roughly one-half are less. This indicates that a sample proportion does not tend to overestimate or underestimate p.
2. Most of the sample proportions are relatively "close" to the value of p. This indicates that a sample proportion is *typically* a good estimate of a population proportion.
3. However, not all of the sample proportions are close to p. From the histogram we see that some of the sample proportions are as low as 0.47 and as high as 0.58. Thus a sample proportion is not always a good estimate of p.

Note also in the simulation that relatively few values of $\hat{p}$ are equal to 0.53, the true value of p. In other words, most of the estimates of p are in error. This leads to the following definition.

Definition 4.4.1 The difference between the value of a population parameter and a sample statistic that results from random fluctuations within the sample is called a *sampling error* (also called an *experimental error*). A difference that occurs when data are improperly collected, recorded, or analyzed is called a *nonsampling error*.

There is nothing we can do to prevent sampling errors. We can, however, analyze them using probability. Nonsampling errors result from human error and thus can be prevented. The next example illustrates one application of the distribution of $\hat{P}$.

Example 4.4.3 Credit Card Fraud By examining the spending habits of one particular consumer, a credit card company observes that during the course of normal transactions, 37% of the charges exceed $150. Out of 50 charges made in one particular month, 27 exceeded $150. Does it appear that these charges were made in the course of normal transactions?

Note that the sample proportion of charges that exceeded $150 is $\hat{p} = \frac{27}{50} = 0.54$. This is larger than the expected rate of 37% in the course of normal transactions. To determine if this is unusually large, we use the idea of the unusual-event principle. We will assume that the transactions are normal and then calculate $P(\hat{P} \geq 0.54)$. If this probability is small (less than 0.05), then we say that $\hat{p} = 0.54$ is unusually large and conclude that the assumption of normal transactions is probably not correct.

To find this probability, let X denote the number of charges out of 50 that exceed $150 in the course of normal transactions. Assuming the charges are independent, then X is $b(50, 0.37)$. Thus $\hat{P} = X/50$ is approximately $N(0.37, 0.37 \cdot 0.63/50) = N(0.37, 0.004662)$

and the z-score of 0.54 is

$$z = \frac{0.54 - 0.37}{\sqrt{0.004662}} \approx 2.49.$$

Therefore,

$$P\left(\hat{P} \geq 0.54\right) \approx P(Z \geq 2.49) = 1 - 0.9936 = 0.0064.$$

Because this probability is less than 0.05, we conclude that the charges were probably not made during the course of normal transactions. This indicates the possibility that some type of fraud is involved. $\square$

The next example illustrates the idea of a *population mean*.

Example 4.4.4 Mean IQ Scores Suppose we want to know the mean IQ score of all college students in the United States, denoted μ. To find the exact value of this number, we would need to measure the IQ of *every* college student in the United States and then calculate

$$\mu = \frac{1}{N} \sum_{i=1}^{N} x_i$$

where $\{x_1, x_2, \ldots, x_N\}$ denotes the set of IQ scores and N is the total number of college students in the United States. The number μ is called the *population mean*, which is another example of a population parameter.

There are several million college students in the United States, so measuring the IQ of every one of them would be impossible. Thus we cannot know the exact value of μ, so it would be more reasonable to estimate μ with a sample mean $\bar{x}$. Suppose we select a random sample of 150 college students from around the United States, measure the IQ of each, and calculate a sample mean of

$$\bar{x} = 115$$

by adding the 150 scores and dividing by 150. This sample mean is an estimate of μ and is an example of a sample statistic. $\square$

To analyze how close this estimate is to the true value of μ, let the random variable X be the IQ score of a randomly selected college student in the United States. Then it seems reasonable to say that

$$E(X) = \mu.$$

That is, the expected value, or mean, of X is the same as the population mean μ. If we choose a random sample of n students, then the sample mean, $\bar{x}$, that we calculate is an observed value of the random variable

$$\bar{X}_n = \frac{X_1 + \cdots + X_n}{n}$$

where X_i denotes the IQ of the ith student. This random variable, as defined in the central limit theorem (Theorem 3.8.2), is also called the *sample mean*. As shown in Example 4.3.5, $\bar{X}_n$ is an unbiased estimator of μ. The central limit theorem says that the distribution of $\bar{X}_n$ is approximately

$$N\left(\mu, \frac{\sigma^2}{n}\right)$$

where $\sigma^2 = \text{Var}(X)$, called the *population variance*. This distribution is called the *sampling distribution of the mean*. The next example illustrates what this distribution means about the quality of a sample mean as an estimate of a population mean.

Example 4.4.5 Simulating Sample Means Suppose, in the population of college students in the United States, that IQ scores have a mean of $\mu = 118$ and variance $\sigma^2 = 15^2$. We simulated collecting 1000 different random samples of IQ scores each of size $n = 150$ and calculated the sample mean of each. This yielded 1000 observed values of $\bar{X}_{150}$ whose histogram is shown in Figure 4.7.

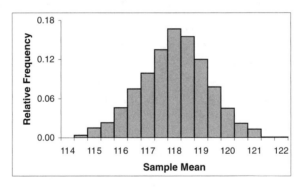

Figure 4.7

Note that the histogram has a bell curve shape as predicted by the central limit theorem. Notice that roughly one-half of the sample means are greater than 118 and one-half are less than 118. This indicates that a sample mean does not tend to overestimate or underestimate the population mean. Also note that most of the sample means are close to 118, but not all. □

Not all sampling distributions are normal. A sample variance

$$s^2 = \frac{1}{n-1} \sum_{i=1}^{n} (x_i - \bar{x})^2$$

is an estimate of a population variance σ^2 and is an observed value of the random variable

$$S^2 = \frac{1}{n-1} \sum_{i=1}^{n} \left(X_i - \bar{X}_n\right)^2$$

where X_i is the random variable representing the ith data value. As shown in Example 4.3.6, if X is normally distributed, then S^2 is an unbiased estimator of σ^2. The distribution of S^2 is called the *sampling distribution of the variance*. The next example illustrates that this distribution is not normal.

Example 4.4.6 **Simulating Sample Variances** Consider a population of thermometers. Ideally each thermometer would read exactly 0°C at the freezing point of water, but they are not perfect, so the actual reading is a random variable. Let X denote this variable and assume that X is $N(0,1)$. Suppose we do not know this population variance $\sigma^2 = 1$, and so we try to estimate it by selecting a sample of 10 thermometers, recording the reading at the freezing point of water of each, and calculating the sample variance.

We simulated doing this 1000 times, yielding 1000 observed values of the random variable S^2. The mean of these 1000 values is 0.996, which is approximately equal to $1 = \sigma^2$, which is expected because S^2 is an unbiased estimator of σ^2. Figure 4.8 shows the histogram of these 1000 values. Note that the shape of this histogram is not a bell curve. It more resembles a chi-square distribution. This observation is discussed in greater detail in Section 4.7. □

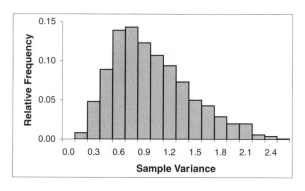

Figure 4.8

A sample standard deviation $s = \sqrt{s^2}$ is an estimate of a population standard deviation σ and is an observed value of the random variable $S = \sqrt{S^2}$. Because S^2 is an unbiased estimator of σ^2, we might expect that S is an unbiased estimator of σ. However, the next example illustrates that this is not the case.

Example 4.4.7 Estimating σ Suppose that in a neighborhood containing three houses, one house owns two boxes of cereal, one owns three boxes, and one owns nine boxes. If we randomly choose a house from this neighborhood, let the random variable X denote the number of boxes of cereal owned by the house. Then X is discrete with range $\{2, 3, 9\}$ and pmf $f(x) = \frac{1}{3}$ for each x. The mean and variance of X are

$$\mu = \frac{1}{3}(2+3+9) = \frac{14}{3} \quad \text{and} \quad \sigma^2 = \frac{1}{3}\left[\left(2-\frac{14}{3}\right)^2 + \left(3-\frac{14}{3}\right)^2 + \left(9-\frac{14}{3}\right)^2\right] = \frac{86}{9}$$

and its standard deviation is $\sigma = \sqrt{86/9} \approx 3.0912$. Suppose we try to estimate σ by randomly selecting two houses *with replacement* and calculating the sample standard deviation s. Table 4.6 shows all nine possible samples of houses and the resulting sample means $\bar{x}$, sample variances s^2, and sample standard deviations s.

Sample		$\bar{x}$	s^2	s
2	2	2	0	0
2	3	2.5	0.5	0.7071
2	9	5.5	24.5	4.9497
3	2	2.5	0.5	0.7071
3	3	3	0	0
3	9	6	18	4.2426
9	2	5.5	24.5	4.9497
9	3	6	18	4.2426
9	9	9	0	0
Mean		14/3	86/9	2.1999

Table 4.6

Notice that the mean of all the sample means is $14/3$, which equals μ, and the mean of all the sample variances is $86/9$, which equals σ^2, illustrating again that the sample mean is an unbiased estimator of μ and the sample variance is an unbiased estimator of σ^2. However, note that the mean of the sample standard deviations is 2.1999, which is much less than $\sigma = 3.0912$. Thus the sample standard deviation tends to *underestimate* σ and is thus a *biased* estimator of σ. □

One may wonder why we sampled *with replacement* in the previous example. In reality we would not want to select the same house twice, so we really sample *without replacement*. However, sampling without replacement means that the selections are technically dependent, which complicates the calculations. In real situations, the sample size is generally a small percentage of the population (less than 5%) so that we can consider the selections as being independent. In other words, we can treat the sampling as being with replacement, even when it is not. Thus it makes sense to only consider sampling with replacement.

Exercises

1. A group of students *spun* a penny 500 times and observed 204 heads and 296 tails (data collected by Daniel Klinge, Jacob Bachman, and Christopher Plucker, 2009).

    a. Calculate the sample proportion of tails.
    b. Assuming that in the population of all spins of a penny the proportion of tails is $p = 0.5$, approximate the probability that the sample proportion is at least as large as that in part a.
    c. What, if anything, do your results say about the assumption that $p = 0.5$?

2. It is often believed that the average body temperature is 98.6°F. A random sample of 110 healthy adults has a mean body temperature of 98.18°F.

    a. Assuming that the population of healthy adults has a mean body temperature of $\mu = 98.6°F$ with a standard deviation of $\sigma = 0.58°F$, approximate the probability that a sample of 110 has a mean less than 98.18 (that is, approximate $P(\bar{X}_{110} \leq 98.18)$).
    b. What, if anything, do your results say about the belief that the average body temperature is 98.6 °F?

3. Let X be $b(n, p)$ and define $\hat{P} = X/n$.

    a. Show that $\text{Var}(\hat{P}) = p(1-p)/n$.
    b. What happens to the value of this variance as $n \to \infty$?
    c. As $n \to \infty$, are observed values of $\hat{P}$ typically closer to its mean p or farther away? Explain.
    d. Based on your results in part c, would a larger sample size typically yield a better estimate of p? Why or why not?

4. Suppose a population has mean μ and variance σ^2.

    a. Show that $\text{Var}(\bar{X}_n) = \sigma^2/n$.

b. What happens to the value of this variance as $n \to \infty$?

c. As $n \to \infty$, are observed values of $\bar{X}_n$ typically closer to its mean μ or farther away? Explain.

d. Based on your results in part c, would a larger sample size typically yield a better estimate of μ? Why or why not?

e. Suppose σ^2 is large. Is it as likely that $\bar{X}_n$ is as close to μ as if σ^2 were small? What does this mean about the quality of a sample mean as an estimate of μ when σ^2 is large compared to when σ^2 is small? Explain.

f. Suppose $\sigma^2 = 0$. What is the variance of $\bar{X}_n$? What does this mean about the quality of a sample mean as an estimate of μ?

5. Suppose that in a population of students, exactly 85% own a car.

a. Approximate the probability that in a sample of $n = 25$ students, more than 95% own a car (that is, approximate $P\left(\hat{P} > 0.95\right)$).

b. Estimate the probability that less than 75% of the sample own a car in the cases where $n = 25, 50, 100$, and 200.

c. Generalize your results in part b. As the sample size gets larger, what happens to the probability that the sample proportion is far from the population proportion?

6. In Example 4.3.6, we showed that we divide by $(n-1)$ rather than n when calculating a sample variance to make S^2 an unbiased estimator of σ^2. In this exercise, we illustrate this point in another way.

a. For each of the nine samples in Example 4.4.7, calculate the "sample variance" using this formula:
$$s^2 = \frac{1}{n} \sum_{i=1}^{n} (x_i - \bar{x})^2.$$

b. Calculate the mean of the nine "sample variances" in part a. Is this mean equal to $\sigma^2 = 86/9$?

c. Based on your results in part b, does this formula for sample variance result in an unbiased estimator of σ^2? Does it appear to over- or underestimate σ^2?

d. For each of the nine samples, calculate the "sample standard deviation" $s = \sqrt{s^2}$ where s^2 is as calculated in part a.

e. Calculate the mean of the nine "sample standard deviations" in part d. Does this formula for the sample variance result in a sample standard deviation that is more or less biased than in the original formula?

7. In this exercise, we examine estimation of a population range and median.

   a. Consider the random variable X in Example 4.4.7. The *range of values* of X is (max value of X) − (min value of X). Calculate the range of values and the median of X. These are called the *population range* and *median*, respectively.

   b. For each of the nine samples in this example, calculate the sample range and median. The sample range can be thought of as an estimate of the range of values of X.

   c. Calculate the mean of the sample ranges and the mean of the medians in part b. Do these equal the population range and median, respectively? Is the sample range an unbiased estimator of the population range? Is the sample median an unbiased estimator of the population median?

   d. Based on these results, is a sample range a good estimate of a population range? Is the sample median a good estimate of the population median?

8. Let X be a discrete random variable with range $\{x_1, x_2, \ldots, x_N\}$ (each value being equally likely) and mean μ. The *mean absolute deviation* (MAD) of X, or *population MAD*, is

$$\text{Population MAD} = \frac{1}{N} \sum_{i=1}^{N} |x_i - \mu|.$$

The mean absolute deviation of a sample of data values $\{x_1, x_2, \ldots, x_n\}$ with mean $\bar{x}$ is

$$\text{Sample MAD} = \frac{1}{n} \sum_{i=1}^{n} |x_i - \bar{x}|.$$

It seems reasonable to assume that a sample MAD is a reasonable estimate of a population MAD, but is this the case?

   a. Calculate the population MAD for the population described in Example 4.4.7.

   b. For each of the nine samples in this example, calculate the sample MAD.

   c. Calculate the mean of the nine sample MADs.

   d. Based on this result, is a sample MAD a good estimate of the population MAD?

4.5 Confidence Intervals for a Proportion

As illustrated in the previous section, a sample proportion $\hat{p}$ is an estimate of a population proportion p. Because $\hat{p}$ is a one-number estimate, it is called a *point estimate*. Whenever we estimate something, we want an idea of how accurate the estimate is.

In the scenario of estimating the proportion of voters who support Marten for state senate, the sample proportion is $\hat{p} = 0.521$. This means that *about* 52% of the voters support Marten. We would like to know how close this estimate is to the true value of the population proportion p. Of course, if we knew *exactly* how close $\hat{p}$ were to p, then we would know the true value of p. Because we cannot know the exact value of p, we cannot know exactly how close our estimate is. However, we can use probability and random variables to get a measure of the *likely* difference between $\hat{p}$ and p. This difference, or measure of accuracy, is called the *margin of error*.

To develop the idea of the margin of error, we first define *critical z-values*.

Definition 4.5.1 Let Z be $N(0,1)$ and let p be a number between 0 and 0.5. A *critical z-value* z_p is a positive number such that

$$P(Z \leq z_p) = 1 - p.$$

Note that the number p referred to in this definition is *not* the population proportion being estimated. The number p in the definition is a probability. Figure 4.9 illustrates the idea of the critical z-value.

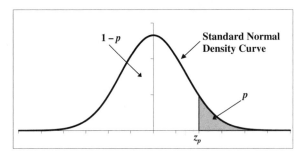

Figure 4.9

Example 4.5.1 **Calculating Critical z-values** Let $p = 0.025$. To find $z_{0.025}$, note that

$$P(Z \leq -z_{0.025}) = \Phi(-z_{0.025}) = 0.025.$$

Examining Table C.1 in Appendix C, we see that $\Phi(-1.96) = 0.025$. Thus we have $z_{0.025} = 1.96$.

Now consider $p = 0.05$. From Table C.1, we see that $\Phi(-1.64) = 0.0505$ and $\Phi(-1.65) = 0.0495$. Taking the average of these two z-scores, we see that $P(Z \leq -1.645) \approx 0.05$, so we take $z_{0.05} = 1.645$. Table C.2 gives the values of z_p for several commonly used values of p.

□

Of particular interest when dealing with critical z-values is the case where α is a number between 0 and 1. Then $p = \alpha/2$ is between 0 and 0.5, so that the the critical z-value $z_{\alpha/2}$ is a positive number such that

$$P(Z \leq z_{\alpha/2}) = 1 - \frac{\alpha}{2} \quad \Rightarrow \quad P(Z > z_{\alpha/2}) = \frac{\alpha}{2}.$$

By symmetry of the standard normal bell curve, this means that

$$P(-z_{\alpha/2} < Z \leq z_{\alpha/2}) = 1 - \alpha.$$

This probability is illustrated in Figure 4.10.

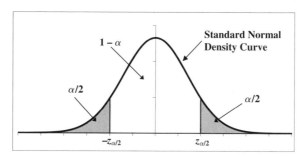

Figure 4.10

Now, to define the margin of error, consider again the scenario of estimating the proportion of voters who support Marten for state senate, p, as first described in Example 4.2.1. Suppose n voters are surveyed. Then the sample proportion $\hat{P} = X/n$, where X is the number in the sample who support Marten, is an unbiased estimator of p.

If we assume that n is small compared to the population size (less than 5%), then X is approximately binomial. Thus by Theorem 4.4.1, $\hat{P}$ is approximately $N(p, p(1-p)/n)$ so that by Theorem 3.4.1 the variable

$$Z = \frac{\hat{P} - p}{\sqrt{p(1-p)/n}}$$

is approximately $N(0,1)$. Then by definition of a critical z-value,
$$P\left(-z_{\alpha/2} \leq Z \leq z_{\alpha/2}\right) = 1 - \alpha$$
$$\Rightarrow \quad P\left(-z_{\alpha/2} \leq \frac{\hat{P} - p}{\sqrt{p(1-p)/n}} \leq z_{\alpha/2}\right) \approx 1 - \alpha.$$

Rewriting this inequality yields
$$P\left(\hat{P} - z_{\alpha/2}\sqrt{\frac{p(1-p)}{n}} \leq p \leq \hat{P} + z_{\alpha/2}\sqrt{\frac{p(1-p)}{n}}\right) \approx 1 - \alpha.$$

Note that p, the quantity being estimated, appears on both ends of this inequality. So we estimate it by replacing it with $\hat{P}$. Thus we have
$$P\left[\hat{P} - z_{\alpha/2}\sqrt{\frac{\hat{P}(1-\hat{P})}{n}} \leq p \leq \hat{P} + z_{\alpha/2}\sqrt{\frac{\hat{P}(1-\hat{P})}{n}}\right] \approx 1 - \alpha.$$

This means the probability that the *random interval*
$$\left[\hat{P} - z_{\alpha/2}\sqrt{\frac{\hat{P}(1-\hat{P})}{n}},\ \hat{P} + z_{\alpha/2}\sqrt{\frac{\hat{P}(1-\hat{P})}{n}}\right]$$

contains the true value of p is approximately $1 - \alpha$. In practical terms, this means that if we were to observe many different values of $\hat{P}$ by calculating sample proportions from many different samples of size n, and calculate such an interval for each, then about $100(1-\alpha)\%$ of them would contain the true value of p. This leads to the definition of the *margin of error* and the *confidence interval*.

Definition 4.5.2 Let $0 < \alpha < 1$ and let x be a number of successes in n observed trials of a Bernoulli experiment with unknown probability of a success p. Define $\hat{p} = x/n$ and let $z_{\alpha/2}$ be a critical z-value. The interval
$$\left[\hat{p} - z_{\alpha/2}\sqrt{\frac{\hat{p}(1-\hat{p})}{n}},\ \hat{p} + z_{\alpha/2}\sqrt{\frac{\hat{p}(1-\hat{p})}{n}}\right]$$

is called a $100(1-\alpha)\%$ *confidence interval estimate for p*. The quantity

$$E = z_{\alpha/2}\sqrt{\frac{\hat{p}(1-\hat{p})}{n}}$$

is called the *margin of error*, and the quantity $\sqrt{\hat{p}(1-\hat{p})/n}$ is called the *standard error of the proportion*. The number α is called the *significance level*, and the number $100(1-\alpha)\%$ is called the *confidence level*.

Requirements

1. The sample must be random.
2. The conditions for a binomial distribution must be satisfied (at least approximately). In other words, the number of trials is fixed; the trials are independent; there are two events, success and failure; and the probability of a success remains the same for all trials.
3. There are at least 5 successes and at least 5 failures observed in n trials. This is to (approximately) verify that the conditions $np \geq 5$ and $n(1-p) \geq 5$ are both met so that $\hat{P}$ is approximately normal.

Confidence intervals are usually written in one of three different but equivalent forms:

$$[\hat{p} - E, \hat{p} + E], \qquad \hat{p} - E \leq p \leq \hat{p} + E, \qquad \text{or} \qquad \hat{p} \pm E.$$

We can think of a confidence interval as giving us a *range* of possible values for p.

Example 4.5.2 Calculating a Confidence Interval Using the data summarized in Table 4.2, calculate a 95% confidence interval estimate for the proportion of all voters who support Marten, and interpret the interval.

We calculate this confidence interval with these steps:

1. Define the population proportion being estimated:

   $p = $ the proportion of all voters who support Marten for state senate.

2. Calculate the sample proportion $\hat{p}$:

$$\hat{p} = \frac{383}{735} \approx 0.521.$$

3. Find the critical value: Because this is a 95% confidence interval, $\alpha = 0.05$, so from Table C.2 we get

$$z_{\alpha/2} = z_{0.05/2} = z_{0.025} = 1.96.$$

4. Calculate the margin of error E:

$$E = z_{\alpha/2}\sqrt{\frac{\hat{p}(1-\hat{p})}{n}} = 1.96\sqrt{\frac{0.521\,(1-0.521)}{735}} \approx 0.0361.$$

5. Calculate the confidence interval:

$$\hat{p} - E \leq p \leq \hat{p} + E \;\Rightarrow\; 0.521 - 0.0361 \leq p \leq 0.521 + 0.0361 \;\Rightarrow\; 0.485 \leq p \leq 0.557.$$

The last, and most important, step is to interpret this confidence interval. We interpret this by saying, "We are 95% confident that the value of p is between 0.485 and 0.557." Notice that the lower limit is less than 0.50. This means that there is a good chance that p is less than 0.50, even though $\hat{p}$ is greater than 0.50. □

An *incorrect* interpretation of the confidence interval in the previous example is to say, "There is a 95% chance that the value of p is somewhere between 0.485 and 0.557." This may seem like an equivalent statement to the correct interpretation, but the words *confident* and *chance* do not mean the same thing. The number p has one value. Either that value is between 0.485 and 0.557, or it is not. There is no probability involved.

The confidence level 95% refers to the *process* of constructing the intervals, not the intervals themselves. It means that if we were to survey many different samples of voters and calculate the corresponding 95% confidence interval using the statistics from each sample, then about 95% of the intervals would contain the true value of p.

In some situations, we may only want an upper or lower bound on p. In these cases, we can calculate a *one-sided* confidence interval estimate for p.

Definition 4.5.3 Let $0 < \alpha < 1$, and let x be a number of successes in n observed trials of a Bernoulli experiment with unknown probability of a success p. Define $\hat{p} = x/n$ and let z_α be a critical z-value. The intervals

$$\left[\hat{p} - z_\alpha\sqrt{\frac{\hat{p}(1-\hat{p})}{n}},\, 1\right] \quad \text{and} \quad \left[0,\, \hat{p} + z_\alpha\sqrt{\frac{\hat{p}(1-\hat{p})}{n}}\right] \tag{4.5}$$

are called $100(1-\alpha)\%$ *one-sided confidence interval estimates for* p.

These intervals could also be expressed in the form

$$\hat{p} - E \leq p \leq 1 \quad \text{and} \quad 0 \leq p \leq \hat{p} + E$$

where $E = z_\alpha \sqrt{\hat{p}(1-\hat{p})/n}$ is the margin of error.

The first interval in Equation (4.5) gives a lower bound for the value of p while the second gives an upper bound. Calculating a one-sided confidence interval is not much different from calculating a one-sided confidence interval, except that the critical value is z_α instead of $z_{\alpha/2}$.

Example 4.5.3 Reporting On-Time Percentage Suppose an airline wants to report an on-time performance of at least $m\%$ with 95% confidence. In a random sample of 650 flights, 587 are on time. Calculate an appropriate confidence interval and give a value of m.

In this problem, we only want a lower limit for the proportion of flights that are on time, so we calculate a one-sided confidence interval. We follow the same basic steps as in Example 4.5.2.

1. Define the population proportion being estimated:

$$p = \text{the proportion of all flights that are on time.}$$

2. Calculate the sample proportion $\hat{p}$:

$$\hat{p} = \frac{587}{650} \approx 0.903.$$

3. Find the critical value: From Table C.2 we see that $z_{0.05} = 1.645$.
4. Calculate the margin of error E:

$$E = z_\alpha \sqrt{\frac{\hat{p}(1-\hat{p})}{n}} = 1.645 \sqrt{\frac{0.903(1-0.903)}{650}} = 0.019.$$

5. Calculate the confidence interval:

$$\hat{p} - E \leq p \leq 1 \Rightarrow 0.903 - 0.019 \leq p \leq 1 \Rightarrow 0.884 \leq p \leq 1.$$

Thus we are 95% confident that p is at least 0.884. So we take $m = 88.4\%$. □

Software Calculations

To calculate a confidence interval estimate of a population proportion p:

Minitab: Select **Stat** → **1-Proportion**. Enter n, the **Number of Trials**, and x, the **Number of Events**. For a two-sided interval, select **Options** and enter the desired **Confidence Level**. Next to **Alternative**, select **not equal**. (The value of the **Test proportion** does not matter.) Checking the box next to **Use test and interval based on normal distribution** tells Minitab to use the same algorithm described in this section where we approximate the binomial distribution with the normal. Otherwise, Minitab will use an exact method. In the output, the confidence interval is labeled CI. For a one-sided interval giving a lower bound for p, follow the same steps except select **greater than** for the **Alternative**. The output gives the lower bound for the interval. For an interval giving an upper bound, select **less than**.

R: The syntax for calculating a two-sided confidence interval at the 95% confidence level is

> binom.test(x, n, alternative = "equal", conf.level = 0.95)

For a one-sided interval giving a lower bound on p, change "equal" to "greater than." For an interval giving an upper bound, change "equal" to "less than." Note that R uses an exact method for calculating the interval whereas the algorithm we present uses an approximation of the binomial distribution with the normal. So the results given by R will be slightly different from those given by our algorithm.

TI-83/84 Calculators: To calculate a two-sided confidence interval, press **STAT** → **TESTS** → **A:1-PropZInt**. Enter x and n and the confidence level, denoted **C-Level**. To calculate a one-sided confidence interval, follow the same steps except use a confidence level of $1 - 2\alpha$, where α is the desired significance level. If you want an interval giving a lower bound for p, take the lower limit to be the limit given in the output and the upper limit to be 1. For an interval giving an upper bound, take the upper limit to be that given in the output and the lower limit to be 0.

Exercises

Directions: In each exercise asking for a confidence interval, (1) define the parameter being estimated, (2) calculate the margin of error, and (3) calculate the confidence interval.

1. Use Table C.1 in Appendix C to estimate the values of z_α and $z_{\alpha/2}$ for the following values of α:
   a. 0.4   b. 0.3   c. 0.005

2. A news report says that in a survey of voters, 65% ± 2% support a particular candidate. Express this confidence interval in the form $\hat{p} - E \leq p \leq \hat{p} + E$.

3. Suppose a university administrator makes the following statement: "The proportion of potential students interested in nursing who would attend the university if we offered a nursing program is approximately 0.45 with a margin of error of 0.43 at the 90% confidence level." Explain why this is practically a meaningless statement.

4. A calculator gives a confidence interval (0.222, 0.444). Find the sample proportion $\hat{p}$ and the margin of error E, and express this confidence interval in the form $\hat{p} \pm E$.

5. A statistics student asks the 25 other students in his class whether they use credit cards and finds that 19 do. Can these data be used to construct a valid 95% confidence interval estimate of the proportion of all college students who use credit cards? Why or why not?

6. A random sample of 87 college students contains 12 who are left-handed (data collected by Jacquelyn Schwartz, 2011). Calculate a 90% confidence interval estimate of the proportion of all college students who are left-handed. It is commonly believed that about 10% of the population is left-handed. Based on this confidence interval, does this belief appear to be reasonable?

7. Evolutionary psychology claims that most people think they look more like their father than their mother. To test this, a student asked 100 fellow students which parent they think they resemble more. A total of 71 students said they looked more like their father (data collected by Eric Hyde, 2011). Construct a 99% confidence interval estimate of the proportion of all students who think they look more like their father. Does the claim of evolutionary psychology appear to be valid?

8. Using the data summarized in Table 4.2, calculate 80%, 90%, and 99% confidence interval estimates of the proportion of all voters who support Marten. What happens to the margin of error and the width of the confidence interval as the confidence level increases?

9. A student observed that 64% of several fellow students said "thank you" when a door was held open for them. Calculate a 95% confidence interval estimate of the proportion of all students who say "thank you" if the sample size is $n = 100, 200$, and 500. What happens to the margin of error and the width of the confidence interval as the sample size increases? What does this say about the quality of $\hat{p}$ as an estimate of p as the sample size increases?

10. In a survey of 1002 adults regarding abortion, 531 said they are pro-choice.

  a. Calculate a 90% confidence interval estimate of the proportion of all adults who are pro-choice. Based on this interval, does it appear that a majority are pro-choice?

b. Repeat part a, but use a 99% confidence level. Does this higher confidence level change the conclusion?

11. A magazine recently changed the design of its layout and asked readers to comment on whether they liked the new design. Of 540 readers who responded, 309 said they liked it.

  a. Use the sample data to construct a 90% confidence interval estimate of the proportion of all readers who like the new design.

  b. Based on the results, does it appear that most readers like the new design? Why or why not?

  c. Do you think that the 540 readers who responded represent a random sample of the population of all readers? Why or why not? Does this cause you to question your conclusion in part b?

12. A student claims to be able to accurately predict the outcome of a flip of a coin. To test the claim, a researcher flips a coin 75 times and the student correctly predicted the outcome 43 times.

  a. Use the sample data to construct a 95% confidence interval estimate of the proportion of all flips of a coin that the student correctly predicts.

  b. Based on the results, does the student's claim appear to be valid? Why or why not?

13. A pollster surveys 500 voters and finds that 54% of them support a particular candidate.

  a. Use the sample data to construct a 95% confidence interval estimate of the proportion of all voters who support the candidate.

  b. Now suppose the pollster surveys an additional 500 voters and finds that out of the $n = 1000$ total voters, 54% support the candidate. Use these data to construct a 95% confidence interval estimate. Compare this confidence interval and the margin of error to those in part a. The information gained from the additional 500 voters decreased the margin of error and the width of the confidence interval. Does this decrease have any real practical use? Explain.

  c. Now suppose the pollster surveys an additional 1000 voters and finds that out of the $n = 2000$ total voters, 54% support the candidate. Use these data to construct a 95% confidence interval estimate. Compare this confidence interval and the margin of error to those in parts a and b. Was there any benefit in surveying the additional 1000 voters? Explain.

14. Suppose that in a recent election, 40% of voters voted for a clean-air proposition. In a survey of 825 voters after the election, 429 said they voted for the proposition.

  a. Use the sample data to construct a 99% confidence interval estimate of the proportion of all voters who say they voted for the proposition.

  b. Given that 40% of voters actually voted for the proposition, what does this confidence interval suggest about the honesty of people when they respond to surveys such as this?

15. An English professor claims that she gets at least 75% of essays graded within 2 days of when they are turned in. In a sample of 200 papers, 168 of them were graded within 2 days.

  a. Use the sample data to construct a 99% one-sided confidence interval that gives a lower bound on the proportion of all essays that are graded within 2 days.

  b. Do the data support the professor's claim? Why or why not?

16. A random sample of 201 newborn piglets contains 90 females (data collected by Alexa Hopping and Brett Troyer, 2011). Calculate a 90% one-sided confidence interval that gives an upper bound on the proportion of all newborn piglets that are female. Based on this interval, does it appear that less than one-half of all newborn piglets are female?

17. To estimate a population proportion p, a statistician collects 50 different random samples from the population. For each sample, she constructs a 90% confidence interval estimate of p. Find the probability that at least 46 of these 50 intervals contain the true value of p. (**Hint:** Let X denote the number of intervals out of the 50 that contain the true value of p. We want to find $P(X \geq 46)$. What type of distribution does X have?)

18. Derive the inequality

$$\hat{P} - z_{\alpha/2}\sqrt{\frac{p(1-p)}{n}} \leq p \leq \hat{P} + z_{\alpha/2}\sqrt{\frac{p(1-p)}{n}}$$

from the inequality

$$-z_{\alpha/2} \leq \frac{\hat{P} - p}{\sqrt{p(1-p)/n}} \leq z_{\alpha/2}.$$

19. The method presented in this section for calculating a confidence interval estimate of p is not the only method of doing so. This method is based on using the normal distribution to approximate probabilities of the sample proportion. Other methods are based on other types of approximations and yield different formulas for the upper and lower limits of a confidence interval.

If we work at a 95% confidence level, collect *many* different samples of the same size, and construct the corresponding confidence interval for each, then theoretically 95% of them should contain the true value of p. However, because formulas for confidence intervals are based on assumptions and simplifications, the observed percentage is not exactly 95%. The "performance" of a method for calculating a confidence interval is measured by how close the observed percentage is to 95%.

The method presented in this section is called the *Wald* method. A similar method called the *adjusted Wald* method uses the following formula: Let x be the number of successes in a sample of size n. Then

$$\hat{p} = \frac{x+2}{n+4} \quad \text{and} \quad E = z_{\alpha/2}\sqrt{\frac{\hat{p}(1-\hat{p})}{n+4}},$$

and the confidence interval is $\hat{p} - E < p < \hat{p} + E$. Another commonly used method called the *Wilson score* method has upper and lower confidence interval limits

$$\frac{\hat{p} + z_{\alpha/2}^2/(2n) \pm z_{\alpha/2}\sqrt{[\hat{p}(1-\hat{p}) + z_{\alpha/2}^2/(4n)]/n}}{1 + z_{\alpha/2}^2/n}$$

where $\hat{p} = x/n$. Both of these methods perform better than the Wald method in the sense that closer to 95% of the resulting intervals contain the true value of p. Other methods include the *Clopper-Pearson "exact"* method and the *mid-P* method. (For a comparison of these and other methods, see A. Agresti and B. A. Coull, "Approximation Is Better than 'Exact' for Interval Estimation of Binomial Proportions," *The American Statistician*, Vol. 52, No. 2, pp. 119–126.)

a. Suppose a random sample of 50 voters finds 20 who support health care reform. Use the adjusted Wald method to calculate a 95% confidence interval estimate of the proportion of all voters who support the reform.

b. Repeat part a, but use the Wilson score method.

c. Suppose a sample of 1000 voters finds 400 who support the reform. Calculate a 95% confidence interval with the Wald method and then with the adjusted Wald method. Are the two confidence intervals much different? What happens to the difference between the two confidence intervals as n gets larger? Why is this so?

4.6 Confidence Intervals for a Mean

A college admission counselor would like to know the mean IQ of all college students in the United States. To find the exact value of this mean, we would need to measure the IQ of every college student in the United States, add them, and divide by the total number of college students. Call this number μ, the population mean. Because there are millions of college students in the United States, this is an impossible endeavor, so it is necessary to estimate the population mean with a sample mean.

Suppose a random sample of 150 college students from across the United States has a mean IQ of $\bar{x} = 115$. In Section 4.4, we illustrated that a sample mean $\bar{x}$ is a point estimate of a population mean μ. As in the previous section, we would like to measure how accurate this estimate is. We do this by calculating a margin of error and then using this to calculate a confidence interval estimate for μ. In this section, we describe how to do this.

A sample mean $\bar{x}$ is an observed value of the random variable $\bar{X}_n$. By the central limit theorem, $\bar{X}_n$ is approximately $N(\mu, \sigma^2/n)$ where μ and σ^2 are the mean and variance, respectively, of the population of interest. Thus by Theorem 3.4.1, the random variable

$$Z = \frac{\bar{X}_n - \mu}{\sigma/\sqrt{n}}$$

is approximately $N(0,1)$.

Now, if $z_{\alpha/2}$ is a critical z-value as in Definition 4.5.1, then

$$P\left(-z_{\alpha/2} \leq Z \leq z_{\alpha/2}\right) = 1 - \alpha$$

$$\Rightarrow P\left(-z_{\alpha/2} \leq \frac{\bar{X}_n - \mu}{\sigma/\sqrt{n}} \leq z_{\alpha/2}\right) \approx 1 - \alpha.$$

Rewriting this second set of inequalities yields

$$P\left(\bar{X}_n - z_{\alpha/2}\frac{\sigma}{\sqrt{n}} \leq \mu \leq \bar{X}_n + z_{\alpha/2}\frac{\sigma}{\sqrt{n}}\right) \approx 1 - \alpha.$$

This means that the probability that the *random* interval

$$\left[\bar{X}_n - z_{\alpha/2}\frac{\sigma}{\sqrt{n}}, \bar{X}_n + z_{\alpha/2}\frac{\sigma}{\sqrt{n}}\right]$$

contains the true value of μ is approximately $1 - \alpha$. This leads us to the definition of the margin of error and confidence interval estimate for μ.

Definition 4.6.1 Let $\bar{x}$ be the mean of a sample of size n taken from a population with known variance σ^2 and unknown mean μ, and let α be between 0 and 1. The interval

$$\left[\bar{x} - z_{\alpha/2}\frac{\sigma}{\sqrt{n}},\ \bar{x} + z_{\alpha/2}\frac{\sigma}{\sqrt{n}}\right]$$

is called a $100(1-\alpha)\%$ *confidence interval estimate* for μ. The quantity

$$E = z_{\alpha/2}\frac{\sigma}{\sqrt{n}}$$

is called the *margin of error*. This type of confidence interval is also called a *Z-interval*.

Requirements

1. The sample is random.
2. The population variance σ^2 is known.
3. The population is normally distributed or $n > 30$.

As in Section 4.5, α is called the *significance level* and $1 - \alpha$ is called the *confidence level*. This confidence interval is usually written in one of three different, but equivalent forms:

$$[\bar{x} - E,\ \bar{x} + E], \qquad \bar{x} - E \leq \mu \leq \bar{x} + E, \qquad \text{or} \qquad \bar{x} \pm E.$$

Example 4.6.1 Calculating a Confidence Interval for μ Consider the scenario of trying to estimate the mean IQ of all college students from the United States where a random sample of 150 college students from across the United States has a mean IQ of $\bar{x} = 115$. Assuming the population variance is $\sigma^2 = 15^2$, calculate a 95% confidence interval estimate for μ, the mean IQ of all college students in the United States.

We calculate a confidence interval for μ with the same steps as a confidence interval for p:

1. Define the population mean being estimated:

    $\mu =$ the mean IQ of all college students in the United States.

2. Find the critical value. Because this is a 95% confidence interval, $\alpha = 0.05$, so from Table C.2 we get $z_{\alpha/2} = 1.96$.

3. Calculate the margin of error E:

$$E = z_{\alpha/2}\frac{\sigma}{\sqrt{n}} = 1.96\frac{15}{\sqrt{150}} \approx 2.40.$$

4. Calculate the confidence interval:

$$\bar{x} - E \leq \mu \leq \bar{x} + E \Rightarrow 115 - 2.40 \leq \mu \leq 115 + 2.40 \Rightarrow 112.6 \leq \mu \leq 117.4.$$

We interpret this interval by saying, "We are 95% confident that the mean IQ of all college students in the United States is between 112.6 and 117.4." □

In practical situations, the population variance σ^2 is rarely known. So we often estimate it by using a sample variance s^2. If $n > 30$, then s^2 is a reasonable estimate of σ^2 and the margin of error is

$$E = z_{\alpha/2}\frac{s}{\sqrt{n}}.$$

For this reason, the Z-interval is often called a *large-sample confidence interval*. If $n \leq 30$ (called a *small sample*), then we need the following theorem to calculate a confidence interval for μ.

Theorem 4.6.1 Let $\bar{X}$ denote the mean and S denote the standard deviation of a sample of size n taken from a normally distributed population with mean μ. The random variable

$$T = \frac{\bar{X} - \mu}{S/\sqrt{n}}$$

has a Student-t distribution with $n - 1$ degrees of freedom. □

A proof of this theorem is given in Appendix A.4. In the proof, the requirement that the population be normally distributed is needed to ensure that $\bar{X}$ is normally distributed. As we saw in the central limit theorem, $\bar{X}$ is approximately normally distributed for $n > 30$ regardless of the population distribution. Thus, Theorem 4.6.1 is approximately true if $n > 30$ regardless of the distribution of the population.

Example 4.6.2 **Illustrating Theorem 4.6.1** To illustrate Theorem 4.6.1, we simulated taking 1000 random samples of size $n = 6$ from a normally distributed population with mean $\mu = 3$. We then took another 1000 samples of size $n = 6$ from an exponential distribution with mean $\mu = \frac{1}{2}$. For each sample, we calculated the quantity

$$t = \frac{\bar{x} - \mu}{s/\sqrt{6}}$$

where $\bar{x}$ and s are the mean and standard deviation of the sample, respectively, and μ is the respective population mean. In the normally distributed population, Theorem 4.6.1 says that these values follow a Student-t distribution with 5 degrees of freedom. The histogram on the left in Figure 4.11 shows these results. Note that the Student-t density curve fits the histogram well, as expected.

In the exponentially distributed population, Theorem 4.6.1 does not hold. The histogram on the right in Figure 4.11 shows the results from this population. Note that the Student-t density curve does not fit the histogram well, indicating that these values are not described by this distribution.

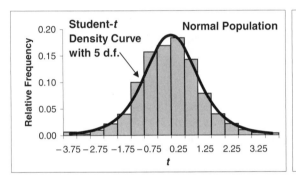
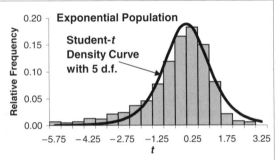

Figure 4.11

□

Now, to derive the confidence interval for μ in the case where σ^2 is unknown, let α be the desired significance level, n be the sample size, and $t_{\alpha/2}(n-1)$ be a critical Student-t value as described in Definition 3.9.6 (for simplicity we call this critical value simply $t_{\alpha/2}$). Then we have

$$P\left(-t_{\alpha/2} \leq \frac{\bar{X} - \mu}{S/\sqrt{n}} \leq t_{\alpha/2}\right) = 1 - \alpha.$$

Rewriting this inequality so that μ is alone in the center yields

$$P\left(\bar{X} - t_{\alpha/2}\frac{S}{\sqrt{n}} \leq \mu \leq \bar{X} + t_{\alpha/2}\frac{S}{\sqrt{n}}\right) = 1 - \alpha.$$

This leads to the following definition.

Definition 4.6.2 Let $\bar{x}$ be the mean and s^2 be the variance of a sample of size n taken from a population with unknown variance σ^2 and mean μ; let α be between 0 and 1; and let $t_{\alpha/2}$ be a critical Student-t value with $n-1$ degrees of freedom. The interval

$$\left[\bar{x} - t_{\alpha/2}\frac{s}{\sqrt{n}},\ \bar{x} + t_{\alpha/2}\frac{s}{\sqrt{n}}\right]$$

is called a $100(1-\alpha)\%$ *confidence interval estimate for μ when σ^2 is unknown*. The quantity

$$E = t_{\alpha/2}\frac{s}{\sqrt{n}}$$

is called the *margin of error*, and the quantity $s/\sqrt{n}$ is called the *standard error of the mean*. This type of confidence interval is also called a *T-interval*.

Requirements

1. The sample is random.
2. The population is normally distributed or $n > 30$.

To help determine which type of interval should be used, Z or T, we give the following suggestions:

1. If $n > 30$ or σ^2 is known, then use a Z-interval.
2. If σ^2 is unknown and the population is normally distributed (at least approximately), then use a T-interval.
3. If $n \leq 30$, σ^2 is unknown and the population is *not* normally distributed, use non-parametric techniques (see Chapter 7).

To use either type of interval, the sample *must* be random.

Example 4.6.3 Calculating a Confidence Interval for μ A random sample of 15 "1-lb" packages of shredded cheddar cheese has a mean weight of $\bar{x} = 1.05$ lb and a standard deviation of $s = 0.02$ lb. Calculate a 99% confidence interval estimate for the mean weight of all such packages.

In this example, we will *not* make any assumptions about the population variance. Instead, we will use the sample variance. The steps are the same as in Example 4.6.1:

1. Define the parameter mean being estimated:

    μ = the mean weight of all 1-lb packages of shredded cheddar cheese.

2. Find the critical value: Note that $\alpha = 0.01$ and that $n = 15$ so that we have 14 degrees of freedom (d.f.). From Table C.3, we get $t_{\alpha/2} = t_{0.005} = 2.977$.
3. Calculate the margin of error E:

$$E = t_{\alpha/2} \frac{s}{\sqrt{n}} = 2.977 \left(\frac{0.02}{\sqrt{15}}\right) \approx 0.0154.$$

4. Calculate the confidence interval:

$$\bar{x} - E \leq \mu \leq \bar{x} + E \Rightarrow 1.05 - 0.0154 \leq \mu \leq 1.05 + 0.0154 \Rightarrow 1.0346 \leq \mu \leq 1.0654.$$

Notice that the lower limit of the interval is greater than 1. This indicates that μ is probably greater than 1, so that the packages are not underfilled, on average. □

Confidence intervals can be used to informally compare means from different populations, as illustrated in the next example.

Example 4.6.4 Comparing Weights of M&Ms A student weighs several yellow and red M&M candies and records the sample means and standard deviations (in grams), as summarized in Table 4.7 (data collected by Frank Ohlinger, 2009). Assuming the samples are random and the populations are normally distributed, calculate 95% confidence interval estimates of the mean weights of each population. Does there appear to be a significant difference between the two population means?

	n	$\bar{x}$	s
Yellow	17	0.8588	0.0508
Red	18	0.8722	0.0338

Table 4.7

First we define the two population means being estimated:

μ_1 = the mean weight of all yellow M&Ms,
μ_2 = the mean weight of all red M&Ms.

Next we calculate 95% confidence intervals for each population mean as in the previous example:

$$\text{Red:} \quad E = 2.120\left(\frac{0.0508}{\sqrt{17}}\right) \approx 0.0261 \Rightarrow 0.8327 \leq \mu_1 \leq 0.8849$$

$$\text{Yellow:} \quad E = 2.110\left(\frac{0.0338}{\sqrt{88}}\right) \approx 0.0168 \Rightarrow 0.8554 \leq \mu_2 \leq 0.8890.$$

Note that these two intervals overlap. This means it is very possible that $\mu_1 = \mu_2$ even though there is a difference in the two sample means. Thus we conclude that there is not sufficient evidence to conclude that the two population means are significantly different. If the two intervals *did not* overlap, we would conclude that there was a significant difference between the two population means. □

The analysis used in the previous example is very informal. *Do not use the comparison of confidence intervals to draw definitive conclusions about the equality (or difference) of population means.* In the next chapter, we get more formal methods for comparing populations.

Software Calculations

To calculate a confidence interval estimate of μ:

Minitab: If you have raw data, enter them in a column. Select **Stat** → **Basic Statistics**. Select **1-Sample Z** if the population standard deviation σ is known, or **1-Sample T** if σ is not known. Select the name of the column containing the data under **Samples in columns**. In the case of the Z-interval, enter σ next to **Standard deviation**. Press **Options** to change the confidence level. For a two-sided interval, select **not equal** next to **Alternative**. For one-sided intervals, select either **less than** or **greater than**. If you only have sample statistics, select **Summarized Data** and enter the necessary statistics.

R: First enter the data $\{x_1, x_2, \ldots, x_n\}$ as avector with the syntax x<-c($x_1, x_2, \ldots, x_n$). The syntax for calculating a two-sided confidence interval at the 95% confidence level is

> t.test(x, alternative = "two.sided", conf.level = 0.95)

For one-sided intervals, change "two.sided" to either "less" or "greater."

TI-83/84 Calculators: If you have raw data, enter them in a list. Then press **STAT** → **TESTS**. Select **7:ZInterval** if the population standard deviation σ is known, or **8:TInterval** if σ is not known. Select **Stats**; then enter the name of the list containing the data and the confidence level. When calculating a Z-interval, enter σ. If you only have sample statistics, select **Stats** and enter the necessary statistics. In all cases, set **Freq** to 1.

Exercises

Directions: In each exercise asking for a confidence interval, (1) define the parameter being estimated, (2) find the critical value, (3) calculate the margin of error, and (4) calculate the confidence interval. Unless otherwise specified, calculate a T-interval.

1. A random sample of 15 G-rated movies has a mean run time of 80.6 min with a standard deviation of 21.2 min (data collected by Meredith Hein and Rachel Dahlke, 2011). Assuming the population is normally distributed, construct a 95% confidence interval estimate of the mean run time of all G-rated movies.

2. A random sample of 10 run times from top-10 finishers in cross-country meets from the GPAC conference has a mean of 1606 sec with a standard deviation of 19.8 sec (data collected by Dawn Martin and Daniel Gibson, 2011). Assuming the population is normally distributed, construct a 90% confidence interval estimate of the mean run time of all top-10 finishers in the GPAC.

3. A student wants to estimate the mean amount of money spent by college students on books this semester. He surveys 50 randomly chosen students and calculates a sample mean of $\bar{x} = \$235$.

a. Calculate a 95% confidence interval estimate of this population mean for each of the following values of the sample standard deviation s: 75, 55, 35, and 15.

b. Based on your results in part a, compare the margin of error for a sample with a small standard deviation to one with a large standard deviation.

c. If you want a good estimate of a population mean, would you want a sample with a large or a small standard deviation? Explain.

4. Suppose we want to estimate a population mean μ in which the population variance σ^2 is known.

a. What happens to the margin of error E as the sample size $n \to \infty$?

b. Based on your results in part a, does a larger sample size provide a better estimate of μ than a smaller sample size?

5. People have claimed that the ratio of a person's overall height to navel height equals the golden ratio (a number approximately equal to 1.618). To test this, a statistic professor has each of 39 students measure his or her overall height and navel height and then calculate the ratio (overall height)/(navel height). The 39 values of this ratio have a mean of $\bar{x} = 1.6494$ and a standard deviation of $s = 0.0474$.

   a. Use these data to calculate a 99% confidence interval estimate for the mean ratio of all students at this university. Assume the sample is a random sample of students at this university.

   b. Do the results support the claim? Why or why not?

   c. Could these results be used to draw any conclusion about the mean ratio of all people in the world? Why or why not?

6. To compare the amounts of sodium and sugar in brand-name cereals to the amounts in generic cereals, a student records the sodium and sugar contents per serving of 74 different brand-name cereals and 62 different generic cereals. The results are shown in the table below (data collected by Rachel Masters, 2009).

	Sodium		Sugar	
	$\bar{x}$	s	$\bar{x}$	s
Brand Name	137.79 mg	60.69 mg	7.77 g	3.36 g
Generic	158.22 mg	73.30 mg	7.84 g	4.49 g

   a. Calculate a 90% confidence interval estimate for the mean amount of sodium in all brand-name cereals. Do the same for the sugar content. Repeat for the generic cereals.

   b. Does there appear to be a significant difference between name-brand and generic cereals in either sodium or sugar content? Why or why not?

7. To compare the number of unpopped kernels in bags of microwave buttered popcorn to the number in bags of microwave kettle corn, a student pops four bags of each kind and

counts the number of unpopped kernels in each bag. The data are shown in the table below (data collected by Adam Wolf, 2009).

Buttered	51	60	123	93
Kettle	14	27	20	57

a. Calculate a 95% confidence interval estimate for the mean number of unpopped kernels in all bags of buttered popcorn. Do the same for the kettle corn. Assume that both populations are normally distributed.

b. Does there appear to be a significant difference between these two populations? Why or why not?

8. Polyethylene livestock water tanks are manufactured by pouring melted resin into a mold. As the resin cools, it shrinks a bit. The percentages of shrink of 30 randomly selected tan-colored and 30 randomly selected black-colored tanks are shown in the table below.

	2.08	2.37	2.97	2.37	2.08	2.37	2.67	2.37	2.67	2.97
Tan	2.37	2.67	2.67	2.97	2.67	2.97	2.67	2.37	2.67	2.37
	2.37	2.37	2.37	2.37	2.37	2.08	2.08	2.67	2.67	2.67
	2.97	3.26	2.67	2.37	2.67	2.37	2.67	2.67	2.67	2.37
Black	2.97	2.67	2.08	2.08	2.67	2.37	2.97	2.67	2.67	2.97
	2.37	2.67	2.37	2.08	2.37	2.97	2.67	2.97	2.67	2.67

a. Calculate a 95% confidence interval estimate for the mean percentage of shrink of all tan-colored tanks. Do the same for the black tanks. Assume that both populations are normally distributed.

b. Does there appear to be a significant difference between the mean percentage of shrink for tan and black tanks? Why or why not?

9. Derive the inequality
$$\bar{X}_n - z_{\alpha/2} \frac{\sigma}{\sqrt{n}} \leq \mu \leq \bar{X}_n + z_{\alpha/2} \frac{\sigma}{\sqrt{n}}$$
from the inequality
$$-z_{\alpha/2} \leq \frac{\bar{X}_n - \mu}{\sigma/\sqrt{n}} \leq z_{\alpha/2}$$

10. As in Section 4.5, we can calculate one-sided confidence interval estimates of μ by replacing the critical value $z_{\alpha/2}$ or $t_{\alpha/2}$ with z_α or t_α in the formulas for the margin of error E. Then the confidence intervals are

$$\bar{x} - E \leq \mu < \infty \quad \text{or} \quad -\infty < \mu \leq \bar{x} + E.$$

   a. Nine supermodels have a mean height of $\bar{x} = 70.3$ in with a standard deviation of $s = 1.4$ in. Calculate a 99% one-sided confidence interval that gives a lower bound on the mean height of all supermodels.

   b. The mean height of women in the general population is about 63.5 in. Does your confidence interval in part a suggest that supermodels are taller than typical women? Why or why not?

11. To apply the results of the central limit theorem to a sample, it is necessary that the members of the sample be chosen independently. Because samples are typically selected without replacement, the selections are technically dependent. However, if n, the sample size, is less than 5% of the population size N ($n < 0.05N$), then treating the selections as independent is a reasonable approximation. In cases where $n \geq 0.05N$, a more valid confidence interval can be obtained by calculating the margin of error E as usual and then calculating the *finite population correction factor*

$$C = \sqrt{\frac{N-n}{N-1}}.$$

The confidence interval is then

$$\bar{x} - C \cdot E \leq \mu \leq \bar{x} + C \cdot E.$$

   a. A student at a university of 650 students wants to estimate the mean number of times students call home each week. He asks 150 students how many times they call home each week and calculates a sample mean of $\bar{x} = 2.3$ times and a standard deviation of $s = 1.85$ times. Calculate a 90% confidence interval estimate of the mean number of times students at this university call home each week, using the finite population correction factor.

   b. Consider two populations, one of size $N_1 = 200$ with mean μ_1 and the other of size $N_2 = 5{,}000{,}000$ with mean μ_2. Suppose that samples of size $n = 100$ are taken from each population and that each sample has a mean of $\bar{x} = 10$ and standard deviation of $s = 1$. Calculate a 90% confidence interval estimate of each population mean, using the finite population correction factor. Which confidence interval is narrower? Based on this, is 10 a better estimate of μ_1 or μ_2?

c. If n is a fixed value, find $\lim\limits_{N \to \infty} C$. What does this say about the necessity of the correction factor when the population size is very large?

d. Find the value of C when $n = N$. What does this say about the accuracy of $\bar{x}$ as an estimate of μ in this case?

12. To monitor the packaging of bags of cement mix, a quality control engineer selects a sample of 5 bags throughout the day and weighs each. The weights (in pounds) of the bags, along with the sample mean from each day, over 6 days of production are shown in the table below.

Day	Weight					Sample Mean
1	61.4	61.7	61.5	61.8	61.6	61.6
2	61.2	62.1	62.3	62.1	61.3	61.8
3	63.4	63.3	63.5	63.6	63.2	63.4
4	62.6	63.2	63.0	62.9	61.8	62.7
5	61.5	62.0	61.9	61.1	61.5	61.6
6	61.2	61.4	62.1	62.0	60.8	61.5

To analyze these sample means, the empirical rule introduced in Section 3.4 states that for data with a normal distribution, virtually all the data values should lie within 3 standard deviations of the population mean. The central limit theorem says that in this scenario, the sample means have a mean of μ, the population mean of the bags of cement, and a standard deviation of $\sigma/\sqrt{5}$ where σ is the population standard deviation of the bags of cement.

We do not know μ or σ, so we estimate them with $\bar{\bar{x}}$ and s, the mean and standard deviation of the 30 bags of cement in the overall sample, respectively ($\bar{\bar{x}}$ could also be calculated by taking the mean of the sample means, hence the double bars). Thus the sample means should be between $\bar{\bar{x}} - 3s/\sqrt{5}$ and $\bar{\bar{x}} + 3s/\sqrt{5}$.

The $\bar{x}$ chart is a graph consisting of the following four components:

1. the six points (day, sample mean);
2. the line $y = \bar{\bar{x}} + 3s/\sqrt{5}$, called the *upper control limit*;
3. the line $y = \bar{\bar{x}} - 3s/\sqrt{5}$, called the *lower control limit*;
4. and the line $y = \bar{\bar{x}}$, called the *centerline*.

This graph is an example of a *control chart*. This packaging process is said to be "in statistical control" if the points appear to randomly fluctuate above and below the centerline and the points are all between the upper and lower control limits.

a. In this set of data, $\bar{\bar{x}} = 62.1$ and $s = 0.813$. Use this information to calculate the upper and lower control limits, and sketch the $\bar{x}$ chart.
b. Does this process appear to be in statistical control? Explain.

4.7 Confidence Intervals for a Variance

Although not as well known as a population proportion or mean, the variance plays an important role in areas such as manufacturing. For example, suppose a manufacturer of butter wants to produce a product that is 80% butterfat. Consider the random experiment of choosing a batch of butter, and define the random variable

$$X = \text{the proportion of butterfat in the batch.}$$

Then we would want $\mu = E(X) = 0.8$. The variance of X, σ^2, is a measure of how close values of X typically are to the mean. If σ^2 is large, it would mean that there are many batches of butter with proportions of butterfat not very close to 0.8. Because we want the batches to be as consistent as possible, we would want σ^2 to be as small as possible. This variance is called the *population variance*.

Suppose that the variance is 0.0025 and that management believes this is too high, so they retool the equipment in an attempt to lower the variance. To determine if the improvements have worked, they need to estimate the new variance. So they select a random sample of 20 batches and measure the proportion of butterfat in each. The data are shown in Table 4.8 (these are 20 observed values of the random variable X).

| 0.755 | 0.863 | 0.825 | 0.767 | 0.825 | 0.836 | 0.844 | 0.792 | 0.827 | 0.821 |
| 0.782 | 0.767 | 0.804 | 0.761 | 0.809 | 0.798 | 0.757 | 0.768 | 0.787 | 0.741 |

Table 4.8

The mean and variance of these data are

$$\bar{x} = 0.8052 \quad \text{and} \quad s^2 = 0.001102.$$

This sample variance s^2 is an estimate of σ^2. The sample variance is an observed value of the random variable

$$S^2 = \frac{1}{n-1} \sum_{i=1}^{n} (X_i - \bar{X})^2$$

which, as illustrated in Section 4.4, is an unbiased estimator of σ^2 if X has a normal distribution. To construct a confidence interval for this estimate, we need the following theorem.

Theorem 4.7.1 *If the population is normally distributed with variance σ^2, then the random variable*

$$C = \frac{(n-1)S^2}{\sigma^2}$$

is $\chi^2(n-1)$. □

The proof of this theorem is beyond the scope of this text, so we omit it. In practical terms, this theorem says that if we collect samples of size n from a population that is normally distributed with (known) variance σ^2, calculate the sample variance of each, multiply by $n-1$, and then divide by σ^2, the resulting values will follow a $\chi^2(n-1)$ distribution.

Example 4.7.1 **Illustrating Theorem 4.7.1** To illustrate Theorem 4.7.1, we simulated taking 1000 random samples, each of size $n=6$, from a population that is $N(5,4)$. For each sample, we calculated the sample variance s^2 and then the quantity $c = (6-1)s^2/4$. This yielded 1000 observed values of the random variable C in Theorem 4.7.1. We then constructed the histogram shown in Figure 4.12 and graphed the $\chi^2(6-1)$ density curve on top. Notice that the density curve follows the tops of the bars, indicating that C does indeed have a $\chi^2(6-1)$ distribution as claimed in the theorem.

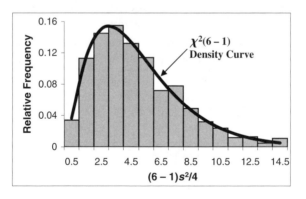

Figure 4.12

□

Now, to calculate the margin of error for s^2 as an estimate of σ^2, let α be the desired significance level and consider the two critical chi-square values $\chi^2_{1-\alpha/2}(n-1)$ and $\chi^2_{\alpha/2}(n-1)$

as described in Definition 3.9.4. By definition, if C is $\chi^2(n-1)$, then

$$P\left(C \le \chi^2_{1-\alpha/2}(n-1)\right) = 1 - (1-\alpha/2) = \alpha/2$$

and $\quad P\left(C \le \chi^2_{\alpha/2}(n-1)\right) = 1 - \alpha/2 \quad \Rightarrow \quad P\left(C > \chi^2_{\alpha/2}(n-1)\right) = \alpha/2.$

These two probabilities are illustrated in Figure 4.13.

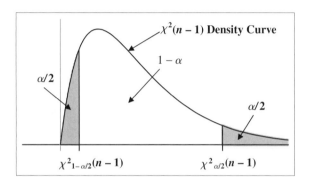

Figure 4.13

If we let $a = \chi^2_{1-\alpha/2}(n-1)$ and $b = \chi^2_{\alpha/2}(n-1)$, we may combine these two probabilities to get

$$P\left(a \le \frac{(n-1)S^2}{\sigma^2} \le b\right) = 1 - \alpha.$$

Rewriting this inequality so that σ^2 is by itself in the center yields

$$P\left(\frac{(n-1)S^2}{b} \le \sigma^2 \le \frac{(n-1)S^2}{a}\right) = 1 - \alpha.$$

This gives us the definition of the confidence interval for σ^2 and σ.

Definition 4.7.1 Let s^2 be the variance of a sample of size n taken from a normally distributed population with unknown variance σ^2, and let α be between 0 and 1. Also, let $a = \chi^2_{1-\alpha/2}(n-1)$ and $b = \chi^2_{\alpha/2}(n-1)$ be critical χ^2 values. The interval

$$\left[\frac{(n-1)s^2}{b}, \frac{(n-1)s^2}{a}\right]$$

is called a $100(1-\alpha)\%$ *confidence interval estimate for* σ^2.

Requirements

1. The sample is random.
2. The population is normally distributed.

This confidence interval could be written in the equivalent inequality form

$$\frac{(n-1)s^2}{b} \leq \sigma^2 \leq \frac{(n-1)s^2}{a}.$$

A confidence interval for σ, the population standard deviation, can be found by taking the square roots of both ends of this interval.

Example 4.7.2 Constructing a Confidence Interval The basic steps for constructing a confidence interval for σ^2 are the same as for other population parameters. We illustrate the process by constructing a 95% confidence interval using the data in Table 4.8, where the sample variance is $s^2 = 0.001102$. We assume that the population is normally distributed (see Exercise 4 of Section 4.10 for a verification of this assumption).

1. Define the population variance being estimated:

   σ^2 = the variance in the proportion of butterfat of all batches of butter.

2. Find the critical values. Note that $\alpha = 0.05$ and that $n = 20$. From Table C.4, we get the critical values

   $$a = \chi^2_{1-0.05/2}(19) = \chi^2_{0.975}(19) = 8.907 \quad \text{and} \quad b = \chi^2_{0.05/2}(19) = \chi^2_{0.025}(19) = 32.852.$$

3. Calculate the confidence interval:

   $$\frac{(n-1)s^2}{b} \leq \sigma^2 \leq \frac{(n-1)s^2}{a} \Rightarrow \frac{(19)(0.001102)}{32.852} \leq \sigma^2 \leq \frac{(19)(0.001102)}{8.907}$$

   $$\Rightarrow 0.000637 \leq \sigma^2 \leq 0.00235.$$

   A confidence interval for σ is then

   $$\sqrt{0.000637} \leq \sigma \leq \sqrt{0.00235} \Rightarrow 0.0252 \leq \sigma \leq 0.0485.$$

4. Interpret the confidence interval. We interpret this in the same way as other types of confidence intervals by saying, "We are 95% confident that the true value of σ^2 is between 0.000637 and 0.00235." Note that the upper confidence interval limit is less than 0.0025, the variance before the machines were retooled. Thus it appears that the variance has indeed decreased. □

One-sided confidence intervals for σ^2 may be found by using the critical values

$$a = \chi^2_{1-\alpha}(n-1) \quad \text{and} \quad b = \chi^2_{\alpha}(n-1).$$

The one-sided intervals are then

$$\frac{(n-1)s^2}{b} \leq \sigma^2 \leq \infty \quad \text{and} \quad 0 \leq \sigma^2 \leq \frac{(n-1)s^2}{a}.$$

For the data in Table 4.8, to find the 95% one-sided confidence interval giving an upper bound on σ^2, the critical value is

$$a = \chi^2_{1-0.5}(19) = \chi^2_{0.95}(19) = 10.12$$

so that the confidence interval is

$$0 \leq \sigma^2 \leq \frac{(19)(0.001102)}{10.12} = 0.00207.$$

As in Example 4.7.2, this upper bound is less than 0.0025, indicating again that the variance has indeed decreased.

Exercises

Directions: In each exercise asking for a confidence interval, (1) define the parameter being estimated, (2) find the critical value, and (3) calculate the confidence interval.

1. Suppose that a sample of 26 steel sheets of a certain type was tested for tensile strength (measured in kilograms per square millimeter), and the variance was $s^2 = 2.41$. Use these data to construct a 90% confidence interval estimate of σ^2, the population variance (assume the population is normally distributed). If the requirements for production of these sheets state that the variance must be less than 3.5, does it appear that the sheets meet specifications?

2. Two baseball players hit 26 baseballs off a tee with a metal bat, recorded the time each ball took to reach home plate, and calculated the average velocity of each. These 26 baseballs had a mean average velocity of 95.54 mph with a standard deviation of 15.58 mph (data collected by Daniel Ullman and Daniel Wilkerson, 2011). Assuming that this sample is random and that the population is normally distributed, calculate a 95% confidence interval estimate of the standard deviation of the average velocity of all baseballs hit with a metal bat.

3. The mass (in grams) of 17 randomly selected yellow M&M candies is shown in the table below (data collected by Frank Ohlinger, 2009). Use these data to calculate a 99% confidence interval estimate of the standard deviation of the mass of all such candies (assume the population is normally distributed).

0.89	0.89	0.85	0.86	0.90	0.83	0.82	0.86	0.94
0.83	0.84	0.88	0.86	0.97	0.82	0.80	0.76	

4. Suppose a manufacturer of bolts produces one such bolt of length 15 mm. Specifications for the bolt require that the variance in lengths be less than 0.004 mm². A random sample of 50 such bolts has a mean length of $\bar{x} = 15.01$ mm with variance $s^2 = 0.0023$ mm². Assuming the population is normally distributed, construct a 99% confidence interval giving an upper bound on the population variance. Does the manufacturing process appear to meet the specifications?

5. A sample of 18 male students was asked how much they spent on textbooks this semester. The sample variance was $s_M^2 = 35.05$. A sample of eight female students was asked the same question, and the sample variance was $s_F^2 = 18.40$. (Data collected by Megan Damron and Spencer Solomon, 2009.) Assume that the amount spent on textbooks is normally distributed for both the populations of male students and of female students.

a. Calculate a 90% confidence interval estimate for σ_M^2, the population variance of the amount spent on textbooks by male students.

b. Calculate a 90% confidence interval estimate for σ_F^2, the population variance of the amount spent on textbooks by female students.

c. Does there appear to be a significant difference between the variances of these two populations? Why or why not?

6. Suppose a random sample of size $n = 10$ is taken from a normally distributed population with variance $\sigma^2 = 1.5$.

a. Use Table C.4 to approximate the probability that the sample variance is less than 2.446. That is, approximate $P(S^2 \leq 2.446)$ where S^2 is the sample variance. (**Hint:** According to Theorem 4.7.1, the random variable $C = (10-1)S^2/1.5$ has a $\chi^2(9)$

distribution. This means that the event $S^2 \le 2.446$ corresponds to the event $C = 9S^2/1.5 \le 9(2.446)/1.5 \approx 14.68$. Thus $P(S^2 \le 2.446) \approx P(C \le 14.68)$.)

b. Now suppose the sample size is $n = 31$. Approximate $P(S^2 \le 0.83955)$.

7. Confidence intervals for μ were expressed in three equivalent ways:

$$[\bar{x} - E, \bar{x} + E], \qquad \bar{x} - E \le \mu \le \bar{x} + E, \qquad \text{or} \qquad \bar{x} \pm E,$$

where E is the margin of error. The confidence interval for σ^2 calculated in Example 4.7.2 could be expressed in two different forms: $0.000637 \le \sigma^2 \le 0.00235$ and $[0.000637, 0.00235]$. It seems reasonable that we could define the margin of error to be one-half the width of these intervals $E = (0.00235 - 0.00637)/2 = 0.008565$ and then express the confidence interval in this form: 0.0014935 ± 0.0008565. Explain why this is not an appropriate form.

8. Derive the inequality

$$\frac{(n-1)S^2}{b} \le \sigma^2 \le \frac{(n-1)S^2}{a}$$

from the inequality

$$a \le \frac{(n-1)S^2}{\sigma^2} \le b.$$

9. One problem with using Table C.4 to find critical chi-square values is that it only goes up to 100 degrees of freedom ($r = 100$). Plus, not every degree up to 100 is listed in the table. One way to approximate critical values for large values of r is with the formulas

$$a = \chi^2_{1-\alpha/2}(r) \approx \frac{1}{2}\left(-z_{\alpha/2} + \sqrt{2r-1}\right)^2 \quad \text{and} \quad b = \chi^2_{\alpha/2}(r) \approx \frac{1}{2}\left(z_{\alpha/2} + \sqrt{2r-1}\right)^2$$

where $z_{\alpha/2}$ is a critical z-value as described in Definition 4.5.1.

a. Use these formulas to approximate a and b for $\alpha = 0.05$ and $r = 100$, and compare the approximations to the exact values given in Table C.4. Compute the relative error in the approximation (the error divided by the exact value). How close are the approximations?

b. Repeat part a for $\alpha = 0.01$ and $r = 15$. How good are these approximations for smaller values of r as compared to larger values?

c. Suppose a sample of size $n = 150$ from a normally distributed population has a sample variance of $s^2 = 2.36$. Calculate a 90% confidence interval estimate for σ^2, the population variance.

10. Suppose a manufacturer of frozen vegetables would like to estimate the variance in the weights of its 16-oz bags of frozen corn. To do this, on each of k days they select a sample of bags and weigh each. If we assume that the variance on each day is the same, σ^2, then an estimate of this value can be obtained by calculating the *pooled variance*

$$s_p^2 = \frac{(n_1 - 1)s_1^2 + (n_2 - 1)s_2^2 + \cdots + (n_k - 1)s_k^2}{(n_1 - 1) + (n_2 - 1) + \cdots + (n_k - 1)},$$

where n_i is the sample size and s_i^2 is the sample variance from day i. The pooled variance is often described as the "weighted average" of the sample variances. Related to the pooled variance is the *pooled standard deviation*, $s_p = \sqrt{s_p^2}$. Calculate the pooled variance and standard deviation of the data shown in the table below.

Day	Weights (oz)				
1	16.09	16.14	15.98	16.21	16.02
2	15.97	16.11	16.11		
3	16.19	15.87	16.19	16.03	
4	16.18	15.95			
5	16.21	16.08	16.18	15.99	16.05

4.8 Confidence Intervals for Differences

To compare two different populations, we may want to estimate the difference between their respective parameters, using a confidence interval. In this section, we describe how to do this to compare proportions and means.

Comparing Proportions

Suppose a statistics professor is testing two different online homework systems for an introduction to statistics class. One section of the class uses the first system while another section uses the second system. Of $n_1 = 95$ students using the first system, $x_1 = 87$ pass the class. Of $n_2 = 82$ students using the second system, $x_2 = 62$ pass the class. Based on these statistics, it appears that the first system is better in terms of the proportion of students who pass. We want to estimate how much better. Specifically, if we let

$$p_1 = \text{the proportion of all students using the first system who pass}$$

and

$$p_2 = \text{the proportion of all students using the second system who pass,}$$

then we want to estimate the difference $(p_1 - p_2)$, using a confidence interval.

To this end, let the random variables $\hat{P}_1$ and $\hat{P}_2$ denote sample proportions from the two populations, and suppose that random samples of size n_1 and n_2 are taken from each population yielding x_1 and x_2 "successes," respectively. Then, assuming that n_1 and n_2 are "large enough," by Theorem 4.4.1 $\hat{P}_1$ and $\hat{P}_2$ are approximately

$$N\left(p_1, \frac{p_1(1-p_1)}{n_1}\right) \quad \text{and} \quad N\left(p_2, \frac{p_2(1-p_2)}{n_2}\right),$$

respectively. Assuming that $\hat{P}_1$ and $\hat{P}_2$ are independent, then by Theorem 3.7.4 the random variable $\hat{P}_1 - \hat{P}_2$ is approximately

$$N\left(p_1 - p_2, \frac{p_1(1-p_1)}{n_1} + \frac{p_2(1-p_2)}{n_2}\right).$$

Thus the random variable

$$Z = \frac{\left(\hat{P}_1 - \hat{P}_2\right) - (p_1 - p_2)}{\sqrt{p_1(1-p_1)/n_1 + p_2(1-p_2)/n_2}}$$

is approximately $N(0,1)$. Now if $z_{\alpha/2}$ is a critical z-value, then we have

$$P\left[-z_{\alpha/2} \leq \frac{\left(\hat{P}_1 - \hat{P}_2\right) - (p_1 - p_2)}{\sqrt{p_1(1-p_1)/n_1 + p_2(1-p_2)/n_2}} \leq z_{\alpha/2}\right] \approx 1 - \alpha.$$

Rewriting this inequality so that $(p_1 - p_2)$ is in the center yields

$$P\left[\left(\hat{P}_1 - \hat{P}_2\right) - p \cdot z_{\alpha/2} \leq (p_1 - p_2) \leq \left(\hat{P}_1 - \hat{P}_2\right) + p \cdot z_{\alpha/2}\right] \approx 1 - \alpha,$$

where

$$p = \sqrt{\frac{p_1(1-p_1)}{n_1} + \frac{p_2(1-p_2)}{n_2}}.$$

This leads to the definition of the confidence interval.

Definition 4.8.1 Let x_1 and x_2 be the number of successes in two independent samples of size n_1 and n_2 taken from two populations with proportions p_1 and p_2. Also, let $\hat{p}_1 = x_1/n_1$ and $\hat{p}_2 = x_2/n_2$ be the sample proportions, let α be between 0 and 1, and let $z_{\alpha/2}$ be a critical z-value. The interval

$$(\hat{p}_1 - \hat{p}_2) - \hat{p} \cdot z_{\alpha/2} \leq (p_1 - p_2) \leq (\hat{p}_1 - \hat{p}_2) + \hat{p} \cdot z_{\alpha/2}$$

where

$$\hat{p} = \sqrt{\frac{\hat{p}_1(1-\hat{p}_1)}{n_1} + \frac{\hat{p}_2(1-\hat{p}_2)}{n_2}}$$

is called a $100(1-\alpha)\%$ *confidence interval estimate for* $(p_1 - p_2)$. The quantity

$$E = \hat{p} \cdot z_{\alpha/2}$$

is called the *margin of error*. This confidence interval is also called a *2-proportion Z-interval*.

Requirements:

1. Both samples are random and independent.
2. Each sample contains at least 5 successes and 5 failures. This is so both sample proportions are approximately normally distributed.

In this example, we have $\hat{p}_1 = 87/95 \approx 0.916$ and $\hat{p}_2 = 62/82 \approx 0.756$ so that

$$\hat{p} = \sqrt{\frac{0.916\,(1-0.916)}{95} + \frac{0.756\,(1-0.756)}{82}} \approx 0.0553.$$

At the 95% confidence level, $z_{0.05/2} = 1.96$ so that the margin of error is $E = (0.0553)(1.96) \approx 0.104$ and the confidence interval is

$$(0.916 - 0.756) - 0.104 \leq (p_1 - p_2) \leq (0.916 - 0.756) + 0.104$$
$$\Rightarrow \quad 0.056 \leq (p_1 - p_2) \leq 0.264.$$

We would interpret this by saying, "We are 95% confident that the difference between the population proportions is between 0.056 and 0.264." Because this interval contains only positive values, it appears that the first system is indeed better than the second.

Software Calculations

To calculate a two-sided confidence interval estimate of the difference $p_1 - p_2$:

Minitab: Select **Stat** → **Basic Statistics** → **2-Proportion**. Select **Summarized Data**; enter n_1 and n_2 as the **Trials** and x_1 and x_2 as the **Events**. Select **Options**, enter the **Confidence level**, set the **Alternative** to **not equal**, and do *not* check **Use pooled estimate of p for the test**.

R: The syntax for calculating a confidence interval at the 95% confidence level is
> prop.test(c(x_1, x_2), c(n_1, n_2), correct=FALSE, conf.level = 0.95).

TI-83/84 Calculators: Press **STAT** → **TESTS** → **B:2-PropZInt**. Enter x_1, n_1, x_2, n_2, and the confidence level, denoted **C-Level**.

Comparing Means

We can do a similar analysis to the above to compare two population means with a confidence interval. Suppose we want to compare the life spans of two brands of lightbulbs. Let X_1 and X_2 denote the life span of a randomly selected bulb of the first and second brands, respectively. Assume that X_1 is $N(\mu_1, 547)$ and X_2 is $N(\mu_1, 478)$. We measure the life spans of $n_1 = 45$ lightbulbs of the first brand and $n_2 = 48$ of the second brand and calculate the sample means $\bar{x}_1 = 905.3$ and $\bar{x}_2 = 890.2$.

Based on these sample statistics, it appears that the first brand lasts longer on average than the second. We want to use these data to quantify just *how much* longer with a confidence interval. To this end, note that by Theorem 3.8.1 the sample mean of the first brand, $\bar{X}_1$, is $N(\mu_1, \sigma_1^2/n_1)$ and the sample mean of the second brand, $\bar{X}_2$, is $N(\mu_2, \sigma_2^2/n_2)$ where σ_1^2 and σ_2^2 are the respective population variances. Then, assuming that X_1 and X_2 are independent, by Theorem 3.7.4 the random variable $\bar{X}_1 - \bar{X}_2$ is

$$N\left(\mu_1 - \mu_2, \frac{\sigma_1^2}{n_1} + \frac{\sigma_2^2}{n_2}\right).$$

For notational simplicity, define $\sigma = \sqrt{\sigma_1^2/n_1 + \sigma_2^2/n_2}$ to be the standard deviation of $\bar{X}_1 - \bar{X}_2$. Thus the random variable

$$Z = \frac{(\bar{X}_1 - \bar{X}_2) - (\mu_1 - \mu_2)}{\sigma}$$

is $N(0,1)$. By definition of the critical value $z_{\alpha/2}$, we have $P(-z_{\alpha/2} \leq Z \leq z_{\alpha/2}) = 1 - \alpha$. Substituting, we get

$$P\left(-z_{\alpha/2} \leq \frac{(\bar{X}_1 - \bar{X}_2) - (\mu_1 - \mu_2)}{\sigma} \leq z_{\alpha/2}\right) = 1 - \alpha.$$

Rewriting this yields

$$P\left[(\bar{X}_1 - \bar{X}_2) - \sigma \cdot z_{\alpha/2} \leq (\mu_1 - \mu_2) \leq (\bar{X}_1 - \bar{X}_2) + \sigma \cdot z_{\alpha/2}\right] = 1 - \alpha.$$

This leads us to the definition of the confidence interval.

Definition 4.8.2 Let $\bar{x}_1$ and $\bar{x}_2$ be the means of two independent samples of size n_1 and n_2 taken from two populations with variances σ_1^2 and σ_2^2 and unknown means μ_1 and μ_2. Let α be between 0 and 1, and let $z_{\alpha/2}$ be a critical z-value. The interval

$$(\bar{x}_1 - \bar{x}_2) - \sigma \cdot z_{\alpha/2} \leq (\mu_1 - \mu_2) \leq (\bar{x}_1 - \bar{x}_2) + \sigma \cdot z_{\alpha/2} \qquad (4.6)$$

where $\sigma = \sqrt{\sigma_1^2/n_1 + \sigma_2^2/n_2}$ is called a $100(1-\alpha)\%$ *confidence interval estimate* for $(\mu_1 - \mu_2)$. The quantity

$$E = \sigma \cdot z_{\alpha/2}$$

is called the *margin of error*. This confidence interval is also called a *2-sample Z-interval*.

Requirements:

1. Both samples are random and independent.
2. Both population variances are known.
3. Both populations are normally distributed or both sample sizes are greater than 30.

In this example, we have

$$\bar{x}_1 - \bar{x}_2 = 905.3 - 890.2 = 15.1 \quad \text{and} \quad \sigma = \sqrt{\frac{547}{45} + \frac{478}{48}} \approx 4.7$$

so that at the 95% confidence level, the margin of error is $E = (4.7)(1.96) \approx 9.2$ and the confidence interval is

$$15.1 - 9.2 \leq (\mu_1 - \mu_2) \leq 15.1 + 9.2 \quad \Rightarrow \quad 5.9 \leq (\mu_1 - \mu_2) \leq 24.3.$$

Thus we would be 95% confident in saying that, on average, the first brand lasts about 6 to 24 hr longer than the second.

In most situations, we do not know the population variances. If both sample sizes are large (for example, $n_1 > 30$ and $n_2 > 30$), then the sample variances s_1^2 and s_2^2 serve as reasonable estimates of σ_1^2 and σ_2^2. This means that

$$\sigma \approx \sqrt{\frac{s_1^2}{n_1} + \frac{s_2^2}{n_2}}$$

and we can calculate an approximate confidence interval using inequality (4.6). For this reason, the 2-sample Z-interval is often called a *large-sample* interval. If the sample sizes are small, then the confidence intervals are calculated using critical t-values as summarized in the box below (see Appendix A.5 for a more in-depth discussion of these techniques). These confidence intervals are called *2-sample T-intervals*.

Small-Sample Confidence Intervals for the Difference of Two Means

If two populations are (approximately) normally distributed and their variances σ_1^2 and σ_2^2 are unknown, then an approximate $100(1-\alpha)\%$ confidence interval for the difference of their means $(\mu_1 - \mu_2)$ using data from two independent samples of the respective populations is

$$(\bar{x}_1 - \bar{x}_2) - E \leq (\mu_1 - \mu_2) \leq (\bar{x}_1 - \bar{x}_2) + E$$

where the margin of error E is calculated according to the following cases:

1. If σ_1^2 and σ_2^2 are assumed to be equal, then

$$E = s_p \cdot t_{\alpha/2} \sqrt{\frac{1}{n_1} + \frac{1}{n_2}}$$

where

$$s_p = \sqrt{\frac{(n_1 - 1) s_1^2 + (n_2 - 1) s_2^2}{n_1 + n_2 - 2}}$$

is the *pooled standard deviation* (this is an estimate of the common population standard deviation) and $t_{\alpha/2}$ is a critical t-value with $(n_1 + n_2 - 2)$ degrees of freedom.

2. If σ_1^2 and σ_2^2 are *not* assumed to be equal, then

$$E = t_{\alpha/2} \sqrt{\frac{s_1^2}{n_1} + \frac{s_2^2}{n_2}}$$

where $t_{\alpha/2}$ is a critical t-value with r degrees of freedom where

$$r = \frac{\left(s_1^2/n_1 + s_2^2/n_2\right)^2}{[1/(n_1-1)]\left(s_1^2/n_1\right)^2 + [1/(n_2-1)]\left(s_2^2/n_2\right)^2}.$$

If r is not an integer, then round it down to the nearest whole number. (**Note:** If software is used to find $t_{\alpha/2}$, there is no need to round r.)

To help decide which type of confidence interval is appropriate, Z or T, we offer these suggestions:

1. If both sample sizes are greater than 30 or both population variances are known, then use a Z-interval.
2. If either sample size is less than 30, the population variances are unknown, and both populations are normally distributed (at least approximately), then use a T-interval.
3. If either sample size is less than 30, the population variances are unknown, and at least one population is *not* normally distributed, then use nonparametric techniques (see Chapter 7).

To use either type of interval, the two samples *must* be random and independent.

Example 4.8.1 Comparing Diets To compare two different diets, call them A and B, volunteers followed each diet for 6 months and their weight losses were recorded. Among the $n_1 = 10$ subjects who followed diet A, the mean weight loss was $\bar{x}_1 = 4.5$ lb with a standard deviation of $s_1 = 6.5$ lb. Among the $n_2 = 10$ subjects who followed diet B, the mean weight loss was $\bar{x}_2 = 3.2$ lb with a standard deviation of $s_2 = 4.5$ lb. Assuming that both populations are normally distributed, construct a 95% confidence interval estimate of the difference of the population mean weight loss ($\mu_1 - \mu_2$). Based on these results, does there appear to be a significant difference between the diets?

We begin by defining the parameters:

$\mu_1 = $ the mean weight loss of all people on diet A

and $\quad \mu_2 = $ the mean weight loss of all people on diet B.

Because we do not know either population variance and both sample sizes are less than 30, we construct 2-sample T-intervals. If we assume that the population variances are equal, then we have $(10 + 10 - 2) = 18$ degrees of freedom so that the critical value is $t_{0.05/2}(18) = 2.101$. The pooled standard deviation is

$$s_p = \sqrt{\frac{(10-1)(6.5^2) + (10-1)(4.5^2)}{10 + 10 - 2}} \approx 5.59.$$

Thus the margin of error is

$$E = 5.59(2.101)\sqrt{\frac{1}{10} + \frac{1}{10}} \approx 5.25$$

and the confidence interval is

$$(4.5 - 3.2) - 5.25 \leq (\mu_1 - \mu_2) \leq (4.5 - 3.2) + 5.25 \quad \Rightarrow \quad -3.95 \leq \mu_1 - \mu_2 \leq 6.55.$$

If we do *not* assume that the population variances are equal, then we have

$$r = \frac{\left(6.5^2/10 + 4.5^2/10\right)^2}{[1/(10-1)]\left(6.5^2/10\right)^2 + [1/(10-1)]\left(4.5^2/10\right)^2} \approx 16$$

degrees of freedom so that the critical value is $t_{0.05/2}(16) = 2.120$ and the margin of error is

$$E = 2.120\sqrt{\frac{6.5^2}{10} + \frac{4.5^2}{10}} = 5.3.$$

The confidence interval is then

$$(4.5 - 3.2) - 5.3 \leq (\mu_1 - \mu_2) \leq (4.5 - 3.2) + 5.3 \quad \Rightarrow \quad -4 \leq (\mu_1 - \mu_2) \leq 6.6.$$

Notice that both confidence intervals are very similar and that they contain 0. This is an indication that $(\mu_1 - \mu_2)$ could be 0, meaning that there is no difference between the two diets. Therefore, we conclude that there is not a "statistically significant" difference between the two diets.

Note that this conclusion does not mean that for an individual person, diet A is as effective as diet B. Also note that the confidence interval indicates that μ_1 could be as much as about 6.5 lb greater than μ_2, meaning that diet A could produce better results than diet B, on average. We need to be very careful about interpreting and applying these statistical results in the real world. *Statistical* significance does not mean the same thing as *practical* significance. □

One important requirement for a confidence interval for the difference of two means is that the data come from independent samples. The next example illustrates what can be done when this requirement is not met.

Example 4.8.2 **Comparing Test Scores** At a large university, freshman students are required to take an introduction to writing class. Students are given a survey on their attitudes toward writing at the beginning and end of the class. Each student receives a score between 0 and 100 (the higher the score, the more favorable the attitude toward writing). The scores of nine different students from the beginning and end of the class are shown in Table 4.9. Based on these data, do the attitudes toward writing appear to increase by the end of the class?

Beginning	77.6	83.2	60.2	93.1	74.6	43.1	86.9	79.3	80.2
End	85.4	79.6	64.2	96.6	79.3	40.5	90.1	89.2	85.6
End − Beginning	7.8	−3.6	4	3.5	4.7	−2.6	3.2	9.9	5.4

Table 4.9

Examining the data, we see that for most students the score at the end is higher than that at the beginning. This indicates that scores do tend to increase. To quantify the amount of increase, we may be tempted to form a confidence interval for the difference in the population means ($\mu_E - \mu_B$). However, because both samples of data come from the *same* nine students, the two samples are *not* independent, so the requirements for this type of confidence interval are not met. Note that each student gives two numbers, a beginning score and an ending score. Because of this, we say these data come in *pairs*.

Instead, we form a confidence interval for the mean difference of the scores for all the students. The third row of Table 4.9 shows the differences of interest for these nine students. We want to form a confidence interval for

$$\mu_d = \text{the mean of (end − beginning) for all freshmen students.}$$

Note that this is a confidence interval for a single population mean, so we use a T-interval as described in Section 4.6.

The mean of the differences in Table 4.9 is $\bar{x}_d = 3.59$ (this is a point estimate of μ_d), and the standard deviation is $s_d = 4.36$. Assuming the population of these differences is normally distributed, then at the 95% confidence level the margin of error is

$$E = t_{\alpha/2}(n-1)\frac{s}{\sqrt{n}} = 2.306\left(\frac{4.36}{\sqrt{9}}\right) \approx 3.35.$$

and the confidence interval is

$$\bar{x}_d - E \leq \mu \leq \bar{x}_d + E \Rightarrow 3.59 - 3.35 \leq \mu_d \leq 3.59 + 3.35 \Rightarrow 0.24 \leq \mu_d \leq 6.94.$$

Because this confidence interval includes only positive values, we conclude that the scores do appear to increase, on average. ⊔

Software Calculations

To calculate a two-sided confidence interval estimate of the difference $\mu_1 - \mu_2$:

Minitab: If you have raw data, enter the two samples in separate columns. To calculate a T-interval, select **Stat** $\to$ **Basic Statistics** $\to$ **2-Sample t**. Select **Samples in different columns**, and enter the names of the columns containing the data. If you assume that the two populations have the same variance, check the box next to **Assume equal variances**. Select **Options**, enter the **Confidence level**, set the **Alternative** to **not equal**. (The value of **Test difference** does *not* matter.) If you only have sample statistics, choose **Summarized data** and enter the necessary statistics. To calculate a confidence interval for μ_d where the data come in pairs, follow the same steps except choose **Paired t** instead of **2-Sample t**.

R: Enter the two samples of data $\{x_1, x_2, \ldots, x_n\}$ and $\{y_1, y_2, \ldots, y_m\}$ into vectors with the syntax x<-c($x_1, x_2, \ldots, x_n$) and y<-c($y_1, y_2, \ldots, y_m$). The syntax for calculating a confidence interval at the 95% confidence level where you do not assume the two populations have the same variance is

> t.test(y, z, equal.variances=FALSE, conf.level=0.95)

If you assume the two populations have the same variance, change FALSE to TRUE. To calculate a confidence interval for μ_d where the data come in pairs, enter the values of the differences in a list and use the procedures for constructing a confidence interval estimate of a single population mean in Section 4.6.

TI-83/84 Calculators: If you have raw data, enter the two samples in different lists. For a Z-interval where the population standard deviations σ_1 and σ_2 are known, press **STAT** $\to$ **TESTS** $\to$ **9:2-SampZInt**. Select **Data**, enter σ_1 and σ_2, the names of the lists containing the data, and the confidence level. For a T-interval, select **0:2-SampTint**, choose **Data**, and enter the necessary information. Always set **Freq1** and **Freq2** to 1. If you assume the two populations have the same variance, select **Pooled: Yes**; otherwise, select **No**. For either type of interval, if you only have sample statistics, choose **Stats** and enter the necessary statistics. To calculate a confidence interval for μ_d where the data come in pairs, enter the values of the differences in a list and use the procedures for constructing a confidence interval estimate of a single population mean in Section 4.6.

Exercises

Directions: In each exercise asking for a confidence interval, (1) define the parameters, (2) find the degrees of freedom (where appropriate), (3) calculate the critical value, and (4) calculate the confidence interval. When calculating a confidence interval for the difference of means, calculate a T-interval unless otherwise specified. In all exercises, use a 95% confidence level unless otherwise specified.

1. A random sample of 419 plain M&Ms contains 62 blue candies while a sample of 142 peanut M&Ms contains 31 blue candies (data collected by Rebekah Brisbois, 2009). Calculate a confidence interval estimate of the difference in the proportions of blue candies for plain and peanut M&Ms. Do the two types of M&Ms appear to contain the same proportion of blue candies?

2. To test for side effects of a new allergy medication, researchers gave the medicine to 1500 volunteers and a placebo to 1550 other volunteers. Of those who received the medicine, 39 reported drowsiness, and of those who received a placebo, 19 reported drowsiness. Researchers would like to compare the proportions of each group that experienced drowsiness. Calculate a confidence interval estimate of the difference of the population proportions. Do those who receive the medicine appear to have a higher proportion of drowsiness? Based on this evidence, is drowsiness a major concern? Explain. is very small

3. Using sign-in records from the university weight room, a student randomly chooses several users and records the person's gender and the time of day the person used the weight room (morning or evening). The results are summarized in the table below (data collected by Joshua Ostrem, 2011). Let p_1 be the proportion of all morning users who are male and p_2 be the proportion of all evening users who are male. Calculate a 95% confidence interval estimate of the difference $(p_1 - p_2)$. Based on this result, is there much difference between morning and evening in terms of the proportion of male users?

	Morning	Evening
Males	228	503
Females	255	495

4. A city is considering the construction of a new middle school. To judge attitudes toward the proposed location, residents living near and far from the location were asked if they favored the location. The data are shown in the table below.

	Near	Far
Sample Size	125	135
Number in Favor	94	108

a. Calculate the sample proportion in favor of the location for both those near and those far from the location. Based on these numbers, does there appear to be much difference between the two populations of citizens?

b. Construct a confidence interval estimate of the difference of the two population proportions. Based on this interval, does there appear to be much difference between the two groups of citizens?

c. Combine the two samples into one large sample, and construct a confidence interval estimate of the proportion of *all* residents of the city who favor the location. The members of the city council want at least 70% support for the location before they proceed with the plans. Based on this confidence interval, do they appear to have this level of support?

5. Explain why a difference in population proportions must be between 0 and 1 in absolute value.

6. Four bottles of one brand of bottled water have a mean volume of 510.75 mL with a standard deviation of 1.25 ml. Four bottles of another brand have a mean volume of 505 mL with a standard deviation of 0.816 mL (data collected by Brittany Singleton and Kelsie Elder, 2009). Assuming that both populations are normally distributed and have the same variance, calculate a confidence interval estimate of the difference of the population means of the two brands. Does there appear to be a significant difference between the population means? 7.58), no

7. Two baseball players hit 30 baseballs off a tee with a metal bat, recorded the time each ball took to reach home plate, and calculated the average velocity of each. Then they repeated this using a wood bat. The results (in miles per hour) are summarized in the table below (data collected by Daniel Ullman and Daniel Wilkerson, 2011). Assuming both samples are random and the populations are normally distributed, calculate a 95% confidence interval estimate of the difference of the mean average velocity of the two types of bats (do not assume equal variances). A baseball commentator claims that baseballs hit with metal bats travel on average about 10 mph faster than those hit with wood bats. Based on this confidence interval, does this claim appear to be reasonable?

	n	$\bar{x}$	s
Metal	30	95.83	15.678
Wood	30	81.16	10.279

8. A biologist wants to compare the mean weights of bluegill and catfish in a lake. She captures 23 bluegill and records a mean weight of 1.25 lb with a standard deviation of

0.15 lb. She also captures 18 catfish and records a mean weight of 4.36 lb with a standard deviation of 1.02 lb. Assume that the weights of both populations are normally distributed.

   a. Based on the sample data, is it reasonable to assume that both populations have the same variance? Explain.

   b. Construct a 2-sample T-interval estimate of the difference of the two population means (do not assume the variances are equal).

9. A sample of 47 education majors at a certain university has a mean grade point average (GPA) of 3.46 with a standard deviation of 0.336 while a sample of 53 noneducation majors has a mean GPA of 3.38 with a standard deviation of 0.449 (data collected by Nicole Sempek and Allison Chrismer, 2009).

   a. Construct a 2-sample Z-interval estimate of the difference in the mean GPA of education and noneducation majors at this university.

   b. Construct a 2-sample T-interval estimate, assuming the population variances are equal. Repeat this while assuming that the variances are *not* equal.

   c. Is there much difference between these three different confidence intervals? Based on these results, does there appear to be a difference between the mean GPAs of education and noneducation majors at this university? Explain.

10. To compare two generic brands of size AA batteries, a father put one battery into a toy train and recorded the number of minutes the train ran. He then repeated this, using a total of five batteries of each brand. The data are given in the table below (data collected by Todd Searls, 2012).

Generic A	624	647	658	682	724
Generic B	600	673	690	712	922

    a. Assuming both populations are normally distributed, but that they do not have the same variance, construct a confidence interval estimate of the difference of the mean life spans of these two brands of batteries. Construct this interval by hand, using the procedures in the box on page 305.

    b. Repeat part a, but use available software to calculate the interval.

    c. Explain why there is so great a difference between the intervals in parts a and b. (**Hint**: See the note at the bottom of the box.)

11. A student counts the number of chocolate chips in cookies of two different brands, call them brands A and B, and records the data shown in the table below (data collected by Emily Brandt, 2010). Assuming both populations are normally distributed and have the same variance, use the data to construct a confidence interval estimate of the difference of the mean number of chocolate chips in the two brands at the 90% confidence level. Does there appear to be a significant difference between the brands?

Brand	Number of Chocolate Chips
A	28 19 23 19 17 23 24 25 23 27
	18 16 24 23 20 24 24 20 24 23
B	27 28 18 20 30 24 21 23 27 24
	30 23 22 25 17 18 22 25 29 30

12. A student conducts an experiment where she starts a stopwatch and asks a fellow student to tell her when 30 sec has passed. She then stops the watch and records how much time has actually passed. The times (in seconds) for 15 male and 15 female students are shown in the table below (data collected by Courtney Cale, 2011).

Males	18.6 28.7 21.6 30.5 24.0 37.8 15.5 30.6
	16.3 51.4 47.3 38.9 26.6 21.4 16.4
Females	30.6 29.7 41.5 25.0 23.3 26.5 37.0 18.8
	38.4 22.7 28.3 40.9 36.1 19.8 30.7

Construct a confidence interval estimate of the difference of the population means, assuming that both of these samples are random, both populations are normally distributed, and the two populations have the same variance. Does there appear to be a significant difference between the abilities of males and females to estimate when 30 sec has passed?

13. A statistics professor asks his students to draw a line 8 cm long without using a ruler. They each measure the line with a ruler and calculate the error in millimeters. They then draw another line, measure it, and calculate this error. Then they each report their first and second errors. The table below shows the errors for 18 different students. Assuming these 18 students are a random sample of all students, calculate a confidence interval for the mean of (first error − second error) for all students (assume these differences are normally distributed). Does the second error appear to be smaller than the first, on average?

First Error	2	2	9	4	6	3	15	2	1.5	2	0	5	1	2	2	5	10	4
Second Error	7	1	2	2	2	9	5	3	10	7	5	2	3	2	0	1	5	3

14. To test the effectiveness of a training program, the 400-m track times (in seconds) of eight athletes were recorded before and after three weeks of training, as shown in the table below (data collected by Andrew Messersmith, 2009). Construct a 95% confidence interval estimate of the mean of the differences (Before − After). Based on this result, does the training program appear to have a significant effect on track times?

Before	53.8	54.2	52.9	53.0	55.1	58.2	56.7	52.0
After	52.7	53.8	51.1	52.7	54.2	55.4	53.3	49.8

15. Do shoes increase vertical jump height? To analyze this question, 10 male athletes at a university jumped three times with shoes and three times without shoes, and the best height (in inches) for each was recorded in the table below (data collected by Todd May, 2010). Construct a 95% confidence interval estimate of the mean of the differences (With − Without). Based on this result, does it appear that shoes significantly affect vertical jump height?

With	15.5	18.5	22.5	17.5	16.5	17.5	20.5	15.5	19.5	18.5
Without	15.5	17.5	22.5	15.5	15.5	18.5	21.5	14.5	17.5	17.5

16. For the 2-sample T-interval in the case where the population variances are *not* assumed to be equal, some texts use the following number of degrees of freedom:

$$r = \text{the smaller of } (n_1 - 1) \text{ and } (n_2 - 1).$$

   a. Use this approach to construct a 95% confidence interval estimate of the difference of the population means, using the data in Example 4.8.1.

   b. Compare the confidence interval found in part a to the confidence interval found in Example 4.8.1, where we did not assume equal population variances. Which is wider? Also compare the number of degrees of freedom used to calculate each interval. Which is larger?

   c. Generalize your results from part b. Show that if r is calculated as described in the box on page 305, then r is greater than the smaller of $(n_1 - 1)$ and $(n_2 - 1)$. (**Hint:** Assume that $n_1 \geq n_2$. Start with the formula for r and multiply both the numerator and denominator by $(n_2 - 1)$; use the fact that $(a + b)^2 > a^2 + b^2$ for any $a, b > 0$; and show that $r > n_2 - 1$.)

   d. What does the result of part c mean in general about the width of a confidence interval calculated using this approach compared to the method in the box? Will a higher or lower percentage of confidence intervals calculated using this approach contain the true value of $(\mu_1 - \mu_2)$? Explain your reasoning.

17. In the procedures for the 2-sample T-interval where the population variances are not assumed to be equal, we calculate r and then round it down if it is not an integer.

   a. By examining Table C.3, state what happens to the size of the critical t-value as the degrees of freedom r increases.

   b. What would happen to the width of a resulting confidence interval if we rounded the calculated value of r up rather than down?

   c. Why do you think we specify to round down rather than up?

Directions for Exercises 18 and 19: As in previous sections, we may construct a one-sided confidence interval estimate of the difference of two proportions or means by replacing the critical value $z_{\alpha/2}$ or $t_{\alpha/2}$ with z_α or t_α (using the same degrees of freedom). Use this approach to find one-sided confidence intervals in Exercises 18 and 19.

18. A manufacturer of bolts has retooled some of its machinery to reduce the number of defective bolts rejected during inspection. Before the retooling, a sample of 175 bolts contained 19 rejects. After the retooling, a sample of 154 bolts contained 6 rejects.

   a. If p_1 and p_2 denote the population proportion of rejected bolts before and after retooling, respectively, calculate a 95% one-sided confidence interval estimate of $(p_1 - p_2)$ that gives a lower bound on this difference.

   b. If the company's goal is to reduce the percentage of defects by at least 2 percentage points, does the company appear to have met its goal? Explain.

19. An SAT tutoring service claims that they raise SAT math scores by an average of 20 points. A sample of 25 students from the area who did not use the service received a mean score of 533 with a standard deviation of 118. A sample of 27 students who did use the service received a mean score of 600 with a standard deviation of 80. If μ_1 denotes the mean score of all students who use the service and μ_2 denotes the mean score of all students who do not use the service, calculate a 95% one-sided confidence interval that gives a lower bound on the difference $(\mu_1 - \mu_2)$. Assume that both populations are normally distributed and have the same variance. Do these data support the claim?

4.9 Sample Size

The first step in choosing a sample for a statistical study is to determine the sample size. Many people believe that the larger the population size, the larger the sample size. However, we will see that as long as the population is "large," its exact size does not matter.

Consider again the scenario of estimating the proportion of voters who support Marten for state senate. In the sample of $n = 785$ voters, the sample proportion of voters who support Marten is $\hat{p} = 0.521$. At the 95% confidence level, this yielded a margin of error of $E = 0.0361$. Suppose Marten would like to obtain a more precise estimate. In other words, he would like a smaller margin of error. How many voters must be surveyed?

To begin to answer this question, note that the formula for the margin of error when estimating a population proportion is

$$E = z_{\alpha/2}\sqrt{\frac{\hat{p}(1-\hat{p})}{n}}$$

where α is the desired significance level and $\hat{p}$ is the sample proportion. Algebraically solving this equation for n yields

$$n = \frac{z_{\alpha/2}^2 \hat{p}(1-\hat{p})}{E^2}. \tag{4.7}$$

In this formula for n, E is the *desired* margin of error, which is referred to as the *maximum error of the estimate*. Also, $\hat{p}$ is an *estimate* of the population proportion p. It may be obtained in at least three different ways:

1. Use an estimate from a previous survey.
2. Use knowledge of the circumstances.
3. If it is completely unknown, use $\hat{p} = 0.5$.

Note that when defining the margin of error for estimating p in Section 4.5, we assumed that the population size was large compared to the sample size. This same assumption needs to hold for Equation (4.7) to give a valid sample size. As long as this assumption is valid, the exact size of the population does not matter. The only factors affecting the necessary sample size are the desired margin of error, the significance level, and the estimate of p. The case of a "small" population is dealt with in Exercise 14 of this section.

Example 4.9.1 Calculating a Sample Size for a Proportion How many voters must be surveyed so that the margin of error for estimating p, the proportion of people who support Marten, is no more than 0.025 at the 95% confidence level?

We want to know n so that $E = 0.025$ at the 95% confidence level. At this confidence level, $z_{\alpha/2} = 1.96$. From the previous survey, we got a sample proportion of 0.521, so we take $\hat{p} = 0.521$. Using Equation (4.7), we get

$$n = \frac{(1.96)^2(0.521)(0.479)}{(0.025)^2} \approx 1533.9.$$

Because n must be an integer and larger sample sizes yield better results, we always round n up. Therefore, we conclude that we must survey at least 1534 voters. □

In the formula for the margin of error, the term $\hat{p}(1-\hat{p})$ appears. It can be shown that $\hat{p}(1-\hat{p}) \leq \frac{1}{4}$ regardless of the value of $\hat{p}$ (see Exercise 4 of this section). This means that

$$E = z_{\alpha/2}\sqrt{\frac{\hat{p}(1-\hat{p})}{n}} \leq z_{\alpha/2}\sqrt{\frac{1/4}{n}} = \frac{z_{\alpha/2}}{2\sqrt{n}}.$$

This gives the following approximation for E:

$$E \approx \frac{z_{\alpha/2}}{2\sqrt{n}}. \tag{4.8}$$

Some statisticians use this approximation as the definition of the margin of error. When polling results are presented in the media, the given margin of error is often calculated by using this formula at the 95% confidence level.

Example 4.9.2 Determining a Sample Size from a Margin of Error A magazine article reported that 60% of surveyed voters did not vote in the last Presidential election with a "4% margin of error." Assuming this margin of error was calculated using Equation (4.8) at the 95% confidence level, find the sample size used in the survey.

The 4% margin of error means that $E = 0.04$ so that

$$0.04 = \frac{0.98}{\sqrt{n}}.$$

Solving this for n yields $n = 600.25$. Thus about 600 voters were surveyed. □

We can calculate the necessary sample size for estimating a population mean in a similar fashion. The margin of error for estimating a mean μ when the population variance σ^2 is known is

$$E = z_{\alpha/2}\frac{\sigma}{\sqrt{n}}.$$

Solving this for n yields

$$n = \frac{z_{\alpha/2}^2 \sigma^2}{E^2}. \tag{4.9}$$

In this formula, σ^2 is an *estimate* of the population variance. If no reasonable estimate of σ^2 is known, it may be necessary to take a preliminary sample to try to estimate it. Another simple estimate is

$$\sigma \approx \frac{\text{(largest possible data value)} - \text{(smallest possible data value)}}{4}. \qquad (4.10)$$

The idea behind this approximation is that in a set of data, "most" of the values lie within 2 standard deviations of the mean (remember the empirical rule). In other words, the smallest possible value is about 2σ below the mean, and the largest possible value is about 2σ above. So the difference between the smallest and largest values is about 4σ.

The definition of the margin of error for estimating μ was based on the central limit theorem, which requires the members of the sample to be chosen independently. As long as the population size is large relative to the sample size, this assumption is reasonable. This assumption needs to hold for Equation (4.9) to give a valid sample size. As long as this assumption is valid, the exact size of the population does not matter. The only factors affecting the necessary sample size are the desired margin of error, the significance level, and the estimate of σ^2.

Example 4.9.3 Estimating IQ Scores A student would like to estimate the mean IQ score of all statistics professors. How many professors must be surveyed so that the margin of error is no more than 3 units at the 90% confidence level?

The desired margin of error is $E = 3$ and $z_{\alpha/2} = 1.645$. Because IQ scores in the general population are normalized to have a variance of 15^2, we will assume that in the population of statistics professors the variance is also 15^2. Thus using Equation (4.9), we have

$$n = \frac{(1.645)^2 (15)^2}{3^2} \approx 67.7.$$

Therefore, we must survey at least 68 professors. □

In some cases, we may want the margin of error to be less than some fraction of σ. That is, we want $E = m\sigma$ where $m > 0$. To find the desired sample size, we need

$$E = m\sigma = z_{\alpha/2} \frac{\sigma}{\sqrt{n}} \quad \Rightarrow \quad n = \left(\frac{z_{\alpha/2}}{m}\right)^2.$$

For instance, if we want the margin of error to be less than one-half of σ at the 95% confidence level, we have

$$n = \left(\frac{1.96}{0.5}\right)^2 \approx 15.4.$$

So we would need a sample of at least 16.

When we estimate a population variance σ^2, determining the sample size is a bit more difficult. Note that we did not define a margin of error for a confidence interval estimate of σ^2. Instead, we find the sample size so that the *width* of the confidence interval is less than some fraction of σ^2.

The confidence interval for σ^2 is

$$\left[\frac{(n-1)s^2}{b}, \frac{(n-1)s^2}{a}\right]$$

where s^2 is the variance of a sample of size n taken from a normally distributed population, and $a = \chi^2_{1-\alpha/2}(n-1)$ and $b = \chi^2_{\alpha/2}(n-1)$ are critical χ^2 values. The width of this interval is

$$\text{Width} = \frac{(n-1)s^2}{a} - \frac{(n-1)s^2}{b} = (n-1)s^2\left(\frac{1}{a} - \frac{1}{b}\right).$$

Now, $s^2 \approx \sigma^2$ so that

$$\text{Width} \approx (n-1)\sigma^2\left(\frac{1}{a} - \frac{1}{b}\right)$$

(technically the expected value of a sample variance equals σ^2 so $E(\text{width})$ equals this quantity on the right). So for this width to be less than $m\sigma^2$, we need

$$(n-1)\left(\frac{1}{a} - \frac{1}{b}\right) < m.$$

Because a and b both depend on the significance level α *and* the value of n, solving this inequality for n is not easy. Instead, we use trial and error by selecting different values of n, finding the corresponding values of a and b from Table C.4, and then choosing the smallest value of n for which this inequality holds.

For $m = 0.70$ at the 95% confidence interval, the calculations in Table 4.10 show that the sample size must be at least 71.

n	$a = \chi^2_{0.975}(n-1)$	$b = \chi^2_{0.025}(n-1)$	$(n-1)\left(\frac{1}{a} - \frac{1}{b}\right)$
51	34.76	71.42	0.845
61	43.19	83.30	0.762
71	51.74	95.02	0.699
81	60.39	106.6	0.649

Table 4.10

Exercises

1. Derive the equation $n = z_{\alpha/2}^2 \hat{p}(1-\hat{p})/E^2$ from the equation $E = z_{\alpha/2}\sqrt{\hat{p}(1-\hat{p})/n}$.

2. A banker would like to estimate the proportion of all college students who use credit cards. Based on her experience, she believes that the proportion is approximately 0.75, but she wants to do a survey to verify this. How many students must be surveyed so that the margin of error is no more than 0.015 at the 90% confidence level?

3. A political pollster wants to estimate the proportion of voters who support a candidate for President. How many voters must be surveyed so that the margin of error is no more than 0.04 at the 95% confidence level?

4. Show that $\hat{p}(1-\hat{p}) \leq \frac{1}{4}$ for all values of $\hat{p}$ between 0 and 1.

5. A biologist has sampled 15 large-mouth bass from a lake and calculated a sample mean length of $\bar{x} = 12.5$ in with a standard deviation of $s = 1.3$ in. How many *more* fish must be sampled so that a 90% confidence interval estimate of the mean length of all fish in the lake has a margin of error of less than 0.25 in?

6. A manufacturer of batteries has updated its manufacturing process and wants to determine if the changes have increased the mean life span of its size AA batteries. Under the old process, the mean life span in a pocket radio was $\mu = 8.9$ hr with a standard deviation of $\sigma = 0.2$ hr.

   a. How many new batteries must be sampled so that the margin of error for the mean life span under the new process is no more than 0.02 hr at the 95% confidence level? (Assume that the change in the standard deviation is minimal.)

   b. Suppose a sample of $n = 400$ new batteries was tested, yielding a sample mean life span of $\bar{x} = 9.05$ hr with a sample standard deviation of $s = 0.18$ hr. Calculate a 95% confidence interval estimate of the mean life span of the new batteries. Does the new process appear to produce longer-lasting batteries?

7. A student would like to estimate the mean GPA of all education majors at a large university. She believes that most of the GPAs will lie between 2.5 and 3.8.

   a. Use the given information and Equation (4.10) to estimate the standard deviation σ of the GPAs of education majors.

   b. Use the estimate of σ in part a to find the sample size necessary so that a confidence interval estimate of the mean GPA of all education majors has a margin of error of less than 0.1 at the 99% confidence level.

8. Find the necessary sample size to estimate a population mean μ so that the margin of error is less than 10% of the population standard deviation σ at the 90% confidence level.

9. In Example 4.9.3, we assumed that the variance of the IQ scores in the population of statistics professors was the same as in the general population, 15^2. How reasonable of an assumption do you think this is? Do you think that the variance in the population of statistics professors is higher or lower than in that the general population? Does this mean that the calculated sample size in this example is too large or too small? Explain.

10. Assume we have a large population and we want to estimate a population proportion with a margin of error of $E = 0.02$.

   a. Find the necessary sample size at the 80%, 90%, 95%, and 99% confidence levels (use $\hat{p} = 0.5$).

   b. Generalize your results in part a. What happens to the necessary sample size as the confidence level increases?

   c. Now suppose we use a 95% confidence level, but change the desired margin of error. Find the necessary sample size for $E = 0.015, 0.01$, and 0.005.

   d. Generalize your results in part d. What happens to the necessary sample size as the margin of error decreases?

11. The owners of a small coffee shop want to start selling a gourmet brand of pastry. To determine if this would be profitable, they would like to estimate the proportion of their customer base that would be interested in purchasing the pastries. They would like to have a margin of error of $E = 0.01$ at the 99% confidence level.

   a. Find the necessary sample size.

   b. The owners decide that the sample size in part a is far too large. Give at least two recommendations as to how they could change their requirements to lower the necessary sample size.

12. Suppose we have a large population and we want to estimate a population proportion p with a margin of error of $E = 0.1$ at the 95% confidence level, but we do know know an estimate of p.

   a. Find the value of $\hat{p}$ so that the sample size n given by Equation (4.7) is maximized.

   b. Generalize your results in part a. For a fixed margin of error and confidence level, find the value of $\hat{p}$ so that the sample size n given by Equation (4.7) is maximized.

c. Explain why your results in part b mean that if no estimate of p is known, we use $\hat{p} = 0.5$.

13. Suppose we want to estimate a population variance σ^2 so that the width of the corresponding confidence interval is no more than $1.52\sigma^2$ at the 95% confidence level. Find the necessary sample size.

14. When a population proportion is estimated, the formula for the margin of error E derived in Section 4.5 and the resulting formula for n in Equation (4.7) are based on the assumption that n is less than 5% of the population size so that the number of successes in the sample, X, is approximately binomial. When n is greater than 5% of the population size (or the population is "small"), X is *not* approximately binomial. Rather, it has a hypergeometric distribution. In this case, an appropriate formula for the margin of error is

$$E = z_{\alpha/2}\sqrt{\frac{\hat{p}(1-\hat{p})}{n}\left(\frac{N-n}{N-1}\right)} \qquad (4.11)$$

and the formula for the sample size is

$$n = \frac{k}{1 + (k-1)/N} \qquad (4.12)$$

where N is the population size and

$$k = \frac{z_{\alpha/2}^2 \hat{p}(1-\hat{p})}{E^2}.$$

Note that k is simply the sample size in the case of a large population as given in Equation (4.7). In the formula for k, $\hat{p}$ is an estimate of p and E is the desired margin of error.

a. In a survey of $n = 100$ students on a campus of $N = 1000$ students, 65 support the construction of a new gym. Use Equation (4.11) to construct a 90% confidence interval estimate of the proportion of all students on the campus who support the construction.

b. In an office building containing $N = 800$ employees, management is considering moving the vending machines and wants to estimate the proportion of employees who support the move. If they want a margin of error of $E = 0.05$ at the 90% confidence level, use Equation (4.12) to find the necessary sample size (use $\hat{p} = 0.5$).

c. Repeat part b, except use $N = 2000, 5000, 10{,}000, 100{,}000$, and then $1{,}000{,}000$. Compare the value of n to k as N gets larger.

d. If we were to use Equation (4.7) to find the sample size in a small population, would we get an over- or underestimate of the necessary sample size?

e. The term $\left(\frac{N-n}{N-1}\right)$ in Equation (4.11) can be thought of as a correction for a "finite population." If n has a fixed value, find the limit of this correction term as $N \to \infty$. What does this imply about the necessity of this correction term as the population gets larger?

15. In this exercise, we derive a formula to find the necessary sample sizes so that a confidence interval for a *difference* of two population proportions has a desired margin of error. To simplify things, we assume that both sample sizes are equal (call the common value n). Then the margin of error as given in Section 4.8 is

$$E = z_{\alpha/2}\sqrt{\frac{\hat{p}_1(1-\hat{p}_1)}{n} + \frac{\hat{p}_2(1-\hat{p}_2)}{n}}.$$

a. Solve this equation for n.

b. Let E denote the desired margin of error and $\hat{p}_1$ and $\hat{p}_2$ denote estimates of the population proportions (taken to be 0.5 if no estimate is known). Use the result of part a to find the sample size n so that the margin of error is no more than 0.04 at the 95% confidence level.

16. In this exercise, we derive a formula to find the necessary sample sizes so that a confidence interval for a *difference* of two population means has a desired margin of error in the case where both population variances are known. As in the previous exercise, we assume that both sample sizes are equal so that the margin of error as given in Section 4.8 is

$$E = z_{\alpha/2}\sqrt{\frac{\sigma_1^2}{n} + \frac{\sigma_2^2}{n}}$$

where σ_1^2 and σ_2^2 are the (known) population variances.

a. Solve this equation for n.

b. Let E denote the desired margin of error. Use the result of part a to find the sample size n so that the margin of error is no more than 0.75 at the 99% confidence level where the population variances are $\sigma_1^2 = 1.5$ and $\sigma_2^2 = 5.6$.

4.10 Assessing Normality

Many of the formulas for calculating confidence intervals we have seen so far have the requirement that the population be normally distributed. Given a sample of values from a population, one way to determine if the population is normally distributed is to construct a histogram of the data. If the histogram has an approximate bell curve shape, then we have an indication that the population is normally distributed.

This is fairly easy to do; however, judging whether the histogram has a bell curve shape is somewhat subjective, especially if the sample size is very small. In this section, we present a slightly more objective way of determining if a population is normally distributed through the construction of a graph called a *normal quantile plot*, also called a *probability plot* or *Q-Q plot*.

We motivate the basic idea behind this type of graph with the following example.

Example 4.10.1 **Thermometers** Suppose that a manufacturer of thermometers is interested in testing the accuracy of its product. Particularly, they are interested in the readings given by the thermometers at the freezing point of water (measured in degrees Celsius). Ideally each would read exactly 0°C, but the thermometers are not perfect; so some give positive readings and others give negative readings. A sample of 10 thermometers was randomly selected, and their readings at the freezing point of water were recorded in the first row of Table 4.11.

Reading (z)	−2.21	−1.33	−0.81	−0.55	−0.29	0.04	0.35	0.48	0.85	1.27
p	—	0.1	0.2	0.3	0.4	0.5	0.6	0.7	0.8	0.9
π_p	—	−1.28	−0.84	−0.52	−0.25	0	0.25	0.52	0.84	1.28

Table 4.11

We want to determine if these readings come from a population that is normally distributed. Specifically, let the random variable Z denote the reading of a randomly selected thermometer. We want to determine if Z is $N(0, 1)$. For each data value z, define

$$p = \text{the proportion of the sample that is less than } z.$$

The second data value, -1.33, is greater than one out of the 10 data values, so $p = 0.1$. This means that -1.33 is the 10th percentile of the data. This data value is an approximation of the 10th percentile of Z. The other values of p are shown in the second row of Table 4.11.

Next, for each p, we find the 100pth percentile of Z if it were indeed $N(0, 1)$, denoted π_p. This is done by searching Table C.1 for a value of z such that $\Phi(z) = p$ (in more technical terms, we are evaluating $\Phi^{-1}(p)$). From the table we see that $\Phi(-1.28) = 0.1003$, thus $\pi_{0.1} \approx -1.28$. The other values of π_p are shown in the third row of Table 4.11.

Now we compare the first and third rows of Table 4.11. The first row contains estimates of the percentiles of Z, and the third row contains the actual percentiles if Z were $N(0,1)$. Note that these estimates are very close to the actual values. This is an indication that Z is $N(0,1)$.

Lastly, we draw a graph by plotting π_p versus the reading, as in Figure 4.14. Because these values are approximately the same, the points should form a straight line through the origin. This is exactly what we see in the graph. This graph is called the *normal quantile plot* of the data. □

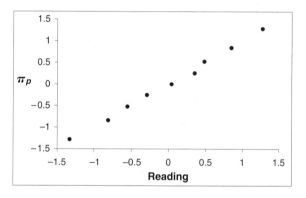

Figure 4.14

Usually, quantile plots are drawn with software. Different software use different algorithms for constructing the plots. Some are more complex than others, but the basic interpretation of all of them is the same: a straight-line pattern indicates that the population is normally distributed. The example above illustrates the idea of the quantile plot in the case where the population is $N(0,1)$. In the box below, we describe one simple algorithm for constructing a quantile plot in more-general cases.

Constructing a Normal Quantile Plot

Purpose: To determine if a set of data is from a normally distributed population.
1. Arrange the data values in increasing order: $x_1 \leq x_2 \leq \cdots \leq x_n$.
2. For each $k = 1, 2, \ldots n$, define
$$p_k = \frac{k}{n+1}.$$

3. Calculate $z_k = \Phi^{-1}(p_k)$ for each k where Φ is the standard normal cdf.
4. Plot the point (x_k, z_k).
5. If the points form a straight-line pattern, then conclude that the population appears to be normal. If the points do not form a straight line or exhibit some other type of nonlinear pattern, then conclude that the population is *not* normal.

Example 4.10.2 Normally Distributed Temperatures The second row of Table 4.12 gives the average daily temperatures in the month of November for the city of Lincoln, Nebraska, for nine different years (data collected by Brandon Metcalf, 2009). Determine if the population of all such temperatures is normally distributed.

k	1	2	3	4	5	6	7	8	9
x_k	32.6	33.1	36.0	36.8	37.7	38.1	42.3	43.2	46.5
p_k	0.1	0.2	0.3	0.4	0.5	0.6	0.7	0.8	0.9
z_k	−1.28	−0.84	−0.52	−0.25	0.00	0.25	0.52	0.84	1.28

Table 4.12

We construct a normal quantile plot of these data, using the algorithm in the box that begins on page 325. The calculations are shown in the third and fourth rows of the box. The quantile plot is shown in Figure 4.15. Note that the points form a straight-line pattern, so we conclude that the population appears to be normal.

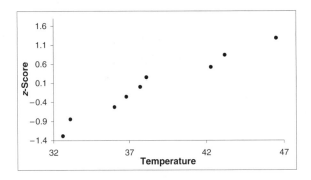

Figure 4.15

□

To better visualize the straight-line pattern that the dots in a quantile plot should resemble, if the population is normally distributed, then we add a straight line to the plot with the following algorithm.

1. Calculate the sample mean and standard deviation of the data, $\bar{x}$ and s.
2. For each k, calculate the following quantity:

$$y_k = \frac{x_k - \bar{x}}{s}$$

(these are simply the z-scores of the data values if the population is $N(\bar{x}, s^2)$).
3. Plot the points (x_k, y_k) on the quantile plot and connect them with a straight line.
4. If the points from the quantile plot lie close to this straight line, then conclude that the population appears to be normal.

The resulting line for the data in Example 4.10.2 is shown in Figure 4.16. Note that the points lie close to this line, so again we conclude that the population is normally distributed. Also note that because these points always form a straight line (see Exercise 1 in this section), we do not really need to calculate y_k for every k. We simply need to calculate two values, plot the points, and connect them with a straight line.

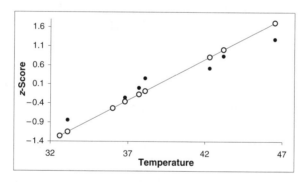

Figure 4.16

Example 4.10.3 **Nonlinear Pattern** Determine if the sample data shown in Table 4.13 come from a population that is normally distributed.

5.88	6.59	7.12	8.56	10.51	12.36	13.12	13.98	14.93
6.29	6.59	7.92	9.07	11.93	12.48	13.68	14.31	
6.41	7.10	8.21	9.75	12.14	12.75	13.85	14.88	

Table 4.13

The quantile plot for these data is shown on the left side of Figure 4.17. One could argue that the points lie close to the straight line. However, there is a definite nonlinear pattern to the points. Thus we conclude that the population is *not* normally distributed. A relative frequency histogram of the data is shown on the right side of the figure. Judging from this, it appears that the population may be uniformly distributed.

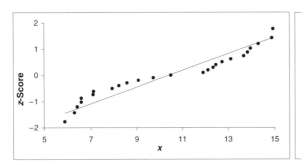
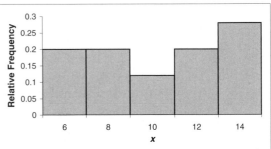

Figure 4.17

□

We conclude this section with the following observation. The central limit theorem says, informally, that the sum of many independent identically distributed random variables is approximately normally distributed. We generalize this observation in the "fuzzy central limit theorem."

If the population is influenced by many small, random, unrelated effects, then

The population may be normally distributed.

Using the fuzzy central limit theorem, we may reason that a population is normally distributed even in the absence of data. Consider the weights of people. Weights are influenced by diet, heredity, socioeconomic status, and many other factors. Thus it may be reasonable to simply assume that weights are normally distributed.

Software Calculations

To construct a normal quantile plot:

Minitab: Enter the sample data in a column. Then select **Graphs** → **Probability Plot**. Choose **Single**. Under **Variables**, enter the name of the column containing the data. Minitab uses a slightly different algorithm for constructing the plot than the one we present, but the interpretation of it is the same.

R: Enter the sample data $\{x_1, x_2, \ldots, x_n\}$ into a vector with the syntax x<-c($x_1, x_2, \ldots, x_n$). Then enter the code
> qqnorm(x, datax=TRUE)

The straight line that the points should follow if the data were normally distributed is plotted with the code
> qqline(x, datax=TRUE)

TI-83/84 Calculators: Enter the sample data by pressing **STAT** → **EDIT** → **1:Edit**. Then press **2ND** → **STAT PLOT** and select the last **Type** of graph. Next to **Xlist** enter the name of the list containing the data. Set the **Freq** to 1. Then press **ZOOM** → **9:ZoomStat**. The calculator uses a slightly different algorithm for constructing the plot than the one we present, but the interpretation of it is the same.

Exercises

1. Explain why the points (x_k, y_k) where $y_k = (x_k - \bar{x})/s$ will always form a straight line regardless of the distribution of the population. (**Hint**: For a given set of data, $\bar{x}$ and s are constants.)

2. The table below shows the scores on a statistics exam for 11 randomly selected students. Draw a normal quantile plot of these data. Based on the graph, does it appear reasonable to assume that these data come from a normally distributed population?

52.3	56.8	57.8	60.7	70.9	89.9	91.7	92.6	92.7	92.8	93.9

3. A professor performs an experiment by asking her students to draw a line 6 cm long without using a ruler. The students then measure the length of the lines and report the error. The errors (in millimeters) are shown in the table below. Determine if these data come from a population that is normally distributed.

2	2	9	4	6	3	15	2	1.5	2	0	5	1	2	2	5	10	4

4. In Example 4.7.2, we used the data from Table 4.8 to construct a confidence interval estimate of the variance in the proportion of butterfat in batches of butter. One of the requirements for the formula for this confidence interval is that the population be normally distributed. Use the data in Table 4.8 to determine if the population is normally distributed. Does the requirement of normality appear to be met?

5. To compare one name brand of size AA batteries to two generic brands, a father put one battery into a toy train and recorded the number of minutes the train ran. He then repeated this using a total of five batteries of each brand. The data are given in the table below (data collected by Todd Searls, 2012). Draw a normal quantile plot of the data from each brand. Based on the graphs, does it appear reasonable to assume that any of these populations of batteries are normally distributed?

Name Brand	850	1002	1007	1088	1100
Generic A	624	647	658	682	724
Generic B	600	673	690	712	922

6. In Exercise 11 of Section 4.8, we constructed a confidence interval for the difference of the mean number of chocolate chips in two different brands of cookies. Construct a normal quantile plot for each sample in the table given in this exercise. Does the assumption that both populations are normally distributed appear to be reasonable?

7. In Exercise 12 of Section 4.8, we constructed a confidence interval for the difference of mean estimated elapsed times of males and females, assuming that both populations were normally distributed. Construct a normal quantile plot for each sample of data in the table given in this exercise, and determine if this assumption of normality was reasonable.

8. In Exercise 13 of Section 4.8, we constructed a confidence interval for the mean of (first error − second error) for all students. Construct a normal quantile plot for the differences in the table given in this exercise. Is it reasonable to assume that these differences are normally distributed as assumed in the exercise?

9. A normal quantile plot can be used to determine if a population has a normal distribution. In some situations, we may want to determine if a population has a distribution other than normal. For instance, we might claim that a population has an exponential distribution. In these cases, we can use the following algorithm to generate a plot similar to a normal quantile plot:

1. Randomly collect sample data and arrange them in increasing order: $x_1 \leq x_2 \leq \cdots \leq x_n$.

2. For each $k = 1, 2, \ldots n$, define $p_k = k/(n+1)$. Then x_k is an estimate of the $(100p_k)$th percentile of the population.

3. Let $f(x)$ be the claimed pdf of the population, $F(x)$ be the corresponding cdf, and F^{-1} be the inverse cdf. Calculate $\pi_k = F^{-1}(p_k)$ for each k. Then π_k is the $(100p_k)$th percentile of the population if the population does have the claimed distribution.

4. Plot the points (x_k, π_k). If the points lie close to a straight line through the origin with slope 1, then conclude that the population appears to have the claimed distribution.

The resulting plot is called a *probability plot* or a *Q-Q plot*. In this exercise, we will generate a Q-Q plot to determine if a population has an exponential distribution.

a. The pdf for an exponential distribution is $f(x) = \lambda e^{-\lambda x}$, and its cdf is $F(x) = 1 - e^{-\lambda x}$ for $x > 0$. Find $F^{-1}(y)$ by letting $y = 1 - e^{-\lambda x}$ and solving for x.

b. The table below gives 15 data values from a population that is believed to have an exponential distribution with parameter $\lambda = \frac{1}{2}$. Use the algorithm to generate a Q-Q plot of the data.

0.1	0.5	1.7	2.2	3.4
0.2	0.7	1.7	2.5	4.5
0.3	1.2	2.1	2.9	5.5

c. On the plot, draw a straight line through the origin with slope 1. Do the points lie close to this line? Does the population appear to have an exponential distribution?

CHAPTER 5

Hypothesis Testing

Chapter Objectives

- Introduce hypothesis testing
- Stress the connection between random variables and hypothesis testing
- Discuss how critical values are used and how P-values are calculated
- Introduce tests for claims about a proportion or mean
- Introduce tests for comparing two parameters
- Introduce chi-square tests
- Introduce the basics of ANOVA
- Stress the requirements of the data for each type of test

CHAPTER 5 Hypothesis Testing

5.1 Introduction

In the media, we may hear statements such as

"More than 65% of voters support this bill," or

"The average home value in the United States is less than $150,000."

To determine if this first statement is true, we would have survey *every* voter and calculate the proportion who support the bill. For the second statement, we would have to find the value of *every* home in the United States and calculate the mean. Both of these endeavors are virtually impossible, so we may never know for certain if either statement is true.

These statements are made about a population proportion and a population mean, respectively. We can estimate these parameters by collecting sample data and calculating a sample proportion and a sample mean. If the sample proportion is greater than 0.65 and the sample mean is less than 150,000, then we are tempted to say that the statements are true. However, we must remember that sample statistics are *estimates* of population parameters. Just because a sample proportion is greater than 0.65, we cannot conclude for certain that the population proportion is greater than 0.65. In this case, the data appear to support the claim, but we need to determine how strong the support is. This is the motivation behind the idea of *hypothesis testing*.

> **Definition 5.1.1** *Hypothesis testing* is a formal approach for determining if data from a sample support a claim about a population.

There are many different types of claims about populations. A claim may be about the value of a population parameter or the distribution of a population, or it may be a statement comparing two different populations. Regardless of the type of claim, hypothesis testing consists of five basic steps:

1. State the null and alternative hypotheses.
2. Calculate the test statistic.
3. Find the critical value (or calculate the *P*-value).
4. State the technical conclusion.
5. State the final conclusion.

In this section, we present an overview of these steps. In subsequent sections, we discuss specific types of claims in greater detail.

The Process

To illustrate the process of hypothesis testing, consider this claim recently made by two students regarding fellow students at their university:

> More than one-half of the students are from out of state.

To test this claim, the two students surveyed 95 fellow students and found that 59 are from out of state (data collected by Creighton Pearse and Marcus Brees, 2011). In these data, the sample proportion is

$$\hat{p} = \frac{59}{95} \approx 0.621.$$

This number is called the *sample statistic* for the hypothesis test. Because this proportion is greater than 0.5, the data do appear to support the claim. Informally, hypothesis testing allows us to determine just how strong the support is.

Step 1: State the null and alternative hypotheses.

The *null* and *alternative hypotheses*, denoted H_0 and H_1, respectively, are two competing mathematical statements. The first step in stating the hypotheses is to define the population parameter about which the claim is being made:

p = the proportion of *all* students at the university who are from out of state.

We stress the word *all* in this definition because the claim is being made about the population of all students at the university, not just those in the sample. The null and alternative hypotheses are statements about the equality or inequality of the value of this parameter. Examples of these statements include

$$p = 0.50, \quad p > 0.25, \quad p \leq 0.75, \quad \text{and} \quad p \neq 0.10.$$

In the advanced theory of hypothesis testing, H_0 and H_1 could take any of these forms. In common applications of hypothesis testing, H_0 takes the form

$$H_0: p = p_0$$

where p_0 is some given number. The null hypothesis can be thought of as a statement of the status quo. It is a statement about the value of the parameter we would expect to be true if there were no data indicating otherwise. The alternative hypothesis is a statement about the value of the parameter suggested by the data.

In this example, we might expect that $p = 0.50$, and the data indicate that $p > 0.50$. Therefore, we test the hypotheses

$$H_0: p = 0.50 \qquad H_1: p > 0.50.$$

Step 2: Calculate the test statistic.

The test statistic is a number calculated using the sample statistic(s) and the number in the null hypothesis. This number is an observed value of a random variable with a certain type of distribution. Each type of hypothesis test has a different method for calculating the test statistic, the details of which we explain further in later sections. For a claim about a single population proportion such as this, the test statistic has a standard normal distribution (so it is a z-score) and is calculated using the formula

$$z = \frac{\hat{p} - p_0}{\sqrt{p_0(1 - p_0)/n}}$$

where $\hat{p}$ is the sample proportion, p_0 is the number used in H_0, and n is the sample size. In other types of claims, the test statistics have other types of distributions (such as the Student-t or the χ^2).

In this claim, $p_0 = 0.5$, $\hat{p} = 0.621$, and $n = 95$. Thus the test statistic is

$$z = \frac{\hat{p} - p_0}{\sqrt{p_0(1 - p_0)/n}} = \frac{0.621 - 0.5}{\sqrt{0.5(1 - 0.5)/95}} \approx 2.36.$$

Step 3: Find the critical value.

To interpret the test statistic, we compare the test statistic to an "appropriate" critical value. The critical value comes from the same distribution as the test statistic. To find the appropriate critical value, we first choose a significance level α (or a confidence level $(1 - \alpha)$). For a claim about a single population proportion, Table 5.1 summarizes how to find the appropriate critical value.

In this example, we choose to use a 95% confidence level, so $\alpha = 0.05$. From Table C.2, we see the critical value is $z_{0.05} = 1.645$.

Claim About a Single Population Proportion

Form of H_1	Critical Value	Critical Region
$p > p_0$	z_α	$[z_\alpha, \infty)$
$p < p_0$	z_α	$(-\infty, -z_\alpha]$
$p \neq p_0$	$z_{\alpha/2}$	$(-\infty, -z_{\alpha/2}] \cup [z_{\alpha/2}, \infty)$

Table 5.1

Step 4: State the technical conclusion.

The technical conclusion is one of these two statements:

$$\text{Reject } H_0. \quad \text{or} \quad \text{Do not reject } H_0.$$

We reject H_0 if the test statistic falls into the critical region, and we do not reject H_0 otherwise. Table 5.1 summarizes the critical region for a test about a single population proportion. In this example, the critical region is

$$[1.645, \infty).$$

Because $z = 2.36$, the test statistic does fall in the critical region. Thus we come to the technical conclusion: Reject H_0.

Step 5: State the final conclusion.

The technical conclusion does not mean much to anyone other than the person doing the analysis. So we want to state the final conclusion in terms that anyone can understand.

The wording of the final conclusion depends on the statement of the original claim and the technical conclusion. In mathematical notation, claims tested in this text take one of the following forms:

$$p = p_0, \quad p > p_0, \quad p < p_0, \quad \text{or} \quad p \neq p_0.$$

The first of these forms is said to "contain equality" while the other three do not. Table 5.2 gives suggested wordings of the final conclusion.

In this example, the claim in mathematical notation is $p > 0.50$. Therefore we come to the final conclusion: The data support the claim. Thus the sample evidence is strong enough to indeed support the claim.

	Does the Claim Contain Equality?	
Technical Conclusion	Yes	No
Reject H_0	There is sufficient evidence to reject the claim.	The data support the claim.
Do Not Reject H_0	There is not sufficient evidence to reject the claim.	The data do not support the claim.

Table 5.2

Example 5.1.1 Polling Results A politician claims that 75% of voters support his clean-water initiative. A random sample of 350 voters contains 240 who support the initiative. Use these data to test the claim at the 90% confidence level.

We use the five steps illustrated above.

1. The parameter about which the claim is made is

$$p = \text{the proportion of all voters who support the initiative.}$$

The claim in mathematical notation is $p = 0.75$. We take this to be the status quo, so it becomes H_0. Because the claim is of equality and any sample proportion much higher or lower than 0.75 would appear to contradict the status quo, we choose H_1 to be $p \neq 0.75$. So we test the hypotheses

$$H_0\text{: } p = 0.75, \qquad H_1\text{: } p \neq 0.75.$$

2. To calculate the test statistic, we note that $\hat{p} = \frac{240}{350} \approx 0.686$, $p_0 = 0.75$, and $n = 350$ so that

$$z = \frac{\hat{p} - p_0}{\sqrt{p_0(1-p_0)/n}} = \frac{0.686 - 0.75}{\sqrt{0.75(1-0.75)/350}} = -2.77.$$

3. We are told to use the 90% confidence level, so $\alpha = 0.10$. The appropriate critical value is $z_{0.10/2} = z_{0.05} = 1.645$.
4. The critical region is $(-\infty, -1.645] \cup [1.645, \infty)$. Because $z = -2.77$ lies in this region, the technical conclusion is to reject H_0.
5. Because the original claim does contain equality, the final conclusion is that there is sufficient evidence to reject the claim. □

The *P*-Value Method

In the examples above, we came to the technical conclusion by comparing the test statistic to a critical value. This approach is called the *traditional* method. A different but equivalent approach for coming to the technical conclusion is called the *P-value* method. In this method, we find the area under the appropriate density curve to either the left or the right of the test statistic. The appropriate density curve is that of the distribution from which the critical value comes. For a test about a single population proportion, the appropriate density curve is the standard normal density curve. Table 5.3 summarizes how to find the *P*-value, and Figure 5.1 illustrates the different types of tests.

Form of H_1	*P*-Value	Type of Test
$p > p_0$	Area to the right of z	Right-tail
$p < p_0$	Area to the left of z	Left-tail
$p \neq p_0$	Twice the area to the left or right of z, whichever area is smaller	Two-tail

Table 5.3

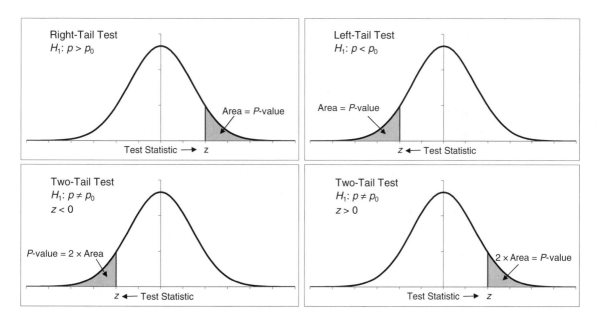

Figure 5.1

In the *P*-value method, we use these criteria for coming to the technical conclusion:

1. Reject H_0 if *P*-value $\leq \alpha$.
2. Do not reject H_0 if *P*-value $> \alpha$.

The final conclusion is worded the same in both the traditional and *P*-value methods.

Example 5.1.2 Calculating *P*-Values In the claim about the proportion of students who are from out of state, the alternative hypothesis is H_1: $p > 0.5$. Thus the *P*-value is the area under the standard normal density curve to the right of the test statistic $z = 2.36$. Using Table C.1, we see the *P*-value is $1 - 0.9909 = 0.0091$. Because this *P*-value is less than the significance level $\alpha = 0.05$, we reject H_0. This is the same conclusion reached by using the traditional method.

In the claim about the voters who support the clean water initiative, the alternative hypothesis is H_1: $p \neq 0.75$, so this is a two-tail test. Because the test statistic is $z = -2.77$, which is negative, the *P*-value is twice the area to the left of -2.77. This area is 0.0028, so the *P*-value is $2(0.0028) = 0.0056$. This *P*-value is less than the significance level $\alpha = 0.10$, so we reject H_0 as we did with the traditional method. □

Exercises

1. A teacher claims that the mean weight of all kindergarten students is less than 50 lb. She weighs 500 kindergarten students and calculates a sample mean of $\bar{x} = 48.6$. She then concludes that this proves her claim. Explain what is wrong with this conclusion.

2. A student purchases a 48-oz bag of candy and claims that the mean weight of all candies in this particular bag is greater than 0.05 oz. She weighs all 1000 candies in the bag, calculates a mean weight of 0.048 oz, and concludes that this disproves her claim. Is this the correct conclusion? Explain.

3. A politician claims that more than one-half of her constituents support her for reelection. A poll of 850 constituents finds that 45% support her. Explain why it would be pointless to do a hypothesis test of this claim using these data.

4. In each of the following claims, define the parameter about which the claim is being made, and state the null and alternative hypotheses.

   a. More than 75% of citizens of the United States own a home.

   b. Less than 25% of introduction to statistics students spend more than 1 hr studying for their tests.

   c. The mean IQ score of professors is greater than 120.

d. The standard deviation of scores on the SAT is 100.

e. The mean weight of 12-oz packages of shredded cheddar cheese is greater than 12 oz.

5. Given below are three different alternative hypotheses for claims about a single population proportion along with the value of the test statistic z and the significance level α. For each, find the appropriate critical value, state the critical region, calculate the P-value, and state the technical conclusion.

   a. H_1: $p < 0.45$, $z = -1.24$, $\alpha = 0.10$
   b. H_1: $p \neq 0.95$, $z = 2.67$, $\alpha = 0.05$
   c. H_1: $p > 0.12$, $z = 0.87$, $\alpha = 0.01$

6. Given below are four different claims about a population parameter and the technical conclusion from the resulting hypothesis test. Use Table 5.2 to state the final conclusion.

   a. Claim: $p = 0.15$, technical conclusion: reject H_0
   b. Claim: $\mu < 3.5$, technical conclusion: do not reject H_0
   c. Claim: $p > 0.10$, technical conclusion: reject H_0

7. Use the data in Example 5.1.1 to test the claim that less than 75% of voters support the clean-air initiative using the P-value method at the 90% confidence level. Does the final conclusion for this claim contradict the final conclusion for the original claim in the example? Explain why or why not.

8. Explain your answers to the following questions about testing a claim about a single population proportion.

   a. If the test statistic is negative, which is larger, $\hat{p}$ or p_0?
   b. If the test statistic is 0, which is larger, $\hat{p}$ or p_0?
   c. If you are doing a left-tail test and get a P-value greater than 0.5, which is larger, $\hat{p}$ or p_0?
   d. If you are doing a two-tail test and get a P-value greater than 0.5, what are the possible values of the test statistic?
   e. Is it possible to get a P-value greater than 1?

f. When we calculate P-values using Table C.1, the answers always have exactly four decimal places. When a piece of software calculates P-values, it uses an algorithm that can give more than four decimal places. If the P-value is extremely small, it is often given in scientific notation. If someone claims to have gotten a P-value of 2.67 using a calculator, what is the mistake he or she most likely made? What is the likely value of the P-value rounded to four decimal places?

9. Consider a claim about a single population proportion where the null hypothesis is H_0: $p = 0.25$. Suppose a sample of size $n = 150$ is surveyed.

   a. Calculate the value of the test statistic z in the cases where the sample proportion is $\hat{p} = 0.26, 0.23, 0.35, 0.12$, and 0.25.

   b. Generalize the results in part a. If $\hat{p} > p_0$, what is the sign of z? What about the case where $\hat{p} < p_0$? What if $\hat{p} = p_0$? If $\hat{p}$ is farther in value from p_0, does z get larger or smaller in absolute value?

5.2 Testing Claims About a Proportion

In this section, we derive the formula for the test statistic for a claim about a single population proportion, and we discuss the rationale behind the steps in hypothesis testing.

The Hypotheses

Consider again the claim that more than one-half of the students are from out of state. The hypotheses for this claim are

$$H_0: p = 0.50 \quad \text{and} \quad H_1: p > 0.50.$$

We can think of these hypotheses as two competing mathematical statements. For the sake of argument, we *assume* that H_0 is a true statement. In this claim, this means we assume that the population proportion is $p = 0.50$. If this assumption were true, then the observed sample proportion of $\hat{p} = 0.621$ would be higher than expected. To determine the significance of this larger than expected sample proportion, we need to determine the likelihood of this value occurring by chance, *assuming* H_0 is true. If this likelihood is small (less than α), then it is an indication that the assumption is incorrect (i.e., we would reject H_0). If this likelihood is not small (greater than α), then the opposite is true (i.e., we would not reject H_0).

The Test Statistic

To find this likelihood, suppose the null hypothesis is of the form

$$H_0: p = p_0$$

where p is the population proportion of interest. Define the random variable

$$X = \text{the number of successes in a sample of size } n.$$

Then, assuming that H_0 is true, X is $b(n, p_0)$ (at least approximately). The sample proportion $\hat{P} = X/n$ by Theorem 4.4.1 is approximately

$$N(p_0, p_0(1 - p_0)/n)$$

so that

$$Z = \frac{\hat{P} - p_0}{\sqrt{p_0(1 - p_0)/n}} \tag{5.1}$$

is approximately $N(0, 1)$. Now let α be the desired significance level and z_α be the critical z-value as described in Definition 4.5.1. If Z is $N(0, 1)$, then by definition of z_α

$$P(Z \leq z_\alpha) = 1 - \alpha$$

$$\Rightarrow \quad P\left(\frac{\hat{P} - p_0}{\sqrt{p_0(1 - p_0)/n}} \leq z_\alpha\right) \approx 1 - \alpha. \tag{5.2}$$

This fraction gives us the test statistic. The associated hypothesis test is called the *1-proportion Z-test*, the steps of which are summarized in the box on the next page.

> **1-Proportion Z-Test for a Claim About a Single Population Proportion**
>
> **Purpose**: To test a claim about a single population proportion where the null hypothesis is of the form H_0: $p = p_0$. Let
>
> - x be the number of "successes" in a sample of size n and
> - $\hat{p} = x/n$ be the sample proportion.
>
> The test statistic is
>
> $$z = \frac{\hat{p} - p_0}{\sqrt{p_0(1 - p_0)/n}}.$$
>
> The critical value is a z-value, and the P-value is an area under the standard normal density curve.
>
> | Form of H_1 | Critical Region | P-Value |
> |---|---|---|
> | $p > p_0$ | $[z_\alpha, \infty)$ | Area to the right of z |
> | $p < p_0$ | $(-\infty, -z_\alpha]$ | Area to the left of z |
> | $p \neq p_0$ | $(-\infty, -z_{\alpha/2}] \cup [z_{\alpha/2}, \infty)$ | Twice the area to the left or right of z, whichever area is smaller |
>
> **Requirements**
> 1. The sample is random.
> 2. The conditions for a binomial distribution must be met (at least approximately).
> 3. The conditions $np_0 \geq 5$ and $n(1 - p_0) \geq 5$ are both met. This is necessary so that the sample proportion is approximately normally distributed.

The Technical Conclusion

Now, Equation (5.2) shows that if H_0 is true, then the probability of observing a test statistic z that is less than z_α is large (approximately equal to $1 - \alpha$). Thus if $z \leq z_\alpha$, we conclude that the observed sample proportion could have occurred by chance and there is probably nothing wrong with the assumption. So we do not reject H_0.

On the other hand, if H_0 is true, then the probability that z is greater than z_α is small (approximately equal to α). Thus if $z > z_\alpha$, we conclude that the observed sample proportion is unlikely to have occurred by chance, and our assumption is probably incorrect. So we reject H_0.

In summary, we reject H_0 if the test statistic lies in the interval $[z_\alpha, \infty)$ (called the *critical region*), and we do not reject H_0 otherwise.

The Final Conclusion

To understand the rationale behind the wording of the final conclusion, note that if the original claim does not contain equality, then it becomes the alternative hypothesis H_1. If we reject H_0, then we are informally saying that the data support H_1 more than H_0, so we conclude that the data support the claim. If we do not reject H_0, then the opposite is the case and we conclude that the data do not support the claim.

If the original claim does contain equality, then it becomes H_0. If we reject H_0, then we are rejecting the claim and we conclude that there is sufficient evidence to reject the claim. If we do not reject H_0, then we are not rejecting the claim and we conclude that there is not sufficient evidence to reject the claim.

The P-Value Method

To motivate the P-value method, note that by definition of z_α, the area under the standard normal bell curve to the right of z_α is α, as illustrated in Figure 5.2. If the test statistic z lies in the critical region, then the area to the right of z (called the *P-value*) is less than α. Thus we reject H_0 if the P-value is less than α and do not reject H_0 otherwise.

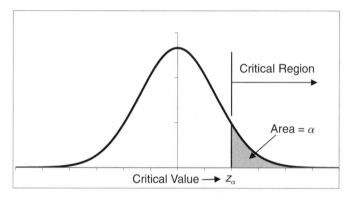

Figure 5.2

Because the P-value is the area to the right of z, we interpret the P-value as follows:

> The P-value is the probability of obtaining a value of the test statistic *at least as extreme* as that observed, assuming H_0 is true.

Informally, the P-value is the probability of observing a sample proportion $\hat{p}$ at least as extreme as that observed, assuming H_0 is true.

Alternative Hypothesis of the Form H_1: $p < p_0$

The above discussion applies to the case where the alternative hypothesis is of the form H_1: $p > p_0$. When the alternative hypothesis is of the form H_1: $p < p_0$, the same principles apply except that the critical value is $-z_\alpha$ and the P-value is the area to the left of the test statistic z.

Example 5.2.1 **Colors of M&Ms** A student claims that fewer than 25% of plain M&M candies are red. A random sample of 195 candies contains 37 red candies. Use these data to test the claim at the 0.05 significance level.

1. The parameter about which the claim is made is

$$p = \text{the proportion of all M\&M candies that are red.}$$

2. The claim in mathematical notation is $p < 0.25$, so we test the hypotheses

$$H_0: p = 0.25, \quad H_1: p < 0.25$$

3. The sample proportion is $\hat{p} = \frac{37}{195} \approx 0.190$ so that the test statistic is

$$z = \frac{\hat{p} - p_0}{\sqrt{p_0(1-p_0)/n}} = \frac{0.190 - 0.25}{\sqrt{0.25(1-0.25)/195}} \approx -1.93.$$

4. The critical value is $-z_{0.05} = -1.645$. The P-value is the area to the left of $z = -1.93$, which is 0.0268.
5. Because $z < -z_\alpha$ and P-value $< \alpha$, the technical conclusion is to reject H_0.
6. Because the original claim does not contain equality, the final conclusion is that the data support the claim. □

Alternative Hypothesis of the Form H_1: $p \neq p_0$

To understand the logic behind the two-tail test, first consider the right-tail test where the alternative hypothesis is of the form $H_1: p > p_0$. Observe that Equation (5.2) shows that if H_0 is true, then

$$P(z \leq z_\alpha) \approx 1 - \alpha \quad \Rightarrow \quad P(z > z_\alpha) \approx \alpha.$$

Thus if H_0 is really a true statement, then the probability of rejecting H_0 is approximately α. Rejecting H_0 when it is really a true statement is called a *Type I error*.

If the alternative hypothesis is of the form $H_1: p \neq p_0$, then a sample proportion $\hat{p}$ much smaller *or* larger than p_0 would cause us to reject H_0 in favor of H_1. In other words, a large negative test statistic z or a large positive test statistic z would cause us to reject H_0. To make the probability of a Type I error equal to α in this case, consider the critical value $z_{\alpha/2}$. Then for Z as defined in Equation (5.1), if H_0 is really a true statement,

$$P(-z_{\alpha/2} < Z < z_{\alpha/2}) \approx 1 - \alpha$$
$$\Rightarrow \quad P\left[(Z \leq -z_{\alpha/2}) \cup (Z \leq z_{\alpha/2})\right] \approx \alpha.$$

Thus the appropriate critical region is $(-\infty, -z_{\alpha/2}] \cup [z_{\alpha/2}, \infty)$.

To understand the P-value in the two-tail test, note that if the test statistic z is less than $-z_{\alpha/2}$, then the area to the left of z is less than $\alpha/2$, so that twice this area is less than α. This is illustrated in Figure 5.3. The same holds true if z is greater than $z_{\alpha/2}$.

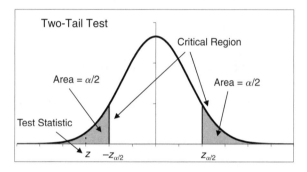

Figure 5.3

Type II Errors

The null hypothesis is a mathematical statement that is either true or false; we do not know which it is, but it is one or the other. If H_0 really is a true statement, then the correct technical conclusion is to not reject H_0. It would be an error to get the opposite technical conclusion. As stated above, this error is called a Type I error, and the probability of committing this type of error is α, the significance level.

If H_0 really is a false statement, the correct technical conclusion is to reject H_0. If we do not reject H_0, this is an error, called a *Type II error*. The probability of committing this type of error is denoted β while the probability of *not* committing this type of error is $(1 - \beta)$. This probability, $(1 - \beta)$, is called the *power* of the test. Table 5.4 summarizes these ideas.

The next example illustrates the idea of a type II error and how to calculate β and the power of a test.

	The True State of Nature	
Technical Conclusion	H_0 Is True	H_0 Is False
Reject H_0	Type I error α	Correct decision $1-\beta$
Do Not Reject H_0	Correct decision $1-\alpha$	Type II error β

Table 5.4

Example 5.2.2 Calculating Power In a certain population of voters, exactly 60% support a particular ballot measure. That is, $p = 0.60$ where p is the proportion of voters who support the measure. Not knowing this, a politician claims that more than 50% of voters support the measure. The resulting hypotheses for this claim are

$$H_0: p = 0.50 \quad \text{and} \quad H_1: p > 0.50.$$

In this case, H_0 is a false statement. Suppose we decide to survey $n = 100$ voters and then conduct the hypothesis test at the 0.05 significance level. Hopefully we will reject H_0, but this is not guaranteed to happen. The probability that we do reject H_0 is $(1-\beta)$, the power of the test.

To find $(1-\beta)$, note that the critical value is $z_{0.05} = 1.645$. Because this is a right-tail test, we will reject H_0 if $z > 1.645$. The formula for z is

$$z = \frac{\hat{p} - 0.50}{\sqrt{0.50(1-0.50)/100}} = \frac{\hat{p} - 0.50}{0.05}.$$

Solving this for $\hat{p}$ yields

$$\hat{p} = 0.05z + 0.50.$$

If $z > 1.645$, then

$$\hat{p} > 0.05(1.645) + 0.50 \approx 0.582.$$

Thus we will reject H_0 if $\hat{p} > 0.582$. The number 0.582 is called the *critical sample proportion*, and the interval $(0.582, 1]$ is the corresponding critical region. To find the probability that $\hat{p} > 0.582$, note that because the true population proportion is $p = 0.60$, then in a sample of $n = 100$ the sample proportion $\hat{P}$ is approximately

$$N\!\left(p, \frac{p(1-p)}{n}\right) = N(0.60, 0.0024).$$

Therefore, the probability that we reject H_0 is

$$P(\hat{P} > 0.582) \approx P\left(Z > \frac{0.582 - 0.60}{\sqrt{0.0024}}\right) = P(Z > -0.37) = 0.6423.$$

Thus the power of the test is $(1 - \beta) = 0.6423$ so that the probability of a type II error is $\beta = 0.3577$. □

In general, to find the power of a test about a single population proportion:

1. Find the appropriate critical z-value(s) (for a left-tail test, this is $-z_\alpha$; for a right-tail test, this is z_α; and for a two-tail test, these are $-z_{\alpha/2}$ and $z_{\alpha/2}$).
2. Find the corresponding critical value(s) of $\hat{p}$ with the formula

$$\hat{p} = z\sqrt{\frac{p_0(1-p_0)}{n}} + p_0$$

where z is a critical value, p_0 is the proportion in H_0, and n is the sample size.
3. Find the probability that the sample proportion is in the appropriate critical region, using the fact that $\hat{P}$ is approximately $N(p, p(1-p)/n)$, where p is the true value of the population proportion. This probability is the power of the test, $(1 - \beta)$.

The next example illustrates these steps for a two-tail test.

Example 5.2.3 Detecting a Change In the past, 75% of employees of a very large company reported that they were satisfied with their jobs. Suppose that for various reasons this percentage has changed to 70%. Based on empirical evidence, management believes that the percentage has changed, but they do not know if it has increased or decreased. They decide to test the claim that the percentage has changed by surveying a sample of $n = 150$ workers and using a 0.05 significance level. Find the probability that the test will detect the change in the percentage.

Formally, the claim to be tested is that p, the proportion of workers who are now satisfied with their jobs, is different from 0.75. The resulting hypotheses are

$$H_0\colon p = 0.75 \quad \text{and} \quad H_1\colon p \neq 0.75.$$

The change will be "detected" if we reject the false null hypothesis. The probability of this is $(1 - \beta)$, the power of the test. We use the steps outlined above to find this power.

1. Because this is a two-tail test, the critical values are $-z_{0.05/2} = -1.96$ and $z_{0.05/2} = 1.96$.
2. The corresponding critical values of $\hat{p}$ are

$$\hat{p} = -1.96\sqrt{\frac{0.75(1-0.75)}{150}} + 0.75 \approx 0.681,$$

and

$$\hat{p} = 1.96\sqrt{\frac{0.75(1-0.75)}{150}} + 0.75 \approx 0.819.$$

3. We will reject H_0 if $\hat{p} < 0.681$ or $\hat{p} > 0.819$. Now, $\hat{P}$ is approximately

$$N(0.7,\ 0.7(1-7)/150) = N(0.7,\ 0.0014)$$

so that

$$P\left(\hat{P} < 0.681\right) = P\left(Z < \frac{0.681 - 0.7}{\sqrt{0.7(1-0.7)/150}}\right) \approx P(Z < -0.51) = 0.3050.$$

By similar calculations, $P\left(\hat{P} > 0.819\right) \approx 0.0007$. Thus the probability of rejecting H_0 is approximately

$$0.3050 + 0.0007 = 0.3057.$$

Informally, we could interpret this result by saying that there is about a 30% chance that the change will be detected with the hypothesis test. □

To calculate the power of a test, we need to know the value of the population proportion p, which we rarely know (for if we did know it, there would be no reason to collect data or do a hypothesis test). Therefore, we cannot know the exact power of a test. However, we can make the following observations (see Exercise 18 of this section):

1. The farther p_0 is from p, the higher the power.
2. The larger the sample size n is, the higher the power.
3. The larger the significance level α is, the higher the power.

Ideally, we want $\alpha = \beta = 0$ so that the probability of either a Type I or Type II error is 0 (stated another way, we want a 100% confidence level and a power of 1). This, however, is impossible. What is often done in practice is to choose α and n according to the seriousness of the claim being tested. For instance, if the claim is regarding the proportion of patients who have a heart attack as the result of taking a certain drug, a small value of α and a large value of n are appropriate (say, $\alpha = 0.01$ and $n = 500$).

> # Software Calculations
>
> To perform a hypothesis test regarding a single a population proportion p:
>
> **Minitab**: Select **Stat** $\to$ **1-Proportion**. Enter n, the **Number of Trials**, and x, the **Number of Events**. Select **Options** and then select the inequality in H_1 next to **Alternative**. Enter p_0 next to **Test proportion**. Checking the box next to **Use test and interval based on normal distribution** tells Minitab to use the same algorithm described in this section where we approximate the binomial distribution with the normal. Otherwise, Minitab will use an exact method.
>
> **R**: The syntax for performing a hypothesis test where H_1 contains a $\neq$ sign is
> > binom.test(x, n, p_0, alternative = "two.sided")
>
> If H_1 contains a $<$ sign, change "two.sided" to "less." If H_1 contains a $>$ sign, change "two.sided" to "greater." Note that R uses an exact method for calculating the P-value whereas the algorithm we present uses an approximation of the binomial distribution with the normal. So the results given by R will be slightly different from those given by our algorithm.
>
> **Excel**: Excel does not have a built-in function for doing the 1-proportion test, but the P-value can be found by first finding the test statistic z, using the formula in the text. Then let $z_E = -|z|$. The P-value for a one-tail test is calculated with the syntax NORM.DIST($z_E, 0, 1, 1$). For a two-tail test, double this probability.
>
> **TI-83/84 Calculators**: Press **STAT** $\to$ **TESTS** $\to$ **5:1-PropZTest**. Enter p_0, x, n, and the inequality in H_1.

Exercises

Directions: In exercises asking to test a claim, use the five-step hypothesis-testing procedure. Give either the P-value or the critical region. Unless otherwise specified, use a 0.05 significance level.

1. A drug company claims that less than 10% of users of its allergy medicine experience drowsiness. In a random sample of 75 users, 3 reported drowsiness. Use the data to test the claim. Repeat the test, but use a 0.01 significance level. Does this change the conclusion?

2. Better Bread Company believes that more than 75% of consumers prefer its classic white bread over its main competitor's. In a blind taste test involving 150 consumers, 115 preferred the bread of Better Bread Company.

a. Use the data to test the claim.

b. Now test the claim that more than 70% prefer its classic white bread over the competitor's.

c. If the company were to advertise its white bread, which of these two statements would be more accurate? "More than 75% of customers prefer our bread," or "More than 70% prefer our bread." Explain.

3. In a random sample of 1008 voters who voted in the 2006 Nebraska governor's election, 24% were 65 years and older. A reporter writes a story about this election and says, "Less than 25% of all voters in this election were aged 65 years and older." Determine if the data support this statement. Would it be appropriate to include this statement in the article? Explain.

4. A high-sugar cereal is defined as one in which the ratio of total carbohydrates to sugar is greater than $4:1$. A student claims that most cereals are high-sugar. A sample of 120 types of cereal contains 70 that are high-sugar (data collected by Hannah McNeiley, 2009). Assuming this is a random sample of all cereals, use these data to test the claim.

5. The ghost crab (*Ocypode quadrata*) has one claw that is larger than the other. In a random sample of 72 ghost crabs, 41 have a larger left claw (data collected by Nicholas Lee, 2010). Test the claim that most ghost crabs have a larger left claw. Based on this evidence, do ghost crabs appear to have a genetic disposition to having a larger left claw?

6. A billiards table contains six pockets, two "behind" the original rack of balls and four "in front" of the rack. A student claims that in a game of billiards, the majority of the balls are pocketed in the pockets behind the rack. He observed a total of 130 pocketed balls and found that 75 were pocketed behind the original rack (data collected by Tyler Walker, 2009). Use these data to test the claim.

7. It seems reasonable to believe that a head football coach has a higher chance of getting fired if he has a losing record. To examine this, a student surveyed 39 randomly selected head football coaches at the Football Bowl Subdivision (FBS) level who had been fired and found that 30 of them had a winning percentage less than 0.500 (data collected by George Toman, 2009). Use this to test the claim that of all the fired head football coaches at the FBS level, a majority have a winning percentage less than 0.500. Does this result mean that having a losing record causes a coach to get fired?

8. When a fair coin is flipped, the probability of getting tails is 0.5. This means that if we were to flip a coin *many* times, the proportion of tails would be 0.5. If we were to *spin* a penny, we might expect to get similar results. However, is this the case? To determine this, a group of students spun a penny 500 times and observed 204 heads and 296 tails

(data collected by Daniel Klinge, Jacob Bachman, and Christopher Plucker, 2009). Use these data to test the claim that in the population of all spins of a penny, the proportion of tails is 0.5. Does it appear that spinning a penny produces the same results as flipping a penny?

9. A university cafeteria ice cream machine offers three types of ice cream: vanilla, chocolate, and twist. Two students observed 56 people serve themselves ice cream and noted that 35 of them chose vanilla while 21 of them chose chocolate or twist (data collected by Danielle Masur and Travis Besel, 2009). Use these data to test the claim that of those at this cafeteria who eat ice cream, more choose vanilla over chocolate or twist.

10. When calculating probabilities involving genders of children, we assumed that $P(\text{girl}) = P(\text{boy}) = 0.5$. At a particular hospital, there were 1712 baby girls and 1753 boys born between the years 2004 and 2010 (data collected by Angie Ramaekers, 2011). Assuming that these births represent a random sample of all births, test the claim that $P(\text{girl}) = P(\text{boy}) = 0.5$. Does our assumption appear to be reasonable?

11. A student claims that a type of candy has more red pieces than blue pieces. A random sample of 250 candies contains 65 red and 55 blue candies. Use these data to test the claim. (**Hint**: We are really only interested in the population of red and blue candies. Consider the 65 red and 55 blue candies to be a sample of size $n = 120$ of this population. If there are more red than blue in the population, what does this mean about the proportion of red?)

12. An opponent of the use of seat belts claims that less than 25% of car crashes occur within 5 mi of home. Because of this, he contends that you don't really need to wear a seat belt on short trips. In a survey of 120 car crashes, it was found that 62 occurred within 5 mi of the home of the driver.

   a. Use these data to test the claim at the 0.01 significance level. Does the opponent's contention appear to be reasonable?

   b. Was there any point in doing the hypothesis test in part a? Explain.

13. Suppose H_1: $p > 0.5$ is the alternative hypothesis for a claim that $p > 0.5$ and that a sample yields the sample statistic $\hat{p} = 0.52$.

   a. Calculate the test statistic z and the P-value for the cases where $n = 500, 1000, 1500, 2000,$ and 2500. State the final conclusion at the 95% confidence level.

   b. Generalize your results in part a. If H_1: $p > p_0$ is the alternative hypothesis and $\hat{p} > p_0$, find $\lim_{n \to \infty} z$ and $\lim_{n \to \infty} P\text{-value}$.

c. If $\hat{p} > p_0$, would you be more confident that the sample data support the claim if n were large or small? Why?

14. All the P-values we have calculated to this point have been approximations of binomial probabilities using the normal distribution. In this exercise, we look at how to calculate *exact* P-values where we do not use the normal as an approximation of the binomial. When we calculate an exact P-value, the test is generically called an *exact binomial test*.

  a. The P-value for the left-tail test calculated in Example 5.2.1 is the approximate probability of obtaining 37 successes or fewer in 195 trials, assuming H_0: $p = 0.25$ is true. This P-value is really an approximation of $P(X \leq 37)$ where X is $b(195, 0.25)$. The exact P-value in this test is defined to be $P(X \leq 37)$. Use available software to calculate this exact P-value. Compare this exact value to the approximation in Example 5.2.1.

  b. The P-value for a right-tail test is the probability of obtaining the number of successes at least as large as that observed, assuming H_0 is true. In more precise terms, the exact P-value is $P(X \geq x)$ where x is the number of observed successes and X is $b(n, p_0)$. Calculate the exact P-value for a test where H_1: $p > 0.3$, $n = 350$, and $x = 120$. Also calculate the P-value using the normal as an approximation of the binomial, and compare the two P-values.

  c. For a two-tail test, the definition of the exact P-value is a bit more complicated. When using the normal distribution to calculate a P-value, we calculated a normal probability and then multiplied it by 2. This calculation was based, in part, on the fact that the normal distribution is symmetric. The binomial distribution, however, is not symmetric. To illustrate the definition of an exact two-tail P-value, consider a test where the hypotheses are H_0: $p = 0.3$, H_1: $p \neq 0.3$ and a sample of $n = 100$ yields $x = 24$ successes. Observe that 24 successes is less than what we would expect if H_0 were true; so to calculate the exact P-value, first we calculate $p_1 = P(X \leq 24)$ where X is $b(100, 0.3)$. Then we find a number x_2 such that the probability $p_2 = P(X \geq x_2)$ is as close to p_1 as possible, but *less than* p_1. The exact P-value is $p_1 + p_2$. Calculate this exact P-value and give the value of x_2. (Note that some software do not require p_2 to be less than p_1, so they may give slightly different exact P-values.)

15. Another method for doing hypothesis testing is called the *confidence interval* method. In this method, we calculate an "appropriate" confidence interval estimate of p, using the

data and the given significance level. If p_0 is in the interval, then we do not reject H_0. Otherwise, we reject H_0. The appropriate confidence interval is described below. (See Section 4.5 for descriptions on how to calculate these confidence intervals.)

a. For a left-tail test, calculate a one-sided confidence interval that gives an upper bound on p.

b. For a right-tail test, calculate a one-sided confidence interval that gives a lower bound on p.

c. For a two-tail test, calculate a two-sided confidence interval estimate of p.

Redo Examples 5.1.1 and 5.2.1, using the confidence interval method. In each case, give the appropriate confidence interval. Are the results any different from those found with the P-value or traditional method?

16. Consider Example 5.2.2.

a. Suppose the claim is that more than 45% support the measure, so that $p_0 = 0.45$. Find the power of the test. What does this say about the power as p_0 gets farther in value from p?

b. Suppose that $p_0 = 0.5$, but $n = 500$. Find the power of the test. What does this say about the power as n gets larger?

c. Suppose that $p_0 = 0.5$ and $n = 100$, but $\alpha = 0.2$. Find the power of the test. What does this say about the power as α gets larger?

17. Suppose that in a population of college students, exactly 20% enjoy statistics class. Not knowing this, a student claims that less than 30% enjoy statistics class and plans to test the claim by surveying $n = 200$ students and using a 0.01 significance level. Find the power of the test.

18. In Example 5.2.3, find the probability of detecting the change if the claim were that the proportion is now less than 0.75 (so that the alternative hypothesis is $H_1: p < 0.75$). Would this hypothesis test be more or less likely to detect the change than the original test?

5.3 Testing Claims About a Mean

In this section, we introduce techniques for testing claims about a population mean. The logic behind this type of test is identical to that of the 1-proportion Z-test. The main difference between testing a claim about a proportion and one about a mean is the calculation of the test statistic.

To motivate the need for a hypothesis test about a mean, consider the following claim: The mean IQ of college students in the United States is greater than 110. The parameter in this claim is

$$\mu = \text{the mean IQ of all college students in the United States.}$$

To determine with absolute certainty whether this claim is true, we would need to calculate μ by testing the IQ score of every single college student in the United States, add them up, and then divide by the number of college students. This is virtually impossible, so it is necessary to collect data from a sample and use them to test the claim about the population with a hypothesis test.

Using the procedures in Section 5.1, the null and alternative hypotheses for this claim are

$$H_0: \mu = 110, \qquad H_1: \mu > 110.$$

Suppose that a random sample of 150 college students from across the United States has a mean IQ of $\bar{x} = 112.5$. This sample mean is greater than expected *if* H_0 were true. We need to determine the significance of this result.

When confidence interval estimates of μ are calculated, there are two types of intervals: a Z-interval for the case where σ^2, the population variance, is known and a T-interval for the case where σ^2 is unknown. Likewise hypothesis tests about μ fall into two different categories corresponding to the same two cases.

To find a formula for the test statistic in the case where σ^2 is known, suppose the null hypothesis is of the form $H_0: \mu = \mu_0$, the population has a variance of σ^2, and a sample of size n has a mean of $\bar{x}$. Assuming H_0 is true, the central limit theorem tell us that the sample mean $\bar{X}$ is approximately $N(\mu_0, \sigma^2/n)$. The probability that the sample mean $\bar{X}$ is greater than the observed value $\bar{x}$ is then

$$P(\bar{X} > \bar{x}) \approx P\left(Z > \frac{\bar{x} - \mu_0}{\sigma/\sqrt{n}}\right).$$

The fraction on the right-hand side gives us the formula for the test statistic. The associated hypothesis test is called the Z-test. The procedures for this test are summarized in the box on the next page.

Z-Test for a Claim About a Single Population Mean with σ^2 Known

Purpose: To test a claim about the mean of a single population where the null hypothesis is of the form $H_0: \mu = \mu_0$ and the population has a variance of σ^2. Let $\bar{x}$ be the mean of a sample of size n.

The test statistic is

$$z = \frac{\bar{x} - \mu_0}{\sigma/\sqrt{n}}.$$

The critical value is a z-value, and the P-value is an area under the standard normal density curve.

Form of H_1	Critical Region	P-Value
$\mu > \mu_0$	$[z_\alpha, \infty)$	Area to the right of z
$\mu < \mu_0$	$(-\infty, -z_\alpha]$	Area to the left of z
$\mu \neq \mu_0$	$(-\infty, -z_{\alpha/2}] \cup [z_{\alpha/2}, \infty)$	Twice the area to the left or right of z, whichever area is smaller

Requirements
1. The sample is random.
2. Either the population is normally distributed or $n > 30$. This is so the sample mean is approximately normally distributed.

Example 5.3.1 Students' IQ Score Consider again the claim that the mean IQ of college students in the United States is greater than 110. Typically IQ scores have a variance of $\sigma^2 = 15^2$, so we will assume that the population of college students has this same variance. Use the sample data where $n = 150$ and $\bar{x} = 112.5$ to test the claim at the 0.05 significance level.

The population parameter and hypotheses have been defined above. The test statistic is

$$z = \frac{\bar{x} - \mu_0}{\sigma/\sqrt{n}} = \frac{112.5 - 110}{15/\sqrt{150}} \approx 2.04.$$

The critical region is $[z_{0.05}, \infty) = [1.645, \infty)$, and the P-value is the area to the right of 2.04, which is 0.0207. Because z is in the critical region and the P-value < 0.05, we reject H_0 and conclude that the data support the claim. □

The drawback of the Z-test is that it requires knowledge of the population variance (or standard deviation). This is rarely known, so we often estimate it with a sample variance (or sample standard deviation). To find the test statistic in this case, we use Theorem 4.6.1, which states that if the population is normally distributed, then the random variable

$$T = \frac{\bar{X} - \mu}{S/\sqrt{n}}$$

has a Student-t distribution with $(n-1)$ degrees of freedom, where μ is the population mean and S denotes the sample standard deviation. By arguments similar to that in the Z-test, the appropriate test statistic is

$$t = \frac{\bar{x} - \mu_0}{s/\sqrt{n}}$$

where s is the observed sample standard deviation. The resulting hypothesis test is called the *T-test*. The procedures for this test are summarized in the box below.

T-Test for a Claim About a Single Population Mean with σ^2 Unknown

Purpose: To test a claim about the mean of a single population where the null hypothesis is of the form H_0: $\mu = \mu_0$. Let

- $\bar{x}$ be the mean of a sample of size n, and
- s be the sample standard deviation.

The test statistic is

$$t = \frac{\bar{x} - \mu_0}{s/\sqrt{n}}.$$

The critical value is a t-value with $(n-1)$ degrees of freedom, and the P-value is an area under the corresponding Student-t density curve.

Form of H_1	Critical Region	P-Value
$\mu > \mu_0$	$[t_\alpha, \infty)$	Area to the right of t
$\mu < \mu_0$	$(-\infty, -t_\alpha]$	Area to the left of t
$\mu \neq \mu_0$	$(-\infty, -t_{\alpha/2}] \cup [t_{\alpha/2}, \infty)$	Twice the area to the left or right of t, whichever area is smaller

Requirements

1. The sample is random.
2. Either the population is normally distributed or $n > 30$. This is so the sample mean is approximately normally distributed.

We do not have a table that gives areas under the Student-t density curve. Thus we cannot find the exact P-value of a T-test by using tables. The P-value is most often found with software. However, we can approximate the P-value using Table C.3 in Appendix C as illustrated in the next example.

Example 5.3.2 **Approximating P-Values** Consider a claim where the alternative hypothesis H_1: $\mu > 5$, and suppose that a sample of $n = 19$ yields the test statistic $t = 1.65$. Because this is a right-tail test, the P-value is the area under the Student-t density curve with $19 - 1 = 18$ degrees of freedom to the right of 1.65. From Table C.3, we note that

$$t_{0.05}(18) = 1.734 \quad \text{and} \quad t_{0.10}(18) = 1.330.$$

This means that the area to the right of 1.734 is 0.05 and the area to the right of 1.330 is 0.10. Because $t = 1.65$ is between 1.734 and 1.330, we conclude that the area to the right of 1.65 (the P-value) is between 0.05 and 0.10. (Using software, we find that the exact value is 0.0581.)

Now consider a claim where the alternative hypothesis is H_1: $\mu \neq 7.2$, and suppose that a sample of $n = 9$ yields the test statistic $t = -3.11$. This is a two-tail test, and because the test statistic is negative, the P-value is twice the area to the left with $9 - 1 = 8$ degrees of freedom. From Table C.3, we see that

$$t_{0.005}(8) = 3.355 \quad \text{and} \quad t_{0.01}(8) = 2.896.$$

By symmetry of the Student-t density curve, this means that the area to the left of -3.355 is 0.005 and the area to the left of -2.896 is 0.01. Because $t = -3.11$ is between -3.355 and -2.896, the area to the left of -3.11 is between 0.005 and 0.01. Therefore, the P-value is between

$$2(0.005) = 0.01 \quad \text{and} \quad 2(0.01) = 0.02.$$

(Using software, the exact area to the left is 0.00722, so that the P-value is $2(0.00722) = 0.01444$.) □

Example 5.3.3 **Weights of Shredded Cheese** A manufacturer of cheese claims that the mean weight of all its 12-oz packages of shredded cheddar is greater than 12 oz. They collect a random sample of $n = 36$ packages, weigh each, and calculate a sample mean of $\bar{x} = 12.05$ and a sample standard deviation of $s = 0.15$. Use these data to test the claim at the 0.05 significance level.

We use the same five-step hypothesis-testing procedure outlined in Section 5.1.

1. The parameter about which the claim is made is

   μ = the mean weight of all 12-oz packages of shredded cheddar.

   The claim is $\mu > 12$, so we test the hypotheses

   $$H_0: \mu = 12, \qquad H_1: \mu > 12.$$

2. Because we do not claim to know the population variance, we use the T-test. The test statistic is then

   $$t = \frac{\bar{x} - \mu_0}{s/\sqrt{n}} = \frac{12.05 - 12}{0.15/\sqrt{36}} = 2.$$

3. Because $n = 36$, we have $36 - 1 = 35$ degrees of freedom, so the appropriate critical value is $t_{0.05}(35) = 1.690$. The P-value is the area to the right of $t = 2$ under the Student-t density curve with 35 degrees of freedom. From Table C.3, we see that this area is between 0.025 and 0.05. (Using software, the P-value is 0.027.)
4. The critical region is $[1.690, \infty)$. Note that $t = 2$ lies in this region, and the P-value is less than 0.05, so the technical conclusion is to reject H_0.
5. Because the original claim does not contain equality, the final conclusion is that the data support the claim. This means that *on average*, it appears as if the bags weigh more than 12 oz. It does *not* mean that every bag weighs more than 12 oz. □

When testing a claim about a mean, one has to decide which test to use, the Z- or T-test. We offer the following suggestions:

1. If the population variance is known, then use the Z-test.
2. If the population variance is not known, but the sample size is large (that is, $n > 30$), then the sample standard deviation s provides a good estimate of the population standard deviation σ, so use the Z-test with s in place of σ.
3. If the population variance is not known and the sample size is small (that is, $n \leq 30$), then use the T-test.
4. If using software, then use the T-test regardless.

Software Calculations

To perform a hypothesis test regarding a single population mean μ:

Minitab: If you have raw data, enter it in a column. Select **Stat** → **Basic Statistics**. Select **1-Sample Z** if the population standard deviation σ is known, or **1-Sample T** if σ is not known. Select the name of the column containing the data under **Samples in columns**. In the case of the Z-test, enter σ next to **Standard deviation**. Check the box next to **Perform hypothesis test**, and enter the claimed population mean μ_0 next to **Hypothesized mean**. Press **Options**, set the confidence level, and select the inequality in H_1 next to **Alternative**. If you only have sample statistics, select **Summarized Data** and enter the necessary statistics.

R: First enter the data $\{x_1, x_2, \ldots, x_n\}$ as a vector with the syntax x<-c($x_1, x_2, \ldots, x_n$). The syntax for calculating the P-value for a two-tail test is
> t.test(x, alternative = "two.sided")

For a one-tail test, change "two-sided" to either "less" or "greater" depending on the inequality in H_1.

Excel: Excel does not have built-in functions for doing the Z- or T-tests, but the P-values can be found by first finding the test statistic z or t, using the formulas in the text. Then let $z_E = -|z|$ or $t_E = -|t|$. The P-value for a one-tail test is calculated with the syntax NORM.DIST($z_E, 0, 1, 1$) or T.DIST(t_E, df, 1) where df is the number of degrees of freedom. For a two-tail test, double the probability.

TI-83/84 Calculators: If you have raw data, enter them in a list. Then press **STAT** → **TESTS**. Select **1:Z-Test** if the population standard deviation σ is known, or **2:T-Test** if σ is not known. Select **Stats**, then enter the claimed population mean μ_0, the name of the list containing the data, the confidence level, and the inequality in H_1. When calculating a Z-interval, enter σ. If you only have sample statistics, select **Stats** and enter the necessary statistics. In all cases, set the **Freq** to 1.

Exercises

Directions: In exercises asking to test a claim, use the five-step hypothesis-testing procedure at the 0.05 significance level. Unless otherwise specified, use a T-test. Give either the P-value or the critical region.

1. Use Table C.3 to estimate the P-values in each of the following T-tests with the given alternative hypothesis, sample size, and test statistic.

   a. H_1: $\mu < 2.4$, $n = 23$, $t = -1.98$
   b. H_1: $\mu \neq 0$, $n = 6$, $t = 2.36$
   c. H_1: $\mu > 16.28$, $n = 14$, $t = 1.30$

2. In Example 5.3.3, the claim is $\mu > 12$ and the sample mean is $\bar{x} = 12.05$. Explain why this sample mean does not *prove* the claim.

3. A thermometer is supposed to read 0°C at the freezing point of water. However, some read less than 0 and others read greater than 0. A manufacturer of thermometers records the readings at the freezing point of water of 75 randomly selected thermometers from its production line and calculates a mean reading of $\bar{x} = 0.12$°C. Assuming the population standard deviation is $\sigma = 1$°C, use a Z-test to test the claim that the mean reading at the freezing point of water of all thermometers from the assembly line is 0°C. Do these thermometers appear to give a reading of 0°C, on average?

4. A sample of 30 homes within a 5-mi radius of a certain lake in Nebraska has a mean area of 2421.7 ft^2 with a standard deviation of 856.8 ft^2. The median area of homes in the Midwest is 1931 ft^2 (data collected by Nathaniel Jensen, 2010). Test the claim that the mean square footage of all homes within a 5-mi radius of this lake is greater than 1931 ft^2. Does it appear that homes around this lake tend to be larger than a typical home? Why might this be?

5. A local gym is offering a new weight loss diet and exercise program. Fifty volunteers follow the program for 6 months, and the change in their weights is recorded (a negative change means the person lost weight). The changes have a mean of $\bar{x} = -7.7$ lb with a standard deviation of $s = 10.3$ lb. Use a Z-test to test the claim that the mean weight change of all people on this program is negative. Does the program appear to be effective?

6. A sample of 40 division I offensive linemen has a mean height of $\bar{x} = 76.875$ in with a standard deviation of $s = 2.115$ in (data collected by Seth Elley, 2010). Use the data to test the claim that the mean height of all division I offensive linemen is greater than 76 in.

7. Two students select four 500-mL bottles of a certain brand of bottled water and weigh the water in each. The masses (in grams) are shown below (data collected by Brittany Singleton and Kelsie Elder, 2009). Assuming the population is normally distributed, use the data to test the claim that the mean mass of water in all bottles of this brand is greater than 500 g. Do the bottles appear to be underfilled on average? (**Note:** 1 mL of water has a mass of 1 g.)

$$505 \quad 506 \quad 505 \quad 504$$

8. A student randomly chooses 10 books on medicine from his university's library and records the copyright date of each, as shown in the table below (data collected by Andrew Asmus, 2009).

1934	1949	1965	1970	1998
1942	1960	1969	1982	2002

a. Generate a normal quantile plot of these data. Does the population of all books on medicine from this library appear to be normally distributed?

b. Suppose someone believes that this library's books on medicine are too old and claims that the mean copyright date on such books is before 1975. Do these data meet the requirements for the T-test so that it can be used to test this claim? If so, use the data to test the claim.

9. To examine the savings when purchasing generic brand products at a large retailer, a student records the prices of 10 different, randomly selected name brand products and the prices of a corresponding generic equivalent, as shown in the table below (data collected by Amanda Schroeder, 2010).

Product	1	2	3	4	5	6	7	8	9	10
Name Brand	7.68	9.56	7.62	1.58	2.98	3.58	6.54	8.97	4.06	6.46
Generic Brand	4.47	2.97	3.34	0.67	3.12	2.98	5.34	6.97	1.76	4.12

a. For each product, calculate the percentage of savings experienced when the generic brand product is purchased instead of the name brand.

b. Use the data to test the claim that average (or mean) savings is greater than 20%. Based on these results, does purchasing generic brand products appear to result in substantial savings?

10. To examine the percentage of pages in magazines devoted to advertising, a student randomly chooses 13 magazines and records the total number of pages in each and the number of pages containing only advertisements, as shown in the table below (data collected by Emma Gaither, 2010). Use the data to test the claim that the mean of the proportion of pages containing only advertisements is greater than 0.25.

Magazine	1	2	3	4	5	6	7	8	9	10	11	12	13
Total	153	213	205	225	241	179	67	178	85	67	67	161	131
Only Ads	58	72	69	88	93	67	20	42	29	15	15	29	23

11. A student claims that the mean height of the starters on the university basketball team is greater than 64 in. He measures the height of all five starters and calculates a mean of 66.4 in and a standard deviation of 1.3 in. Explain why it is *not* necessary to do a hypothesis test in this scenario.

12. Examine the critical t-values in Table C.3, and compare the values as n gets larger to the critical z-values in Table C.2. Explain how this helps justify the use of the Z-test in place of the T-test when n is large.

13. Suppose $H_1: \mu > 12$ is the alternative hypothesis for a claim that $\mu > 12$, which we want to test with a Z-test, and that a sample of size $n = 50$ yields the sample mean $\bar{x} = 12.05$.

   a. Calculate the test statistic z and the P-value for the cases where the sample standard deviation is $s = 0.5, 0.2,$ and 0.1. In each case, state the final conclusion at the 95% and 99% confidence levels.

   b. Generalize your results in part a. If $\mu > \mu_0$ is the claim, n is a fixed number, and $\bar{x} > \mu_0$, find $\lim_{s \to 0} z$ and $\lim_{s \to 0} (P\text{-value})$.

   c. If $\bar{x} > \mu_0$, would you be more confident that the sample data supported the claim if s were large or small? Explain.

14. Suppose that a sample of size $n = 30,000$ yields a mean of $\bar{x} = 15.05$ with a standard deviation of $s = 0.91$.

   a. If μ is the population mean, test the claim that $\mu > \mu_0$ for these values of μ_0: 15.01, 15.02, 15.03, and 15.04.

   b. Because the sample size is so large, is there really any point in doing a hypothesis test with these data? Explain.

15. Consider a hypothesis test where the null hypothesis is $H_1: \mu < 40$ and the sample data yield a P-value of 0.025. Suppose an outlier was discovered that is much larger than the other data values. If this outlier were removed and the P-value were recalculated using the remaining data, would the P-value increase or decrease? Explain your reasoning.

16. As noted in the text, in cases where σ is not known, but n is large (typically $n > 30$), we can simply replace σ with s and perform a Z-test. This typically gives the same results as the T-test, but not always.

a. Redo Example 5.3.3, using the Z-test. Does this give the same result as the T-test?
b. Consider a claim that $\mu = 100$ so that the hypotheses are H_0: $\mu = 100$ and H_1: $\mu \neq 100$. Suppose that a random sample of size $n = 36$ gives the sample statistics $\bar{x} = 105$ and $s = 15$. Use these data to test the claim at the 0.05 significance level, using the T-test and then the Z-test. Do these two tests give the same result?

17. As discussed in Section 5.2, a Type II error is the error of not rejecting a false null hypothesis. The probability of doing this is denoted β. The quantity $(1 - \beta)$ is the probability of correctly rejecting a false null hypothesis and is called the *power of the test*. To calculate the power of a Z-test for a claim about a single population mean, we use the following algorithm:

1. Find the appropriate critical value(s).
2. Find the corresponding critical value(s) of $\bar{x}$ with the formula

$$\bar{x} = \mu_0 + z \frac{\sigma}{\sqrt{n}},$$

where z is a critical value, μ_0 is the mean used in H_0, σ is the population standard deviation, and n is the sample size.
3. Find the probability that the sample mean is in the appropriate critical region determined by the critical values of $\bar{x}$ using the fact that $\bar{X}$ is approximately $N(\mu, \sigma^2/n)$ where μ is the true value of the population mean. This probability is the power of the test, $(1 - \beta)$.

Suppose a population is $N(10, 1)$ and that someone claims that $\mu < 10.5$. She decides to test the claim with a Z-test using a random sample of size $n = 15$. Use the algorithm above to the find power of this test.

5.4 Comparing Two Proportions

The previous sections dealt with testing a claim about a parameter of a single population. Often we want to compare parameters from two different populations. We can test claims about the equality (or difference) of parameters from two different populations in a similar way as in the previous sections.

Suppose researchers are trying to determine if a new allergy medication causes drowsiness, as is often the case. To investigate this, they give the medicine to $n_1 = 1320$ patients and a placebo (a pill with no medicine) to $n_2 = 1350$ patients. They record that $x_1 = 110$ of the patients given the medicine experience drowsiness and that $x_2 = 100$ of the patients given a placebo experience drowsiness.

We use the data to test the claim that a higher proportion of those taking the medicine experience drowsiness than those taking the placebo. The parameters of interest are

p_1 = the proportion of all patients given the medicine who experience drowsiness, and

p_2 = the proportion of all patients given a placebo who experience drowsiness.

The claim is $p_1 > p_2$, so we test the hypotheses

$$H_0: p_1 = p_2, \qquad H_1: p_1 > p_2.$$

From the data, we get the two sample proportions

$$\hat{p}_1 = \frac{110}{1320} \approx 0.0833 \quad \text{and} \quad \hat{p}_2 = \frac{100}{1350} \approx 0.0741.$$

These sample proportions are estimates of p_1 and p_2, respectively. Based on these estimates, it appears that p_1 is possibly higher than p_2. However, we must remember that these are *estimates*, and even if $p_1 = p_2$, it is still possible that a sample proportion estimate of p_1 is higher than one of p_2. We need to determine if the fact that $\hat{p}_1 > \hat{p}_2$, as observed, is significant, or if this difference can be explained by chance.

To find the appropriate test statistic, let the random variables $\hat{P}_1$ and $\hat{P}_2$ denote sample proportions of the two populations. We consider the fact that $\hat{p}_1 > \hat{p}_2 \Rightarrow \hat{p}_1 - \hat{p}_2 > 0$ to be significant if

$$P\left(\hat{P}_1 - \hat{P}_2 > \hat{p}_1 - \hat{p}_2\right)$$

is small (i.e., less than 0.05). To find this probability, let p_1 and p_2 denote the two population proportions and suppose that random samples of size n_1 and n_2 are taken from each population, yielding x_1 and x_2 "successes," respectively. Then by Theorem 4.4.1, $\hat{P}_1$ and $\hat{P}_2$ are approximately

$$N\left(p_1, \frac{p_1(1-p_1)}{n_1}\right) \quad \text{and} \quad N\left(p_2, \frac{p_2(1-p_2)}{n_2}\right),$$

respectively. Assuming that $\hat{P}_1$ and $\hat{P}_2$ are independent, then by Theorem 3.7.4, the random variable $\hat{P}_1 - \hat{P}_2$ is approximately

$$N\left(p_1 - p_2, \frac{p_1(1-p_1)}{n_1} + \frac{p_2(1-p_2)}{n_2}\right).$$

Thus the random variable

$$Z = \frac{\left(\hat{P}_1 - \hat{P}_2\right) - (p_1 - p_2)}{\sqrt{p_1(1-p_1)/n_1 + p_2(1-p_2)/n_2}}$$

is approximately $N(0,1)$. Now, assume that H_0: $p_1 = p_2$ is true, and let p denote the common population proportion, that is, $p = p_1 = p_2$. Then the random variable Z becomes

$$Z = \frac{\left(\hat{P}_1 - \hat{P}_2\right) - (p - p)}{\sqrt{p(1-p)/n_1 + p(1-p)/n_2}} = \frac{\hat{P}_1 - \hat{P}_2}{\sqrt{p(1-p)(1/n_1 + 1/n_2)}}.$$

To calculate probabilities of Z, we need to know the value of p. Because we do not know this value, we need to estimate it with the data. Because larger sample sizes typically give better estimates, we "pool" the two samples to get the estimate

$$p \approx \frac{x_1 + x_2}{n_1 + n_2}.$$

We denote this estimate by $\hat{p}$. Using this value, we get the test statistic

$$z = \frac{\hat{p}_1 - \hat{p}_2}{\sqrt{\hat{p}(1-\hat{p})(1/n_1 + 1/n_2)}}.$$

The associated hypothesis test is called the *2-proportion Z-test*. The procedures for this test are summarized in the box on the next page.

For the claim regarding the allergy medications, we have

$$\hat{p} = \frac{110 + 100}{1320 + 1350} \approx 0.0787$$

so that the test statistic is

$$z = \frac{0.0833 - 0.0741}{\sqrt{0.0787(1 - 0.0787)(1/1320 + 1/1350)}} \approx 0.88.$$

At the 0.05 significance level, the critical region is $[1.645, \infty)$. The P-value is the area to the right of 0.88, which is 0.1894. Because z is not in the critical region (and the P-value is greater than 0.05), we do not reject H_0. Because the original claim does not contain equality, we conclude that the data do not support the claim that a higher proportion of those taking the medicine experience drowsiness than those taking the placebo.

> **2-Proportion Z-Test for a Claim Comparing Two Population Proportions**
>
> **Purpose**: To test a claim comparing proportions from two independent populations where the null hypothesis is of the form H_0: $p_1 = p_2$. The test statistic is
>
> $$z = \frac{\hat{p}_1 - \hat{p}_2}{\sqrt{\hat{p}(1-\hat{p})(1/n_1 + 1/n_2)}}$$
>
> where
> - n_1 and n_2 are the sample sizes taken from the two populations,
> - x_1 and x_2 are the numbers of successes in the two samples,
> - $\hat{p}_1 = \dfrac{x_1}{n_1}$, $\hat{p}_2 = \dfrac{x_2}{n_2}$, and
> - $\hat{p} = \dfrac{x_1 + x_2}{n_1 + n_2}$.
>
> The critical value is a z-score, and the P-value is the area under the standard normal density curve.
>
> | Form of H_1 | Critical Region | P-Value |
> |---|---|---|
> | $p_1 > p_2$ | $[z_\alpha, \infty)$ | Area to the right of z |
> | $p_1 < p_2$ | $(-\infty, -z_\alpha]$ | Area to the left of z |
> | $p_1 \neq p_2$ | $(-\infty, -z_{\alpha/2}] \cup [z_{\alpha/2}, \infty)$ | Twice the area to the left or right of z, whichever area is smaller |
>
> **Requirements**
> 1. Both samples are random and independent.
> 2. In both samples, the conditions for a binomial distribution are satisfied.
> 3. In both samples, there are at least 5 successes and 5 failures. This is so both sample proportions are approximately normally distributed.

We might report this conclusion by saying that there is not a *statistically significant* difference between those who received the medicine and those who received the placebo. This does *not* mean that the medicine does not cause drowsiness. Data alone cannot be used to determine cause-and-effect relationships. Also, we should be very cautious about using results such as this to draw final conclusions about the safety or efficacy of a product. Statistical results should be one component, but other factors such as the severity of the possible side effects and the magnitude of the benefits should also be considered. In short:

Do not make a final definitive conclusion based solely on a P-value.

Example 5.4.1 Polling Results In a survey of voters in the 2008 Texas Democratic primary, 54% of 1167 females voted for Hillary Clinton while 47% of 881 males voted for Clinton (data from www.CNNPolitics.com, March 5, 2008). Use these data to test the claim that of the voters in the 2008 Texas Democratic primary, the proportion of females who voted for Clinton is higher than the proportion of males who voted for Clinton.

We use the same five-step hypothesis-testing procedure outlined in Section 5.1.

1. The parameters about which the claim is made are

    p_1 = the proportion of all females who voted for Clinton

    and p_2 = the proportion of all males who voted for Clinton.

    (Note that we are really only referring to females and males in the 2008 Texas Democratic primary.) The claim is $p_1 > p_2$, so we test the hypotheses

    $$H_0: p_1 = p_2, \qquad H_1: p_1 > p_2.$$

2. To calculate the test statistic, we need the number of successes in each sample. That is, we need to know the numbers of females and males who voted for Clinton. These are $x_1 = 0.54(1167) \approx 630$. Likewise, $x_2 = 0.47(881) = 414$. Then we have

    $$\hat{p} = \frac{630 + 414}{1167 + 881} \approx 0.510$$

    and the test statistic is

    $$z = \frac{0.54 - 0.47}{\sqrt{0.510(1 - 0.510)(1/1167 + 1/881)}} \approx 3.14.$$

3. At the 0.05 significance level, the critical value is 1.645. The P-value is the area to the right of 3.14, which is 0.0008.
4. The critical region is $[1.645, \infty)$, so the technical conclusion is to reject H_0.
5. Because the original claim does not contain equality, the final conclusion is that the data support the claim. Note that this conclusion can only be applied to voters in the 2008 Texas Democratic primary.

> ## Software Calculations
>
> To perform a hypothesis test comparing two population proportions p_1 and p_2:
>
> **Minitab**: Select **Stat** → **Basic Statistics** → **2-Proportion**. Select **Summarized Data**, and enter n_1 and n_2 as the **Trials** and x_1 and x_2 as the **Events**. Select **Options**, enter the **Confidence level**, choose the inequality in H_1 next to **Alternative**, set the **Test difference** to 0, and *do* check **Use pooled estimate of p for the test**. Note that Minitab states the test in terms of the difference $(p_1 - p_2)$, but the results are equivalent to those in the text.
>
> **R**: The syntax for performing a two-tail test is
>
> > prop.test(c(x_1, x_2), c(n_1, n_2), alternative="two.sided", correct=FALSE)
>
> For a one-tail test, change "two.sided" to either "less" or "greater" depending on the inequality in H_1. Note that R uses a different algorithm for computing the test statistic than the one in the text. So it will not give the test statistic z. However, the P-value given by R is the same as that given by the algorithm in the text.
>
> **TI-83/84 Calculators**: Press **STAT** → **TESTS** → **6:2-PropZTest**. Enter x_1, n_1, x_2, and n_2, and select the inequality in H_1.

Exercises

Directions: In exercises asking to test a claim, use the five-step hypothesis-testing procedure. Give either the P-value or the critical region. Unless otherwise specified, use a 0.05 significance level.

1. A student claims that a type of candy has more red pieces than blue pieces. A random sample of 250 candies contains 65 red and 55 blue candies. Explain why it is *not* appropriate to use this data and a 2-proportion Z-test to test this claim.

2. It seems reasonable to believe that students who are athletes take naps more often than nonathletes. To test this, a student surveyed 75 athletes and 82 nonathletes at her university and found that 72% of the athletes take naps on a regular basis, whereas 71% of the nonathletes do (data collected by Hannah Denk, 2009). Use these data to test the claim that the proportion of athletes at this university who take naps on a regular basis is higher than the proportion of nonathletes who do. Does it appear that athletes take naps more often than nonathletes?

3. Using sign-in records from the university weight room, a student randomly chooses several users and records the person's gender and the time of day the person used the weight room (morning or evening). The results are summarized in the table below (data collected by Joshua Ostrem, 2011). Let p_1 be the proportion of all morning users who are male and let p_2 be the proportion of all evening users who are male. Test the claim that $p_1 = p_2$. Based on this result, is there much difference between morning and evening in terms of the proportion of male users?

	Morning	Evening
Males	228	503
Females	255	495

4. To compare older quarters to newer quarters, a student spun several quarters from the 1960s and several from the 2000s and recorded whether each landed heads up or tails up. The results are summarized in the table below (data collected by Ray Spiering, 2011). Test the claim that the proportion of tails when spinning new quarters is greater than the proportion of tails when spinning old quarters. Based on this result, does there appear to be a significant difference between old and new quarters when spinning them?

	Old	New
Heads	351	296
Tails	249	304

5. Two students asked several fellow students with blonde or brunette hair their preference in hair color of the opposite sex. The results are summarized in the table below (data collected by Kayla Asche and Natalie Jeppensen, 2011). Test the claim that the proportion of blondes who prefer blondes equals the proportion of brunettes who prefer brunettes. Based on this result, does it appear that students typically prefer members of the opposite sex with the same hair color as themselves?

	Preference				
	Blonde	Brunette	Black	Red	None
Blonde	21	7	0	2	7
Brunette	12	32	1	0	16

6. In a test of a new pain medicine, $n_1 = 75$ patients were given a placebo while $n_2 = 70$ patients were given the medicine. Within 6 months, $x_1 = 1$ patient who was given the placebo had a heart attack and $x_2 = 5$ patients given the medicine had a heart attack.

a. Use the given data to test the claim that the proportion of heart attacks among those given a placebo equals the proportion among those given the medicine. Does there appear to be a *statistically significant* difference between these two medicines? Does this mean the new medicine is safe?

b. Now suppose it was discovered that three patients given the medicine who had a heart attack were "overlooked" so that the data really should be $n_2 = 73$ and $x_2 = 8$. Now does there appear to be a statistically significant difference between these two medicines?

7. Two treatments for a kind of tumor are tested on both small and large tumors. The results are summarized in the table below, where x denotes the number of tumors successfully treated. Use a 0.2 significance level in the following claims.

	Treatment A		Treatment B	
	n	x	n	x
Small Tumors	90	84	260	224
Large Tumors	250	180	85	54
Total	340	264	345	278

a. Test the claim that the proportion of small tumors successfully treated by treatment A is higher than the proportion of small tumors successfully treated by treatment B.

b. Test the claim that the proportion of large tumors successfully treated by treatment A is higher than the proportion of large tumors successfully treated by treatment B. Based on this result and that in part a, which treatment appears to be better?

c. Test the claim that the proportion of *all* tumors successfully treated by treatment A is *less* than the proportion of *all* tumors successfully treated by treatment B. Based on this result, which treatment appears to be better? Does this agree with your answer in part b? (Results such as these, where the proportions of successes in different samples appear to be reversed when the samples are combined, are called *Simpson's paradox*.)

8. To compare the incidence of drowsiness among three new allergy medications, researchers selected three different groups of 100 people each and treated each group with one of the medications. The numbers of participants in each group who reported drowsiness are summarized in the table below.

Medication	A	B	C
Number	12	18	24

a. Test the claim that the proportion of all people taking medication A who experienced drowsiness is equal to the proportion of all people taking medication B who experienced drowsiness. Is there a statistically significant difference between medications A and B?

b. Repeat part a, but compare medication B to medication C. Is there a statistically significant difference between medications B and C?

c. Repeat part a, but compare medication A to medication C. Is there a statistically significant difference between medications A and C?

d. Generalize your observations: Is the relation of "not significantly different" transitive? That is, if population A is not significantly different from B and B is not significantly different from C, is it true that A is not significantly different from C? Explain.

9. We can informally compare two population proportions by constructing confidence interval estimates of each. If the intervals overlap, we conclude that there is not a significant difference between then. This, however, does not always give the same result as a formal hypothesis test. Suppose that $x_1 = 90$ out of $n_1 = 150$ females say they always wear a seat belt while $x_2 = 75$ out of $n_2 = 150$ males say they always wear a seat belt.

a. Construct a two-sided 90% confidence interval estimate of the proportion of all females who always wear a seat belt. Do the same for the males. Does there appear to be a difference between these two population proportions?

b. Now do a formal hypothesis test on the claim that the proportion of all females who always wear a seat belt equals the proportion of males who do the same. Use a 0.1 significance level. Based on this result, does there appear to be a difference between these two population proportions?

10. A hypothesis test such as the 2-proportion Z-test can be used to detect a difference between two population proportions. One problem with it is that it does not measure *how much* the difference is. One way to solve this is to construct a confidence interval for the difference of the proportions. Consider the following scenario: In a survey of 2200 suburban families, 550 said they do at least one load of laundry a day. In a survey of 2000 rural families, 450 said they do at least one load a day.

a. Use these data to test the claim that the proportion of suburban families who do at least one load of laundry a day is higher than the proportion of rural families who do. Based on this result, does there appear to be a *statistically significant* difference between the two populations?

b. Construct a 95% confidence interval for the difference of the two population proportions as described in Section 4.8. Based on this result, does there appear to be a *practical* difference between the two populations?

11. We can modify the 2-proportion Z-test to test a claim that the difference between two population proportions is some number c, using the test statistic

$$z = \frac{(\hat{p}_1 - \hat{p}_2) - c}{\sqrt{\hat{p}_1(1-\hat{p}_1)/n_1 + \hat{p}_2(1-\hat{p}_2)/n_2}}.$$

The critical region and P-value are found in the same way as in the regular test according to the inequality in H_1. For instance, a political candidate claims that voters in the suburbs support him more than voters in the city by more than 10 percentage points. If p_1 and p_2 denote the proportions of voters in the suburbs and the city, respectively, who support him, then the claim is $p_1 - p_2 > 0.1$ so that $c = 0.1$. The hypotheses for this claim are H_0: $p_1 - p_2 = 0.1$ and H_1: $p_1 - p_2 > 0.1$. A survey of $n_1 = 350$ voters from the suburbs contains $x_1 = 252$ who support him while a survey of $n_2 = 400$ voters from the city contains $x_2 = 220$ who support him. Use the data to test the claim.

12. To determine if caffeine consumption is related to time spent studying, two students asked several fellow students the following two questions: On average, on the night before a test, how many hours do you usually study? and During that time, how much caffeine do you consume? The results are summarized in the table below (data collected by Johnathon Carnoali and Brandon Rehm, 2011). Let p_1 be the proportion of students who consume one or more servings of caffeine in the population of students who study 2+ hr and p_2 be the proportion in the population who study 1–2 hr. Use the procedures in Exercise 11 above to test the claim that $p_1 - p_2 > 0.2$.

	No Caffeine	One or More Servings
2+ Hr Studying	13	21
1–2 Hr Studying	67	19

5.5 Comparing Two Variances

In this section, we present a hypothesis test for the comparison of two population variances called the *F-test*. To illustrate this test, consider the following scenario: The weights of a sample of 28 blue M&Ms have a standard deviation of $s_1 - 0.0555$ g while a sample of 18 red M&Ms has a standard deviation of $s_2 = 0.0338$ g (data collected by Frank Ohlinger,

2009). Based on these data, it appears that blue M&Ms have a larger variance (or standard deviation) than the red ones do. We want to determine if this observed difference in the sample standard deviations is statistically significant or not.

We begin by defining the parameters:

$$\sigma_1^2 = \text{the variance of the weights of all blue M\&Ms}$$

and

$$\sigma_2^2 = \text{the variance of the weights of all red M\&Ms}.$$

Then the hypotheses are

$$H_0:\ \sigma_1^2 = \sigma_2^2, \qquad H_1:\ \sigma_1^2 > \sigma_2^2.$$

To derive the appropriate test statistic, let S_1^2 and S_2^2 denote the two sample variances from samples of size n_1 and n_2, respectively. Assuming both populations are normally distributed, then by Theorem 4.7.1 the random variables

$$\frac{(n_1-1)S_1^2}{\sigma_1^2} \quad \text{and} \quad \frac{(n_2-1)S_2^2}{\sigma_2^2}$$

are $\chi^2(n_1-1)$ and $\chi^2(n_2-1)$, respectively. Then by Theorem 3.9.3, the variable

$$F = \frac{\left[(n_1-1)S_1^2/\sigma_1^2\right]/(n_1-1)}{\left[(n_2-1)S_2^2/\sigma_2^2\right]/(n_2-1)} = \frac{S_1^2/\sigma_1^2}{S_2^2/\sigma_2^2}$$

has an F distribution with (n_1-1) and (n_2-1) degrees of freedom. Now, if $H_0:\ \sigma_1^2 = \sigma_2^2$ were true, then the variable F would simplify to

$$F = \frac{S_1^2}{S_2^2}.$$

Thus the appropriate test statistic is

$$f = \frac{s_1^2}{s_2^2}.$$

In this scenario, the test statistic is

$$f = \frac{0.0555^2}{0.0338^2} \approx 2.696.$$

If $H_0: \sigma_1^2 = \sigma_2^2$ were true, then we would expect the test statistic f to be close to 1. A large value of f suggests that H_0 may not be true. We consider f to be large, causing us to reject H_0, if $P(F > f)$ is small (i.e., less than 0.05).

In this scenario, F has $(28 - 1) = 27$ and $(18 - 1) = 17$ degrees of freedom. In Table C.5, the closest degrees of freedom are 30 and 15, and we see that $f_{0.05}(30, 15) = 2.25$. This is the (approximate) critical value (using software, the exact value is closer to 2.17). This means that

$$P(F > 2.25) \approx 0.05 \quad \Rightarrow \quad P(F > 2.696) < 0.05 \tag{5.3}$$

so that we reject H_0 and conclude that the data do support the claim that the weights of blue M&Ms have a larger variance than do red M&Ms.

The P-value is the area under the F-distribution density curve to the right of the test statistic $f = 2.696$. This is very difficult to approximate using Table C.5. We see from Equation (5.3) that the P-value is less than 0.05, but this is as exact as we can get using the table. Using software, the exact P-value is found to be 0.018.

This test is called the *F-test*, and its procedures are summarized in the box on the next page.

We need to make two comments about the F-test:

1. We require that s_1^2 be the larger of the two variances so that the test statistic f is greater than 1. This means that the critical values are also greater than 1. If we did not make this requirement, then we would have to include more critical values in Table C.5, making it even longer.

2. The requirement of normality is very strict, meaning that if either population is not normal, then this test will not give valid results. Because of this sensitivity, some statisticians recommend against using this test. Many alternative tests for the equality of population variances have been devised. (See Exercise 20 in Section 5.6 for one such alternative.)

Example 5.5.1 **Verifying Assumptions** To compare two different diets, call them A and B, volunteers followed each diet for six months and their weight losses were recorded. Among the $n_1 = 10$ subjects who followed diet A, their mean weight loss was $\bar{x}_1 = 4.5$ lb with a standard deviation of $s_1 = 6.5$ lb. Among the $n_2 = 10$ subjects who followed diet B, their mean weight loss was $\bar{x}_2 = 3.2$ lb with a standard deviation of $s_2 = 4.5$ lb.

To compare these two diets, we could construct a 2-sample T-interval estimate of the population means. To construct such an interval, we first need to determine if it is reasonable to assume that the two populations have the same variance. We can do this using the F-test.

> **F-Test for a Claim Comparing Two Population Variances**
>
> **Purpose**: To test a claim comparing the variances of two independent populations where the null hypothesis is of the form H_0: $\sigma_1^2 = \sigma_2^2$. The test statistic is
>
> $$f = \frac{s_1^2}{s_2^2}$$
>
> where
>
> - s_1^2 and s_2^2 are the sample variances,
> - s_1^2 is the *larger* of the two variances, and
> - n_1 and n_2 are the sample sizes.
>
> The critical value is an F-value from Table C.5 with $n = (n_1 - 1)$ and $d = (n_2 - 1)$ degrees of freedom, and the P-value is an area under the corresponding F-distribution density curve.
>
> | Form of H_1 | Critical Region | P-Value |
> |---|---|---|
> | $\sigma_1^2 > \sigma_2^2$ | $[f_\alpha(n,d), \infty)$ | Area to the right of f |
> | $\sigma_1^2 \neq \sigma_2^2$ | $[f_{\alpha/2}(n,d), \infty)$ | Twice the area to the right of f |
>
> **Requirements**
> 1. Both samples are random and independent.
> 2. Both populations are normally distributed (strict requirement).

Specifically, we want to test the claim that the two populations have the same variance. We follow the basic five-step hypothesis-testing approach:

1. The parameters about which the claim is made are

$$\sigma_1^2 = \text{the variance of the weight loss for all those on diet } A$$
and
$$\sigma_2^2 = \text{the variance of the weight loss for all those on diet } B.$$

The claim is $\sigma_1^2 = \sigma_2^2$, so we test the hypotheses

$$H_0: \sigma_1^2 = \sigma_2^2, \qquad H_1: \sigma_1^2 \neq \sigma_2^2.$$

2. The test statistic is

$$f = \frac{6.5^2}{4.5^2} \approx 2.09.$$

3. At the 0.05 significance level, the critical value is $f_{0.05/2}(9,9) = 4.03$. The P-value is twice the area to the right of $f = 2.09$. Using software, this value is found to be 0.288.

4. The critical region is $[4.03, \infty)$. Because $f = 2.09$ does not lie in this region (and the P-value is greater than 0.05), the technical conclusion is to not reject H_0.
5. Because the original claim does contain equality, the final conclusion is that there is not sufficient evidence to reject the claim. Therefore, it appears to be reasonable to assume that the populations have the same variance. □

Using the F-distribution, we can construct a confidence interval estimate of the ratio of two population variances σ_1^2/σ_2^2. As shown above, the variable

$$F = \frac{S_1^2/\sigma_1^2}{S_2^2/\sigma_2^2}$$

has an F-distribution with $(n_1 - 1)$ and $(n_2 - 1)$ degrees of freedom. Let f_L and f_R be two critical values such that

$$P(f_L \leq F \leq f_R) = 1 - \alpha. \qquad (5.4)$$

Figure 5.4 illustrates the definition of f_L and f_R.

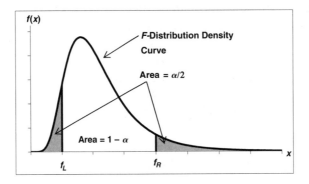

Figure 5.4

Then substituting the definition of F into Equation (5.4) and rewriting the inequality so that σ_1^2/σ_2^2 appears in the middle, we find

$$P\left(\frac{1}{f_R} \cdot \frac{S_1^2}{S_2^2} \leq \frac{\sigma_1^2}{\sigma_2^2} \leq \frac{1}{f_L} \cdot \frac{S_1^2}{S_2^2}\right) = 1 - \alpha.$$

This leads to the definition of the confidence interval.

Definition 5.5.1 Let $0 < \alpha < 1$ and let s_1^2 and s_2^2 be the variances of independent samples of size n_1 and n_2, respectively, from two normally distributed populations with variances σ_1^2 and σ_2^2. Also, let f_L and f_R be numbers such that

$$P(f_L \leq F \leq f_R) = 1 - \alpha$$

where F has an F-distribution with $(n_1 - 1)$ and $(n_2 - 1)$ degrees of freedom. The interval

$$\left[\frac{1}{f_R} \cdot \frac{s_1^2}{s_2^2}, \frac{1}{f_L} \cdot \frac{s_1^2}{s_2^2} \right]$$

is called a $100(1 - \alpha)\%$ *confidence interval estimate for the ratio* σ_1^2/σ_2^2.

The tricky part about calculating such a confidence interval lies in finding f_L and f_R from Table C.5. The next example illustrates how to do this.

Example 5.5.2 Calculating a Confidence Interval Suppose two independent samples of size $n_1 = 16$ and $n_2 = 10$ taken from two normally distributed populations have standard deviations of $s_1 = 3.25$ and $s_2 = 1.52$, respectively. Calculate a 95% confidence interval for σ_1^2/σ_2^2.

In this case, the random variable F has 15 and 9 degrees of freedom. To find f_R, we observe in Figure 5.4 that

$$P(F > f_R) = \frac{0.05}{2} = 0.025.$$

Thus $f_R = f_{0.025}(15, 9) = 3.77$. To find f_L, observe again in Figure 5.4 that

$$P(F < f_L) = 0.025 \quad \Rightarrow \quad P\left(\frac{1}{F} > \frac{1}{f_L}\right) = 0.025.$$

But because F has 15 and 9 degrees of freedom, the variable $1/F$ has 9 and 15 degrees of freedom (see Exercise 18 of Section 3.9). Thus $1/f_L = f_{0.025}(9, 15) = 3.12$ so that $f_L = 1/3.12 \approx 0.321$.

Therefore, the confidence interval is

$$\left[\frac{1}{3.77} \cdot \frac{3.25^2}{1.52^2}, \frac{1}{0.321} \cdot \frac{3.25^2}{1.52^2} \right] = [1.21, 14.2].$$

Based on this confidence interval, we are 95% confident that σ_1^2 is from 1.21 to 14.2 times as large as σ_2^2. Because this confidence interval does not contain 1, we are confident that $\sigma_1^2 > \sigma_2^2$. □

Generalizing the calculations in this example, we get the following formulas for f_R and f_L:

$$f_R = f_{\alpha/2}(n, d) \quad \text{and} \quad f_L = \frac{1}{f_{\alpha/2}(d, n)}$$

where $n = (n_1 - 1)$ and $d = (n_2 - 1)$.

Software Calculations

To perform a hypothesis test comparing two population variances σ_1^2 and σ_2^2:

Minitab: If you have raw data, enter both samples in two different columns. Select **Stat → Basic Statistics → 2 Variances**. Select **Samples in different columns**, and enter the names of the columns containing the data. If you only have sample statistics, select **Summarized data** and enter the necessary statistics. (Note that Minitab asks for the sample variances, which are squares of the sample standard deviations.)

R: R will give both the *P*-value for the hypothesis test and a confidence interval estimate of the ratio σ_1^2/σ_2^2. Enter the two samples of data $\{x_1, x_2, \ldots, x_n\}$ and $\{y_1, y_2, \ldots, y_m\}$ into vectors with the syntax x<−c($x_1, x_2, \ldots, x_n$) and y<−c($y_1, y_2, \ldots, y_m$). The syntax for performing a two-tail test and obtaining a two-sided confidence interval at the 95% confidence level is

> var.test(x, y, ratio = 1, alternative = "two.sided", conf.level = 0.95)

For a one-tail test, change "two.sided" to either "less" or "greater" depending on the inequality in H_1.

Excel: Enter both samples in different columns of a spreadsheet. Select **Data → Data Analysis → F-Test 2-Sample for Variances**. Enter the ranges containing the two samples, and select a blank cell for the **Output Range**. In the output, the *P*-value for a one-tail test is labeled "P(F<=f) one-tail." To find the *P*-value for a two-tail test, simply double the *P*-value in the output. To find the critical value $f_p(n, d)$, use the syntax FINV(p, n, d).

TI 83/84 Calculators: If you have raw data, enter both samples in two different lists. Press **STAT → E:2-SampFTest → Data**. Enter the names of the lists containing the data, set **Freq1** and **Freq2** to 1, and select the inequality in H_1. If you only have sample statistics, select **Stats**, enter the necessary statistics, and select the inequality in H_1.

5.5 Comparing Two Variances

Exercises

Directions: In exercises asking to test a claim, use the five-step hypothesis-testing procedure. Give either the P-value or the critical region. Unless otherwise specified, use a 0.05 significance level.

1. Supervisors at a company that manufactures baseball bats have retooled some machinery in an attempt to reduce variation in the weights of the bats. Before retooling, a sample of 41 bats had a standard deviation of 1.24 oz. After retooling, a sample of 61 bats had a standard deviation of 0.54 oz. Assuming that both populations are normally distributed, test the claim that the variance before retooling is higher than the variance after retooling. Does the retooling appear to have worked?

2. A wine connoisseur records the percentage of alcohol content of 31 different red wines and 33 white wines. The standard deviation of the red wines was 1.24% while the standard deviation of the white wines was 1.55% (data collected by Rachel Brainard, 2011). Assuming that the populations are normally distributed, test the claim that both populations of wine have the same variance. P-value $= 0.2190$

3. A sample of 53 noneducation majors at a university has GPAs with a standard deviation of 0.449 while a sample of 47 education majors has a standard deviation of 0.336 (data collected by Nicole Sempek and Allison Chrismer, 2009). Assuming that both populations are normally distributed, test the claim that the variance of noneducation majors is higher than the variance of education majors.

4. A sample of 23 bluegill from a certain lake has a mean weight of 1.25 lb with a standard deviation of 0.15 lb while a sample of 18 catfish from the same lake has a mean weight of 4.36 lb with a standard deviation of 1.02 lb. Assuming that both populations are normally distributed, use an F-test to determine if the two populations have the same variance at the 99% confidence level.

5. Calculate f_R and f_L used in the confidence interval estimate for the ratio σ_1^2/σ_2^2 when $\alpha = 0.05$ for each of the following values of n_1 and n_2.

   a. $n_1 = 8$, $n_2 = 13$
   b. $n_1 = 10$, $n_2 = 21$
   c. $n_1 = 31$, $n_2 = 61$
   d. Generalize your results. What happens to the values of f_R and f_L as the sample sizes get larger? What does this mean about the width of the confidence interval as the sample sizes get larger (assume s_1^2 and s_2^2 are the same for any sample size)? What does this imply about the quality of s_1^2/s_2^2 as an estimate of σ_1^2/σ_2^2?

6. A student records the sodium content (in milligrams) per serving of several different generic brand and name brand cereals. The results are shown in the table below (data collected by Rachel Masters, 2009).

	n	$\bar{x}$	s
Generic Brand	62	158.22	73.30
Name Brand	74	137.79	60.69

   a. Assuming both populations are normally distributed, construct a 95% confidence interval estimate of the ratio of the two population variances. Based on this result, does there appear to be a significant difference between the variances? Explain.

   b. Now use the F-test to test the claim that the variances in the sodium content of the two types of cereal are equal. Does this agree with your conclusion in part a?

7. A student records the masses (in grams) of several regular M&Ms and several peanut M&Ms. The results are shown in the table below (data collected by Brian Maxson, 2011).

	n	$\bar{x}$	s
Regular	31	0.8889	0.04982
Peanut	31	2.370	0.30772

   a. Assuming both populations are normally distributed, construct a 95% confidence interval estimate of the ratio of the two population variances. Suppose someone claims that the variance of the peanut M&Ms is about 50 times as large as that of the regular M&Ms. Do our results support this claim? Explain.

   b. Now suppose the sample sizes were both 201 instead of 31, but that the sample standard deviations did not change. Construct the resulting 95% confidence interval estimate of the ratio of the two population variances. Compare the width of this interval to that found in part a. Generalize this observation: If the sample sizes increase, what happens to the width of the confidence interval?

8. A student conducts an experiment where she starts a stopwatch and asks a fellow student to tell her when 30 sec has passed. She then stops the watch and records how much time has actually passed. The times (in seconds) for 15 male and 15 female students are shown

in the table below (data collected by Courtney Cale, 2011). Assuming that the populations are normally distributed, perform an F-test to determine if it is reasonable to assume that the two populations have the same variance.

Males	18.6	28.7	21.6	30.5	24.0	37.8	15.5	30.6
	16.3	51.4	47.3	38.9	26.6	21.4	16.4	
Females	30.6	29.7	41.5	25.0	23.3	26.5	37.0	18.8
	38.4	22.7	28.3	40.9	36.1	19.8	30.7	

9. Construct a 95% confidence interval estimate of the ratio of the population variances using the data in Exercise 8 above.

10. The masses (in grams) of several red and blue candies are shown in the table below.

Red	0.7619	0.7791	0.7969	0.8096	0.8147	0.8184	0.8384
	0.7645	0.7967	0.8020	0.8096	0.8160	0.8207	
Blue	0.8559	0.8839	0.8900	0.8935	0.8992	0.9090	0.9270
	0.8644	0.8873	0.8905	0.8988	0.9023	0.9135	

Assuming that both populations of candies are normally distributed, use an F-test to determine if the two populations have the same variance at the 90% confidence level.

11. Construct a 95% confidence interval estimate of the ratio of the population variances, using the data in Exercise 10 above.

5.6 Comparing Two Means

Tests for comparing two population means fall into the same categories as methods for constructing confidence intervals for the difference of two means:

1. The population variances are known (or both samples are large),
2. The population variances are unknown but assumed to be equal, and
3. The population variances are unknown and *not* assumed to be equal.

The first test is called a *2-sample Z-test* while the last two are called *2-sample T-tests*. The theoretical details behind these tests are all very similar to those presented in Section 4.8 and Appendix A.5, so we omit them. These tests are summarized in the box on the next page.

2-Sample Z- and T-Tests for Comparing Two Independent Population Means
Purpose: To test a claim regarding the means of two independent populations where the null hypothesis is of the form H_0: $\mu_1 - \mu_2 = c$. Begin by selecting random and independent samples from each population. Let

- n_1 and n_2 denote the sample sizes,
- $\bar{x}_1$ and $\bar{x}_2$ denote the sample means, and
- s_1^2 and s_2^2 denote the sample variances, respectively.

The test statistic is calculated according to the following cases:

1. If both population variances are known, then the test statistic is

$$z = \frac{(\bar{x}_1 - \bar{x}_2) - c}{\sqrt{\sigma_1^2/n_1 + \sigma_2^2/n_2}}.$$

   The critical value is a z-value, and the P-value is an area under the standard normal density curve.

2. If the population variances are not known but are assumed to be equal, then the test statistic is

$$t = \frac{(\bar{x}_1 - \bar{x}_2) - c}{s_p\sqrt{1/n_1 + 1/n_2}}$$

   where

$$s_p = \sqrt{\frac{(n_1 - 1)s_1^2 + (n_2 - 1)s_2^2}{n_1 + n_2 - 2}}$$

   is the *pooled standard deviation*. The critical value is a t-value with $(n_1 + n_2 - 2)$ degrees of freedom, and the P-value is an area under the Student-t density curve with this number of degrees of freedom.

3. If the population variances are not known and *not* assumed to be equal, then the test statistic is

$$t = \frac{(\bar{x}_1 - \bar{x}_2) - c}{\sqrt{s_1^2/n_1 + s_2^2/n_2}}.$$

   The critical value is a t-value with r degrees of freedom, where r is

$$r = \frac{\left(s_1^2/n_1 + s_2^2/n_2\right)^2}{[1/(n_1 - 1)]\left(s_1^2/n_1\right)^2 + [1/(n_2 - 1)]\left(s_2^2/n_2\right)^2}$$

   If r is not an integer, then round it down to the nearest whole number. (**Note**: If software is used to find the critical value, there is no need to round r.) The P-value is an area under the Student-t density curve with r degrees of freedom.

5.6 Comparing Two Means

In all three cases, the critical region and P-value are found according to the form of H_1 in the same way as for other types of tests.

Requirements
1. Both samples are random.
2. The samples are independent.
3. Both populations are normally distributed, or both sample sizes are greater than 30.

If the variances are unknown, but both sample sizes are large (meaning greater than 30), then the sample variances provide a good estimate of the population variances so that the Z-test can be used where σ_1^2 and σ_2^2 are replaced with s_1^2 and s_2^2. For this reason, the 2-sample Z-test is called a *large-sample* test.

Example 5.6.1 Comparing Diets To compare two different diets, call them A and B, volunteers followed each diet for 6 months, and their weight losses were recorded. Among the $n_1 = 10$ subjects who followed diet A, their mean weight loss was $\bar{x}_1 = 4.5$ lb with a standard deviation of $s_1 = 6.5$ lb. Among the $n_2 = 10$ subjects who followed diet B, their mean weight loss was $\bar{x}_2 = 3.2$ lb with a standard deviation of $s_2 = 4.5$ lb. Assuming both populations are normally distributed and have the same variance, test the claim that the mean weight loss on diet A is higher than that on diet B.

We follow the basic five-step hypothesis-testing approach:

1. The parameters about which the claim is made are

$$\mu_1 = \text{the mean weight loss for all those on diet } A$$
$$\text{and} \quad \mu_2 = \text{the mean weight loss for all those on diet } B.$$

The claim is $\mu_1 > \mu_2$, which is equivalent to $\mu_1 - \mu_2 > 0$, so we test the hypotheses

$$H_0: \mu_1 - \mu_2 = 0, \qquad H_1: \mu_1 - \mu_2 > 0.$$

2. The pooled standard deviation is

$$s_p = \sqrt{\frac{(10-1)6.5^2 + (10-1)4.5^2}{10 + 10 - 2}} = 5.59$$

so that the test statistic is

$$t = \frac{(4.5 - 3.2) - 0}{5.59\sqrt{1/10 + 1/10}} \approx 0.52$$

3. At the 0.05 significance level, the critical value is $t_{0.05}(10 + 10 - 2) = 1.734$. The P-value is the area to the right of $t = 0.52$ under the Student-t density curve with 18 degrees of freedom. Using software, this value is found to be 0.305.
4. The critical region is $[1.734, \infty)$. Because $t = 0.52$ does not lie in this region (and the P-value is greater than 0.05), the technical conclusion is to not reject H_0.
5. Because the original claim does not contain equality, the final conclusion is that the data do not support the claim. Therefore, it appears that the mean weight loss for diet A is not statistically different from that for diet B. □

Example 5.6.2 Nitrate in Water A town has installed a new water filtration system in hopes of reducing the levels of nitrate in the drinking water by an average of more than 15 mg/L. Samples of water from randomly selected houses across the town before and after the system was installed were tested, and the results (in milligrams per liter) are shown in Table 5.5. Assuming that both populations are normally distributed, test the claim that the difference in the mean nitrate levels before and after the system was installed is more than 15. Does the goal appear to have been met?

	n	$\bar{x}$	s
Before	25	55.6	3.49
After	18	36.7	5.67

Table 5.5

We perform a 2-sample T-test and do *not* assume the population variances are equal:

1. The parameters about which the claim is made are

$$\mu_1 = \text{the mean nitrate level before the system was installed}$$
and
$$\mu_2 = \text{the mean nitrate level after the system was installed.}$$

The claim is $\mu_1 - \mu_2 > 15$, so we test the hypotheses

$$H_0: \mu_1 - \mu_2 = 15, \qquad H_1: \mu_1 - \mu_2 > 15.$$

2. The test statistic is

$$t = \frac{(55.6 - 36.7) - 15}{\sqrt{3.49^2/25 + 5.67^2/18}} \approx 2.59$$

and the number of degrees of freedom is

$$r = \frac{\left(3.49^2/25 + 5.67^2/18\right)^2}{[1/(25-1)](3.49^2/25)^2 + [1/(18-1)](5.67^2/18)^2} \approx 26.$$

3. At the 0.05 significance level, the critical value is $t_{0.05}(26) = 1.706$. The P-value is the area to the right of $t = 2.59$ under the Student-t density curve with 26 degrees of freedom. Using software, this value is found to be 0.008.
4. The critical region is $[1.706, \infty)$. Because $t = 2.59$ lies in this region (and the P-value is less than 0.05), the technical conclusion is to reject H_0.
5. Because the original claim does not contain equality, the final conclusion is that the data support the claim. Thus it appears that the goal has been met. □

One important requirement for a 2-sample test is that the data come from two independent samples. In the next example, we collect survey results from each student before taking a class and after taking the class, and we compare the results. At first glance, this appears to be a 2-sample test. However, the two samples (the results before and the results after) are not independent because they come from the same group of students. So a 2-sample test does not apply. Instead, we apply the *paired T-test* as described in the box below.

Paired *T*-Test

Purpose: To test the claim that a set of paired data (x_i, y_i), $i = 1, \ldots, n$, come from a population in which the differences $(x - y)$ have a mean less than, greater than, or equal to 0.

Steps
1. Let μ_d denote the population mean of the differences.
2. The null hypothesis is H_0: $\mu_d = 0$.
3. Calculate the differences $(x_i - y_i)$, $i = 1, \ldots, n$. This forms a sample of size n from the population of all differences.
4. Use the differences from step 3 and the T-test from Section 5.3 to test the claim.

Requirements
1. The data are randomly selected.
2. The differences come from a normally distributed population or $n > 30$.

Example 5.6.3 Comparing Test Scores At a large university, freshmen students are required to take an introduction to writing class. Students are given a survey on their attitudes toward writing at the beginning and end of the class. Each student receives a score between 0 and 100 (the higher the score, the more favorable his or her attitude toward writing). The scores of nine different students from the beginning and end of the class are shown in Table 5.6. Test the claim that the scores improve from the beginning to the end.

Beginning	77.6	86.1	62.1	93.1	74.6	42.0	86.9	79.3	80.2
End	85.4	83.2	64.2	96.6	79.3	40.5	87.2	89.2	85.6
(End − Beginning)	7.8	−2.9	2.1	3.5	4.7	−1.5	0.3	9.9	5.4

Table 5.6

These data are said to come in "matched pairs" (each of the nine students provides two numbers, a beginning score and an end score). Thus these two samples are *not* independent, and it would be inappropriate to use a 2-sample test to compare the mean scores at the beginning to those at the end. Instead, we use a paired T-test.

We look at the difference in the scores (end − beginning). If the score really did improve from beginning to end, then we expect these differences to be positive. Specifically, the mean of these differences would be positive. Define the parameter

$$\mu_d = \text{the population mean of the difference (end − beginning)}.$$

Then we test the claim that $\mu_d > 0$. Assuming this population of differences is normally distributed, we use a T-test as described in Section 5.3.

1. The claim is $\mu_d > 0$, so we test the hypotheses

$$H_0: \mu_d = 0, \qquad H_1: \mu_d > 0.$$

2. The sample mean and standard deviation of the differences in the third row of Table 5.6 are $\bar{x} = 3.256$ and $s = 4.215$. Thus the test statistic is

$$t = \frac{3.256 - 0}{4.215/\sqrt{9}} \approx 2.32.$$

3. At the 0.05 significance level, the critical value is $t_{0.05}(9-1) = 1.860$. The P-value is the area to the right of $t = 2.32$ under the Student-t density curve with 8 degrees of freedom. Using software, this value is found to be 0.025.
4. The critical region is $[1.860, \infty)$. Because $t = 2.47$ does lie in this region (and the P-value is less than 0.05), the technical conclusion is to reject H_0.
5. Because the original claim does not contain equality, the final conclusion is that the data support the claim. Therefore, it appears that the scores do improve from the beginning to the end. □

Software Calculations

To perform a hypothesis test regarding the difference of two population means $\mu_1 - \mu_2$:

Minitab: If you have raw data, enter the two samples in separate columns. To perform a T-test, select **Stat** $\to$ **Basic Statistics** $\to$ **2-Sample t**. Select **Samples in different columns** and enter the names of the columns containing the data. If you assume that the two populations have the same variance, check the box next to **Assume equal variances**. Select **Options**, enter the **Confidence level**, enter c next to **Test difference**, and choose the inequality in H_1 next to **Alternative**. If you only have sample statistics, choose **Summarized data** and enter the necessary statistics. To perform a paired T-test, follow the same steps except choose **Paired t** instead of **2-Sample t**.

R: Enter the two samples of data $\{x_1, x_2, \ldots, x_n\}$ and $\{y_1, y_2, \ldots, y_m\}$ into vectors with the syntax x<−c$(x_1, x_2, \ldots, x_n)$ and y<−c$(y_1, y_2, \ldots, y_m)$. The syntax for performing a two-tail test where the two samples are independent and the two populations are not assumed to have the same variance is

> t.test(x, y, mu=c, alternative="two-tail", paired=FALSE,
        equal.variances=FALSE)

If the two populations are assumed to have the same variance, change "equal.variances=FALSE" to "equal.variances=TRUE." To perform a one-tail test, change "two-tail" to "less" or "greater" depending on the inequality in H_1. To perform a paired T-test, change "paired=FALSE" to "paired=TRUE."

Excel: Enter the two samples of data into different columns. Select **Data** $\to$ **Data Analysis**. Choose **t-Test: Paired Two Sample for Mean**, **t-Test: Two-Sample Assuming Equal Variances**, **t-Test: Two-Sample Assuming Unequal Variances**, or **z-Test: Two-Sample for Means** depending on what type of test you want to perform. For each type of test, enter the ranges containing the samples, enter c next to **Hypothesized Mean Difference**, enter the significance level next to **Alpha**, and select a blank cell for the **Output Range**. For the Z-test, also enter the (known) population variances σ_1^2 and σ_2^2.

TI-83/84 Calculators: The calculator will only perform a hypothesis test where $c = 0$. If you have raw data, enter the two samples in different lists. For a Z-test where the population standard deviations σ_1 and σ_2 are known, press **STAT** → **TESTS** → **3:2-SampZTest**. Select **Data**, enter σ_1 and σ_2, the names of the lists containing the data, and select the inequality in H_1. For a T-test, select **4:2-SampTTest**, choose **Data**, and enter the necessary information. Always set **Freq1** and **Freq2** to 1. If you assume the two populations have the same variance, select **Pooled: Yes**; otherwise, select **No**. For either type of test, if you only have sample statistics, choose **Stats** and enter the necessary statistics. To perform a paired T-test, enter the values of the differences in a list and use the procedures for testing a claim about a single population mean in Section 5.3.

Exercises

Directions: In exercises asking to test a claim, use the five-step hypothesis-testing procedure. Give either the P-value or the critical region. Unless otherwise specified, use a 2-sample T-test and a 0.05 significance level.

1. A sample of 47 education majors at a certain university has a mean GPA of 3.46 and a standard deviation of 0.336 while a sample of 53 noneducation majors has a mean GPA of 3.38 and a standard deviation of 0.449 (data collected by Nicole Sempek and Allison Chrismer, 2009). Use a 2-sample Z-test to test the claim that education majors at this university have a higher mean GPA than noneducation majors.

2. A wine connoisseur recorded the percentage of alcohol content of different red and white wines. The resulting sample statistics are summarized in the table below (data collected by Rachel Brainard, 2011). Assuming that both populations are normally distributed, and that these data come from random samples of each population, use a 2-sample Z-test to test the claim that the mean percentage of alcohol content of all red wines is 1 percentage point higher than the mean percentage of alcohol content of all white wines.

	n	$\bar{x}$	s
Red	31	12.91	1.24
White	33	11.30	1.55

3. A student at a small Christian university wanted to study the mood of students before and after chapel. Before chapel on several different days, he asked 30 fellow students to rate their mood on a scale of 1 to 10, where 1 represents a bad mood and 10 a good mood. The

scores had a mean of 6.70 with a standard deviation of 2.23. He then surveyed a different sample of 30 students after chapel and got a mean score of 6.67 with a standard deviation of 2.14 (data collected by Justin Boutwell, 2009). Assuming that the populations are normally distributed and have the same variance, test the claim that the mean score before chapel is the same as the mean score after chapel. Does there appear to be a significant difference between the moods before and after chapel?

4. Four bottles of brand A of bottled water have a mean volume of 510.75 mL with a standard deviation of 1.25 mL. Four bottles of brand B have a mean volume of 505 mL with a standard deviation of 0.816 mL (data collected by Brittany Singleton and Kelsie Elder, 2009). Assuming that both populations are normally distributed and have the same variance, test the claim that the mean volume of brand A is more than 5 mL higher than that of brand B.

5. At a summer camp, 40 boys and 40 girls between the ages of 9 and 11 years old were shown images of 12 different items used at the camp. They were allowed to study the images for 40 to 45 sec, and then they took a break. After the break, they were given 1 min to write down as many items as they could recall, and the number of items correctly recalled by each was recorded. The boys correctly recalled an average of 9.35 items with a standard deviation of 1.61, while the girls had an average of 10.13 items with a standard deviation of 1.44 (data collected by Tyler Zander, 2011). Assuming that both populations are normally distributed and have the same variance, and that these data come from random samples of each population, test the claim that the mean number of items correctly recalled is the same for boys and girls. Does there appear to be a significant difference between the mean recall times of boys and girls?

6. Two baseball players hit 30 baseballs off a tee with a metal bat, recorded the amount of time each ball took to reach home plate, and calculated the average velocity of each. Then they repeated this, using a wood bat. The results (in miles per hour) are summarized in the table below (data collected by Daniel Ullman and Daniel Wilkerson, 2011). Assuming that both samples are random and the populations are normally distributed, test the claim that the mean average velocity for baseballs hit with a metal bat is 8 mph greater than the mean average velocity for baseballs hit with a wood bat (do not assume equal variances).

	n	$\bar{x}$	s
Metal	30	95.83	15.678
Wood	30	81.16	10.279

7. A soccer fan records the heights (in centimeters) of several randomly chosen European World Cup players and several players from the Americas. The sample statistics are summarized in the table below (data collected by Santiago Keinbaum, 2011).

	n	$\bar{x}$	s
Europe	61	181.26	5.69
Americas	69	179.83	7.19

a. Test the claim that the mean height of all European players equals the mean height of all players from the Americas. Use the 2-sample Z-test and then the 2-sample T-test where we do not assume equal variances. Use software to calculate the P-value for each test.

b. Is there much difference between the P-values from the two tests in part a? Based on this result, if we have large sample sizes, does it really matter which test we use?

8. To compare two types of wind turbines, Micon and Vestas, the outputs of several turbines of each type were recorded (in kilowatt hours) under similar wind conditions. The data are shown in the table below (data collected by Chris Jensen, 2010). Assuming that both populations are normally distributed and that they have the same variance, test the claim that the mean output of Vestas is more than 700 kWh greater than that of Micon.

	n	$\bar{x}$	s
Micon	27	301.7	10.5
Vestas	9	1019.22	9.54

9. A student records the masses (in grams) of several regular M&Ms and several peanut M&Ms. The results are shown in the table below (data collected by Brian Maxson, 2011). Assume that both populations are normally distributed, but their variances are not equal. Use a T-test to test the claim that the mean mass of all peanut M&Ms is 1.25 g higher than that of regular M&Ms.

	n	$\bar{x}$	s
Regular	30	0.8889	0.04982
Peanut	30	2.370	0.30772

10. Polyethylene livestock water tanks are manufactured by pouring melted resin into a mold. As the resin cools, it shrinks a bit. The percentages of shrink of 30 randomly selected tan-colored and 30 randomly selected black-colored tanks are shown in the table below (data collected by Russell D. Bartling, 2010). Assuming that both populations are normally distributed and have the same variance, test the claim that the mean percentage of shrink of tan tanks is the same as that for black tanks. Based on these results, does the color of the tank appear to affect the percentage of shrink?

Tan	2.08	2.37	2.97	2.37	2.08	2.37	2.67	2.37	2.67	2.97
	2.37	2.67	2.67	2.97	2.67	2.97	2.67	2.37	2.67	2.37
	2.37	2.37	2.37	2.37	2.37	2.08	2.08	2.67	2.67	2.67
Black	2.97	3.26	2.67	2.37	2.67	2.37	2.67	2.67	2.67	2.37
	2.97	2.67	2.08	2.08	2.67	2.37	2.97	2.67	2.67	2.97
	2.37	2.67	2.37	2.08	2.37	2.97	2.67	2.97	2.67	2.67

11. Do younger prisoners receive shorter sentences? The table below gives the minimum sentences given to a sample of prisoners at a state penitentiary who did not get life sentences (data collected by Helen Schnackenberg, 2011). Test the claim that the mean minimum sentence for prisoners age 25 years and under is less than the mean sentence for those over age 25 years (do not assume equal variances).

Age 25 Years and Under	1	8	4	20	2	2.5	2	1.5	9	3	12	1.25	4	4	3	2
	3	2	3	3	2	6	4	2.5	5	2	2	2.5	10	4	1	3.5
Over Age 25 Years	12.5	3	9	3	1.5	1.5	4	2	4	10	5	30	15	4	2	12
	3	3	8	26	5	15	6	4	5	8	13.5	4	35	8	10	9

12. Are mothers having their first child at an older age? A researcher asked 33 mothers who had their first child prior to 1993 their age at the birth of their first child. She then asked the same question of 33 mothers who had their first child in 1993 or later. The data are given in the table on the next page (data collected by Sue Kohlwey, 2012). Assuming these data are randomly selected, use the 2-sample Z-test to test the claim that the mean age of mothers who had their first child prior to 1993 is less than the mean age of mothers

who had their first child in 1993 or later. According to these results, does it appear that mothers now are having their first child at an older age?

Prior to 1993	15	17	19	19	20	20	20	21	21	21	21	21	21	22	23	23	23
	23	24	24	24	24	24	25	25	26	26	27	28	28	28	30	32	
1993 or Later	16	18	19	20	20	21	22	22	22	22	22	22	24	25	25	25	26
	26	26	26	27	27	27	27	28	28	29	30	32	33	33	35	35	

13. In Example 5.6.3, we assumed that the population of differences is normally distributed. Construct a quantile plot of the sample of the differences in Table 5.6. Does the assumption of normality appear to be reasonable?

14. To test the effectiveness of a training program, the 400-m track times (in seconds) of eight athletes were recorded before and after 3 weeks of training, as shown in the table below (data collected by Andrew Messersmith, 2009). Use these data to test the claim that the mean of the differences (before − after) is positive. Based on this result, does the training program appear to have a significant effect on track times?

Before	53.8	54.2	52.9	53.0	55.1	58.2	56.7	52.0
After	52.7	53.8	51.1	52.7	54.2	55.4	53.3	49.8

15. Do shoes increase vertical jump height? To analyze this question, 10 male athletes at a university jumped three times with shoes and three times without shoes, and the best height (in inches) for each was recorded in the table below (data collected by Todd May, 2010). Test the claim that in the population of all male athletes at this university, the mean of the differences (with − without) is positive. Based on this result, does it appear that shoes significantly affect vertical jump height?

With	15.5	18.5	22.5	17.5	16.5	17.5	20.5	15.5	19.5	18.5
Without	15.5	17.5	22.5	15.5	15.5	18.5	21.5	14.5	17.5	17.5

16. Twelve male students at a university were asked to report their heights in inches. After they reported their heights, the surveyor accurately measured their heights. The reported and measured heights are given in the table on the next page. Test the claim that in the population of all males at this university, the mean of the differences (reported − measured) is positive. Do these data support the claim that male students tend to exaggerate their

height? Can you think of another reason to explain the discrepancy between reported and measured heights?

Reported	67	73	67.5	70	68	70	71	68	67.5	66	70	71.5
Measured	66.3	72.8	68	69.4	67.3	70.2	70.9	67.6	67.1	65.3	68.8	70.9

17. Two types of thermometers, bimetallic stemmed and infrared, were used to measure the exterior temperatures at several points on an engine. The data (in degrees Celsius) are given in the table below. Use the data to determine if there is a significant difference between the temperature readings of the two types of thermometers.

Bimetallic	87.5	82.9	59.2	49.2	42.7	32.9
Infrared	88.7	83.7	61.9	49.1	43.1	33.2

18. We can informally compare two population means by constructing confidence interval estimates of each and then determining if the intervals overlap. This usually, but not always, gives the same conclusion as a formal hypothesis test of the equality of the means. The table below shows the sodium content per serving (in milligrams) of several different generic brand and name brand cereals (data collected by Rachel Masters, 2009).

	n	$\bar{x}$	s
Generic Brand	62	158.22	73.30
Name Brand	74	137.79	60.69

a. Calculate an appropriate two-sided 90% confidence interval estimate for the mean amount of sodium in all name brand cereals using the techniques in Section 4.8. Do the same for the generic brand cereals. What type of confidence interval did you use, a Z or T? Based on these confidence intervals, does there appear to be a significant difference between the mean amounts of sodium for the different types of cereal?

b. Now use an appropriate hypothesis test to test the claim that the mean amount of sodium in generic brand cereals equals the mean amount of sodium in name brand cereals using a 90% confidence level. What type of 2-sample test did you use, a Z or T? Based on this result, does there appear to be a difference between these two population means? Is this the same conclusion reached in part a?

19. The masses (in grams) of several red and blue candies are shown in the table below.

Red	0.7619	0.7791	0.7969	0.8096	0.8147	0.8184	0.8384
	0.7645	0.7967	0.8020	0.8096	0.8160	0.8207	
Blue	0.8559	0.8839	0.8900	0.8935	0.8992	0.9090	0.9270
	0.8644	0.8873	0.8905	0.8988	0.9023	0.9135	

a. Generate a normal quantile plot of each set of data. Do the two populations appear to be normally distributed?

b. Use an F-test to determine if the two populations have the same variance.

c. Use an appropriate 2-sample T-test to test the claim that the mean weight of blue candies is greater than that of the red candies.

20. As mentioned in Section 5.5, the F-test for the equality of two population variances is very sensitive to nonnormality. A more robust test for the equality of two variances is the *modified Levene's* test (also called the *Levene–Brown–Forsythe* test). The algorithm is as follows:

1. Calculate the median of each sample, m_1 and m_2.

2. For each data value in the first sample x, replace it with $|x - m_1|$. Repeat for the second sample using m_2.

3. Use a 2-sample T-test to test the claim that these transformed values come from populations with equal means and variances (i.e., test the claim that $\mu_1 = \mu_2$ using the values in step 2 assuming equal variances).

4. If we reject H_0 in step 3, then conclude that the two populations have different variances; otherwise, conclude that the variances are equal.

Use the modified Levene's test to test the claim that the two samples in the table below come from populations with equal variances.

Sample 1	5.66	6.62	6.76	7.58	7.66	7.77	7.95	8.38	8.38	8.50	9.13
Sample 2	7.41	7.62	7.63	7.73	7.84	8.15	8.26	8.27	8.46	8.54	8.75

5.7 Goodness-of-Fit Tests

In this section, we present a test for the claim that a random variable has a certain distribution. We have seen several examples of informally testing such a claim. For instance, in Example 3.3.7, we informally tested the claim that the time people wait in line in a grocery store had an exponential distribution by constructing a histogram of the data. The "shape" of the histogram resembled the graph of an exponential pdf, so we concluded that the claim was reasonable. We showed in Section 4.10 how we could determine if a random variable has a normal distribution by constructing a normal quantile plot and looking for a straight-line pattern.

These two approaches are relatively simple. However, they both require a subjective interpretation of a graph. The *chi-square goodness-of-fit test* we present here is a more objective approach. To illustrate the idea behind this test, suppose a gambler claims that a die is "fair." To test the claim, he rolls it 180 times and records the data shown in the second row of Table 5.7.

Outcome	1	2	3	4	5	6
Observed Freq	34	22	27	28	36	33
Expected Freq	30	30	30	30	30	30

Table 5.7

Let the random variable X denote the outcome of a single roll, and define

$$p_i = P(X = i), \quad \text{for } i = 1, \ldots, 6.$$

If the die were indeed fair, then $p_i = \frac{1}{6}$ for each i. In other words, the claim is that X has a uniform distribution. So we test the hypotheses

$$H_0: p_1 = \cdots = p_6 = \frac{1}{6}, \qquad H_1: \text{at least one } p_i \neq \frac{1}{6}.$$

If the die were indeed fair (i.e., if H_0 were true), then we would expect each side to appear $180(\frac{1}{6}) = 30$ times. Comparing the second and third rows of Table 5.7, we see that the observed and expected frequencies are close to each other. We measure the total difference between these quantities with the following calculation:

$$c = \frac{(34-30)^2}{30} + \frac{(22-30)^2}{30} + \cdots + \frac{(33-30)^2}{30} = 4.6. \tag{5.5}$$

A large value of c would indicate that the observed and expected frequencies were not close and would cause us to reject H_0. A small value of c would indicate the opposite. The following theorem gives us a way to determine if c is large or small.

Theorem 5.7.1 *Suppose a discrete random variable X has the range $\{x_1, x_2, \ldots, x_k\}$ and that*

$$p_i = P(X = x_i).$$

Let the random variable X_i denote the number of times x_i appears in n total observed values of X. Then the random variable

$$C = \sum_{i=1}^{k} \frac{(X_i - np_i)^2}{np_i}$$

is approximately $\chi^2(k-1)$. This approximation is reasonable if $np_i \geq 5$ for all i. □

To relate this theorem to the example of testing if the die is fair, note the following:

1. The variable X is the result of one roll of the die. Its range is $\{1, \ldots, 6\}$ so that $k = 6$.
2. The 180 rolls of the die result in $n = 180$ observed values of X.
3. The data in the second row of Table 5.7 are observed values of $X_1, \ldots, X_6$.
4. The number c calculated in Equation (5.5) is an observed value of the variable C.

Example 5.7.1   Justifying Theorem 5.7.1   The proof of Theorem 5.7.1 is far beyond the scope of this text. To justify this theorem, we simulated 1000 trials of rolling a truly fair die 180 times. For each trial, we calculated the value of c as in Equation (5.5). This yielded 1000 observed values of the random variable C. A histogram of these data is shown in Figure 5.5. A graph of the $\chi^2(6-1)$ pdf is plotted on top of the histogram. Note that the curve follows the tops of the bars, indicating that the $\chi^2(6-1)$ distribution is reasonable. □

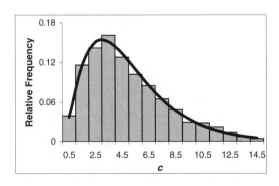

Figure 5.5

Theorem 5.7.1 tells us that to determine if c is large or small, we need to compare it to a critical chi-square value with $(k-1)$ degrees of freedom. If $c > \chi_\alpha^2(k-1)$, we consider c to be large, causing us to reject H_0.

The critical values are given in Table C.4. If we choose a 95% confidence level for the claim regarding the die, then $\alpha = 0.05$, and the appropriate degrees of freedom is $r = 6 - 1 = 5$. Thus the critical value is $\chi_{0.05}^2(5) = 11.07$. We have $c = 4.6 < 11.07$, so we do not reject H_0 and we conclude that there is not sufficient evidence to reject the claim that the die is fair.

The P-value for a chi-square test is the area to the right of c under the $\chi^2(k-1)$ density curve. This is usually found with software. For this claim, the P-value is approximately 0.47. This is greater than $\alpha = 0.05$, so we do not reject H_0. The procedures for this test are summarized in the box below.

Note that if H_0 were true, then by the definition of $\chi_\alpha^2(k-1)$ the probability of rejecting H_0 would be

$$P(C > \chi_\alpha^2(k-1)) = \alpha.$$

Thus the significance level α is the probability of rejecting a true H_0, a Type I error, as in all other hypothesis tests.

Chi-Square Goodness-of-Fit Test

Purpose: To test the claim that a random variable X has some particular distribution.

Steps

1. Collect n observed values of X.
2. Define k disjoint categories of the values of X in such a way that every possible value of X belongs to exactly one category.
3. Define
$$p_i = P(X \in \{\text{category } i\}), \quad \text{for } i = 1, \ldots, k$$
to be the *exact* values of these probabilities.
4. Let π_i be the *claimed* values of $P(X \in \{\text{category } i\})$, $i = 1, \ldots, k$, calculated according to the claimed distribution.
5. The hypotheses are
$$H_0: p_i = \pi_i, \quad \text{for } i = 1, \ldots, k, \qquad H_1: p_i \neq \pi_i, \quad \text{for at least one } i.$$
6. Let O_i be the number of data values in category i, $i = 1, \ldots, k$ (these are the observed frequencies).
7. Calculate $E_i = n\pi_i$, $i = 1, \ldots, k$ (these are the expected frequencies).

8. Calculate the test statistic c:

$$c = \sum_{i=1}^{k} \frac{(O_i - E_i)^2}{E_i}.$$

9. Find the critical value $\chi_\alpha^2(k-1)$, or calculate the P-value, which is the area to the right of c under the $\chi^2(k-1)$ density curve.

10. If $c > \chi_\alpha^2(k-1)$ or the P-value is less than α, then reject H_0.

Requirements
1. The data have been randomly chosen.
2. Each expected frequency E_i is at least 5.

Example 5.7.2 **Colors of M&Ms** In 1996, the colors of plain M&M candies were distributed as follows: 30% brown, 20% yellow, 20% red, 10% orange, 10% green, and 10% blue. Some years later, a class of statistics students tested the claim that the distribution has not changed. They recorded the colors of 1191 randomly selected plain M&M candies, as shown in the first two columns of Table 5.8.

To set up the test, consider the random experiment of choosing a plain M&M candy. Define the random variable X to be 1 if the candy is brown, 2 if it is yellow, etc. The observed frequencies of the colors are the observed frequencies of the values of X. The null hypothesis is thus

$$H_0:\ p_1 = 0.3,\ p_2 = p_3 = 0.2,\ p_4 = p_5 = p_6 = 0.1.$$

The calculation of the test statistic is shown in Table 5.8.

Color	O_i	E_i	$(O_i - E_i)^2/E_i$
Brown	154	$1191(0.3) = 357.3$	115.7
Yellow	201	$1191(0.2) = 238.2$	5.8
Red	156	$1191(0.2) = 238.2$	28.4
Orange	228	$1191(0.1) = 119.1$	99.6
Green	205	$1191(0.1) = 119.1$	62.0
Blue	247	$1191(0.1) = 119.1$	137.4
			$c = 448.9$

Table 5.8

At the 95% confidence level, the critical value is $\chi_{0.05}^2(6-1) = 11.071$. Because $c = 448.9 > 11.071$, we reject H_0 and conclude that the distribution of colors has changed. □

Example 5.7.3 **Dandelions in a Lawn** Consider a 1000 ft² lawn that contains 3000 dandelions, and let X denote the number of dandelions in a randomly chosen 1 ft² section of lawn. It seems reasonable to claim that X has a Poisson distribution with mean $\lambda = 3$.

To test this claim, a student simulated such a lawn by randomly placing 300 dots on a piece of paper with an area of 100 in². He then randomly chose 75 different 1 in² sections of paper and counted the number of dots in each section. The data are summarized in Table 5.9.

Number of Dots	0	1	2	3	4	5	6	7	8
Freq	4	15	14	16	13	6	4	2	1

Table 5.9

Now let p_i denote the probability that i dots appear in a randomly chosen 1 in² section of paper. If X, the number of dots in a 1 in² section of paper, really has a Poisson distribution with mean $\lambda = 3$ as claimed, then, for instance, the claimed value of p_1 is

$$\pi_1 = P(X = 1) = \frac{3^1}{1!} e^{-3} \approx 0.149.$$

Because X could have any positive integer value and the largest observed value of X is 8, we combine all values of X greater than or equal to 8 into one category, $[8, \infty)$, so that

$$\pi_8 = P(X \geq 8) = 1 - P(X \leq 7) \approx 1 - 0.9881 = 0.0119.$$

The values of π_i and the expected frequencies $E_i = 75\pi_i$ are shown in Table 5.10.

x	0	1	2	3	4	5	6	7	≥ 8
π_i	0.050	0.149	0.224	0.224	0.168	0.101	0.050	0.022	0.012
E_i	3.73	11.20	16.80	16.80	12.60	7.56	3.78	1.62	0.89

Table 5.10

Notice that the last three values of E_i are less than 5, violating the requirements for the test. This problem can be easily rectified by combining the corresponding values of X into one category, $[6, \infty)$. The resulting calculations are shown in Table 5.11.

Using these seven different categories, the null hypothesis is

$$H_0: p_0 = 0.050, \, p_1 = 0.149, \ldots, p_6 = 0.084.$$

x	π_i	O_i	E_i	$(O_i - E_i)^2/E_i$
0	0.050	4	3.73	0.02
1	0.149	15	11.20	1.29
2	0.224	14	16.80	0.47
3	0.224	16	16.80	0.04
4	0.168	13	12.60	0.01
5	0.101	6	7.56	0.32
≥ 6	0.084	7	6.29	0.08
				$c = 2.23$

Table 5.11

At the 95% confidence level, the critical value is $\chi^2_{0.05}(7-1) = 12.59$. Because $c = 2.23 < 12.59$, we do not reject H_0 and we conclude that it is reasonable to assume that the number of dots is described by a Poisson distribution with parameter $\lambda = 3$. □

In the previous example, the parameter of the claimed distribution was assumed to be known. In some cases, the parameter(s) of the claimed distribution are not known and must be estimated using the data. In these cases, we "lose a degree of freedom" for each parameter estimated. Thus if we estimate one parameter, the critical value is $\chi^2_\alpha(k-1-1)$. If we estimate two parameters, the critical value is $\chi^2_\alpha(k-1-2)$. The next example illustrates this idea.

Example 5.7.4 **Exponential Distribution** Table 5.12 shows the times spent waiting in line at a grocery store (in minutes) of 30 different customers. We test the claim that X, the amount of time a randomly selected customer spends waiting in line, has an exponential distribution.

0.03	0.21	0.33	0.66	0.91	1.33	1.56	1.84	2.74	3.46
0.04	0.21	0.36	0.87	0.92	1.40	1.60	1.87	2.79	3.90
0.10	0.22	0.47	0.90	0.96	1.51	1.76	2.36	3.24	4.25

Table 5.12

To test this claim with a chi-square test, first we note that X is continuous and can take on any positive value. Thus we divide the values of X into subintervals (or categories) and count the number of data values in each interval, as in the first two rows of Table 5.13.

Interval	$[0,1)$	$[1,2)$	$[2,3)$	$[3,4)$	$[4,\infty)$
Freq	15	8	3	3	1
π_i	0.508	0.250	0.123	0.060	0.058
E_i	15.24	7.5	3.69	1.8	1.74

Table 5.13

Now, to calculate the claimed probabilities π_i, we need to estimate the parameter of the claimed distribution. The mean of these data is $\bar{x} = 1.41$, so the parameter λ is approximately

$$\lambda = \frac{1}{1.41} \approx 0.71.$$

Thus the pdf of X is approximately

$$f(x) = 0.71e^{-0.71x}.$$

We use this pdf to calculate the values of π_i. For instance,

$$\pi_1 = P(0 \leq X < 1) = \int_0^1 0.71e^{-0.71x}\, dx \approx 0.508.$$

The other values of π_i are shown in the third row of Table 5.13. Using these values, we calculate $E_i = 30\pi_i$ as shown in the fourth row. Notice that the last three values of E_i are less than 5, so we combine these intervals into one as shown in Table 5.14.

Interval	π_i	O_i	E_i	$(O_i - E_i)^2/E_i$
$[0,1)$	0.508	15	15.24	0.0038
$[1,2)$	0.250	8	7.5	0.0333
$[2,\infty)$	0.242	7	7.26	0.0093
				$c = 0.0464$

Table 5.14

Using these three intervals, the null hypothesis is

$$H_0\colon p_1 = 0.508, p_2 = 0.250, p_3 = 0.242$$

and the test statistic is $c = 0.0464$. Because we estimated the parameter λ, we have $3 - 1 - 1 = 1$ degree of freedom. At the 95% confidence level, the critical value is $\chi^2_{0.05}(1) = 3.841$. Because $c = 0.0464 < 3.841$, we do not reject H_0 and we conclude that it is reasonable to assume that the wait time is described by an exponential distribution. □

Software Calculations

To perform a χ^2 goodness-of-fit test:

Minitab: Enter the observed frequencies in one column and the expected proportions in another column. Select **Stat → Tables → Chi-Square Goodness-of-Fit Test**. Enter the name of the column containing the observed frequencies next to **Observed counts**. Under **Test**, select either **Equal proportions** or **Specific proportions** depending on what you are claiming. In the latter case, enter the name of the column containing the expected proportions. Note that the given P-value is based on $(k-1)$ degrees of freedom.

R: The syntax is
> chisq.test(c($O_1, \ldots, O_k$), p = ($\pi_1, \ldots, \pi_k$))

where the O's are the observed frequencies and the π's are the claimed proportions. Note that the given P-value is based on $(k-1)$ degrees of freedom.

Excel: Excel does not have a built-in function for performing the χ^2 goodness-of-fit test. However, you may easily calculate the P-value for such a test with the syntax CHIDIST(c, df) where c is the test statistic from the test and df is the number of degrees of freedom.

TI-83/84 Calculators: Enter the observed frequencies in list L_1 and the expected frequencies in list L_2. On a TI-84 calculator, press **STAT → TESTS → D:χ^2GOF-Test**. Enter the names of the lists containing the observed and expected values, and enter the number of degrees of freedom, denoted **df**.

On a TI-83 calculator, at the home screen enter the command $(L_1 - L_2)^2 / L_2 \to L_3$. Then press **LIST → MATH → 5:sum(→ L_3 → ENTER**. This gives the value of the test statistic c. To find the P-value, press **DISTR → 8:χ^2cdf(** and use the following syntax:
χ^2cdf(c, 1E99, df)

where df is the number of degrees of freedom and E is entered by pressing **2ND → ,**

Exercises

Directions: In each of the following exercises, (1) define the parameters, (2) state the null hypothesis, (3) calculate the test statistic, (4) state the number of degrees of freedom, (5) find the critical value or the P-value, and (6) state the technical and final conclusions.

1. A union leader at a factory claims that worker absences on the different days of the week are equally likely. To test this claim, a company manager collects the data shown in the table below.

Day	Mon.	Tues.	Wed.	Thurs.	Fri.
Number of Absences	35	14	13	25	42

   a. Without doing any calculations, do the data appear to support the claim? Explain.
   b. Use a chi-square goodness-of-fit test to test this claim at the 99% confidence level.

2. A student claims that the four different colors of a certain kind of candy occur with equal frequency. He selects a random sample of 60 of these candies and records the frequencies of the colors as shown in the table below (data collected by Jake Harmon, 2009).

Color	Yellow	Orange	Red	Pink
Freq	18	20	12	10

   a. Without doing any calculations, do the data appear to support the claim? Explain.
   b. Use a chi-square goodness-of-fit test to test this claim at the 90% confidence level.
   c. Now suppose the observed frequencies are 10 times the original amounts (that is, 180, 200, 120, and 100, respectively). Use these larger frequencies to test the claim at the 99% confidence level.
   d. Based on your results in parts b and c, if data appear to contradict a claim of equal frequencies, would you be more confident in rejecting the claim if the sum of the frequencies were small or large?

3. When a single card is dealt from a well-shuffled deck, it seems reasonable to assume that each suit will occur with equal frequency. To test this assumption, two students shuffled a full deck, dealt one card, recorded the suit, replaced the card, and repeated this a total of 150 times. The results are summarized in the table below (data collected by Faith Hoppen and Sarah Mack, 2011). Use these data to test the claim that the suits occur with equal frequency. Based on this result, is it reasonable to say that $P(\text{hearts}) = P(\text{diamonds}) = P(\text{clubs}) = P(\text{spades}) = \frac{1}{4}$ when a single card is dealt from a full deck?

Suit	Hearts	Diamonds	Clubs	Spades
Freq	37	49	34	30

4. Demographic records show that historically, of students who attend a certain university, 65% are from the home state of the university, 25% are from another state, and the remainder are international students. In the current year, 510 enrolled students are from the home state, 258 are from another state, and 32 are international students. Does it appear that the demographics have changed? If so, which categories appear to have changed the most? Support your conclusion with an appropriate hypothesis test.

5. A bag has four blue, three red, two green, and one yellow cube. Consider the random experiment of choosing 3 cubes without replacement, and let X denote the number of blue cubes in the sample of 3. Then X has a hypergeometric distribution, and its pmf is shown in the first two rows of the table below.

x	0	1	2	3
$f(x)$	0.167	0.500	0.300	0.033
Freq	47	104	75	14

A group of students does 240 trials of this random experiment and records the number of blue cubes in each trial, as shown in the third row of the table. Use a chi-square test to determine if the pmf of X is as claimed at the 99% confidence level.

6. A biologist claims that in a certain type of corn, $\frac{9}{16}$ of the kernels are smooth and purple (S & P), $\frac{3}{16}$ are wrinkled and purple (W & P), $\frac{3}{16}$ are smooth and yellow (S & Y), and $\frac{1}{16}$ are wrinkled and yellow (W & Y). One particular ear of corn had frequencies of the different types of kernels shown in the table below. Use the data to test the claim at the 95% confidence level.

S & P	W & P	S & Y	W & Y
212	85	62	19

7. A student claims that in a certain brand of candy, 20% of the candies are red, 30% are blue, 40% are green, and 10% are yellow. She randomly selects 30,000 such candies and records the data in the table below.

Color	Red	Blue	Green	Yellow
Freq	5850	9060	11,850	3240

   a. Use a goodness-of-fit test to test the claim with these data. Does there appear to be a *statistically significant* difference between the actual proportions of the colors and the claimed proportions?

b. Calculate the sample proportions of each color. Based on these results, does there appear to be a *practically significant* difference between the actual proportions of the colors and the claimed proportions? Explain.

8. When testing a new type of missile for the army, researchers shot 519 missiles toward a large target area. They then divided the area into 570 squares of $\frac{1}{4}$ km^2 each and counted the number of missiles that landed in each square. The data are shown in the table below. Use these data to test the claim that X, the number of missiles that landed in a randomly selected $\frac{1}{4}$ km^2 section, has a Poisson distribution at the 95% confidence level. Give the estimated value of the parameter of the distribution.

Number of Missiles	0	1	2	3	≥ 4
Freq	225	221	86	32	6

9. Use a chi-square test to determine if the data given in the table below are observed values of a random variable X with a normal distribution at the 99% confidence level. Give the estimated values of the parameters of the distribution. (**Suggestion**: Use five intervals, $(-\infty, 2)$, $[2, 4)$, $[4, 6)$, $[6, 8)$, and $[8, \infty)$.)

0.87	3.39	3.74	4.04	4.34	4.99	5.71	5.92	7.14	8.07
2.42	3.61	3.74	4.06	4.36	5.38	5.85	6.63	7.19	8.20
2.86	3.61	3.91	4.18	4.79	5.43	5.91	6.71	7.21	10.24

10. A student claims to have written a computer program that will generate values of a random variable X with the pdf

$$f(x) = \begin{cases} 2x - 2, & 1 \leq x \leq 2 \\ 0, & \text{otherwise.} \end{cases}$$

The table below shows 30 values generated by the program. Use a chi-square test to determine if the program works as claimed at the 95% confidence level. (**Suggestion**: Use four intervals of width 0.25.)

1.03	1.16	1.26	1.30	1.45	1.55	1.62	1.69	1.85	1.99
1.08	1.18	1.28	1.42	1.46	1.58	1.65	1.76	1.93	1.99
1.15	1.20	1.29	1.44	1.53	1.60	1.66	1.79	1.95	2.00

11. A random sample of 100 candies of a certain type yields 54 red and 46 green candies.

   a. Use a 1-proportion Z-test to test the claim that in the population of all candies, the proportion of red candies equals 0.5. State the P-value.

   b. Use a chi-square goodness-of-fit test to test the claim that in the population, red and green candies occur with equal frequency. Use software to find the P-value. Is this the same P-value as in part a?

   c. Informally, explain why we should expect to get the same P-value in both parts a and b. (**Hint**: Consider the informal interpretation of a P-value.)

12. One pseudorandom generator designed to give values of a random variable that is $U(0, 1)$ has the following algorithm:

   1. $x_0 =$ an arbitrary number between 0 and 1,
   2. $x_{n+1} =$ fractional part of $(9821 * x_n + 0.211327)$.

   Use available software to generate 100 values with this algorithm and then use a goodness-of-fit test to test the claim that these are observed values of a variable that is $U(0, 1)$.

13. One algorithm for generating observed values of a random variable X with an exponential distribution and parameter λ, $\{x_1, \ldots, x_n\}$, is to first generate values of a random variable Y that is $U(0, 1)$, $\{y_1, \ldots, y_n\}$, and then calculate

$$x_i = -\frac{1}{\lambda} \ln(1 - y_i).$$

Use available software to generate 100 values of X with this algorithm (choose your own value of λ), and then use a goodness-of-fit test to test the claim that these are observed values of a variable with an exponential distribution and the chosen value of λ.

5.8 Test of Independence

In this section, we present a test for the independence of two sets of events. At first glance this type of test looks much different than a goodness-of-fit test studied in the previous section. However, the calculations are very similar, and the theory behind both tests relies on Theorem 5.7.1.

To illustrate this test, consider the following scenario: Two students want to determine if the university men's basketball team benefits from home-court advantage. They randomly select 205 games played by the team and record if each one was played at home or away and if the team won or lost, as summarized in Table 5.15 (data collected by Emily Hudgins and Courtney Santistevan, 2009). Such a table is called a *contingency table*.

	Result		
Location	Win	Loss	Total
Home	55	30	85
Away	58	62	120
Total	113	92	205

Table 5.15

If the team does *not* benefit from home-court advantage, then the result should not be related to the location. In other words, they should be *independent*, so we test the following hypotheses:

H_0: The result is independent of the location, $\quad H_1$: They are dependent.

To derive the test statistic, consider the random experiment of choosing a game. The outcome falls into four different, but not disjoint, events: home, away, win, and loss. If H_0 is true, then, for instance, the events home and win are independent. This means that

$$P(\text{home} \cap \text{win}) = P(\text{home}) \cdot P(\text{win}).$$

From the data, we estimate the two probabilities on the right:

$$P(\text{home}) \approx \frac{85}{205} \quad \text{and} \quad P(\text{win}) \approx \frac{113}{205}.$$

If H_0 is true, then in 205 games we would expect the number that are played at home and result in a win to be

$$\begin{aligned}
\text{Expected number at home and win} &= P(\text{home} \cap \text{win}) \cdot 205 \\
&= P(\text{home}) \cdot P(\text{win}) \cdot 205 \\
&\approx \frac{85}{205} \cdot \frac{113}{205} \cdot 205 \\
&= \frac{85 \cdot 113}{205} = 46.854.
\end{aligned}$$

We perform similar calculations to get the expected numbers for the other combinations of events, as shown in Table 5.16.

	Result	
Location	Win	Loss
Home	46.854	38.146
Away	66.146	53.854

Table 5.16

We compare the observed values to the expected values in the same way as a goodness-of-fit test, as shown in Table 5.17.

Event	O	E	$(O-E)^2/E$
Home and win	55	46.854	1.416
Home and loss	30	38.146	1.740
Away and win	58	66.146	1.003
Away and loss	62	53.854	1.232
			$c = 5.391$

Table 5.17

Analogous to Theorem 5.7.1, the test statistic c has a χ^2 distribution. To find the appropriate degrees of freedom, suppose the contingency table has a rows and b columns. Then the data fall into ab different categories, so we might be tempted to think that we have $(ab-1)$ degrees of freedom. We also have $(a+b)$ events, and we have to estimate the probability of each, so we lose some degrees of freedom. However, we do not have to estimate all of these probabilities directly from the data. Because the sum of the probabilities of the events corresponding to the rows of the contingency table must equal 1, we really only need to estimate $(a-1)$ of the probabilities directly from the data. Likewise, we only need to estimate $(b-1)$ of the probabilities associated with the columns. Thus the appropriate number of degrees of freedom is

$$ab - 1 - (a-1) - (b-1) = ab - a - b + 1 = (a-1)(b-1).$$

Chi-Square Test of Independence

Purpose: To test if the row events of a contingency table are independent of the column events. Let

- n = the total number of observations in the table,
- a = the number of rows in the table,
- b = the number of columns in the table,

- O_{ij} = the value in the ith row and jth column of the table, $i = 1, \ldots, a$, $j = 1, \ldots, b$,
- $R_i = \sum_{j=1}^{b} O_{ij}$ = the sum of the observations in the ith row of the table, and
- $C_j = \sum_{i=1}^{a} O_{ij}$ = the sum of the observations in the jth column of the table.

To test the null hypothesis

H_0: The rows are independent of the columns,

the expected frequencies are

$$E_{ij} = \frac{R_i \cdot C_j}{n}$$

and the test statistic is

$$c = \sum_{i=1}^{a} \sum_{j=1}^{b} \frac{(O_{ij} - E_{ij})^2}{E_{ij}}.$$

The critical value is $\chi^2_\alpha[(a-1)(b-1)]$. We reject H_0 if c is greater than the critical value. The P-value is the area to the right of c under the $\chi^2[(a-1)(b-1)]$ density curve.

Requirements
1. The data in the table represent frequency counts and are randomly selected.
2. All expected frequencies are at least 5.

In this scenario, $a = b = 2$ so that we have $(2-1)(2-1) = 1$ degree of freedom. At the 95% confidence level, the critical value is $\chi^2_{0.05}(1) = 3.841$. The P-value is the area to the right of $c = 5.391$ under the $\chi^2(1)$ density curve. By using software, this is found to be 0.02. Because $c = 5.391$ is greater than 3.841, we reject H_0 and conclude that the result is not independent of the location. Thus the team appears to benefit from home-court advantage. The procedures for the test of independence are summarized in the box above.

In Example 5.7.2, we used a goodness-of-fit test to determine if the colors of M&Ms occur with some claimed proportions. In the next example, we use a test of independence to determine if the colors of two different types of M&Ms occur with the same proportions. Such a test is called a *test of homogeneity*.

Example 5.8.1 Comparing M&Ms To determine if regular M&Ms have the same proportions of each color as peanut M&Ms, a student selects a random sample of each type and counts the number of each color, as shown in Table 5.18 (data collected by Rebekah Brisbois,

2009). Use these data to test the claim that the two types have the same proportions of the different colors.

Color (Observed)

Type	Blue	Red	Yellow	Green	Brown	Orange	Total
Regular	62	60	68	98	45	84	417
Peanut	31	10	41	13	12	35	142
Total	93	70	109	111	57	119	559

Table 5.18

Consider the random experiment of choosing an M&M (regular or peanut). If the two types have the same proportions of the different colors, then the color is independent of the type. So we perform a test of independence where the hypotheses are

H_0: The color is independent of the type, H_1: They are dependent.

Color (Expected)

Type	Blue	Red	Yellow	Green	Brown	Orange
Regular	69.376	52.218	81.311	82.803	42.521	88.771
Peanut	23.624	17.782	27.689	28.197	14.479	30.229

Table 5.19

The expected values are shown in Table 5.19, and the test statistic is

$$c = \frac{(62 - 69.376)^2}{69.376} + \cdots + \frac{(35 - 30.229)^2}{30.229} \approx 28.79.$$

We have $(2-1)(6-1) = 5$ degrees of freedom, so at the 95% confidence level, the critical value is $\chi^2_{0.05}(5) = 11.07$. The P-value is the area to the right of $c = 28.79$, which is approximately 0.00003 when found by using software. Because $28.79 > 11.07$, we reject H_0 and conclude that color is not independent of type. Thus it appears that the two types do not have the same proportions of the colors. □

Software Calculations

To perform a test of independence:

Minitab: Enter the first column of the contingency table in a blank column. Enter the next column of the contingency table in the next blank column, and so on. Select **Stat → Tables → Chi-Square Test (Two-Way Table in Worksheet)**. Enter the names of the columns containing the contingency table.

R: First enter the contingency table with the syntax
> tbl<− matrix(c($O_{11}, O_{21} \ldots, O_{ab}$), nrow=$a$ ncol=b)

where the O's are the observed frequencies read top to bottom and left to right from the contingency table and a and b are the number of rows and columns of the table, respectively. The syntax for the test is then
> chisq.test(tbl, correct=FALSE)

Excel: Excel does not have a built-in function for performing a test of independence. However, you may easily calculate the P-value for such a test with the syntax CHIDIST(c, df) where c is the test statistic and df is the number of degrees of freedom.

TI-83/84 Calculators: Enter the contingency table as a matrix by pressing **2ND → MATRIX → EDIT**. Scroll down to the desired matrix and press **ENTER**. Enter the size of the matrix and then the entries of the contingency table. Then press **STAT → TESTS → C:χ^2-Test**. Next to **Observed**, enter the name of the matrix containing the contingency table by pressing **2ND → MATRIX → NAMES**. Scroll down to the desired matrix and press **Enter**. Next to **Expected**, enter the name of the matrix where you want the expected frequencies to be stored (you do *not* have to calculate these values).

Exercises

Directions: In each exercise asking for a test of independence, (1) calculate the test statistic, (2) state the number of degrees of freedom, (3) find the critical value or the P-value, and (4) state the technical and final conclusions. Unless otherwise specified, use a 0.05 significance level.

1. In Example 5.8.1, we concluded that the two types of M&Ms do not have the same proportions of the colors. To assess where the differences are, calculate the sample proportions of each color for each type, using the data in Table 5.18. Compare the proportions.

Which colors appear to have equal proportions in both types? Which colors appear to have different proportions?

2. Several different pollsters ask 1196 voters whether they plan to vote in the next election. The responses and the genders of the pollsters are summarized in the table below. Use the data to test the claim that the response is independent of the gender of the pollster. Based on the data, does it appear that gender influences the response?

	Gender of Pollster		
Response	Male	Female	Total
Yes	500	350	850
No	248	98	346
Total	748	448	1196

3. To compare three different brands of cereal with marshmallows, a student counted the number of marshmallow and grain pieces in random samples of each brand. The data are shown in the table below (data collected by Katie Bangert, 2010). Test the claim that these three brands of cereal have the same proportions of marshmallow and grain pieces. Based on these results, which brand (if any) appears to have the highest proportion of marshmallow pieces?

	Name Brand	Generic Brand A	Generic Brand B	Total
Marshmallow	331	467	413	1211
Grain	1989	1387	1629	5005
Total	2320	1854	2042	6216

4. To test for side effects of a new allergy medicine, researchers give the medicine to 1500 volunteers and a placebo to 1550 other volunteers, and then they record the numbers who experience drowsiness as summarized in the table below. Test the claim that experiencing drowsiness is independent of treatment. Does it appear that those taking the medicine experience drowsiness more frequently than those taking the placebo?

	Drowsiness		
	Yes	No	Total
Medicine	39	1461	1500
Placebo	19	1531	1550
Total	58	2992	3050

5. The colors of the Nebraska Cornhuskers are scarlet and cream. A student surveyed 132 residents of Lincoln, Nebraska (home of the Cornhuskers), and asked each if he or she is a Cornhusker fan and what color his or her car is. The data are summarized in the table below. Use these data to test the claim that car color is independent of whether the person is a Cornhusker fan. Does it appear that Cornhusker fans tend to prefer any one specific color?

	Color of Car				
Fan	Red/Maroon	White/Cream/Gray	Black/Blue	Other	Total
Yes	24	31	21	17	93
No	4	18	8	9	39
Total	28	49	29	26	132

6. Is the probability of getting tails the same when flipping a coin as when spinning a coin? To answer this question, several students spun and flipped different types of coins and recorded the number of heads and tails obtained, as summarized in the table below (data collected by Austin Velasquez, Aleks Strehlke, Danielle Cornish, Jamie Crouse, Elizabeth Young, Jordan Ahl, David Eads, and Reed Shoaff, 2010). For each type of coin, test the claim that the proportions of heads and tails are the same when flipping as when spinning. If you were playing a coin game and were given the choice of coin and whether to flip or spin it, what would you choose and would you bet heads or tails? Explain.

	Dime		Nickel		Quarter		Half-Dollar		Dollar	
	Flip	Spin	Flip	Spin	Flip	Spin	Flip	Spin	Flip	Spin
Heads	242	278	263	320	241	254	253	249	247	238
Tails	258	222	237	180	259	246	247	251	253	262

7. When babies first crawl, some begin with an "army crawl" while others start on their hands and knees. The table below summarizes data on gender and first crawling styles of 151 babies (data collected by Anna Huff, 2011). Test the claim that gender and how a baby first crawls are independent. Does there appear to be a difference between genders and crawling styles?

	Boy	Girl
Army Crawl	41	17
Hands and Knees	47	46

8. To determine if the location of a person's home is related to dog ownership, two students asked several people from the Midwest where they live—in a rural area, on an acreage (a large plot of land near a city), or in a city; and whether they own a dog. The results are shown in the table below (data collected by Steven Musselwhite and Colton Schneider, 2010). Assuming this is a random sample of people from the Midwest, test the claim that dog ownership is independent of the location of a person's home. Does it appear that one type of home location is more often associated with dog ownership?

	Rural	Acreage	Town
Yes	43	43	26
No	21	17	30

9. It is often said that "sex sells," and as a result many television commercials are "sexually suggestive." To determine if the time of day affects the frequency of sexually suggestive commercials, two students recorded the number of such commercials on five different networks at three different times of day, as summarized in the table below (data collected by Seth Elley and Brenton Whitaker, 2011). Assuming that the data represent a random sample of all commercials during these time periods, use the data to determine if there is a significant difference in the proportions of sexually suggestive commercials during these time periods. Suppose a parent says, "I won't let my kids watch TV in the evening because there is more sex on TV than at other times of day." What, if anything, do our results say about the validity of this claim?

	9–10 AM	2–3 PM	7–8 PM
Suggestive	23	31	40
Not	165	169	161

10. The observed values in a contingency table are discrete whereas the χ^2 distribution used to calculate the critical value and the P-value is continuous. To correct for this difference in continuity, some statisticians use the *Yates' correction for continuity*. This is done by calculating the expected frequencies, degrees of freedom, and critical value as normal, but the test statistic is calculated as

$$c = \sum_{i=1}^{a} \sum_{j=1}^{b} \frac{(|O_{ij} - E_{ij}| - 0.5)^2}{E_{ij}}.$$

When this correction should be used (or even if it should be used at all) is a matter of debate. It is often used for 2×2 contingency tables or in any case where an expected frequency is less than 5.

	Disease		
Smoke	Yes	No	Total
Yes	560	308	868
No	240	100	340
Total	800	408	1208

Suppose the table above gives the results of an investigation into the link between smoking and the occurrence of a certain disease.

a. Use these data to test the claim that smoking is independent of the disease, using the standard test of independence. State the test statistic, the critical value, and the technical conclusion.

b. Now test the claim, using Yates' correction for continuity. Does this lead to the same technical conclusion?

c. Based on the results in parts a and b, does Yates' correction for continuity increase the likelihood of rejecting H_0 or decrease it?

11. Another approach to testing for independence using a 2×2 contingency table is *Fisher's exact test*. To illustrate this test, consider the following scenario: To investigate a complaint that female employees have been denied promotion more often than male employees, a company examines the records of 24 employees who applied for promotion. The resulting data are summarized in the table below.

	Gender		
Promoted?	Females	Males	Total
Yes	5	9	14
No	8	2	10
Total	13	11	24

Now consider these 24 applicants as a "population" in which 14 were promoted and 10 were not promoted. Suppose we choose the 11 male applicants without replacement and let the random variable X denote the number of these 11 applicants who were promoted.

Assuming that gender is independent of being promoted, then X has a hypergeometric distribution with $n = 11$, $N = 24$, $N_1 = 14$, and $N_2 = 10$ so that the pmf of X is

$$f(x) = P(X = x) = \frac{\binom{14}{x}\binom{10}{11-x}}{\binom{24}{11}}, \quad x = 0, 1, \ldots, 11.$$

The data contain the observed value $X = 9$, and $f(9)$ is the probability of observing this value. We need to calculate the probability of observing a value of X at least this extreme. To do this, follow these steps:

1. Calculate $f(x)$ for $x = 0, 1, \ldots, 11$.
2. Add the values of $f(x)$ from step 1 that are less than or equal to $f(9)$.
3. The result of step 2 is the P-value for the two-tail test of the null hypothesis H_0: gender is independent of being promoted. This P-value is interpreted in the same way as every other P-value.

Calculate the P-value for these data. At the 95% confidence level, do the data appear to back up the complaint that female employees have been denied promotion more often than male employees?

5.9 One-Way ANOVA

In Section 5.6, we introduced hypothesis tests for comparing two population means. In this section, we introduce a test for comparing more than two population means. The theory behind this test is well beyond the scope of this text, so we do not discuss it. Instead, we focus on understanding the basic ideas, terminology, and mechanics of the calculations.

To illustrate this test, consider the following scenario: To test the claim that different colors of M&Ms have the same mean mass, a student weighs several M&Ms and records the data in Table 5.20 (data collected by Frank Ohlinger, 2009).

Color	Mass (g)					$\bar{x}$	s
Yellow	0.89	0.89	0.85	0.86	0.90	0.878	0.0217
Red	0.85	0.92	0.85	0.91	0.94	0.894	0.0416
Blue	0.88	0.81	0.89	0.84	0.86	0.856	0.0321
Orange	0.80	0.85	0.87	0.87	0.86	0.850	0.0292

Table 5.20

To set up the hypotheses, we define the parameters μ_1 to be the mean mass of all yellow M&Ms, μ_2 to be the mean mass of all red M&Ms, and so on. Then the claim is $\mu_1 = \mu_2 = \mu_3 = \mu_4$, so we test the hypotheses

$$H_0: \mu_1 = \mu_2 = \mu_3 = \mu_4, \qquad H_1: \text{At least one mean is different.}$$

We begin the test by analyzing the sample means. Note that the four sample means are not all the same and that the mean for red is the largest. This is an indication that μ_2 is larger than the rest, suggesting that we should reject H_0. These differences between the sample means are referred to as *variance between the samples*. The larger this variance is, the more confident we are in rejecting H_0.

However, when comparing sample means, we also need to consider the sample standard deviations. Note that the standard deviation for red is the largest. This is an indication that the sample mean 0.894 for red is possibly not a good estimate of μ_2. This makes us less confident in rejecting H_0. The sample standard deviations are referred to as *variance within the samples*. The larger this variance is, the less confident we are in rejecting H_0.

To put these ideas together, we calculate the ratio

$$F = \frac{\text{variance between the samples}}{\text{variance within the samples}}.$$

The larger this ratio is, the more confident we are in rejecting H_0. The resulting test is called *analysis of variance*, or ANOVA for short. To formally define this ratio, we use the following notation. Let

1. k = the number of samples = the number of populations being compared,
2. n_i = the size of the ith sample,
3. x_{ij} = the jth value in the ith sample,
4. $N = n_1 + \cdots + n_k$ = the total number of data values,
5. $\bar{x}_i$ = the mean of the ith sample,
6. s_i = the standard deviation of the ith sample, and
7. $\bar{\bar{x}}$ = the mean of all the data values = the mean of the k sample means.

Using this notation, we define the following quantities, which are measures of different types of variance within the data.

1. **Sum of squares treatment.** This is a measure of the total variance between the sample means.

$$\text{SS(treatment)} = \sum_{i=1}^{k} n_i (\bar{x}_i - \bar{\bar{x}})^2$$

This is also called the SS(factor) or SS(between groups).

2. **Sum of squares error.** This is a measure of the total variance within the samples.

$$\text{SS(error)} = \sum_{i=1}^{k} (n_i - 1)\, s_i^2$$

This is also called the SS(within samples) or SS(within groups).

3. **Total sum of squares.** This is a measure of the total variance within all the data.

$$\text{SS(total)} = \sum_{i=1}^{k} \sum_{j=1}^{n_i} (x_{ij} - \bar{\bar{x}})^2$$

In simpler terms, SS(total) is calculated by subtracting $\bar{\bar{x}}$ from each data value, squaring each difference, and then adding them up.

The word *treatment* (or *factor*) is a generic term meaning a characteristic that distinguishes the different populations (or groups) from one another. In this example, the treatments are the four colors. Because each population is distinguished by only one factor, this type of hypothesis test is called *one-way ANOVA*, or *single-factor ANOVA*. In the next section, we discuss a similar test where each population is distinguished by two factors, called *two-way ANOVA*.

Example 5.9.1 **Calculating Sums of Squares** We calculate SS(error), SS(treatment), and SS(total) for the data in Table 5.20. First note that $k = 4$ and that $n_i = 5$ for each i so that $N = 20$. The values of $\bar{x}_i$ and s_i are given in the table. Then we have

$$\bar{\bar{x}} = \frac{0.89 + 0.89 + \cdots + 0.86}{20} = 0.8695,$$

$$\text{SS(treatment)} = 5(0.878 - 0.8695)^2 + \cdots + 5(0.850 - 0.8695)^2 = 0.006175,$$

$$\text{SS(error)} = (5-1)(0.0217^2) + \cdots + (5-1)(0.0292^2) = 0.01632, \text{ and}$$

$$\text{SS(total)} = (0.89 - 0.8695)^2 + \cdots + (0.86 - 0.8695)^2 = 0.022495.$$

Notice that these calculations are not terribly complicated, just tedious. They are most often done with software. □

Notice that in Example 5.9.1

$$0.022495 = 0.006175 + 0.01634.$$

This illustrates the following property that is true in all cases:

$$\text{SS(total)} = \text{SS(treatment)} + \text{SS(error)}. \tag{5.6}$$

This property means that we do not need to calculate all three of these quantities directly from the data. We can calculate two of them and then use this property to find the third.

The following theorem, which we present without proof, is the key to analyzing the results.

Theorem 5.9.1 *If all k populations are normally distributed and have a common variance σ^2 and the k samples are independent, then the following results hold:*

1. *If H_0: $\mu_1 = \cdots = \mu_k$ is true, then the quantities*

$$\frac{\text{SS(treatment)}}{\sigma^2} \quad \text{and} \quad \frac{\text{SS(error)}}{\sigma^2}$$

   are observed values of independent random variables that are $\chi^2(k-1)$ and $\chi^2(N-k)$, respectively.

2. *Whether or not H_0 is true, the quantity*

$$\frac{\text{SS(error)}}{N-k}$$

   is an unbiased estimate of σ^2.

3. *The quantity*

$$F = \frac{\text{SS(treatment)}/(k-1)}{\text{SS(error)}/(N-k)}$$

   is an observed value of a random variable with an F-distribution with $(k-1)$ and $(N-k)$ degrees of freedom. □

The proof of the first part of this theorem is beyond the scope of this text. The second part is a result of the first part and properties of the χ^2 distribution. The third part is a result of the first part and the definition of the F-distribution. This theorem leads to the following definitions and the test statistic for the one-way ANOVA test:

1. Mean squares treatment: $\text{MS(treatment)} = \dfrac{\text{SS(treatment)}}{k-1}$.

2. Mean squares error: $\text{MS(error)} = \dfrac{\text{SS(error)}}{N-k}$.

3. Mean squares total: $\text{SS(total)} = \dfrac{\text{SS(total)}}{N-1}$.

4. Test statistic for one-way ANOVA: $F = \dfrac{\text{MS(treatment)}}{\text{MS(error)}}$.

Example 5.9.2 Calculating Mean Squares Here we calculate the mean squares and the test statistic for the data in Table 5.20. We have $k=4$, $N=20$, $\text{SS(treatment)} = 0.006175$, $\text{SS(error)} = 0.01632$, and $\text{SS(total)} = 0.022495$ so that

$$\text{MS(treatment)} = \dfrac{0.006175}{4-1} = 0.002058, \quad \text{MS(error)} = \dfrac{0.01632}{20-4} = 0.00102,$$

$$\text{MS(total)} = \dfrac{0.022495}{20-1} = 0.001184, \quad \text{and} \quad F = \dfrac{0.002058}{0.00102} = 2.018.$$

These quantities are often summarized in an *ANOVA table* as shown in Table 5.21.

Source	SS	df	MS	F
Treatment	0.006175	3	0.002058	2.018
Error	0.01632	16	0.00102	
Total	0.022495	19	0.001184	

Table 5.21

Now we want to find the appropriate critical value. As noted earlier, the larger the test statistic F is, the more confident we are in rejecting H_0. Because F has an F-distribution with $(k-1)$ and $(N-k)$ degrees of freedom (assuming H_0 is true), the appropriate critical value is $F_\alpha(k-1, N-k)$, found in Table C.5. If F is greater than this value, we reject H_0. The P-value is the area under the corresponding density curve to the right of the test statistic. These ideas are summarized in the box on the next page.

In this example, we have 3 and 16 degrees of freedom. In Table C.5, the closest degrees of freedom are 3 and 15. We see that $f_{0.05}(3, 15) = 3.29$ (using software, the exact critical value with 3 and 16 degrees of freedom is 3.24). The critical region is $[3.29, \infty)$. (The

P-value is the area to the right of 2.018. By using software, this is found to be 0.152.) Because $F = 2.018$ does not lie in this region, we do not reject H_0. Because the original claim contains equality, the final conclusion is that there is not sufficient evidence to reject the claim.

One-Way ANOVA Test of Equality of Population Means

To test the null hypothesis
$$H_0: \mu_1 = \cdots = \mu_k,$$
the test statistic is
$$F = \frac{\text{MS(treatment)}}{\text{MS(error)}},$$
and the critical region is $[f_\alpha(k-1, N-k), \infty)$. The P-value is the area to the right of F under the F-distribution density curve with $(k-1)$ and $(N-k)$ degrees of freedom.

Requirements

1. The populations are normally distributed. (This is a "loose" requirement, meaning that the test gives valid results as long as no population is "too far" from normal.)
2. The populations have the same variance. (This, too, is a loose requirement.)
3. The samples are random and independent.

The requirement of normality can be easily verified by constructing a normal quantile plot of each sample. The requirement of equal variances can be verified by comparing the sample variances (or standard deviations). One rule of thumb states that if the sample sizes are equal (or nearly equal), as long as the largest sample variance is no more than nine times the smallest (or the largest standard deviation is no more than three times the smallest), the requirement is met.

One might wonder why we need a test like ANOVA to compare multiple population means. We might argue that we could simply compare all the possible pairs of means. That is, we could test each of the six subhypotheses

$$H_0: \mu_1 = \mu_2 \qquad H_0: \mu_1 = \mu_3 \qquad H_0: \mu_1 = \mu_4$$

$$H_0: \mu_2 = \mu_3 \qquad H_0: \mu_2 = \mu_4 \qquad H_0: \mu_3 = \mu_4$$

individually using a 2-sample T-test. If we rejected any of these null hypotheses, then we would reject the overall null hypothesis $H_0: \mu_1 = \mu_2 = \mu_3 = \mu_4$. The problem with this approach is that the probability of a Type I error in each individual subtest is 0.05 (at the 95% confidence level). The probability of making a Type I error in *at least one* of the subtests is much greater than 0.05 (assuming the tests are independent, this probability is

0.265). Thus our confidence in the result of the overall test is much less than 95%. ANOVA gives us a way of doing all six comparisons at once so that the probability of a Type I error is 0.05, as desired.

Putting the various definitions together, we get these somewhat simplified formulas:

$$\text{MS(treatment)} = \frac{\sum_{i=1}^{k} n_i (\bar{x}_i - \bar{\bar{x}})^2}{k-1} \quad \text{and} \quad \text{MS(error)} = \frac{\sum_{i=1}^{k} (n_i - 1) s_i^2}{N-k}.$$

We use these formulas in the next example.

Example 5.9.3   Comparing Corn Seed A seed company plants four types of new corn seed on several plots of land and records the yield (in bushels per acre) of each plot, as shown in Table 5.22. Test the claim that the four types of seed produce the same mean yield.

Type	Yield				$\bar{x}$	s
A	68.88	59.80	65.57		64.75	4.60
B	62.75	64.95	65.57	71.33	66.15	3.66
C	72.36	82.87	76.95		77.39	5.27
D	60.51	57.17	60.50	55.71	58.47	2.42

$$\bar{\bar{x}} = 66.07$$

Table 5.22

Define the parameters μ_1 to be the mean yield of type A, μ_2 to be the mean yield of type B, and so on. We test the hypotheses

$$H_0: \mu_1 = \mu_2 = \mu_3 = \mu_4, \quad H_1: \text{At least one mean is different.}$$

From the table, we see that $n_1 = n_3 = 3$ and $n_2 = n_4 = 4$ so that $N = 14$. Thus we have

$$\text{MS(treatment)} = \frac{3(64.75 - 66.07)^2 + \cdots + 4(58.47 - 66.07)^2}{4-1} \approx 206.93,$$

$$\text{MS(error)} = \frac{(3-1)(4.60^2) + \cdots + (4-1)(2.42^2)}{14-4} \approx 15.55, \quad \text{and}$$

$$F = \frac{206.93}{15.55} \approx 13.31.$$

The critical value is $f_{0.05}(3, 10) = 3.71$, and the critical region is $[3.71, \infty)$. The P-value is the area to the right of $F = 13.31$, which is 0.0008 when found by using software. Thus we reject H_0 and conclude that the data do not support the claim of equal means. Therefore, there appears to be a statistically significant difference between the four types of seeds. Whether this difference is significant in a practical sense is another issue. □

In the previous example, we concluded that the four population means are not all equal. We may want to know which one is different from the others. We could try to answer this by comparing every combination of pairs, using 2-sample T-tests. However, as mentioned above, the result does not have the desired level of confidence. One simple way around this problem is to decrease the significance level in each of the individual tests (this increases the confidence). For instance, instead of using a significance level of $\alpha = 0.05$, we could use a significance level of $\beta = 0.05/6 \approx 0.0083$ (we divide by six because there are 6 pairs of means).

Another approach, called the *Bonferroni multiple-comparison method*, is to construct confidence intervals for the differences of the means. This method is summarized in the box below and is based on the basic idea of decreasing the significance level for each interval. This method is conservative in that the probability of a Type I error in the overall test is a bit less than the desired level of α, but it is reasonably close. Bonferroni's method is only one method for comparing population means. Other methods include Fisher's least significant difference method and Tukey's honestly significant difference method.

Bonferroni Multiple-Comparison Method

To determine which pairs of k population means are significantly different, let
- α be the desired significance level,
- $c = \binom{k}{2} =$ the number of pairs of means,
- $\beta = \alpha/c$, and
- MS(error) = the mean square error from the ANOVA test of equality of the means.

For each $i \neq j$, the bounds for a confidence interval estimate of $(\mu_i - \mu_j)$ are

$$(\bar{x}_i - \bar{x}_j) \pm t_{\beta/2}\sqrt{\text{MS(error)}\left(\frac{1}{n_i} + \frac{1}{n_j}\right)}$$

where $t_{\beta/2}$ is a critical t-value with $(N-k)$ degrees of freedom. If this interval does *not* contain 0, then conclude that μ_i and μ_j are significantly different.

Example 5.9.4 **Multiple Comparisons** Here we perform the Bonferroni multiple-comparison method on the data in Example 5.9.3. Because there are four means, $c = \binom{4}{2} = 6$ so that at the 95% confidence level, $\beta = 0.05/6 \approx 0.0083$. The critical t-value is $t_{0.0083/2}(10)$. Using software, this is found to be 3.277.

The confidence interval estimate of $(\mu_1 - \mu_2)$ is then

$$(64.75 - 66.15) \pm 3.277\sqrt{15.55\left(\frac{1}{3} + \frac{1}{4}\right)}$$

$$\Rightarrow \quad -11.27 \leq (\mu_1 - \mu_2) \leq 8.47.$$

Because this interval contains 0, we conclude that there is not a significant difference between types A and B. The results for the other pairs are shown in Table 5.23.

i	j	$(\bar{x}_i - \bar{x}_j)$	Lower	Upper
1	2	−1.40	−11.27	8.47
1	3	−12.64	−23.19	−2.09
1	4	6.28	−3.59	16.15
2	3	−11.24	−21.11	−1.37
2	4	7.68	−1.46	16.82
3	4	18.92	9.05	28.79

Table 5.23

Note that the only intervals that do not contain 0 are those involving μ_3 (corresponding to type C). Thus we are (approximately) 95% confident in concluding that type C is significantly different from types A, B, and D, and that no other pair is significantly different. □

Software Calculations

To perform a one-way ANOVA test:

Minitab: Enter each sample of data in a separate column. Select **Stat** → **ANOVA** → **One-Way (Unstacked)**. Enter the names of the columns containing the samples.

R: First create a data frame to store the data. To create the data frame for the data in Table 5.20, enter the following code:

> MMData<− data.frame(Weight=c(1:20), Color=factor(1:20))

Next type **fix("MMData")** and enter the data into the data editor. Close the data editor window, and ignore any warnings about added factor levels. The randomized block design ANOVA is performed with the following command:

> summary(aov(Weight~Color, data=MMdata))

Excel: First enter the samples in consecutive columns. Then select **Data → Data Analysis → Anova: Single Factor**. Next to **Input Range**, select the range containing the samples, and select a blank cell for the **Output Range**.

TI-83/84 Calculators: Enter each sample of data in a separate list. Then press **STAT → TESTS → H:ANOVA**. This will paste the ANOVA(command at the home screen. Then enter the names of the lists containing the samples, separated by commas, and press **ENTER**.

Exercises

Directions: In all exercises asking for an ANOVA test, use available software to find the values of MS(treatment), MS(error), and F. Give the critical value or the P-value. In all exercises, use an $\alpha = 0.05$ significance level.

1. A student surveys several people regarding their attitudes toward bipolar disorder. Each person answers several questions and then receives an overall score between 0 and 10, with a higher score indicating greater prejudice. Respondents are also asked their ages and are assigned to one of four different categories. The scores for five respondents in each category are shown in the table below (data collected by Sara Agee, 2010). Use these data to test the claim that the four age groups have the same mean score. Based on this result, does there appear to be a difference in the level of prejudice between the age groups?

Category	Score				
18–26	1.27	0.13	3.00	0.27	0.60
27–45	0.40	0.20	0.40	0.40	0.87
46–64	0.67	1.20	1.00	0.20	1.00
65+	0.36	0.87	0.27	0.20	0.80

2. Two students ask several fellow students how many minutes per day they typically spend on social networks. Each student is also assigned to one of three age categories. The resulting data are shown in the table on the next page (data collected by Jan Hudak and Adam Dumas, 2009). Assuming that all three populations are normally distributed, use the data to test the claim that the three populations have the same mean. Does it appear that one age group spends more or less time, on average, than the other groups?

Age	Minutes Spent														
≤ 17	10	15	30	40	45	60	60	60	60	60	60	120	120	180	200
18	5	10	10	15	15	30	30	45	60	60	120	120	180	180	180
≥ 19	10	15	20	30	45	60	60	60	60	120	120	120	180	180	200

3. The table below gives the lengths (in minutes) of samples of movies rated G, PG, PG-13, and R (data collected by Meredith Hein and Rachel Dahlke, 2011). Assuming that all three populations are normally distributed, use the data to test the claim that the three populations have the same mean. Which type of movie, if any, appears to be shorter than the rest? Why might this be?

G			PG			PG-13			R		
25	75	88	87	97	104	95	116	130	93	98	123
63	76	97	94	99	114	97	117	130	95	100	125
68	82	98	95	100	120	99	117	153	95	103	129
74	83	106	96	103	143	105	122	157	96	109	138
74	83	117	96	104	152	111	124	178	97	119	148

4. Show that the quantity $SS(total)/\sigma^2$ is an observed value of a random variable that is $\chi^2(N-1)$. (**Hint:** Use Equation (5.6), Theorem 5.9.1, and Exercise 12 of Section 3.9.)

5. Show that if six hypothesis tests are performed, each at the $\alpha = 0.05$ significance level, then probability of a Type I error in *at least one test* is approximately 0.265, assuming the results are mutually independent.

6. If X is the random variable representing the quantity SS(error), show that $E[X/(N-k)] = \sigma^2$. (**Hint:** Use the fact that X/σ^2 is $\chi^2(N-k)$, and apply the results of Exercise 11 of Section 3.9.)

7. A student records the copyright dates of books on various subjects in his university library, as shown in the table below (data collected by Andrew Asmus, 2009).

Subject	Copyright Date				
Energy	1951	1958	1972	1979	1989
	1991	1992	1997	1999	2008
Medicine	1934	1942	1949	1960	1965
	1969	1970	1982	1998	2002
Law	1950	1957	1966	1966	1968
	1971	1973	1980	1984	1987
History	1938	1952	1954	1969	1972
	1998	1999	1999	2002	

a. Construct a normal quantile plot of each sample. Is it reasonable to assume that each population is normally distributed?

b. Calculate the standard deviation of each sample. According to the rule of thumb, is it reasonable to assume that each population has the same variance?

c. Test the claim that the four populations have the same mean copyright data. Based on this result, does it appear that books in any one of these subjects are older or younger on average than in any other subject?

8. We may construct a confidence interval for a single population mean μ_i, using the results of an ANOVA test with the following formula:

$$\bar{x}_i \pm t_{\alpha/2} \sqrt{\frac{\text{MS(error)}}{n_i}}$$

where $t_{\alpha/2}$ is a critical t-value with $(N-k)$ degrees of freedom. We may use these intervals to informally compare population means. If two intervals do not overlap, then conclude that the two means are significantly different. Otherwise, conclude that they are not different.

Construct confidence intervals for each of the population means in Example 5.9.3, using this formula, and compare the intervals. Based on these intervals, which mean(s) appear to be different from the rest?

9. Use the procedures in Exercise 8 above to find confidence interval estimates of the population means, using the data in the table in Exercise 3. Which type of movie, if any, appears to be shorter than the rest? 112.8), G-rated is shorter

10. Independent samples are taken from three different populations, as shown in the table below.

Sample	Value					$\bar{x}$	s
A	9.90	10.80	6.80	9.98	9.44	9.384	1.525
B	10.99	12.19	11.35	8.78	11.06	10.874	1.264
C	13.00	11.88	14.00	10.67	12.86	12.482	1.261

a. Test the claim that the three populations have the same mean.

b. Use the Bonferroni multiple-comparison method to compare each pair of means. Which means appear to be significantly different? (The critical t-value is $t_{0.017/2}(12) = 2.779$.)

c. If μ_1 is *not* significantly different from μ_2, and μ_2 is *not* significantly different from μ_3, can we conclude that μ_1 is *not* significantly different from μ_3? Explain. (**Hint:** Use the results of part b.)

11. Levene's test for the equality of two population variances, as described in Exercise 20 of Section 5.6, can be extended to test the equality of more than two variances. The algorithm for three populations is as follows.

   1. Calculate the median of each sample, m_1, m_2, and m_3.
   2. For each data value x in the first sample, replace it with $|x - m_1|$. Repeat for the second sample, using m_2, and for the third, using m_3.
   3. Perform a one-way ANOVA to test the claim that these three samples of modified values come from populations with equal means.
   4. If we reject H_0 in step 3, then we conclude that the three populations have different variances; otherwise, we conclude that the variances are equal.

Use this algorithm to test the claim that the three samples in the table below come from populations with equal variances.

Sample 1	5.66	6.62	6.76	7.58	7.66	7.77	7.95	8.38	8.38	8.50	9.13
Sample 2	7.41	7.62	7.63	7.73	7.84	8.15	8.26	8.27	8.46	8.54	8.75
Sample 3	4.27	4.82	5.60	6.12	6.36	7.98	8.04	8.67	9.78	9.89	9.99

5.10 Two-Way ANOVA

In this section, we present two methods of testing for equality of population means where each data value is categorized by two factors. Many of the concepts are very similar to the previous section, and the theory is even more complicated, so we omit the details. These two methods are distinguished by the way the statistical experiment is performed and the data are collected. The two types of experiments are called *randomized block designs* and *a × b factorial experiments*.

Randomized Block Designs

To illustrate the first method, consider a scenario where a statistics professor is comparing four different delivery methods for her introduction to statistics class: face-to-face, online, hybrid, and video. These delivery methods are generically called the *treatments*.

She divides the population of students into three groups according to their overall GPA: high, middle, and low. These groups are called *blocks*. She randomly chooses four students from each block and randomly assigns each one to a class, using one of the delivery methods. At the end of the semester, she records each student's overall grade, as in Table 5.24. Such an experiment has what is called a *randomized block design*.

	Delivery Method			
GPA	Face-to-Face	Online	Hybrid	Video
High	88.5	91.3	89.4	85.6
Middle	82.4	81.8	87.5	78.4
Low	79.4	80.6	81.4	81.7

Table 5.24

Using these data, we analyze two questions: (1) Do the four treatments have the same population mean? (2) Do the blocks have any effect on the scores? To this end, we make the following definitions:

1. Let b = the total number of blocks and $i = 1, 2, \ldots, b$ denote the block number.
2. Let k = the total number of treatments and $j = 1, 2, \ldots, k$ denote the treatment number.
3. Let x_{ij} = the data value from block i and treatment j.
4. Let μ_j = the population mean of the jth treatment.

We assume that each data value x_{ij} is an observed value of a random variable X_{ij},

$$X_{ij} = \mu_j + \beta_i + \epsilon_{ij}$$

where

The F-Test for a Randomized Block Design

Let

- $T = \sum_{i=1}^{b}\sum_{j=1}^{k} x_{ij}$ = the sum of all the data values,

- $T_j = \sum_{i=1}^{b} x_{ij}$ = the sum of all values from treatment j,

- $B_i = \sum_{j=1}^{k} x_{ij}$ = the sum of all values from block i, and

- $A = T^2/(b\,k)$.

Sum of Squares	**Degrees of Freedom**

- SS(treatment) = $\sum_{j=1}^{k} \dfrac{T_j^2}{b} - A$ $k-1$

- SS(block) = $\sum_{i=1}^{b} \dfrac{B_i^2}{k} - A$ $b-1$

- SS(total) = $\sum_{i=1}^{b}\sum_{j=1}^{k} x_{ij}^2 - A$ $bk-1$

- SS(error) = SS(total) − SS(treatment) − SS(block) $(b-1)(k-1)$

For the hypothesis H_0: $\mu_1 = \mu_2 = \cdots = \mu_k$, the test statistic is

$$F_1 = \frac{\text{SS(treatment)}/(k-1)}{\text{SS(error)}/[(b-1)(k-1)]}$$

and the critical value is $f_\alpha\,[(k-1),(b-1)(k-1)]$.

For the hypothesis H_0: $\beta_1 = \beta_2 = \cdots = \beta_b$, the test statistic is

$$F_2 = \frac{\text{SS(block)}/(b-1)}{\text{SS(error)}/[(b-1)(k-1)]}$$

and the critical value is $f_\alpha\,[(b-1),(b-1)(k-1)]$.

For each hypothesis, the critical region is $[f, \infty)$ where f is the critical value, and the P-value is the area to the right of the test statistic.

- β_i is the *block effect* (this is a measure of the effect that block i has on the value), and
- ϵ_{ij} is a random variable representing experimental error. (We further assume that each ϵ_{ij} is $N(0, \sigma^2)$ where σ^2 is a common variance, and that all these random variables are mutually independent.)

Using this notation, we simultaneously test the following two null hypotheses:

H_0: $\mu_1 = \mu_2 = \cdots = \mu_k$ (the treatment means are all the same)
H_B: $\beta_1 = \beta_2 = \cdots = \beta_b$ (the block effects are all the same).

Using theory beyond the scope of this text, it can be shown that the appropriate sums of squares and test statistics are calculated as described in the box on the previous page. Each of these sums of squares is an observed value of a random variable that is χ^2 with the indicated degrees of freedom. We define the associated mean squares by dividing the sum of squares by the degrees of freedom.

Example 5.10.1 Calculating Sums of Squares We calculate the sums of squares and the test statistics for the data in Table 5.24. Here we have $k = 4$ and $b = 3$. The values of T, T_j, and B_i are shown in Table 5.25.

GPA	Face-to-Face	Online	Hybrid	Video	Sum
		Delivery Method			
High	88.5	91.3	89.4	85.6	$B_1 = 354.8$
Middle	82.4	81.8	87.5	78.4	$B_2 = 330.1$
Low	79.4	80.6	81.4	81.7	$B_3 = 323.1$
Sum	$T_1 = 250.3$	$T_2 = 253.7$	$T_3 = 258.3$	$T_4 = 245.7$	$T = 1008$

Table 5.25

Then we have

$$A = \frac{1008^2}{4 \cdot 3} = 84{,}672$$

$$\text{SS(treatment)} = \frac{250.3^2 + 253.7^2 + 258.3^2 + 245.7^2}{3} - 84{,}672 \approx 28.387$$

$$\text{SS(block)} = \frac{354.8^2 + 330.1^2 + 323.1^2}{4} - 84{,}672 = 138.665$$

$$\text{SS(total)} = 88.5^2 + 91.3^2 + \cdots + 81.7^2 - 84{,}672 = 201.04$$

$$\text{SS(error)} = 201.04 - 28.3867 - 138.665 = 33.988$$

$$F_1 = \frac{28.3867/(4-1)}{33.988/[(3-1)(4-1)]} \approx 1.67$$

$$F_2 = \frac{138.665/(3-1)}{33.988/[(3-1)(4-1)]} \approx 12.24$$

Note that as with one-way ANOVA, these calculations are not terribly complicated, just very tedious. They are most often performed with software. The P-values for both hypotheses are found by using software. These calculations are summarized in the ANOVA table shown in Table 5.26.

At the 95% confidence level, the critical value for the hypothesis $H_0: \mu_1 = \cdots = \mu_4$ is $f_{0.05}(3, 6) = 4.76$ and the critical region is $[4.76, \infty)$. Because F_1 is not in the critical region, we do not reject this hypothesis. Therefore, we conclude that there is not a statistically significant difference between the treatment means. In simpler terms, the delivery methods appear to produce the same scores, on average.

For the hypothesis $H_B: \beta_1 = \beta_2 = \beta_3$, the critical value is $f_{0.05}(2, 6) = 5.14$ and the critical region is $[5.14, \infty)$. Because F_2 is in this critical region, we reject this hypothesis. Therefore, we conclude that there is a statistically significant difference in the block effects. In simpler terms, the GPA of the student appears to affect the scores.

Source	SS	df	MS	F	P-Value	Critical F
Treatments	28.387	3	9.462	1.67	0.271	4.76
Blocks	138.665	2	69.333	12.24	0.008	5.14
Error	33.988	6	5.665			
Total	201.04	11				

Table 5.26

One might be tempted to do a one-way ANOVA to test the equality of the four treatment means where we consider each student as simply one in a sample of three (if we treat the columns of Table 5.24 as the samples, then this table looks very similar to those used in Section 5.9). However, this completely ignores the possibility of block effects. Ignoring block effects may hide the differences between the treatments (see Exercise 4). Because we rejected the equality of the block effects, taking them into account certainly appears to be justified. If we had not rejected the equality of the block effects, we would want to do a one-way ANOVA test on the data.

$a \times b$ Factorial Experiments

One drawback of the block design is that it does not account for the possibility of an *interaction* between the factors. The next two-way ANOVA test accounts for this. Consider this scenario: Suspecting that the delivery method affects different levels of students differently, the statistics professor randomly chooses 16 students from each GPA level, randomly assigns four to each delivery method, and records their scores at the end of the semester, as shown in Table 5.27.

5.10 Two-Way ANOVA

	Delivery Method			
GPA	Face-to-Face	Online	Hybrid	Video
High	79.4	91.2	86.8	83.1
	84.7	91.0	87.1	92.4
	92.0	93.1	89.0	93.3
	90.4	88.6	90.2	80.7
Middle	84.1	72.0	86.1	78.1
	85.0	85.2	90.4	81.7
	77.1	77.6	87.2	80.1
	84.0	75.2	88.6	79.8
Low	83.7	75.5	76.8	81.2
	86.6	74.5	80.7	83.6
	83.1	80.1	75.5	78.8
	81.3	79.9	80.6	78.5

Table 5.27

Such an experiment is called a 3×4 *factorial experiment with four replications per treatment*. We have two factors, GPA and delivery method (generically called factors A and B); three *levels* of factor A, four *levels* of factor B, and four *replications* of each combination of the two factors. To begin to analyze the data, we calculate the mean of each combination and graph the results, as shown in Figure 5.6.

	Mean Scores			
GPA	F-to-F	Online	Hybrid	Video
H	86.6	91.0	88.3	87.4
M	82.6	77.5	88.1	79.9
L	83.7	77.5	78.4	80.5

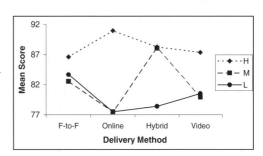

Figure 5.6

Notice in the graph that when we move right from the face-to-face method to the online method, the mean score for the high-GPA students increases while the mean score for the other students decreases. This suggests that the online delivery method interacts differently with high-GPA students than with other GPA levels. If there were *no* difference in the interactions, we would expect the lines in the graph to be roughly parallel.

We want to answer three questions with these data: (1) Is there any difference between the population mean scores of the delivery methods? (2) Does the GPA level affect the scores? (3) Does the interaction of the delivery method and GPA level affect the scores? To this end, we make the following definitions:

1. Let a = the total number of levels of factor A and $i = 1, 2, \ldots, a$ denote the corresponding factor number.
2. Let b = the total number of levels of factor B and $j = 1, 2, \ldots, b$ denote the corresponding factor number.
3. Let r = the number of repetitions of each combination of the two factors, and let $k = 1, 2, \ldots, r$ denote the repetition number.
4. Let x_{ijk} = the data value from the ith level of factor A, the jth level of B, and the kth repetition.

We assume that each data value x_{ijk} is an observed value of a random variable X_{ijk},

$$X_{ijk} = \mu_j + \alpha_i + \gamma_{ij} + \epsilon_{ijk}$$

where

- μ_j is the population mean of the jth level of factor B (the population mean of the jth column of the data),
- α_i is the effect of the ith level of factor A (this is a measure of the effect that level i has on the value),
- γ_{ij} is the *interaction effect* of the ith level of factor A and the jth level of factor B,
- ϵ_{ijk} is a random variable representing experimental error. (We further assume that each ϵ_{ijk} is $N(0, \sigma^2)$, where σ^2 is a common variance and that all are mutually independent.)

Using this notation, we simultaneously test the following three null hypotheses:

H_B: $\mu_1 = \mu_2 = \cdots = \mu_b$ (the factors of B all have the same mean),
H_A: $\alpha_1 = \alpha_2 = \cdots = \alpha_a$ (the effect of the factors of A are all the same),
H_{AB}: $\gamma_{11} = \gamma_{12} = \cdots = \gamma_{ab}$ (the interaction effects are all the same).

The appropriate sums of squares and degrees of freedom are shown in the box on the next page. Here, SS(B), SS(A), and SS(AB) are the sums of squares for H_B, H_A, and H_{AB}, respectively. The mean squares are defined as in the other ANOVA tests. The corresponding test statistics are found by dividing the mean squares by the mean square error. The critical values, critical regions, and P-values are found in the same way as for the other ANOVA tests.

> **The F-Test for an $a \times b$ Factorial Experiment**
>
> Let
>
> - $B_j = \sum_{i=1}^{a} \sum_{k=1}^{r} x_{ijk}$ = the sum of all ar values at the jth level of B,
> - $A_i = \sum_{j=1}^{b} \sum_{k=1}^{r} x_{ijk}$ = the sum of all br values at the ith level of A,
> - $(AB)_{ij} = \sum_{k=1}^{r} x_{ijk}$ = the sum of all r values at the ith level of A and the jth level of B,
> - $S = \sum_{i=1}^{a} \sum_{j=1}^{b} \sum_{k=1}^{r} x_{ijk}$ = the sum of all abr values, and
> - $T = S^2/(abr)$.
>
> **Sum of Squares**   **Degrees of Freedom**
>
> - $SS(B) = \dfrac{1}{ar} \sum_{j=1}^{b} B_j^2 - T$ $b-1$
> - $SS(A) = \dfrac{1}{br} \sum_{i=1}^{a} A_i^2 - T$ $a-1$
> - $SS(AB) = \dfrac{1}{r} \sum_{i=1}^{a} \sum_{j=1}^{b} AB_{ij}^2 - T - SS(B) - SS(A)$ $(a-1)(b-1)$
> - $SS(\text{total}) = \sum_{i=1}^{a} \sum_{j=1}^{b} \sum_{k=1}^{r} x_{ijk}^2 - T$ $abr - 1$
> - $SS(\text{error}) = SS(\text{total}) - SS(B) - SS(A) - SS(AB)$ $ab(r-1)$

Using software, the resulting calculations for the data in Table 5.27 are shown in Table 5.28. Based on these results, we draw the following conclusions:

- For H_B, the P-value is 0.176, indicating that we should not reject H_B. Thus it appears that the population means of the delivery methods are not significantly different. This is the same conclusion reached with the block design.
- For H_A, the P-value is 0.000, indicating that we should reject H_A. This means that there does appear to be a significant difference between the GPA levels. This, again, is the same conclusion reached with the block design.

- For H_{AB}, the P-value is 0.004, indicating that we should reject H_{AB}. This means that there does appear to be a significant interaction between the two factors.
- Because there appears to be an interaction, we should be very careful about interpreting the effects of factor A or B without also considering both.

Source	SS	df	MS	F	P-Value	Critical F
B	68.17	3	22.72	1.74	0.176	2.87
A	599.06	2	299.53	22.93	0.000	3.26
AB	312.51	6	52.08	3.99	0.004	2.36
Error	470.29	36	13.06			
Total	1450.02	47				

Table 5.28

What do these results mean in a practical sense to the professor trying to decide which delivery method to use? Because there is not a significant difference between the population means, one method does not appear to be better overall than any other. However, because there does appear to be a significant interaction, she might consider assigning the different GPA levels to different delivery methods. Examining Figure 5.6, we see that high-level students appear to do best with the online method, middle-level students appear to do best with the hybrid method, and low-level students appear to do best with the face-to-face method. She might consider dropping the video method altogether.

Software Calculations

To perform a two-way ANOVA test:

Minitab: Title a blank column "Data," a second "A," and a third "B." For a randomized block design, enter each data value in column Data. Enter the corresponding row number in column A and the column number in column B. For example, for the data value 78.4 in Table 5.25, enter 2 in column A and 4 in column B. For a factorial experiment, enter the data in column Data, the corresponding level of factor A in column A, and the level of factor B in column B. For example, for the data value 93.3 in Table 5.27, enter 1 in column A and 4 in column B. Select **Stat** → **ANOVA** → **Two-Way**. Select column Data for the **Response**, column A for the **Row factor**, and column B for the **Column factor**.

R: First create a data frame to store the data. To create the data frame for the randomized block design data in Table 5.25, enter the following code:

> ScoresData<−data.frame(Score=c(1:12), GPA=factor(1:12),
                      Method=factor(1:12))

For the factorial experiment data in Table 5.27, change the 12 to 48 because there are 48 scores. Then type **fix("ScoresData")**. For each data value, enter the score in column **Score**; in column **GPA**, enter H for high, M for middle, or L for low; and in column **Method**, enter F for face-to-face, O for online, H for hybrid, or V for video. Close the data editor window and ignore any warnings about added factor levels. The randomized block design ANOVA is performed with the following code:

> summary(aov(Score~GPA+Method, data=ScoresData))

For a factorial experiment, change + to ∗.

Excel: For either a randomized block design or a factorial experiment, enter the data in a worksheet as they appear in Table 5.25 or Table 5.27, respectively (including the row and column labels). Then select **Data → Data Analysis**. For a randomized block design, select **Anova: Two-Factor Without Replication**, and for a factorial experiment, select **Anova: Two-Factor With Replication**. Next to **Input Range**, select the range containing the data and labels, and select a blank cell for the **Output Range**. In the former case, check the box next to **Labels**; in the latter case, enter the number of replications per treatment next to **Rows per sample**.

Exercises

Directions: In all problems asking for an ANOVA test, give the values of the appropriate test statistics F and the critical values (or the P-values). In all exercises, use a 0.05 significance level.

1. A new variety of seed corn is being tested under various soil types and treatment methods. Three fields with different soil types (I, II, and III) are divided into four quadrants. In one quadrant, the seeds are irrigated; in another, they are given fertilizer; in still another, they are given both; and in the last, they are given no treatment. The table below lists the yields (in bushels per acre) of each combination. Does there appear to be a significant difference between the treatment methods? Does the soil type appear to affect the yields? Based on these results, which treatment method and soil type appear to produce the highest yield?

Soil Type	Treatment			
	Irrigation	Fertilizer	Both	None
I	68.2	65.3	72.5	60.7
II	65.3	60.2	69.4	61.2
III	73.6	72.1	76.1	70.1

2. Three different blends of gasoline were tested in four different automobiles. The table below lists the recorded miles per gallon (mpg) of each combination. Does there appear to be a significant difference between the mean mpg of the four automobiles? Does there appear to be a significant difference between the effects of the three blends of gasoline? Based on these results, does any one blend appear to be better than any other?

	Automobile			
Gasoline	A_1	A_2	A_3	A_4
G_1	14.3	15.9	14.8	15.1
G_2	15.2	14.6	17.2	16.2
G_3	17.5	16.7	16.9	17.1

3. The owner of a lawn mower repair shop is trying three different methods of sharpening lawn mower blades on three different types of blades. He records the time (in minutes) it takes to sharpen each type while using each method, as shown in the table below. Does there appear to be a significant difference between the means of the three methods? Does the type of blade appear to affect the time?

	Method		
Type	I	II	III
Regular	5.4	4.9	5.2
Mulching	5.2	5.4	5.3
Riding Mower	6.9	6.5	6.2

4. Consider again the scenario of comparing different delivery methods for the introduction to statistics class, but suppose the data are as given in the table below.

	Delivery Method			
GPA	Face-to-Face	Online	Hybrid	Video
High	88.5	95.6	89.4	85.6
Middle	82.4	88.6	87.5	78.4
Low	79.4	84.2	81.4	81.7

a. Perform a two-way ANOVA test on the data. Does there appear to be a difference in the treatment means?

b. Now perform a one-way ANOVA test, treating each group of three students using each delivery method as simply a random sample. Based on this result, does there appear to be a difference in the treatment means?

c. Which test, the one- or two-way ANOVA, appears to be better able to detect the difference in the treatment means?

5. A bicycle manufacturer has updated its procedures for final assembly. Each of three different crews uses the old procedures for 3 days and then the new procedures for 3 days. The outputs (in bicycles per hour) are shown in the table below. Does there appear to be an interaction between the crew and the procedure? Does there appear to be a difference in the mean outputs of the two procedures? Does the crew appear to affect the output? Based on these results, do the new procedures appear to increase output?

Crew	Old			New		
I	16.8	17.4	15.4	18.1	19.4	18.3
II	16.5	15.3	16.9	17.5	17.8	17.6
III	16.0	14.8	18.1	16.6	16.4	17.6

6. A student surveys several people regarding their attitudes toward bipolar disorder. Each person answers several questions and then receives a score between 0 and 10, with a higher score indicating greater prejudice. Respondents are also asked their education level and whether they have had an encounter with a person who has bipolar disorder. Scores of three randomly selected respondents from each combination are shown in the table below (data collected by Sara Agee, 2010). Does there appear to be an interaction between education level and having an encounter? Does having an encounter appear to affect the scores? Does there appear to be a difference in scores between the education levels?

Encounter	Education Level				
	High School	Associates	Bachelor	Masters	Higher
Yes	1.80	0.27	0.93	1.33	0.87
	1.60	1.13	0.47	0.60	0.80
	1.20	0.47	0.80	1.20	1.00
No	1.47	1.20	1.13	1.20	0.93
	1.20	0.60	1.60	1.13	1.27
	1.33	1.13	1.20	0.73	1.20

7. To compare the prices of diapers sold in two different package sizes at three different stores, a student randomly selects three regular-size and three jumbo-size packages of size 5 diapers at each store and records the price per diaper of each package, as shown in the table below (data collected by Kay Zrust, 2012). Does there appear to be a significant difference between the mean prices of size 5 diapers at these three stores? Does the package size appear to have an effect? Does there appear to be an interaction between the store and the package size?

	Store		
Package Size	A	B	C
Regular	0.292	0.296	0.288
	0.355	0.370	0.405
	0.256	0.238	0.213
Jumbo	0.225	0.242	0.242
	0.322	0.380	0.322
	0.221	0.200	0.228

CHAPTER 6

Simple Regression

Chapter Objectives

- Introduce the basics of regression
- Define covariance and correlation
- Introduce the method of least squares and the simple linear model
- Present hypothesis tests for claims about the simple linear model
- Discuss methods of nonlinear regression
- Introduce basic concepts of multiple regression

6.1 Introduction

Often in life, we look for relationships. For instance, we might ask the following questions:

- Is weight related to heart disease?
- Is time spent studying for a standardized exam related to scores?
- Is the number of home runs hit in a ball park related to the altitude of the park?

Note that all these quantities (weight, time, scores, and so on) are random variables. In more general terms, these questions have the form

Is there a relationship between two random variables?

It is extremely important to understand that we are only asking a question about the existence of a *relationship* between variables. We are *not* asking if one *causes* the other. Probability and statistics cannot in and of themselves determine cause and effect. Cause and effect is an entirely different subject altogether. Identifying possible relationships is the goal of *regression*. In this chapter, we present some of the basic concepts of regression.

To begin to illustrate these ideas, consider the following scenario: To determine if there is a relationship between shoe length and the height of a person, the author had 10 of his students measure their shoe length (to the nearest 0.25 in) and height (to the nearest 0.5 inch). The resulting data are shown in Table 6.1.

Shoe Length	9	10	10.5	11	11.5	11.75	12	12.5	12.75	13
Height	62	64	64.5	69	70	73	72	75	74	77

Table 6.1

The first step in analyzing the data is to graph them. Note that these data come in *pairs*, because each student has a shoe length and a height, and that each pair is independent of the others. We plot height versus shoe length in the *scatterplot* shown in Figure 6.1.

The graph shows that as shoe length increases, height also tends to increase. Also note that the points form roughly a straight line. For this reason, we say that the variables shoe length and height have a *positive linear correlation*. *Correlation* simply means "relationship," *linear* means the relationship is described by a straight line, and *positive* means the line has a positive slope.

Figure 6.2 illustrates other types of relationships between variables X and Y. In a *negative linear correlation*, the points resemble a straight line with a negative slope. In this type

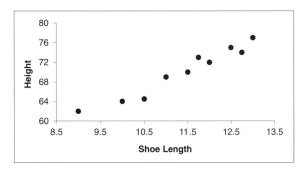

Figure 6.1

of relationship, as X increases, Y decreases. When there is *no correlation*, there does not appear to be any sort of relationship between the variables. The points do not form a pattern. As X increases, we cannot say that Y increases or decreases. In the last graph, the points do form a pattern, but it is not a straight-line pattern. In this case, we say that these variables have a *nonlinear correlation*.

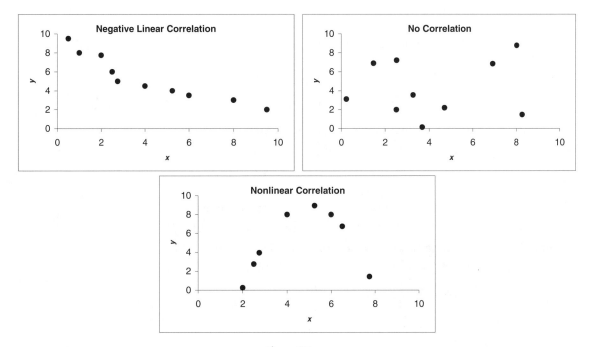

Figure 6.2

Once we have established graphically that there is a correlation between two variables, we ask two questions:

1. How strong is the relationship?
2. If we know the value of one variable, can we predict the value of the other?

Techniques for answering these questions are discussed in the next sections.

6.2 Covariance and Correlation

To begin to measure the strength of a relationship between two random variables, we define the *covariance*.

> **Definition 6.2.1** Let X and Y be two random variables with respective means μ_X and μ_Y. The *covariance of X and Y*, denoted $\text{Cov}(X, Y)$, is
> $$\text{Cov}(X, Y) = E[(X - \mu_X)(Y - \mu_Y)].$$

We need to make two comments regarding the covariance:

1. The covariance is a generalization of the variance because
$$\text{Cov}(X, X) = E[(X - \mu_X)(X - \mu_X)] = E\left[(X - \mu_X)^2\right] = \text{Var}(X).$$

2. Using linearity properties of the expected value, we get the following alternate form of the covariance (see Exercise 1 in this section):
$$\text{Cov}(X, Y) = E(XY) - \mu_X \mu_Y.$$

The definition of covariance applies to both discrete and continuous random variables. We use an example involving discrete variables to illustrate one interpretation of the covariance.

Example 6.2.1 Calculating Covariance for Discrete Random Variables Consider the discrete random variables X and Y whose joint pmf is given in Table 6.2. We have

$$\mu_X = 1\left(\frac{1}{9}\right) + 2\left(\frac{1}{3}\right) + 3\left(\frac{5}{9}\right) = \frac{22}{9},$$

$$\mu_Y = 1\left(\frac{5}{9}\right) + 2\left(\frac{1}{3}\right) + 3\left(\frac{1}{9}\right) = \frac{14}{9}, \text{ and}$$

$$E(XY) = 1 \cdot 1\left(\frac{1}{9}\right) + 2 \cdot 1\left(\frac{2}{9}\right) + 3 \cdot 1\left(\frac{2}{9}\right) + 2 \cdot 2\left(\frac{1}{9}\right) + 3 \cdot 2\left(\frac{2}{9}\right) + 3 \cdot 3\left(\frac{1}{9}\right) = 4.$$

Thus the covariance is

$$\text{Cov}(X, Y) = 4 - \left(\frac{22}{9}\right)\left(\frac{14}{9}\right) = \frac{16}{81} \approx 0.1975.$$

	$f(x, y)$	x = 1	2	3	$P(Y = y)$
y	1	$\frac{1}{9}$	$\frac{2}{9}$	$\frac{2}{9}$	$\frac{5}{9}$
	2	0	$\frac{1}{9}$	$\frac{2}{9}$	$\frac{1}{3}$
	3	0	0	$\frac{1}{9}$	$\frac{1}{9}$
	$P(X = x)$	$\frac{1}{9}$	$\frac{1}{3}$	$\frac{5}{9}$	

Table 6.2

□

What does $\text{Cov}(X, Y) \approx 0.1975$ tell us about the relationship between X and Y? By examining Table 6.2, we see that when X is small ($X = 1$), the only possible value for Y is 1, so Y must also be small. However, when X is large ($X = 3$), Y could be large ($Y = 3$), but it could be smaller. Thus there is a positive correlation between X and Y; this is indicated by the positive covariance. However, this relationship is somewhat weak; this is indicated by the relatively small value of the covariance.

The following theorem, whose proof we leave as an exercise, relates covariance to independence. The converse of this theorem is not true in general. That is, if $\text{Cov}(X, Y) = 0$, we cannot conclude that X and Y are independent (see Exercise 3 of this section).

Theorem 6.2.1 *If X and Y are independent, then*

$$\text{Cov}(X, Y) = 0.$$

□

One difficulty with using the covariance is that its units are (units of X)·(units of Y). Thus if we change the units on either variable, the strength of the relationship between the variables does not change, but the value of the covariance does change (see Exercise 4 of this section). It also makes it difficult to compare the strength of one relationship to another. One way to get around these problems is to use the closely related *correlation coefficient*.

Definition 6.2.2 Let X and Y be random variables with standard deviations σ_X and σ_Y, respectively. The *correlation coefficient of X and Y*, denoted $\rho(X, Y)$, is

$$\rho(X, Y) = \frac{\text{Cov}(X, Y)}{\sigma_X \sigma_Y} = \frac{E(XY) - \mu_X \mu_Y}{\sigma_X \sigma_Y}.$$

Remember that the units of the standard deviation of a variable are the same as the units of the variable. Thus $\rho(X, Y)$ is dimensionless. The correlation coefficient also has the following three properties, which we present without proof.

Theorem 6.2.2 *For any variables X and Y,*

1. $|\rho(X, Y)| \leq 1$.
2. *If X and Y are independent, then $\rho(X, Y) = 0$.*
3. $|\rho(X, Y)| = 1$ *if and only if $P(Y = mX + b) = 1$ for some constants m and b.* □

The second property is a direct result of Theorem 6.2.1. The third property is especially useful. It says that if $|\rho(X, Y)| = 1$ (or is close to 1), then the relation between X and Y can be described (or approximated) with a straight line. We will see an application of this in future sections.

Example 6.2.2 Calculating the Correlation Coefficient Consider again the random variables X and Y in Example 6.2.1. We calculated $\mu_X = 22/9$, $\mu_Y = 14/9$, and $\text{Cov}(X, Y) = 16/81$. The standard deviations are

$$\sigma_X = \sqrt{E[(X - \mu_X)^2]} = \sqrt{(1 - 22/9)^2 \cdot 1/9 + (2 - 22/9)^2 \cdot 1/3 + (3 - 22/9)^2 \cdot 5/9} \approx 0.685$$

$$\sigma_Y = \sqrt{E[(Y - \mu_Y)^2]} = \sqrt{(1 - 14/9)^2 \cdot 5/9 + (2 - 14/9)^2 \cdot 1/3 + (3 - 14/9)^2 \cdot 1/9} \approx 0.685$$

so that the correlation coefficient is

$$\rho(X, Y) = \frac{16/81}{0.685 \cdot 0.685} \approx 0.421. \qquad \square$$

In most cases, we do not know the joint pmf (or pdf) of X and Y, so we need a way of estimating $\rho(X, Y)$ by using data (i.e., observed values of X and the corresponding values of Y). Suppose we have observed n pairs of data values, $\{(x_1, y_1), \ldots, (x_n, y_n)\}$. We can use the data to estimate the parameters of X by

$$\mu_X \approx \bar{x} = \frac{1}{n}\sum_{i=1}^{n} x_i \quad \text{and} \quad \sigma_X \approx \sqrt{\frac{1}{n}\sum_{i=1}^{n}(x_i - \bar{x})^2}.$$

Similar estimates for the parameters of Y can be obtained by replacing x_i with y_i. We can also make the approximation

$$E(XY) \approx \frac{1}{n}\sum_{i=1}^{n} x_i y_i.$$

Putting these estimates together, we get an estimate of $\rho(X, Y)$ called the *sample correlation coefficient*.

Definition 6.2.3 The *sample correlation coefficient*, denoted r, is

$$r = \frac{(1/n)\sum x_i y_i - \bar{x}\bar{y}}{\sqrt{(1/n)\sum(x_i - \bar{x})^2}\sqrt{(1/n)\sum(y_i - \bar{y})^2}} \tag{6.1}$$

where all sums are taken from $i = 1$ to n.

The sample correlation coefficient is also called the *Pearson product-moment correlation coefficient* (or simply the *Pearson correlation coefficient*) in honor of the British statistician Karl Pearson, who did pioneering work in probability and statistics in the early 20th century. An equivalent form of Equation (6.1) is

$$r = \frac{n\sum x_i y_i - \sum x_i \sum y_i}{\sqrt{n\sum x_i^2 - \left(\sum x_i\right)^2}\sqrt{n\sum y_i^2 - \left(\sum y_i\right)^2}}. \tag{6.2}$$

Example 6.2.3 **Calculating a Sample Correlation Coefficient** Consider the data on shoe length and height given in Table 6.1. We calculate the value of r by using Equation (6.2) in Table 6.3. Thus we have

$$r = \frac{10(8043) - (114)(700.5)}{\sqrt{10(1314.375) - (114)^2}\sqrt{10(49{,}034.25) - (700.5)^2}} \approx 0.974.$$

x_i	y_i	$x_i y_i$	x_i^2	y_i^2
9	62	558	81	3844
10	64	640	100	4096
10.5	64.5	677.25	110.25	4160.25
11	69	759	121	4761
11.5	70	805	132.25	4900
11.75	73	857.75	138.0625	5329
12	72	864	144	5184
12.5	75	937.5	156.25	5625
12.75	74	943.5	162.5625	5476
13	77	1001	169	5929
Sum = 114	700.5	8043	1314.375	49,304.25

Table 6.3

Note that these calculations are very tedious, so they are most often done with software. If the random variable X denotes the shoe length of a randomly selected person and Y denotes the height, these calculations show that $\rho(X, Y) \approx 0.974$. □

Figure 6.3 shows the scatterplots of several different sets of data and their corresponding values of r. First note that the sign of r indicates the type of relationship, positive or negative. Second, note that the "closer" the points lie to a straight line, the larger the value of r. This gives us an important interpretation of the correlation coefficient:

The correlation coefficient r measures the strength of a linear relationship.

If r is close to ± 1, it indicates that there is a linear relationship between X and Y. That is, it appears that $Y = mX + b$ for some constants m and b. In Section 6.3, we will see how to estimate the values of the constants m and b.

Testing $H_0: \rho = 0$

Next we discuss a hypothesis test for a claim about ρ. We begin with a generalization of the normal distribution in the case of two variables, called the *bivariate normal distribution* or the *two-dimensional normal distribution*.

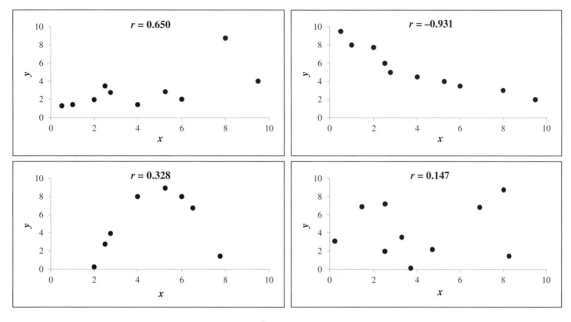

Figure 6.3

Definition 6.2.4 Let

$$h(x, y) = \frac{1}{1-\rho^2}\left[\left(\frac{x-\mu_X}{\sigma_X}\right)^2 - 2\rho\left(\frac{x-\mu_X}{\sigma_X}\right)\left(\frac{y-\mu_Y}{\sigma_Y}\right) + \left(\frac{y-\mu_Y}{\sigma_Y}\right)^2\right].$$

Two variables X and Y are said to have a *bivariate normal distribution* if their joint pdf is

$$f(x, y) = \frac{1}{2\pi\sigma_X\sigma_Y\sqrt{1-\rho^2}}\, e^{-h(x,y)/2}.$$

It can be shown that marginal pdfs of a bivariate normal distribution are

$$f_X(x) = \frac{1}{\sqrt{2\pi}\sigma_X}\exp\left\{-\frac{1}{2}\left(\frac{x-\mu_X}{\sigma_X}\right)^2\right\} \quad \text{and} \quad f_Y(y) = \frac{1}{\sqrt{2\pi}\sigma_Y}\exp\left\{-\frac{1}{2}\left(\frac{y-\mu_Y}{\sigma_Y}\right)^2\right\}.$$

This means that both X and Y are normally distributed with means μ_X and μ_Y and standard deviations σ_X and σ_Y, respectively. Another property of the bivariate normal distribution is that for any fixed value of X, the values of Y follow a normal distribution (technically we say that the conditional distribution of Y given a value of X is normal), and vice versa.

The parameter ρ in the joint pdf is the correlation coefficient defined above. Note that if $\rho = 0$, then the joint pdf is

$$f(x, y) = \frac{1}{2\pi\sigma_X\sigma_Y} \exp\left\{ -\frac{1}{2}\left(\frac{x - \mu_X}{\sigma_X}\right)^2 - \frac{1}{2}\left(\frac{y - \mu_Y}{\sigma_Y}\right)^2 \right\},$$

which is simply the product of the marginal pdfs. Therefore, X and Y are independent by definition. This, along with Theorem 6.2.2, establishes the following result.

Theorem 6.2.3 *Two random variables X and Y with a bivariate normal distribution are independent if and only if $\rho(X, Y) = 0$.* □

The sample correlation coefficient r calculated according to Equation (6.2) is based on data. Different sets of data from the same random variables X and Y may yield different values of r. Thus r is an observed value of a random variable. Call this variable R. The following theorem, whose proof is beyond the scope of this text, gives the distribution of R.

Theorem 6.2.4 *If X and Y have a bivariate normal distribution and assuming that $\rho(X, Y) = 0$, then the random variable*

$$T = R\sqrt{\frac{n - 2}{1 - R^2}}$$

has a Student-t distribution with $(n - 2)$ degrees of freedom

This theorem forms the basis of the hypothesis test described in the box on the next page.

Example 6.2.4 **Testing $\rho = 0$** Consider again the $n = 10$ pairs of data relating shoe length and height in Table 6.1. We found the correlation coefficient to be $r = 0.974$. To test the claim that $\rho = 0$, the hypotheses are

$$H_0: \rho = 0, \quad H_1: \rho \neq 0,$$

T-Test of $\rho = 0$ for Bivariate Random Variables

Purpose: To test the null hypothesis
$$H_0 \colon \rho(X, Y) = 0$$
where X and Y have a bivariate normal distribution. Let $(x_1, y_1), \ldots, (x_n, y_n)$ be n pairs of observed values and let

$$r = \frac{n \sum x_i y_i - \sum x_i \sum y_i}{\sqrt{n \sum x_i^2 - \left(\sum x_i\right)^2} \sqrt{n \sum y_i^2 - \left(\sum y_i\right)^2}}$$

be the sample correlation coefficient. The test statistic is

$$t = r \sqrt{\frac{n-2}{1-r^2}}.$$

The critical value is a t-value with $(n-2)$ degrees of freedom, and the P-value is the area under the corresponding density curve. Both the critical value and the P-value are found in a manner similar to all other hypothesis tests according to the form of the alternative hypothesis H_1.

---

and the test statistic is

$$t = 0.974 \sqrt{\frac{10-2}{1-(0.974)^2}} \approx 12.16.$$

At the 0.05 significance level, the critical value is $t_{0.05/2}(10-2) = 2.306$. The P-value is twice the area to the right of $t = 12.16$ under the Student-t density curve with $(10-2)$ degrees of freedom. This is approximately 0. Because $12.16 > 2.306$ (and the P-value is less than 0.05), we reject H_0 and conclude that there is a statistically significant linear relationship between shoe length and height.

To test the claim that $\rho > 0$, the hypotheses are

$$H_0 \colon \rho = 0, \quad H_1 \colon \rho > 0.$$

The test statistic is the same as before, $t = 12.16$. The critical value is $t_{0.05}(10-2) = 1.860$. The P-value is the area to the right of $t = 12.16$, which is approximately 0. Again, we reject H_0 and conclude that there is a positive linear relationship between shoe length and height. □

Software Calculations

To calculate a sample correlation coefficient and test a claim about ρ:

Minitab: Enter the x and y values in separate columns. Select **Stat** → **Regression** → **Regression**. Next to **Response** and **Predictors**, enter the names of the columns containing the x and y values, respectively. In the output, the square of the sample correlation coefficient r is labeled "R-Sq" and is given as a percentage. To find r, take the square root of R-Sq. The test statistic and P-value for the alternative hypothesis H_1: $\rho \neq 0$ are given in the second row under the headings "T" and "P," respectively.

R: To input the shoe data, enter the following code:
> ShoeData<– data.frame(Length=c(9,10,...), Height=c(62,64,...))

where the ellipses indicate the remaining data. Then enter the following code:
> summary(lm(Height~Length, data=ShoeData))

In the output, the square of the sample correlation coefficient r is labeled "Multiple R-Squared." To find r, take the square root of this number. The test statistic and P-value for the alternative hypothesis H_1: $\rho \neq 0$ are given in the second row under the headings "t value" and "Pr(> |t|)," respectively.

Excel: Enter the x and y values in separate columns. Select **Data** → **Data Analysis** → **Regression**. Enter the ranges containing the x and y values and select a blank cell for the **Output Range**. In the output, the sample correlation coefficient r is labeled "Multiple R." The test statistic and P-value for the alternative hypothesis H_1: $\rho \neq 0$ are given in the last line of the output under the headings "t-stat" and "P-value."

TI-83/84 Calculators: Enter the x values in a list and the y values in a separate list. Then press **STAT** → **TESTS** → **F:LinRegTTest**. Enter the names of the lists containing the x and y values, set **Freq** to 1, and select the inequality in H_1. Leave **RegEQ** blank.

Exercises

1. Show that $\text{Cov}(X, X) = E(XY) - \mu_X \mu_Y$. (**Hint**: Distribute $(X - \mu_X)(Y - \mu_Y)$ and use linearity properties of the expected value operator.)

2. Show that if X and Y are independent, then $\text{Cov}(X, Y) = 0$. (**Hint**: Begin with $\text{Cov}(X, X) = E(XY) - \mu_X \mu_Y$ and use Theorem 3.7.1.)

6.2 Covariance and Correlation

3. Suppose the points $\{(-2, 2), (0, 0), (2, 2)\}$ are plotted on the plane. Consider the random experiment of choosing one of these points at random. Then the sample space consists of these three points, and each outcome is equally likely. Define the random variables

$$X = \text{the first coordinate}, \quad Y = \text{the second coordinate}.$$

Then, for instance, $X(-2, 2) = -2$ and $Y(-2, 2) = 2$.

   a. Find $P(X = 2, Y = 2)$, $P(X = 2)$, and $P(Y = 2)$. Are X and Y independent or dependent? Explain.

   b. Now calculate $E(XY)$, $E(X)$, $E(Y)$, and $\text{Cov}(X, Y)$.

   c. Generalize your observations. If $\text{Cov}(X, Y) = 0$, can we conclude that X and Y are independent?

4. Assume the units of the variables X and Y in Example 6.2.1 are feet. Suppose we change the units to inches so that the range of each variable is $\{12, 24, 36\}$ but that the probabilities do not change. Calculate $\text{Cov}(X, Y)$ and $\rho(X, Y)$, using these new units. Do the new units change the covariance? Do they change the correlation coefficient?

5. Calculate the correlation coefficient r for the shoe length and height data, except interchange the values of x and y. Compare this value to the value found in Example 6.2.3. Is there any difference? Explain why or why not.

6. Suppose that X and Y are random variables such that $Y = mX + b$ for some constants $m \neq 0$ and b.

   a. Show that $\sigma_Y = |m|\sigma_X$. (**Hint:** Find $\sigma_Y^2 = \text{Var}(Y) = \text{Var}(mX + b)$ by treating mX and the constant b as independent random variables. Apply Theorem 3.7.2 and Theorem 2.6.2.)

   b. Use the definition of covariance to show that $\text{Cov}(X, Y) = m\sigma_X^2$.

   c. Use the results of parts a and b and the definition of the correlation coefficient to show that $\rho(X, Y) = \pm 1$.

7. Consider the random variables X and Y with the pmf given in the table below.

	$f(y, x)$	$x=1$	$x=2$	$x=3$	$P(Y = y)$
	0	0.3	0.05	0	0.35
y	1	0	0.3	0.05	0.35
	2	0	0	0.3	0.3
	$P(X = x)$	0.3	0.35	0.35	

a. Without doing any calculations, do you think $\text{Cov}(X, Y)$ is more or less than than that calculated in Example 6.2.1? Explain.

b. Calculate $\text{Cov}(X, Y)$ to test your conjecture in part a.

8. Calculate $\rho(X, Y)$ where X and Y have the pmf given in the table below. Find a relationship of the form $Y = mX + b$ for some constants m and b.

	$f(y, x)$	1	2	3	$P(Y = y)$
	0	0	0	$\frac{1}{4}$	$\frac{1}{4}$
y	1	0	$\frac{1}{2}$	0	$\frac{1}{2}$
	2	$\frac{1}{4}$	0	0	$\frac{1}{4}$
	$P(X = x)$	$\frac{1}{4}$	$\frac{1}{2}$	$\frac{1}{4}$	

(column header: x)

9. Calculate $\rho(X, Y)$ where X and Y have the pmf

$$f(x, y) = \frac{x^3 y}{54}, \quad \text{for } x = 1, 2, \ y = 1, 2, 3.$$

10. Show that

$$\frac{1}{n}\sum(x_i - \bar{x})^2 = \frac{1}{n}\sum x_i^2 - \frac{1}{n^2}\left(\sum x_i\right)^2.$$

11. Use the results of Exercise 10 above and the definitions of $\bar{x}$ and $\bar{y}$ to derive Equation (6.2) from Equation (6.1).

Directions: For each set of data in Exercises 12–14, draw a scatterplot, calculate the correlation coefficient r, and test the claim that $\rho = 0$ at the 0.05 significance level.

12. The table below gives the diameter of the trunk at chest height (in inches) and volume of wood (in cubic inches) in several pine trees. Does there appear to be a linear relationship between diameter and volume?

Diameter x	32	29	24	45	20	30	26	40	24	18
Volume y	185	109	95	300	30	125	55	246	60	15

13. Suppose a biologist records the number of pulses per second of the chirps of a cricket at different temperatures (in degrees Fahrenheit). The data collected are shown in the table below. Does there appear to be a linear relationship between temperature and pulses per second?

Temperature x	72	73	89	75	93	85	79	97	86	91
Pulses/sec y	16	16.2	21.2	16.5	20	18	16.75	19.25	18.25	18.5

14. The table below gives the number of people per physician and male life expectancy (in years) for various countries around the world (data from *World Almanac Book of Facts*, 1992, Pharos Books). Based on the hypothesis tests, does there appear to be a linear relationship between people per physician and life expectancy? Create a graph of the data. Do the points appear to follow a straight line, or could there be a nonlinear curve that better "fits" the data?

Country	People per Physician x	Life Expectancy y
Spain	275	74
United States	410	72
Canada	467	73
Romania	559	67
China	643	68
Taiwan	1010	70
Mexico	1037	67
South Korea	1216	66
India	2471	57
Morocco	4873	62
Bangladesh	6166	54
Kenya	7174	59

15. Use Equation (6.2) to calculate the sample correlation coefficient for the data in the table below where $k \neq 0$ is a constant.

x	1	2	4
y	k	$2k$	$4k$

16. This exercise illustrates that we should not blindly accept the results of a hypothesis test. Consider the data in the table below.

x	0.50	1.00	2.00	2.50	2.75	4.00	5.25	6.00	6.50	7.75
y	-11.25	-7.00	0.25	2.75	3.94	8.00	8.94	8.00	6.75	1.44

a. Calculate the correlation coefficient r for these data.

b. Test the claim that $\rho = 0$. Based on this result, does there appear to be a linear relationship between X and Y?

c. Create a scatterplot of these data. Based on the graph, does there appear to be a relationship between X and Y? Does this relationship appear to be linear?

17. A confidence interval for ρ may be obtained with the following algorithm:

   1. Calculate r.
   2. Calculate
   $$z = \frac{1}{2} \ln \frac{1+r}{1-r} \quad \text{and} \quad k = \frac{z_{\alpha/2}}{\sqrt{n-3}}$$
   where $z_{\alpha/2}$ is a critical z-value.
   3. Calculate $l = \tanh(z - k)$ and $u = \tanh(z + k)$, where
   $$\tanh(x) = \frac{e^x - e^{-x}}{e^x + e^{-x}}$$
   is the hyperbolic tangent function.
   4. The approximate $100(1 - \alpha)\%$ confidence interval estimate of ρ is then $l \leq \rho \leq u$.

Suppose a set of $n = 150$ data points has a sample correlation coefficient of $r = 0.748$. Construct a 95% confidence interval estimate of ρ.

18. Another way to test the null hypothesis $H_0: \rho = 0$ against the alternative $H_0: \rho \neq 0$ is to calculate *critical values* for r. To illustrate how this is done, suppose we have $n = 10$ pairs of data and we want to test the claim at the 95% confidence level. According to T-test described in the box in this section, the critical value, call it t_c, is $t_c = t_{0.05/2}(8) = 2.306$ and the test statistic is

$$t = r \sqrt{\frac{n-2}{1-r^2}}.$$

We would reject H_0 if $|t| > t_c$. Using elementary algebra, we can show that $|t| > t_c$ if and only if

$$|r| > \sqrt{\frac{t_c^2}{t_c^2 + n - 2}} = \sqrt{\frac{2.306^2}{2.306^2 + 10 - 2}} \approx 0.632.$$

The number 0.632 is the critical value for this particular scenario. This tells us how large r must be to conclude that there is a statistically significant linear relationship between the two random variables.

a. Show that $|t| > t_c$ if and only if $|r| > \sqrt{\dfrac{t_c^2}{t_c^2 + n - 2}}$.

b. Calculate the critical values at the 95% and 99% confidence levels for $n = 5, 15, 25,$ and 37.

c. What happens to the value of the critical value as n increases? If we had a smaller set of data, would we need a larger or smaller value of r to conclude that there was a significant linear relationship between the two random variables? Explain.

6.3 Method of Least Squares

Consider again the height and shoe length data. In Example 6.2.4, we tested the claim that $\rho = 0$ and rejected it. This means that it is reasonable to assume that height Y and shoe length X are described by a linear relationship $Y = mX + b$. Suppose we now want to predict the height of a person if we know her shoe length. In other words, we want to predict a value of Y given a value of X. This is called *regression*. To perform regression, we need to know the values of m and b. In this section, we present a method for estimating theses values.

To this end, consider the data points $(x_i, y_i), i = 1, \ldots, n$, in Figure 6.4 that appear to indicate a linear relationship. Graphically, we want to find a straight line that describes this relationship. This line is called the *regression line*, and the equation that describes it, $y = \hat{m}x + \hat{b}$, is called the *linear regression equation* (the "hats" indicate that $\hat{m}$ and $\hat{b}$ are estimates of m and b, respectively). The process of finding this straight line is called *linear regression*.

The variable X is called the *predictor variable* whereas Y is called the *response variable*. The value $\hat{y}_i = \hat{m}x_i + \hat{b}$ is called the *predicted* value of y_i.

To find formulas for $\hat{m}$ and $\hat{b}$, we note that it seems reasonable to want the differences between y_i and $\hat{y}_i$ to be as small as possible. These differences are illustrated with the

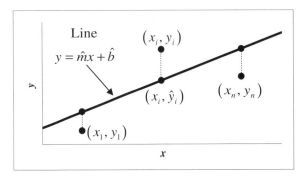

Figure 6.4

dashed lines in Figure 6.4. In mathematical terms, we want to find $\hat{m}$ and $\hat{b}$ that minimize the quantity

$$S = \sum_{i=1}^{n}(y_i - \hat{y}_i)^2 = \sum_{i=1}^{n}\left[y_i - \left(\hat{m}x_i + \hat{b}\right)\right]^2.$$

The requirement that $\hat{m}$ and $\hat{b}$ minimize S is called the *least-squares criterion*. Because there are two variables ($\hat{m}$ and $\hat{b}$), a necessary condition for optimality is that the partial derivatives with respect to each of these variables be zero. This gives the equations

$$\frac{\partial S}{\partial \hat{m}} = \sum 2\left(y_i - \hat{m}x_i - \hat{b}\right)(-x_i) = 0$$
$$\frac{\partial S}{\partial \hat{b}} = \sum 2\left(y_i - \hat{m}x_i - \hat{b}\right)(-1) = 0$$

where all summations are from 1 to n. Rewriting these equations and solving for $\hat{m}$ and $\hat{b}$ yield the formulas

$$\hat{m} = \frac{n\sum x_i y_i - \sum x_i \sum y_i}{n\sum x_i^2 - (\sum x_i)^2} \quad \text{and} \quad \hat{b} = \bar{y} - \hat{m}\bar{x}, \tag{6.3}$$

where $\bar{y}$ and $\bar{x}$ are the means of the x and y values, respectively (see Appendix A.6 for a derivation of these formulas). An alternative, but more complicated, formula for $\hat{b}$ is

$$\hat{b} = \frac{\sum x_i^2 \sum y_i - \sum x_i y_i \sum x_i}{n\sum x_i^2 - (\sum x_i)^2}.$$

The resulting equation $y = \hat{m}x + \hat{b}$ is called the *least-squares* or *best-fit* regression equation.

Example 6.3.1 Calculating the Linear Regression Equation Consider again the data relating shoe length and height given in Table 6.1. In Table 6.4, we calculate $\hat{m}$ and $\hat{b}$, using Equations (6.3).

x_i	y_i	$x_i y_i$	x_i^2	
9	62	558	81	
10	64	640	100	$\hat{m} = \dfrac{10(8043) - (114)(700.5)}{10(1314.375) - (114)^2}$
10.5	64.5	677.25	110.25	
11	69	759	121	≈ 3.878
11.5	70	805	132.25	
11.75	73	857.75	138.0625	$\hat{b} = 70.05 - 3.878(11.4)$
12	72	864	144	
12.5	75	937.5	156.25	≈ 25.84
12.75	74	943.5	162.5625	
13	77	1001	169	
Sum = 114	700.5	8043	1314.375	
Mean = 11.4	70.05			

Table 6.4

Thus the regression equation is $y = 3.878x + 25.84$. This equation is graphed along with the data in Figure 6.5. Notice that the regression equation describes the linear relationship very well. Graphically, we say the line "fits" the data well.

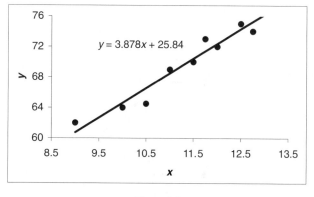

Figure 6.5

Suppose a crime scene investigator finds a shoe print outside a window that measures 11.25 in long and would like to estimate the height of the person who made the print. This is easy to do with the regression equation:

$$\hat{y} = 3.878(11.25) + 25.84 \approx 69.47.$$

Thus we would say that the "best predicted" height of a person with a shoe length of 11.25 in is 69.47 in, based on the data. In the next section, we will see a method for obtaining a prediction interval estimate of this height. □

The next example illustrates that a regression equation is not always appropriate for making predictions.

Example 6.3.2 No Correlation Table 6.5 shows the price (in dollars) of several name brand items and their generic equivalents at a large retail store (data collected by Amanda Schroeder, 2010). Also shown are the percent savings if the generic product is purchased instead of the name brand.

Name Brand x	7.68	9.56	7.62	1.58	2.98	3.58	6.54	8.97	4.06	6.46
Generic Brand	4.47	2.97	3.34	0.67	3.12	2.98	5.34	6.97	1.76	4.12
Percent y	41.8	68.9	56.2	57.6	−4.7	16.8	18.3	22.3	56.7	36.2

Table 6.5

Suppose a name brand product costs $1.99 and we would like to predict the percent savings if we were to purchase the generic equivalent. The graph of percent savings (y) versus name brand price (x) along with the linear regression equation and the value of the correlation coefficient r are shown in Figure 6.6.

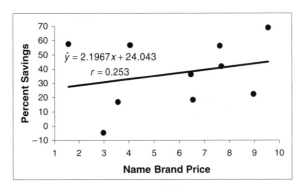

Figure 6.6

We might be tempted to simply plug $x = 1.99$ into the regression equation and predict the savings to be $\hat{y} = 2.1967(1.99) + 24.043 \approx 28.4\%$. However, note that the value of r is rather small, indicating that there may not be a linear correlation. Testing the claim that $\rho = 0$ yields the hypotheses H_0: $\rho = 0$ and H_1: $\rho \neq 0$. The test statistic is

$$t = 0.253 \sqrt{\frac{10-2}{1-(0.253)^2}} \approx 0.74.$$

At the 0.05 significance level, the critical value is $t_{0.05/2}(10-2) = 2.306$. Because $t = 0.74 < 2.306$, we do not reject H_0 and we conclude that there is not a linear correlation between the price of the name brand product and the percent savings. Therefore, it would *not* be appropriate to use the linear regression equation to make predictions.

To find a reasonable prediction for the percent of savings, note that because there does not appear to be a linear relationship (or any other type of relationship), the appropriate value of m in the linear relationship $Y = mX + b$ is $m = 0$. From Equation (6.3), we see that an estimate of b is then

$$\hat{b} = \bar{y} - \hat{m}\bar{x} = \bar{y} - 0\bar{x} = \bar{y}$$

so that the relationship is described by $Y = \bar{y}$. Thus the best predicted value of Y is $\bar{y}$. From the data, we get $\bar{y} = 37.01$. This means that the best predicted percent savings is about 37% regardless of the price of the name brand product. □

The previous example illustrates the first of two important cautions when doing linear regression:

1. If there is no linear correlation, do not use a linear regression equation to make predictions.

2. Only use a linear regression equation to make predictions within the range of the x values of the data. For values outside this range, the trend of the data may change.

Software Calculations

To calculate the coefficients in the least-squares regression equation, follow the steps in Section 6.2. On the next page we describe ways of graphing the regression equation on top of the data:

Minitab: Select **Graph** → **Scatterplot** → **With Regression**. Under **Y variables** and **X variables**, enter the names of the columns containing the y and x values, respectively.

R: To graph the regression equation on top of the shoe data, first enter the data as described in Section 6.2. Then enter the following two lines of code:
> plot(ShoeData$Length, ShoeData$Height)
> abline(lm(Height~Length, data=ShoeData))

Excel: Create a scatterplot of the data using the **Chart Wizard**. Right-click on one of the data points in the scatterplot and select **Add Trendline**. Under **Type**, select **Linear**. Click on the **Options** tab and select **Display equation on chart**.

TI-83/84 Calculators: First follow the steps in Section 6.2. Next to **RegEQ**, enter a function to store the regression equation by pressing **VARS** → **Y-VARS** → **1:Function** and select one of the functions. Select **Calculate**, and the regression equation will be calculated and stored at the chosen function. Next create a scatterplot of the data by pressing **2ND** → **STAT PLOT** and selecting one of the plots. Turn the plot **ON**, choose the first **Type**, and enter the names of the lists containing the x and y values. Then press **ZOOM** → **9:ZoomStat**.

Exercises

1. The table below gives the diameter of the trunk at chest height (in inches) and volume of wood (in cubic inches) in several pine trees. Find the linear regression equation. If a tree has a diameter of 26 in, find the best predicted volume.

Diameter x	32	29	24	45	20	30	26	40	24	18
Volume y	185	109	95	300	30	125	55	246	60	15

2. Suppose a biologist records the number of pulses per second of the chirps of a cricket at different temperatures (in degrees Fahrenheit), as shown in the table below. Find the linear regression equation and the best predicted number of pulses per second if the temperature is 80°F.

Temperature x	72	73	89	75	93	85	79	97	86	91
Pulses/sec y	16	16.2	21.2	16.5	20	18	16.75	19.25	18.25	18.5

6.3 Method of Least Squares

3. The data in the table below give the shoe length and height of all 26 students in the author's introduction to statistics class. Assuming these students represent a random sample of all students at this university, graph the data, calculate the correlation coefficient, determine if there is a linear correlation between shoe length and height in the population of students at this university, and find the linear regression equation that predicts height in terms of shoe length. Find the best predicted height if a student has a shoe length of 12.25 in.

Shoe Length	Height	Shoe Length	Height	Shoe Length	Height
11.75	73	9	64	10	64
12	72	12	70	10.5	70
12.5	75	10.75	67	13	77
12.75	74	12.5	74	10	65
11.75	74.5	11	68	12.75	76
11	69	11.75	71	12	73
11.5	70	10	68	12.5	75
9	62	10.5	64.5	11.5	68
10	64	10	64		

4. In this exercise, we explore the effects of an outlier on the regression equation. Consider the data in the table below, which gives shoe length and height data for a group of students.

Shoe Length	10.75	11	11.75	10.5	10.5	10	10	13	12.75	12	12.5
Height	67	68	71	57	64.5	64	64	77	76	73	75

a. Create a scatterplot of these data and identify any outlier(s).

b. Calculate the linear regression equation with the outlier(s); then calculate it with the outlier(s) removed.

c. Based on your observations in part b, is the linear regression equation sensitive to outliers? Explain.

5. Graph the data in the table below. Does there appear to be a linear correlation? Support your answer with an appropriate hypothesis test. Find the best predicted value of Y if $x = 3$.

x	0	2	4	6	8	10
y	26	20	12	50	25	30

6. In some cases we may suspect that the relationship between two variables X and Y is described by $Y = mX$. To estimate the value of m by using the least-squares criterion, we want to find a number $\hat{m}$ that minimizes

$$S = \sum_{i=1}^{n}(y_i - \hat{y}_i)^2 = \sum_{i=1}^{n}(y_i - \hat{m}x_i)^2.$$

a. Take the derivative of S with respect to $\hat{m}$, set it equal to 0, and solve it for $\hat{m}$ to find a formula for $\hat{m}$ in terms of the x and y coordinates.

b. Use the formula in part a to estimate the value of m for the data in the table below.

x	5	10	15	20	25	30	35	40
y	1.02	1.86	3.00	3.94	4.95	5.82	6.95	7.80

7. Suppose we suspect that the relationship between X and Y is described by $Y = mX + b_0$, where b_0 is some specified value. To estimate the value of m by using the least-squares criterion, we want to find a number $\hat{m}$ that minimizes

$$S = \sum_{i=1}^{n}[y_i - (\hat{m}x_i + b_0)]^2.$$

a. Take the derivative of S with respect to $\hat{m}$, set it equal to 0, and solve it for $\hat{m}$ to find a formula for $\hat{m}$ in terms of the x and y coordinates and b_0.

b. Use the data in the table below and the formula in part a to estimate the value of m if the relationship is $Y = mX + 15$.

x	10	15	20	25	30	35
y	57.4	67.6	71.6	104.5	106.9	136.0

8. Suppose we think that a random variable Y has a constant value b. To estimate the value of b using the least-squares criterion, we want to find a number $\hat{b}$ that minimizes the number

$$S = \sum_{i=1}^{n}\left(y_i - \hat{b}\right)^2$$

where $\{y_i : i = 1, \ldots, n\}$ is a set of observed values of Y. Take the derivative of S with respect to $\hat{b}$, set it equal to 0, and show that $\hat{b} = \bar{y}$, the mean of the observed y values.

9. Show that the graph of the least-squares regression equation $y = \hat{m}x + \hat{b}$ always goes through the point $(\bar{x}, \bar{y})$.

10. In the method of least squares, we estimate m and b by minimizing the sum $S = \sum (y_i - \hat{y}_i)^2$. This method is often used because squares are easy to work with (from the standpoint of calculus) resulting in simple formulas for $\hat{m}$ and $\hat{b}$. Also, as we will see in the next two sections, the resulting values have important statistical properties.

Another method for estimating m and b is to minimize the sum of the absolute values of the differences rather than the squares of the differences. That is, we find $\hat{m}$ and $\hat{b}$ that minimize the sum

$$S_A = \sum_{i=1}^{n} |y_i - \hat{y}_i| = \sum_{i=1}^{n} \left| y_i - \left(\hat{m}x_i + \hat{b}\right) \right|.$$

Consider the height and shoe length data given in Table 6.4.

  a. Calculate the value of S_A using $\hat{m} = 3.878$ and $\hat{b} = 25.84$ found using Equations (6.3). Also calculate the value of S.

  b. Now calculate the values of S_A and S, using $\hat{m} = 4.086$ and $\hat{b} = 23.88$.

  c. Based on your calculations, do values of $\hat{m}$ and $\hat{b}$ that minimize S also minimize S_A? Would calculating $\hat{m}$ and $\hat{b}$ by minimizing S_A result in the same regression equation as if we minimized S? Explain.

11. Yet another method for estimating m and b is called *Chebyshev's criterion*. This method is based on the idea that the linear regression equation should make the largest of the differences $|y_i - \hat{y}_i|$ as small as possible. In more technical terms, this criterion says that the regression equation should minimize the number

$$C = \text{Maximum}\,\{|y_i - \hat{y}_i| : i = 1, 2, \ldots, n\}.$$

Consider the height and shoe length data given in Table 6.4.

  a. Calculate the value of C, using $\hat{m} = 3.878$ and $\hat{b} = 25.84$ found by using Equations (6.3).

  b. Now calculate the value of C, using $\hat{m} = 4$ and $\hat{b} = 24.25$.

  c. Based on your calculations, do values of $\hat{m}$ and $\hat{b}$ that minimize S also minimize C? Would calculating $\hat{m}$ and $\hat{b}$ by minimizing C result in the same regression equation as if we minimized S? Explain.

6.4 The Simple Linear Model

In Section 6.3, we saw how to estimate m and b in the linear relationship $Y = mX + b$. We also saw how a regression equation $\hat{y} = \hat{m}x + \hat{b}$ can be used to estimate a value of Y given a value of X. In this section, we introduce techniques for constructing confidence intervals for these estimates and performing hypothesis tests.

We begin by defining the *simple linear model* that forms the basis for the theory of this section.

> **Definition 6.4.1** Two random variables X and Y are said to be described by a *simple linear model* if
> $$Y = mX + b + \epsilon$$
> where m and b are constants and ϵ is a random variable independent of X that is $N(0, \sigma^2)$ where σ^2 is a constant.

Note that in the model, X and Y do not follow a perfect linear relationship $Y = mX + b$. They follow a linear relationship plus some "noise" modeled by the variable ϵ. This noise could be due to random fluctuations, the effects of other variables not being considered, or other causes. Also note that we do not require X and Y to have a bivariate normal distribution.

For any fixed value of X, say $X = x_i$, the linear model implies the following three points, whose proof we leave as an exercise and which are illustrated in Figure 6.7:

1. Y has a normal distribution,
2. $E(Y) = mx_i + b$, and
3. $\text{Var}(Y) = \sigma^2$, the variance of ϵ.

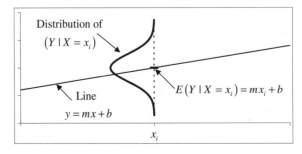

Figure 6.7

Associated with the simple linear model are *residuals*.

Definition 6.4.2 For a set of data $(x_1, y_1), \ldots, (x_n, y_n)$, the *residuals* are

$$(y_i - \hat{y}_i) = \left(y_i - \hat{m}x_i - \hat{b}\right) \quad \text{for } i = 1, \ldots, n$$

where $\hat{m}$ and $\hat{b}$ are the least-squares estimates of m and b as calculated in Section 6.3.

In more informal terms, residuals are (observed y values) − (predicted y values). Residuals can be thought of as observed values of the random variable ϵ in the simple linear model. They can be used to verify that the simple linear model applies.

Example 6.4.1 Calculating Residuals Consider again the data relating shoe length and height given in Table 6.1. The regression equation for these data is $y = 3.878x + 25.839$. In Figure 6.8, we calculate the residuals and show a scatterplot of residuals versus x.

x	y	$\hat{y}$	Residual
9	62	60.7428	1.2572
10	64	64.621	−0.621
10.5	64.5	66.5601	−2.0601
11	69	68.4992	0.5008
11.5	70	70.4383	−0.4383
11.75	73	71.40785	1.59215
12	72	72.3774	−0.3774
12.5	75	74.3165	0.6835
12.75	74	75.28605	−1.28605
13	77	76.2556	0.7444
		Mean =	−0.00048

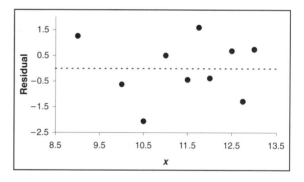

Figure 6.8

First, note that the mean of the residuals is close to 0, as expected if the simple linear model holds. Second, note that there is no "pattern" in the graph of the residuals. This indicates that ϵ is independent of X. Further analysis of the residuals can be done by constructing a normal quantile plot as in Figure 6.9. This quantile plot indicates that ϵ is normally distributed. This analysis leads us to conclude that shoe length and height are described by a simple linear model. □

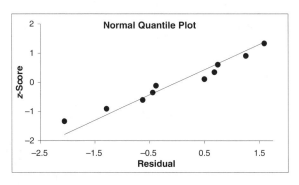

Figure 6.9

Now, to estimate σ^2, the variance of ϵ, we begin with the formula for the sample variance

$$s^2 = \frac{1}{n-1} \sum_{i=1}^{n} (x_i - \bar{x})^2$$

where x_i are observed values of the random variable and $\bar{x}$ is the mean of these values. Taking the residuals as observed values of ϵ and assuming their mean is 0, we find

$$s^2 = \frac{1}{n-1} \sum_{i=1}^{n} [(y_i - \hat{y}_i) - 0]^2 = \frac{1}{n-1} \sum_{i=1}^{n} \left(y_i - \hat{m}x_i - \hat{b}\right)^2$$

as an estimate of σ^2. Modifying this formula slightly gives us the following *unbiased* estimate of σ^2:

$$s_e^2 = \frac{1}{n-2} \sum_{i=1}^{n} \left(y_i - \hat{m}x_i - \hat{b}\right)^2.$$

This leads to the following definition:

Definition 6.4.3 Let X and Y be described by a simple linear model. The *standard error of estimate* is

$$s_e = \sqrt{\frac{1}{n-2} \sum_{i=1}^{n} \left(y_i - \hat{m}x_i - \hat{b}\right)^2}.$$

Using the standard error of estimate, we get the following interval estimate of the value of Y given a value of X, called a *prediction interval*, the derivation of which is beyond the scope of this text.

Definition 6.4.4 Let X and Y be described by a simple linear model. Given a value of X, say, $X = x_0$, a $100(1-\alpha)\%$ *prediction interval estimate* for the corresponding value of Y is

$$\hat{y} - E < Y < \hat{y} + E$$

where $\hat{y} = \hat{m}x_0 + \hat{b}$, the margin of error E is

$$E = t_{\alpha/2} s_e \sqrt{1 + \frac{1}{n} + \frac{(x_0 - \bar{x})^2}{\sum(x_i - \bar{x})^2}}$$

and $t_{\alpha/2}$ is a critical t-value with $(n-2)$ degrees of freedom.

Example 6.4.2 Calculating a Prediction Interval Consider again the shoe length and height data. Using the values of the residuals in Figure 6.8, we find the standard error of estimate to be

$$s_e = \sqrt{\frac{1}{10-2}[(1.2572)^2 + \cdots + (0.7444)^2]} \approx 1.225.$$

When $X = 11.25$, we found that $\hat{y} = 3.878(11.25) + 25.839 \approx 69.467$. To find the corresponding 95% confidence interval estimate of Y, note that $\bar{x} = 11.4$ so that

$$\sum_{i=1}^{10}(x_i - \bar{x})^2 = (9 - 11.4)^2 + \cdots + (13 - 11.4)^2 = 14.775.$$

The critical value is $t_{0.05/2}(10-2) = 2.306$. Thus the margin of error is

$$E = 2.306(1.225)\sqrt{1 + \frac{1}{10} + \frac{(11.25 - 11.4)^2}{14.775}} \approx 2.96$$

and the prediction interval is

$$69.47 - 2.96 < Y < 69.47 - 2.96 \quad \Rightarrow \quad 66.51 < Y < 72.43.$$

We interpret this prediction interval by saying, "we are 95% confident that with a shoe length of 11.25 in, the height of a person is between 66.51 and 72.43 in, based on the data." □

As mentioned earlier, $\hat{m}$ is an estimate of m in the linear model $Y = mX + b + \epsilon$. We can calculate a confidence interval estimate of m in Definition 6.4.5 and perform hypothesis tests regarding m using the procedures in the box that follows.

Definition 6.4.5 Let X and Y be described by a simple linear model $Y = mX + b + \epsilon$. A $100(1-\alpha)\%$ confidence interval estimate of m is

$$\hat{m} - E < m < \hat{m} + E$$

where the margin of error E is

$$E = t_{\alpha/2} \frac{s_e}{\sqrt{\sum(x_i - \bar{x})^2}}$$

and $t_{\alpha/2}$ is a critical t-value with $(n-2)$ degrees of freedom. The quantity $s_e/\sqrt{\sum(x_i - \bar{x})^2}$ is called the *standard error of the coefficient*.

T-Test of the Slope of a Simple Linear Model

Let X and Y be described by a simple linear model $Y = mX + b + \epsilon$. To test the null hypothesis

$$H_0: m = m_0,$$

the test statistic is

$$t = \frac{1}{s_e}(\hat{m} - m_0)\sqrt{\sum(x_i - \bar{x})^2},$$

the critical value is a t-score with $(n-2)$ degrees of freedom, and the P-value is the area under the corresponding density curve. Both the critical value and the P-value are found in a manner similar to all other hypothesis tests according to the form of the alternative hypothesis H_1.

Example 6.4.3 Estimating m For the shoe length and height data, we found that $\hat{m} = 3.878$. To form a 95% confidence interval estimate of m, note that the critical value is $t_{0.05/2}(10-2) = 2.306$. In Example 6.4.2, we found that $s_e = 1.225$ and $\sum(x_i - \bar{x})^2 = 14.775$ so that the margin of error is

$$E = 2.306 \frac{1.225}{\sqrt{14.775}} \approx 0.735$$

and the confidence interval is

$$3.878 - 0.735 < m < 3.878 + 0.735 \quad \Rightarrow \quad 3.143 < m < 4.613.$$

We interpret this interval by saying, "We are 95% confident that m is between 3.143 and 4.613." To test the claim that $m > 0$, the hypotheses are

$$H_0: m = 0, \quad H_1: m > 0.$$

The test statistic is

$$t = \frac{1}{1.225}(3.878 - 0)\sqrt{14.775} \approx 12.168.$$

The critical value is $t_{0.05}(10 - 2) = 1.860$. The P-value is the area under the Student-t density curve with $(10 - 2)$ degrees of freedom to the right of $t = 12.168$, which is approximately 0. Therefore, we reject H_0 and conclude that the data support the claim that $m > 0$. □

Through much exhausting algebraic manipulation, it can be shown that when $m_0 = 0$, the test statistic for the T-test described on the previous page is equal to the test statistic for the test that $\rho = 0$ as described in the T-test in the box at the end of Section 6.2. That is, one can show that

$$r\sqrt{\frac{n-2}{1-r^2}} = \frac{\hat{m}}{s_e}\sqrt{\sum(x_i - \bar{x})^2}.$$

(Compare the value of the test statistic in the previous example to that found in Example 6.2.4. They are, within rounding error, the same.) Thus we can think of the hypotheses $H_0: \rho = 0$ and $H_0: m = 0$ as being equivalent. Intuitively, this should make sense. If X and Y are bivariate random variables, then independence is equivalent to having $\rho = 0$. If they also are described by a simple linear model $Y = mX + b + \epsilon$, then independence means that $m = 0$.

Software Calculations

To calculate the standard error of estimate s_e and perform the T-test for the slope of the simple linear model, follow the steps in Section 6.2. Below we describe ways of calculating a confidence interval for the slope and prediction intervals:

Minitab: In the output, the standard error of estimate s_e is denoted "S". To calculate a prediction interval, enter the value(s) of x_0 in a blank column. Select **Stat → Regression → Regression**. Enter the names of the columns containing the **Response** and **Predictor** variables. Select **Options** and enter the name of the column containing the value(s) of x_0 under **Prediction intervals for new observations**. Also enter the desired confidence level. A normal quantile plot of the residuals can be constructed by selecting **Graphs** and checking the box next to **Normal plot of residuals**.

R: In the output, the standard error of estimate s_e is denoted "Residual standard error." To calculate a 95% confidence interval estimate of the slope, enter the following two lines of code:

> linreg=lm(Height~Length, data=ShoeData)
> confint(linreg, level=0.95)

To calculate a prediction interval at the 95% confidence level using the shoe data for the case where $X = 11.25$, enter the following code:

> predict(linreg, data.frame(Length=11.75), level=0.95, interval="prediction")

To generate a quantile plot of the residuals, enter the following code:

> qqnorm(resid(linreg), datax=TRUE)

Excel: In the output, the standard error of estimate s_e is denoted "Standard Error." A confidence interval estimate of the slope is given in the last row of the output under the headings "Lower 95%" and "Upper 95%." The confidence level can be changed in the Regression window. Also the residuals can be calculated and plotted by checking the boxes next to **Residuals** and **Residual Plots**.

TI-83/84 Calculators: In the output, the standard error of estimate s_e is denoted "s". To calculate a confidence interval for the slope, press **STAT** → **TESTS** → **G:LinRegTInt**. Input the necessary information. The confidence interval is given in the second row of the output. Note that s_e is also given.

Exercises

1. For the data in the table below, the linear regression equation is $y = 0.0316x + 2.9979$. Use this to create a graph of the residuals. Does it appear that the variables X and Y are described by a simple linear model? Explain.

x	0	2	4	6	8	10
y	3.000	3.061	3.122	3.186	3.250	3.316

2. The table below gives the weights (in pounds) and highway miles per gallon (mpg) of several cars. The linear regression equation for the data is $y = -0.01x + 60.6$. Calculate the standard error of estimate s_e.

Weight x	3250	3425	2400	2250
MPG y	26	28	37	38

3. The table below gives the amounts of nitrogen applied to different cornfields (in pounds per acre) and the resulting yields (in bushels per acre). The linear regression equation is $y = 0.56x + 68.8$, and the standard error of estimate is $s_e = 10.93$. Calculate a 95% prediction interval estimate of the yield if 150 lb/acre of nitrogen is applied.

Nitrogen x	0	60	120	180	240
Yield y	78	90	140	162	210

4. The table below gives the budgets of different advertising campaigns done by a company (in thousands of dollars) and the resulting number of units sold (in thousands of units). The linear regression equation for the data is $y = 2.638x + 5.838$, and the standard error of estimate is $s_e = 6.726$. Test the claim that $m > 2.5$. In an attempt to get more money for advertising, someone in the advertising department claims that for each additional $1000 spent, more than 2500 additional units will be sold. Do the data support the claim? Why or why not?

Budget x	1	4	9	11	19	21
Units Sold y	5.7	23.4	21.7	36.7	62.2	56.8

5. The table below gives the diameter of the trunk at chest height (in inches) and the volume of wood (in cubic inches) in several pine trees. The linear regression equation is $y = 10.894x - 191.75$, and the standard error of estimate is $s_e = 20.385$. Find a 95% confidence interval estimate of the slope m.

Diameter x	32	29	24	45	20	30	26	40	24	18
Volume y	185	109	95	300	30	125	55	246	60	15

6. Explain why for a fixed set of data and confidence level, the minimum margin of error E for a prediction interval for Y occurs when $x_0 = \bar{x}$.

7. If X and Y are described by a simple linear model $Y = mX + b + \epsilon$, and X takes on a certain value, say, $X = x_0$, show that Y is $N(mx_0 + b, \sigma^2)$ where σ^2 is the variance of ϵ. (**Hint:** Use Exercise 14 of Section 3.5.)

8. In Example 6.4.2 we calculated a prediction interval for the height of a person Y whose shoe length is $x_0 = 11.25$ in. If we were to measure the heights of many people whose shoe length is 11.25 in, the heights would not all be the same. Thus the heights are observed

values of a random variable, call it Z. (Technically, this random variable is Y given $X = x_0$, denoted $(Y|X = x_0)$.) To form a confidence interval for the mean of this variable $E(Z)$, we use a variation of the prediction interval given in Definition 6.4.4. The confidence interval is

$$\hat{y} - E < E(Z) < \hat{y} + E$$

where the margin of error E is

$$E = t_{\alpha/2} s_e \sqrt{\frac{1}{n} + \frac{(x_0 - \bar{x})^2}{\sum (x_i - \bar{x})^2}}$$

and $t_{\alpha/2}$ is a critical t-value with $(n - 2)$ degrees of freedom. Find a 95% confidence interval estimate of $E(Z)$. Compare the width of this confidence interval to the width of the prediction interval in Example 6.4.2.

9. In Definition 6.4.5, we defined a confidence interval for the slope m in the simple linear model $Y = mX + b + \epsilon$. We can define a confidence interval estimate for the y intercept b as

$$\hat{b} - E < b < \hat{b} + E$$

where the margin of error E is

$$E = t_{\alpha/2} s_e \sqrt{\frac{\sum x_i^2}{n \sum (x_i - \bar{x})^2}}$$

and $t_{\alpha/2}$ is a critical t-value with $(n - 2)$ degrees of freedom. Calculate at 95% confidence interval an estimate of b for the shoe length and height data. (**Hint:** See Table 6.3 for the value of $\sum x_i^2$.)

10. When the temperature of a fixed volume of gas decreases, the pressure of the gas also decreases. *Absolute zero* can be thought of as the temperature at which the pressure is 0. In terms of regression, we can estimate absolute zero with the y intercept of a linear regression equation for data of the form (x_i, y_i), where x_i is the pressure of a fixed volume of gas at temperature y_i. The table below gives the pressure (in bars) and corresponding temperature (in degrees Celsius) of a gas at a fixed volume (data collected by Zachary Klatt, 2012).

Pressure x	1.05	1.05	1.09	1.12	1.15	1.18	1.20
Temperature y	18.3	20.0	30.0	40.0	50.0	60.0	70.0

a. Find the linear regression equation for these data and calculate the standard error of estimate s_e.

b. Use the procedures in Exercise 9 to find a 99% confidence interval estimate of the y intercept.

c. The accepted value of absolute zero is $-273°C$. Does the confidence interval in part b contain this value?

11. In the prediction interval described in Definition 6.4.4, if we replace x_0 with the more general x, $\hat{y}$ with $\hat{m}x + \hat{b}$, and E with its definition, we get the following description of the confidence interval:

$$\hat{m}x + \hat{b} - t_{\alpha/2} s_e \sqrt{1 + \frac{1}{n} + \frac{(x - \bar{x})^2}{\sum (x_i - \bar{x})^2}} < Y < \hat{m}x + \hat{b} + t_{\alpha/2} s_e \sqrt{1 + \frac{1}{n} + \frac{(x - \bar{x})^2}{\sum (x_i - \bar{x})^2}}$$

where $t_{\alpha/2}$ is a critical t-value with $(n - 2)$ degrees of freedom. This defines two curves:

$$y_L(x) = \hat{m}x + \hat{b} - t_{\alpha/2} s_e \sqrt{1 + \frac{1}{n} + \frac{(x - \bar{x})^2}{\sum (x_i - \bar{x})^2}}$$

and

$$y_U(x) = \hat{m}x + \hat{b} + t_{\alpha/2} s_e \sqrt{1 + \frac{1}{n} + \frac{(x - \bar{x})^2}{\sum (x_i - \bar{x})^2}}.$$

These curves are called *prediction bands* for Y. The first curve y_L forms a lower bound for the prediction interval for Y when $X = x$ while the second curve y_U forms an upper bound. Consider the data relating shoe length and height given in Table 6.1.

a. Use the calculations in Example 6.4.2 to construct the prediction bands, and graph them along with the data and the regression line.

b. When $X = 11.25$, verify that these bands give a prediction interval for Y of $66.51 < Y < 72.43$, as found in Example 6.4.2.

c. The width of the prediction band is $y_U(x) - y_L(x)$. This is simply the width of the prediction interval for Y when $X = x$. Graph the curve $y = y_U(x) - y_L(x)$ over the interval $9 \leq x \leq 13$, which gives the width of the prediction band at each value of x in the data.

d. Describe what happens to the width of the prediction band as x increases from 9 to 13. Where is the width minimized? Does this agree with the results of Exercise 6 above?

e. Suppose we use the linear regression equation to predict a value of Y for $X = 9.5$. Then we predict a value of Y for $X = 11.25$. Based on your answer to part d, which of these predictions is probably more accurate? Explain.

12. In Exercise 8, we defined a confidence interval for $E(Y|X = x_0)$. We can use this idea to define *confidence bands* in the same way as we defined prediction bands in Exercise 11. Calculate these confidence bands and graph them along with the data and the prediction bands. Which are narrower, the confidence bands or the prediction bands?

6.5 Sums of Squares and ANOVA

In this section, we introduce an analysis of variance (ANOVA) technique for testing the hypothesis H_0: $m = 0$ where m is the slope in the simple linear model $Y = mX + b + \epsilon$. This test is equivalent to the T-test described in Section 6.4 in the sense that both tests will yield the same conclusion and P-value; but the test statistic in this test is an F-value rather than a t-value. The theory behind this technique is beyond the scope of this book, but it is very similar to the ANOVA test for the equality of population means discussed in Section 5.9.

We begin by defining different types of *deviation* and the associated *sums of squares*. Let

- $(x_i, y_i), \ldots, (x_n, y_n)$ be n observed values of X and Y,
- $\bar{y} = \frac{1}{n} \sum y_i$ be the mean of the y values,
- $y = \hat{m}x + \hat{b}$ be the linear regression equation, and
- $\hat{y}_i = \hat{m}x_i + \hat{b}$ be the predicted value of y_i corresponding to x_i.

We define the following types of deviation:

- $(y_i - \bar{y})$ is called the *total deviation of y_i from $\bar{y}$*,
- $(\hat{y}_i - \bar{y}_i)$ is called the *deviation of y_i from $\bar{y}$ explained by the regression equation* (or simply the *explained deviation*), and
- $(y_i - \hat{y}_i)$ is called the *unexplained deviation*.

Figure 6.10 illustrates these types of deviation. We see from the figure that

$$\begin{align} \text{(Total deviation)} &= \text{(explained deviation)} + \text{(unexplained deviation)} \\ (y - \bar{y}) &= (\hat{y} - \bar{y}) + (y - \hat{y}) \end{align}$$

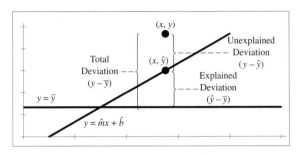

Figure 6.10

If we square these deviations and add them for all the data points, we get amounts of *variation* (measures of the total deviation). The total variation is called the *total sum of squares*, defined as

$$SS_{Tot} = \sum (y_i - \bar{y})^2.$$

The explained variation is called the *regression sum of squares*, defined as

$$SS_{Reg} = \sum (\hat{y}_i - \bar{y})^2.$$

The unexplained variation is called the *residual sum of squares*, defined as

$$SS_{Res} = \sum (y_i - \hat{y}_i)^2.$$

Analogous to the relationship between the deviations, we get

(Total variation) = (explained variation) + (unexplained variation)
$$\sum (y_i - \bar{y})^2 = \sum (\hat{y}_i - \bar{y})^2 + \sum (y_i - \hat{y}_i)^2$$

(proving this relationship is true is not a trivial matter). Rewriting this equation, we get

(Explained variation) = (total variation) − (unexplained variation)
$$\sum (\hat{y}_i - \bar{y})^2 = \sum (y_i - \bar{y})^2 - \sum (y_i - \hat{y}_i)^2$$
$$SS_{Reg} = SS_{Tot} - SS_{Res}$$

Before we continue with the ANOVA test, we make the following connection between these sums of squares and the sample correlation coefficient r. After much tedious algebraic manipulation of the alternative form of the formula for r given in Equation (6.2), we get the following relation:

$$r^2 = \frac{\sum (y_i - \bar{y})^2 - \sum (y_i - \hat{y}_i)^2}{\sum (y_i - \bar{y})^2}.$$

The quantity r^2 is called the *coefficient of determination*. Rewriting this equation in terms of the sums of squares, we see that

$$r^2 = \frac{SS_{Tot} - SS_{Res}}{SS_{Tot}}. \tag{6.4}$$

The quantity $SS_{Tot} - SS_{Res}$ can be thought of as measuring the amount of deviation from $\bar{y}$ "explained" by the linear regression equation. The quantity SS_{Tot} can be thought of as measuring the total amount of deviation from $\bar{y}$ within the data themselves. Thus we get the following interpretation of r^2:

r^2 **is the proportion of the total variation in the** y**-values from** $\bar{y}$ **explained (or accounted for) by the regression equation.**

Informally, we can think of r^2 as a measure of how well the regression line "fits" the data. The closer r^2 is to 1, the better the line fits the data. We can calculate r^2 by first calculating r with Equation (6.2) and then squaring it, or we can calculate the sum of squares first and then use Equation (6.4).

Example 6.5.1 **Calculating** r^2 In Example 6.2.3 we found that $r = 0.974$ for the shoe length and height data. The coefficient of determination is then

$$r^2 = (0.974)^2 \approx 0.949.$$

We interpret this by saying, "94.9% of the total variation in the y values from $\bar{y}$ is explained by the regression equation." Graphically, this means that the regression line fits the data very well. In the next example, we illustrate that Equation (6.4) gives the same value of r^2. □

Now, returning to the ANOVA test, the following theorem forms its foundation:

Theorem 6.5.1 *Assuming H_0: $m = 0$ is true, then SS_{Reg}, SS_{Res}, and SS_{Tot} are observed values of random variables that have χ^2 distributions with 1, $(n-2)$, and $(n-1)$ degrees of freedom, respectively. The variables corresponding to SS_{Reg} and SS_{Res} are independent, and the quantity*

$$f = \frac{SS_{Reg}}{SS_{Res}/(n-2)}$$

is an observed value of a random variable that has an F-distribution with 1 and $(n-2)$ degrees of freedom. □

As in the one-way ANOVA discussed in Section 5.9, the associated *mean squares* are defined by dividing the sums of squares by their corresponding degrees of freedom. The box below gives procedures for using the mean squares to test the hypothesis H_0: $m = 0$ versus H_1: $m \neq 0$. Notice that this test is *only* for a two-tail test and that the P-value and critical value are not found in the same way as for a typical two-tail test.

F-test of the Slope of a Simple Linear Model

Let X and Y be described by a simple linear model $Y = mX + b + \epsilon$. To test the hypotheses

$$H_0: m = 0 \quad \text{versus} \quad H_1: m \neq 0,$$

the test statistic is

$$f = \frac{\text{MS}_{\text{Reg}}}{\text{MS}_{\text{Res}}} = \frac{\text{SS}_{\text{Reg}}}{\text{SS}_{\text{Res}}/(n-2)}.$$

The critical value is $f_\alpha(1, n-2)$, the critical F-score with 1 and $(n-2)$ degrees of freedom. The P-value is the area under the corresponding density curve to the right of the test statistic.

Example 6.5.2 Calculating Sums of Squares Consider again the shoe length and height data. The linear regression equation is $y = 3.878x + 25.839$. The calculations for SS_{Reg}, SS_{Res}, and SS_{Tot} are shown in Table 6.6.

x_i	y_i	$\hat{y}_i$	$(y_i - \bar{y})^2$	$(\hat{y}_i - \bar{y})^2$	$(y_i - \hat{y}_i)^2$
9	62	60.74	64.80	86.63	1.58
10	64	64.62	36.60	29.48	0.39
10.5	64.5	66.56	30.80	12.18	4.24
11	69	68.50	1.10	2.41	0.25
11.5	70	70.44	0.00	0.15	0.19
11.75	73	71.41	8.70	1.84	2.54
12	72	72.38	3.80	5.41	0.14
12.5	75	74.32	24.50	18.20	0.47
12.75	74	75.29	15.60	27.41	1.65
13	77	76.26	48.30	38.50	0.55
		Sum =	234.23	222.22	12.01

Table 6.6

From the table, we see that $SS_{Reg} = 222.22$, $SS_{Res} = 12.01$, and $SS_{Tot} = 234.23$. Because there are $n = 10$ pairs of data, the associated degrees of freedom are 1, $(10-2)$, and $(10-1)$, respectively. This information is organized in the ANOVA table shown in Table 6.7.

	df	SS	MS	f	P-Value
Regression	1	222.22	222.22	148.15	0.000
Residual	8	12.01	1.50		
Total	9	234.23			

Table 6.7

To test the claim that the slope m in the simple linear model is not 0, the hypotheses are H_0: $m = 0$ and H_1: $m \neq 0$. The test statistic is

$$f = \frac{222.22}{1.50} \approx 148.15.$$

At the 95% confidence level, the critical value is $f_{0.05}(1, 8)$. From Table C.5, this is found to be 5.32. The P-value is the area to the right of $f = 148.15$ under the F-distribution density curve with 1 and 8 degrees of freedom. This area is approximately 0. Because f is larger than the critical value (and the P-value is less than 0.05), we reject H_0 and conclude that the data support the claim. This test statistic and P-value are also shown in the ANOVA table.

Note that

$$r^2 = \frac{SS_{Tot} - SS_{Res}}{SS_{Tot}} = \frac{234.23 - 12.01}{234.23} \approx 0.949,$$

which is the same value of r^2 found in Example 6.5.1. Also note that in Example 6.4.3, we did a T-test of the hypothesis H_0: $m = 0$ where the test statistic was $t = 12.168$. Observe that $t^2 = 12.168^2 \approx 148.15 = f$. This illustrates that t^2 and f are, within rounding error, equal. (See Exercise 8 of Section 3.9.8 for a justification of this.) □

Software Calculations

To generate the ANOVA results in Minitab and Excel, follow the steps in Section 6.2. The P-value for the hypothesis H_1: $m \neq 0$ is labeled P in Minitab and *Significance F* in Excel.

R: To generate the ANOVA results, first enter the data as described in Section 6.2. Then enter the following code:

> anova(lm(Height Length, data = ShoeData))

Exercises

1. The table below gives the weights (in pounds) and highway miles per gallon (mpg) of several cars. The linear regression equation for the data is $y = -0.01x + 60.6$. Calculate SS_{Reg}, SS_{Res}, SS_{Tot}, and r^2. Use the F-test to test the claim that $m \neq 0$.

Weight x	3250	3425	2400	2250
MPG y	26	28	37	38

2. The table below gives the amounts of nitrogen applied to different cornfields (in pounds per acre) and the resulting yields (in bushels per acre). The linear regression equation is $y = 0.56x + 68.8$. Calculate SS_{Reg}, SS_{Res}, SS_{Tot}, and r^2. Use the F-test to test the claim that $m \neq 0$.

Nitrogen x	0	60	120	180	240
Yield y	78	90	140	162	210

3. The table below gives average rebounds per game (RPG) and average points scored per game (PPG) of 11 players on a university basketball team (data collected by Alexa Hopping, 2012).

RPG x	3.1	2.8	6.7	2.5	1.5	1.4	4.8	1.4	3.7	4.1	3.6
PPG y	8.6	4.5	7.1	15.8	1.6	5.2	8.0	2.8	4.0	14.5	5.0

a. Calculate the linear regression equation for these data.
b. Calculate SS_{Reg}, SS_{Res}, SS_{Tot}, and r^2. Based on the r^2 value, does the regression line fit the data very well?
c. Use the F-test to test the claim that $m = 0$.
d. Based on these results, is RPG a good predictor of PPG?

4. Show that yet another formula for r is

$$r = \hat{m} \cdot \frac{\sqrt{n \sum x_i^2 - \left(\sum x_i\right)^2}}{\sqrt{n \sum y_i^2 - \left(\sum y_i\right)^2}}.$$

(**Hint:** Multiply the formula for $\hat{m}$ in Equation (6.3) by this fraction, and simplify to make it look like Equation (6.2).)

5. It can be shown that the following two properties hold:

$$\sum (y_i - \hat{y}_i) = 0 \quad \text{and} \quad \sum x_i(y_i - \hat{y}_i) = 0.$$

a. Use the calculations in Table 6.6 to show that these two properties hold (within rounding error) for the shoe length and height data.

b. Use these two properties to prove the following equation is true in general:

$$\sum (y_i - \bar{y})^2 = \sum (\hat{y}_i - \bar{y})^2 + \sum (y_i - \hat{y}_i)^2$$

In other words, show that

$$\text{SS}_{\text{Tot}} = \text{SS}_{\text{Reg}} + \text{SS}_{\text{Res}}.$$

(**Hint:** Start with

$$\sum_{i=1}^{n} (y_i - \bar{y})^2 = \sum_{i=1}^{n} (y_i - \bar{y} + \hat{y}_i - \hat{y}_i)^2 = \sum_{i=1}^{n} [(\hat{y}_i - \bar{y}) + (y_i - \hat{y}_i)]^2,$$

expand the right-hand side, use the fact that $\hat{y}_i = \hat{m}x_i + \hat{b}$, and rewrite so that you get the terms $\sum (\hat{y}_i - \bar{y})^2$, $\sum (y_i - \hat{y})^2$, $\sum (y_i - \hat{y}_i)$, and $\sum x_i (y_i - \hat{y}_i)$.)

6.6 Nonlinear Regression

In this section, we introduce a technique for describing relations between random variables that are not linear. To illustrate this technique, consider the data shown in the first two rows of Table 6.8.

x	−2	−1	1	2	3
y	−0.24	3.47	4.36	−0.56	−3.92
x^2	4	1	1	4	9

Table 6.8

Suppose we suspect that the relationship between the two variables is described by $Y = aX^2 + b$, where a and b are constants. Notice that this relation is of the form "Y is a constant times X^2 plus a constant." Because of this, we say that Y is *linear with respect to* X^2. To estimate the values of a and b, we *transform* the data by squaring each x value while retaining the y values and then fit a straight line to the transformed data, using the methods of Section 6.3. The transformed data are shown in the third row of Table 6.8. The

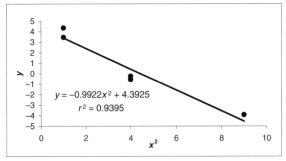 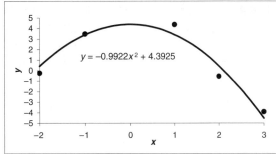

Figure 6.11

graph of the transformed data along with the resulting regression line is shown in the left half of Figure 6.11.

From the linear regression equation, we see the relation is described by $y = -0.9922x^2 + 4.3925$. The graph of this model along with the original data is shown in the right half of Figure 6.11. Notice that this graph captures the trend of the data very well.

The coefficient of determination $r^2 = 0.9395$ for the linear model fit to the transformed data is calculated according to the methods in Section 6.3. This number can be thought of as measuring how well the model $y = -0.9922x^2 + 4.3925$ fits the original data. The closer r^2 is to 1, the better the model fits.

For some types of models, both the x and y values must be transformed. Consider a *power* model of the form $y = ax^b$. To find the values of a and b, we take the natural logarithm of both sides of the model to get

$$\ln y = \ln(ax^b) = \ln a + \ln x^b = \ln a + b \ln x.$$

Thus a straight line fit to the graph of $\ln y$ versus $\ln x$ will have a slope of b and a y intercept of $\ln a$. Several commonly used nonlinear models and their appropriate transformations are shown in Table 6.9.

Type	Model	Transformation
Logarithmic	$y = a + b \ln x$	$y = a + b \ln x$
Power	$y = ax^b$	$\ln y = \ln a + b \ln x$
Exponential	$y = ae^{bx}$	$\ln y = \ln a + bx$
Logistic	$y = \dfrac{L}{1 + e^{a+bx}}$	$\ln\left(\dfrac{L-y}{y}\right) = a + bx$

Table 6.9

Example 6.6.1 Fitting Nonlinear Models Consider the data in Table 6.10 that give the number of people per physician (x) and the male life expectancy in years (y) for several countries around the world (data from *World Almanac Book of Facts*, 1992, Pharos Books). We will fit power and exponential models to the data and compare the two.

	x	y	$\ln x$	$\ln y$
Spain	275	74	5.617	4.304
United States	410	72	6.016	4.277
Canada	467	73	6.146	4.290
Romania	559	67	6.326	4.205
China	643	68	6.466	4.220
Taiwan	1010	70	6.918	4.248
Mexico	1037	67	6.944	4.205
South Korea	1216	66	7.103	4.190
India	2471	57	7.812	4.043
Morocco	4873	62	8.491	4.127
Bangladesh	6166	54	8.727	3.989
Kenya	7174	59	8.878	4.078

Table 6.10

To fit the power model, we graph $\ln y$ versus $\ln x$ and fit a straight line as in the top left corner of Figure 6.12. The slope of this line is the value of b in the power model. The y intercept of this line is the value of $\ln a$ in the model. From the graph we see that $b = -0.0823$ and

$$\ln a = 4.7675 \quad \Rightarrow \quad a = e^{4.7675} \approx 117.62.$$

Therefore, the power model is $y = 117.62 x^{-0.0823}$. This model is graphed on top of the original data in the lower left corner of Figure 6.12.

We fit the exponential model in a similar way by graphing $\ln y$ versus x and fitting a straight line as in the top right corner of Figure 6.12. From the graph we see that $b = -0.00003$ and $a = e^{4.2561} \approx 70.53$. Therefore, the model is $y = 70.53 e^{-0.00003x}$.

Comparing these two models, we see that the power model captures the trend of the data much better than the exponential model does. Comparing the r^2 values, we see that the value for the power model is much larger than the value for the exponential model. Based on these r^2 values, we would say that the power model is the "better-fitting" model.

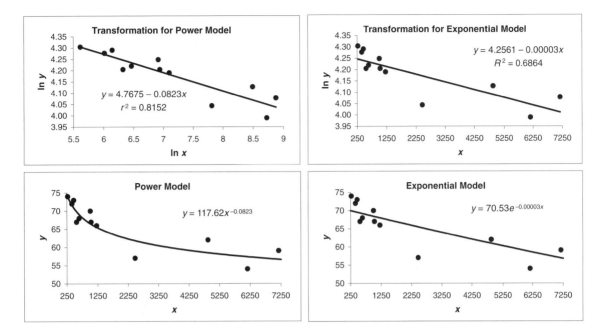

Figure 6.12

Now, suppose a country has 3500 people per physician ($x = 3500$) and we would like to predict the life expectancy of a male (the y value). Because the power model fits the data better, we will use this model to predict the value of y by plugging in $x = 3500$:

$$y = 117.63(3500)^{-0.0823} \approx 60.10 \text{ yr.}$$

We certainly could plug $x = 3500$ into the exponential model, but because this model does not fit the data as well, it may not give as good a prediction. □

Example 6.6.2 Logistic Model The logistic model is often used to model a population whose size is constrained due to limited food supply or other factors. The number L in the model is the maximum size of the population, called the *carrying capacity*. The first two rows of Table 6.11 give the number of bacteria in a petri dish y (in hundreds) at the end of each hour x.

These data are graphed in the right half of Figure 6.13. Notice that the population begins to stop increasing when it approaches 620; so this population is constrained, and we can use a logistic model to predict values of the population. We take L to be 621. To find the

x	0	2	4	6	8	10	12	14	16	18
y	10.3	27	80.2	176.2	330.8	440	560.4	610.8	618.3	620
$\ln\left(\frac{621-y}{y}\right)$	4.08	3.09	1.91	0.93	−0.13	−0.89	−2.22	−4.09	−5.43	−6.43

Table 6.11

values of a and b, we graph $\ln[(621 - y)/y]$ versus x and fit a straight line as in the left half of Figure 6.13. The slope of the line is the value of b, and the y intercept is the value of a.

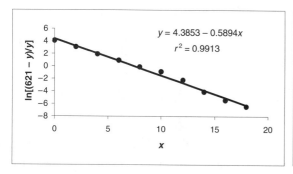
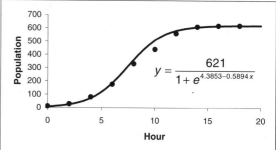

Figure 6.13

The resulting model is graphed in the right half of Figure 6.13. We see that graphically this model fits the data very well. To predict the size of the population at $x = 5$, for instance, we evaluate

$$y = \frac{621}{1 + e^{4.3853 - 0.5894(5)}} \approx 119.$$

Thus there were approximately 11,900 bacteria in the petri dish at the end of hour 5. □

Software Calculations

To fit nonlinear models:

Minitab: Enter the x and y values in blank columns. Title other blank columns to hold the transformed data. Then enter appropriate formulas to transform the data by selecting **Calc** $\to$ **Calculator**. To illustrate this, suppose we want to fit a power model and the column containing the x values is titled "X" and the column containing the transformed values is titled "Transformed X." In the **Calculator** window, enter "Transformed X" next to **Store result in variable**, and under **Expression**, enter the formula **LOGE(X)**. Calculate a linear regression equation for the transformed data, and use the resulting slope and y intercept to calculate the parameters of the nonlinear model.

R: The function **nls** can be used to fit virtually any type of nonlinear model. To illustrate how to use this function, consider the life expectancy data in Table 6.10. Enter these data by setting up a data frame with the code

> LifeExpData<− data.frame(x=c(1:12), y=c(1:12))

Next type **fix("LifeExpData")** and use the data editor to enter the data. To fit a power model, enter the following code:

> summary(nls(y~a*x^b, data=LifeExpData, start=list(a=100, b=-0.001)))

The result gives the values of a and b, so no additional calculations are needed. The numbers 100 and -0.001 are simply educated guesses of the values of a and b. If R returns an error, try using different guesses. The formula y~a*x^b can be changed to any nonlinear model desired. Note that R uses a different algorithm than the one presented in the text, so its results may differ slightly from those in the text.

Excel: Enter the data in blank columns. In separate columns, enter appropriate formulas to transform the data. Create a scatterplot of the transformed data, using the **Chart Wizard**. Right-click on one of the data points in the scatterplot and select **Add Trendline**. Under **Type**, select **Linear**. Click on the **Options** tab and select **Display equation on chart**. Use the slope and y intercept of the resulting linear equation to calculate the parameters of the nonlinear model. Excel also has built-in formulas to fit several types of nonlinear models in the **Add Trendline** window.

TI-83/84 Calculators: Enter the x and y values in lists L1 and L2, respectively. Highlight the label for list L3 and enter an appropriate formula to transform the x values. Do the same for list L4 to transform the y values. For instance, to fit a power model, enter the following formulas:

$$L3 = \ln(L1) \qquad L4 = \ln(L2)$$

Calculate a linear regression equation for the transformed data, and use the resulting slope and y intercept to calculate the parameters of the nonlinear model. The calculator also has built-in formulas to fit several types of nonlinear models in the **STAT** → **CALC** menu.

Exercises

1. For each set of data below, fit a model of the given form by transforming the data appropriately and fitting a straight line to the transformed data. Graph the resulting model on top of the original data, and analyze how well the model fits the data.

a. Model: $y = ax^2 + b$

x	1	2	3	4	5	6
y	16.3	23.1	37.4	46.9	58.7	91.0

b. Model: $y = a\sin(x) + b$

x	1	2	3	4	5	6
y	1.34	1.61	−0.98	−3.80	−4.55	−2.30

c. Model: $y = a\dfrac{x^2+1}{\ln(x)} + b$

x	2	3	4	5	6	7
y	3.30	5.63	9.52	14.31	19.84	26.061

6.6 Nonlinear Regression

2. For each data set below, fit an exponential, a power, and a logarithmic model, and determine which model best fits the data.

a.

x	1	2	3	4	5	6
y	1.66	2.41	6.04	9.89	17.31	31.54

b.

x	1	2	3	4	5	6
y	2.68	5.61	8.71	9.83	11.16	11.03

3. Fit a logarithmic model to the life expectancy data in Table 6.10 and calculate the r^2 value. Compare this model to the power and exponential models found in Example 6.6.1. Which model appears to best fit the data?

4. The table below gives the population y (in thousands) of a city in year x, where $x = 0$ corresponds to the year 2000. Fit an exponential model to these data. Assuming this trend of growth continues, predict the population in the year 2020. How reasonable of an assumption do you think this is?

x	0	1	2	3	4	5	6	7	8	9
y	100	110	120	132	145	156	168	180	199	210

5. Derive the transformation for the logistic model. (**Hint:** Start by solving for e^{a+bx}.)

6. The table below contains data on the population of foxes in a forest y over a period of several years x. Using $L = 235$, fit a logistic equation to the data. How well does the model fit the data?

x	0	1	2	3	4	5	6	7	8	9	10
y	50	85	110	130	175	200	215	221	228	232	234

7. Consider the bacteria population modeled in Example 6.6.2. Suppose we take the carrying capacity to be $L = 625$ instead of 621. Fit a logistic model to the data using $L = 625$, find the r^2 value, and graph the resulting model on top of the original data. How well does this model fit the data compared to the model where $L = 621$?

8. In the text, we fit a nonlinear model by transforming the data appropriately, fitting a linear model to the transformed data, and using the resulting slope and y intercept to find the parameters of the nonlinear model. Another approach to fitting a nonlinear model is

to use the *least-squares criterion* introduced in Section 6.3. According to this criterion, we choose values of the parameters that minimize the quantity

$$S = \sum_{i=1}^{n} (y_i - \hat{y}_i)^2$$

where $\{y_1, \ldots, y_n\}$ are the y values from the data and $\{\hat{y}_1, \ldots, \hat{y}_n\}$ are the predicted values of y using the model and the corresponding x values. For some types of models, the least-squares criterion yields the same values of the parameters as the method in the text. For other types, it does not.

a. Consider the logistic model in Example 6.6.2, where we calculated the parameters $a = 4.3853$ and $b = -0.5894$. Calculate the value of S for this set of parameters.

b. Now calculate S for the parameters $a = 3.9497$ and $b = -0.5024$, and compare it to that found in part a. For this logistic model, would the least-squares criterion give the same values of the parameters as the method in the text? Explain.

c. Consider the nonlinear model $y = ax^3$. Using the least-squares criterion, derive a formula for the parameter a in terms of the data values (x_i, y_i), $i = 1, \ldots, n$. (**Hint**: $\hat{y}_i = ax_i^3$. Take the derivative of S with respect to a and set it equal to 0.)

d. Consider the data in the table below. Fit the model $y = ax^3$ to the data, using the method in the text. Then use the formula in part c. Do these two methods give the same result?

x	−1	−0.5	0.5	1
y	−6	−0.25	0.25	6

6.7 Multiple Regression

In the previous sections, we dealt with using a single predictor variable X to predict the value of a variable Y. In this section, we discuss using two or more predictor variables $X_1, X_2, \ldots, X_k$. This topic is called *multiple regression*. The calculations involved with multiple regression are very tedious, so we do not discuss the formulas (see Exercise 11 of this section for one description). Instead, we focus on interpreting the outputs from software.

Consider the problem of predicting the selling price of a house. The selling price is affected by many factors including the age of the house, living area, number of bedrooms, and so on. Table 6.12 lists the selling price (in dollars), living area (in square feet), acres of land,

and the number of bedrooms of 10 homes in a neighborhood. In this case, the response variable is selling price Y and the predictor variables are area X_1, acres X_2, and number of bedrooms X_3.

Selling Price (y)	Area (x_1)	Acres (x_2)	Bedrooms (x_3)
100,000	2205	2.5	3
93,500	2155	0.8	3
95,650	2600	1.1	4
75,025	1900	0.35	3
95,000	1200	2.5	2
80,250	2050	1.8	3
85,250	2250	0.9	4
121,250	2490	1.8	3
94,575	2390	1.6	2
109,000	3100	1.0	4

Table 6.12

We assume that these variables are related by an equation of the form

$$Y = m_1 X_1 + m_2 X_2 + m_3 X_3 + b + \epsilon$$

where m_1, m_2, m_3, and b are constants (generically these are called *coefficients*) and ϵ is a random variable that is $N(0, \sigma^2)$ and independent of the predictor variables. The goal is to find estimates of these constants yielding the *multiple regression equation*

$$y = \hat{m}_1 x_1 + \hat{m}_2 x_2 + \hat{m}_3 x_3 + \hat{b}$$

that minimizes the quantity

$$S = \sum_{i=1}^{n} (y_i - \hat{y}_i)^2.$$

This is the same criterion used in the case of a single predictor variable, but the formulas for the estimates are much more complicated. Different software programs output the results of the calculations in different formats. Below we display the outputs generated by Microsoft Excel and describe the different components.

Estimates of the Coefficients

Table 6.13 shows the estimates of the coefficients for the data in Table 6.12.

	Coefficients	Standard Error	t Stat	P-Value	Lower 95%	Upper 95%
Intercept	41510.9	25776.2	1.610	0.158	−21561.2	104583.0
Area	20.9	9.8	2.145	0.076	−2.9	44.8
Acres	10339.6	5958.4	1.735	0.133	−4240.2	24919.4
Bedrooms	−2641.1	7158.3	−0.369	0.725	−20156.8	14874.5

Table 6.13

- **Coefficients.** These are the values of $\hat{b}$, $\hat{m}_1$, $\hat{m}_2$, and $\hat{m}_3$, respectively. This yields the multiple regression equation

$$\hat{y} = 20.9x_1 + 10{,}339.6x_2 - 2641.1x_3 + 41{,}510.9.$$

- **Standard Error.** The standard error s_i is used in calculating confidence interval estimates of the constants, using the formula

$$\hat{m}_i - t_{\alpha/2}s_i < m_i < \hat{m}_i + t_{\alpha/2}s_i$$

where $t_{\alpha/2}$ is a critical t-value with $(n - k - 1)$ degrees of freedom, n is the number of observations, and k is the number of predictor variables ($n = 10$ and $k = 3$ in this case). The limits of the corresponding confidence intervals are shown in the columns titled Lower 95% and Upper 95%.

- **t Stat.** This is the test statistic for the hypotheses

$$H_0\colon m_i = 0, \qquad H_1\colon m_i \neq 0$$

in the presence of the other predictor variables. The corresponding P-value is given next to the test statistic. A small P-value indicates that the variable is *statistically significant*. These P-values indicate that the most significant variable is area, that the second most significant is acres, and that bedrooms is not significant at all. Note that these interpretations are valid only if we consider all three of these predictor variables together. If we consider other sets of variables (as done below), the results may change.

ANOVA Results

With multiple regression, we get a set of ANOVA results very similar to those described in Section 6.5. These outputs are shown in Table 6.14.

The number in the column titled F is the test statistic for the hypotheses

$$H_0\colon m_1 = m_2 = m_3 = 0, \qquad H_1\colon \text{at least one is not 0.}$$

	df	SS	MS	F	*Significance F*
Regression	3	897207121.4	299069040.5	2.467	0.160
Residual	6	727309128.6	121218188.1		
Total	9	1624516250			

Table 6.14

The number under *Significance F* is the corresponding *P*-value. This *P*-value measures the "overall significance" of the set of predictor variables. Because this *P*-value is greater than 0.05, we do not reject H_0 and we conclude that this set of predictor variables is not statistically significant. In practical terms, this means that this set of variables does not do a good job of predicting the selling price.

Regression Statistics

The last set of outputs is shown in Table 6.15.

Regression Statistics	
Multiple R	0.743
R Square	0.552
Adjusted R Square	0.328
Standard Error	11009.913
Observations	10

Table 6.15

- **Multiple R.** This is the multiple regression equivalent of the sample correlation coefficient r defined in Section 6.2. When there is only one predictor variable, this value is r. The multiple R value can be calculated by taking the square root of the next statistic, R squared.
- **R Squared.** This is called the *multiple coefficient of determination*. It is the multiple regression equivalent of the coefficient of determination discussed in Section 6.5. It can be thought of as measuring the percentage of variation of the y values from $\bar{y}$ explained by the multiple regression equation. It is calculated using sums of squares according to Equation (6.4). Exercise 10 of this section describes another way to calculate this statistic.
- **Adjusted R Square.** This is calculated with the formula

$$\text{Adjusted } R^2 = 1 - \left[\frac{n-1}{n-(k+1)}\right](1-R^2).$$

The adjusted R^2 value takes into account the number of observations (the more observations, the higher the adjusted R^2 value) and the number of predictor variables (the more predictor variables, the lower the adjusted R^2 value). The higher the adjusted R^2 value is, the better the overall quality of the model.
- **Standard Error.** This is an estimate of the standard deviation of the random variable ϵ in the multiple regression model. This is also called the *standard error of estimate.*

Interpreting the Results

Extracting meaning from all the multiple regression results is very complicated. The important question is, What set of predictor variables is "best" at making predictions? There are many different ways of trying to answer this question. A detailed discussion of these methods is beyond the scope of this text. Below, we describe one simple method where we choose different sets of predictor variables and compare them with different statistics.

In Table 6.16, we compare the R^2 values, adjusted R^2 values, and P-values measuring the overall significance of all four different sets of two or three predictor variables.

Predictor Variables	R^2	Adjusted R^2	P-Value
Area, Acres	0.542	0.411	0.065
Area, Bedrooms	0.327	0.135	0.249
Acres, Bedrooms	0.209	−0.017	0.440
Area, Acres, Bedrooms	0.552	0.328	0.160

Table 6.16

Note that including the variable Bedrooms typically results in lower R^2 values. This indicates that the number of bedrooms is not a good predictor of the selling price. Also note that the R^2 value for the set of all three predictor variables is the highest, but the adjusted R^2 value is lower than that for Area and Acres. This indicates that the set of three variables gives better predictions (a higher R^2 value); but the additional variable makes it more complicated, so it is less desirable as a model (a lower adjusted R^2 value). Also note that the combination of Area and Acres has the lowest P-value, indicating that this set of variables is the "most significant" of those considered.

Based on this simple analysis, we would conclude that the variables Area and Acres are the "best" combination at predicting the selling price. The multiple regression equation for this set of variables is

$$\text{Selling price} = 18.87(\text{Area}) + (11{,}239.21 \text{ Acres}) + 36{,}669.58.$$

If a house has 2000 ft² and sits on 1.5 acres of land, this regression equation predicts that the selling price is

$$\text{Selling price} = 18.87(2000) + 11{,}239.21(1.5) + 36{,}669.58 \approx \$91{,}265.$$

However, this set of predictor variables has rather small R^2 and adjusted R^2 values and a relatively large P-value. This indicates that although this set of variables is the best, it may not be very good at predicting selling price. Thus we must be careful about using this predicted price of \$91,265. We may want to collect data on other variables such as age of the house, values of neighboring homes, and so on, and consider them in our analysis.

Example 6.7.1 **Verifying Assumptions** In the multiple regression model, it was assumed that ϵ is normally distributed. We can verify this by calculating the residuals $(y_i - \hat{y}_i)$ and analyzing them. The multiple regression equation that predicts Selling Price in terms of Area, Acres, and Bedrooms is

$$\hat{y} = 20.9 x_1 + 10{,}339.6 x_2 + -2641.1 x_3 + 41{,}510.9.$$

For the first house in the data, the predicted selling price (also called the *fitted value*) is

$$\hat{y}_1 = 20.9(2205) + 10{,}339.6(2.5) + -2641.1(3) + 41{,}510.9 \approx 105{,}521$$

so that the residual is $(100{,}000 - 105{,}521) = -5521$. The other fitted values and residuals are shown in Table 6.17.

y_i	Area	Acres	Bedrooms	$\hat{y}_i$	$(y_i - \hat{y}_i)$
100,000	2,205	2.5	3	105,521	−5,521
93,500	2,155	0.8	3	86,899	6,601
95,650	2,600	1.1	4	96,660	−1,010
75,025	1,900	0.35	3	76,916	−1,891
95,000	1,200	2.5	2	87,158	7,842
80,250	2,050	1.8	3	95,044	−14,794
85,250	2,250	0.9	4	87,277	−2,027
121,250	2,490	1.8	3	104,240	17,010
94,575	2,390	1.6	2	102,723	−8,148
109,000	3,100	1.0	4	106,076	2,924

Table 6.17

A graph of residuals versus fitted values along with a normal quantile plot of the residuals is shown in Figure 6.14. Notice that there is no pattern to the residuals and the quantile plot shows a straight line. Thus we conclude that the assumptions are indeed met. □

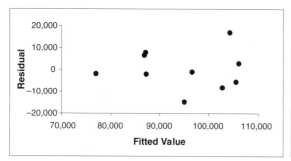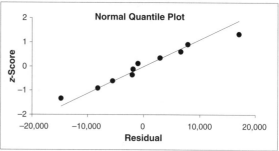

Figure 6.14

Example 6.7.2 Polynomial Model We can use multiple regression techniques to fit polynomial models to data. Suppose we want to fit a model of the form $Y = m_1 X + m_2 X^2 + b$ to the data in the first two rows of Table 6.18.

y	16.3	22.8	25.1	28.9	31.2	35.7
x	5.2	5.8	6.3	6.4	7.2	8.5
x^2	27.04	33.64	39.69	40.96	51.84	72.25

Table 6.18

We can approach this as though we have two "predictor" variables, X and X^2. The third row of the table shows values of X^2. These data yield the multiple regression equation

$$y = 25.84x - 1.47x^2 - 78.16.$$

This model is graphed along with the original data in Figure 6.15. Note that the model fits the data very well. □

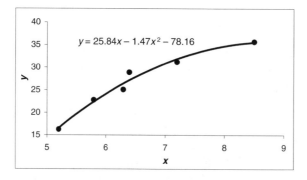

Figure 6.15

Software Calculations

To perform multiple regression:

Minitab: Enter the x and y values in blank columns. Select **Stat** → **Regression** → **Regression**. Next to **Response**, enter the names of the columns containing the y values. Next to **Predictor**, enter the names of the columns containing the desired x values. A normal quantile plot of the residuals can be constructed by selecting **Graphs** and checking the box next to **Normal plot of residuals**.

R: To perform multiple regression on the house selling price data in Table 6.12, set up a data frame with this code:

> HData<−data.frame(SP=c(1:10), Area=c(1:10), Acres=c(1:10), Beds=c(1:10))

Next type **fix("HData")** and use the data editor to enter the data. To use all three predictor variables, enter the following two lines of code:

> multireg<−lm(SP~Area+Acres+Beds, data=HData))
> summary(multireg)

Any combination of predictor variables can be considered by simply changing the variables on the right-hand side of the formula. In the output, the test statistic F for the null hypothesis that all the coefficients are 0 is labeled **F-statistic** and the associated P-value is given to the right.

To generate a quantile plot of the residuals, enter this code:

> qqnorm(resid(multireg), datax=TRUE)

Excel: Enter the data, along with labels, in adjacent columns. Select **Data** → **Data Analysis** → **Regression**. Enter the ranges containing the y values and all the desired x values (include the labels). Check the box next to **Labels**, and select a blank cell for the **Output Range**.

Exercises

Directions: Use available software to perform the required calculations.

1. Using the data in Table 6.12, find the regression equation that predicts Selling Price in terms of Area. Repeat this process, using the predictor variable Acres and then Bedrooms. Which one of these single-variable predictors is best at predicting the selling price based on the R^2 values? Is any one of them better than the best combination of multiple predictor variables considered in Table 6.16?

2. In an attempt to predict the final grade of students in an introduction to statistics class, the professor gives each student a 20-point pretest at the beginning of the year. The table below gives the final grade, pretest score, ACT score, and year (1 = freshman, 2 = sophomore, etc.) of 10 students.

Grade	84.5	82.3	69.2	65.1	80.1	85.9	88.1	90.7	87.2	92.7
Pretest	9	8	18	10	6	8	16	11	15	19
Year	1	2	2	4	3	3	1	4	4	3
ACT	25	20	18	17	20	22	30	28	27	31

a. Find the regression equation that predicts Grade in terms of Pretest Score. Repeat using Year and then ACT. Which of these single-variable predictors is best at predicting the final grade based on the R^2 values? Does the pretest score alone appear to be a good predictor of the final grade? Explain.

b. Consider all four different combinations of two or three predictor variables. Determine which combination is best at predicting the final grade, using the methods described in this section. Based on your results, does it seem worthwhile to give the pretest as a way of predicting the final grade? Does the year of the student appear to affect the final grade? Explain.

c. Use the multiple regression equation that predicts Grade in terms of Pretest and ACT to predict the grade of a student who has a pretest score of 18 and an ACT score of 28.

3. The table below gives the poverty level, unemployment rate, high school graduation rate, and divorce rate (all in percentages) of 10 randomly selected states in 2007 (data collected by Matthew Schranz, 2011). Determine which combination of predictor variable(s) is best at predicting poverty level. Based on these data, does the divorce rate appear to be related to the poverty level at all?

Poverty Level	Unemployment Rate	High School Graduation Rate	Divorce Rate
14.3	4.3	69.6	3.9
9.3	3.8	71.9	3.7
8.9	3.8	86.5	2.6
12.8	5.3	81.9	3.8
5.8	3.5	81.7	3.8
15.5	5.0	68.6	4.0
12.8	5.3	73.8	3.9
9.4	2.7	82.5	3.1
10.0	4.0	88.6	3.6
11.0	4.5	88.5	2.9

4. Table 6.19 gives the population growth rate (PGR), land area (km^2), gross domestic product (GDP) (per capita), amount of renewable water (km^3), median age (yr), and unemployment rate (UR) (%) of several countries (data from CIA world factbook online as collected by Danny Frastaci, 2011). Our goal is to find a set of predictor variables(s) that best predicts the population growth rate.

   a. For each of the individual predictor variables, find the single-regression equation that predicts the population growth rate, and calculate the R^2 value. Which, if any, of these single-variable predictors appear to be related to the population growth rate?

   b. Consider different combinations of the single-variable predictors that are related to the population growth rate as identified in part a. Calculate the multiple regression equation for each combination along with the R^2 value and the adjusted R^2 value. Then use all five predictor variables. Which combination is best?

Country	PGR	Area	GDP	Water	Age	UR
Yemen	2.65	527,968	2,700	4.1	18.1	35.0
Kenya	2.46	580,367	1,600	30.2	18.9	40.0
Sudan	2.48	1,861,484	2,300	154.0	18.5	18.7
Egypt	1.96	1,001,450	6,200	86.8	24.3	9.0
Venezuela	1.49	912,050	12,700	1,233.2	26.1	8.5
Pakistan	1.57	796,095	2,500	233.8	21.6	15.4
Australia	1.15	7,741,220	41,000	398.0	37.7	5.2
Mexico	1.10	1,964,375	13,900	457.2	27.1	5.4
Brazil	1.13	8,514,877	10,800	8,233.0	29.3	6.7
Paraguay	1.28	406,752	5,200	336.0	25.4	6.9
New Zealand	0.88	267,710	27,700	397.0	37.0	6.5
Chile	0.84	756,102	15,400	922.0	32.1	7.1
France	0.50	643,801	33,100	189.0	39.9	9.3
China	0.49	9,596,961	7,600	2,829.6	35.5	4.3
Italy	0.42	301,340	30,500	175.0	43.5	8.4
Japan	−0.28	377,915	34,000	430.0	44.8	5.0
Hungary	−0.17	93,028	18,800	120.0	40.2	11.2
United States	0.96	9,826,675	47,200	3,069.0	36.9	9.6

Table 6.19

5. Explain why, in the multiple regression model $Y = m_1 X_1 + m_2 X_2 + m_3 X_3 + b + \epsilon$, Y cannot be a discrete random variable.

6. Consider the house Selling Price data in Table 6.12. For the multiple regression equation that gives Selling Price in terms of Area and Acres, calculate the residuals and plot them as in Example 6.7.1. Do the assumptions about ϵ appear to be met for this set of predictor variables?

7. The table at the top of the next page gives the average points scored per game (PPG), total number of turnovers, average minutes per game played, and free-throw percentage of 11 college basketball players over the course of a season (data collected by Alex Hopping, 2012).

PPG	Turnovers	Min/Game	Free-Throw %
8.6	75	19.8	0.797
4.5	60	18.7	0.532
7.1	22	22.7	0.780
15.8	48	29.1	0.767
1.6	16	9.0	0.500
5.2	20	12.7	0.750
8.0	56	21.6	0.726
2.8	13	8.6	0.706
4.0	19	15.8	0.583
14.5	62	25.3	0.848
5.0	30	16.0	0.577

a. Find the regression equation that predictes PPG in terms of turnovers, minutes per game, and free-throw percentage.

b. Calculate the residuals for the regression equation in part a, plot them, and create a normal quantile plot of them as in Example 6.7.1. Do the assumptions about ϵ appear to be met for this set of predictor variables? workbook Project 1 Hopping for the solutions.

8. Fit a polynomial model of the form $Y = m_1 X + m_2 X^2 + m_3 X^3 + b$ to the data in the table below.

x	1	2	3	4	5
y	4.28	4.57	5.83	8.76	12.11

9. Use the definition of r^2 in Equation (6.4) and the ANOVA results in Table 6.14 to calculate the multiple coefficient of determination for the house selling price data in Table 6.12. Verify that this is the same value given in Table 6.15.

10. The multiple coefficient of determination can be calculated by using the following formula:

$$R^2 = \frac{S_{yy} - SS_{Res}}{S_{yy}}$$

where

$$S_{yy} = \sum_{i=1}^{n} y_i^2 - \frac{1}{n}\left(\sum_{i=1}^{n} y_i\right)^2$$

and SS_{Res} is the residual sum of squares given in the ANOVA results. Use this formula to calculate R^2 for the regression equation that predicts house Selling Price in terms of Area, Acres, and Bedrooms, and confirm that it is the same as that given in Table 6.15 (use the value of SS_{Res} given in Table 6.14).

11. One way of describing the formulas for the coefficients in a multiple regression equation is with the matrix equation

$$\mathbf{m} = (\mathbf{A}^T \mathbf{A})^{-1} \mathbf{A}^T \mathbf{b}$$

where

- $\mathbf{m} = (\hat{m}_1, \ldots, \hat{m}_k, \hat{b})$ is the vector of coefficients,
- $\mathbf{A}$ is an $n \times (k+1)$ matrix where the first k columns are the observed values of the predictor variables and the last column contains all 1s,
- $\mathbf{b}$ is the vector of observed values of the response variable, and
- $\mathbf{A}^T$ is the transpose of matrix $\mathbf{A}$.

For instance, to fit a multiple regression equation of the form $\hat{y} = \hat{m}_1 x_1 + \hat{m}_2 x_2 + \hat{b}$ using the data in Table 6.20, the matrices are

$$\mathbf{m} = \begin{bmatrix} \hat{m}_1 \\ \hat{m}_2 \\ \hat{b} \end{bmatrix}, \quad \mathbf{A} = \begin{bmatrix} 9 & 25 & 1 \\ 18 & 18 & 1 \\ 15 & 27 & 1 \\ 19 & 31 & 1 \end{bmatrix}, \quad \text{and} \quad \mathbf{b} = \begin{bmatrix} 84.5 \\ 69.2 \\ 87.2 \\ 92.7 \end{bmatrix}.$$

Use available software to calculate $\mathbf{m} = (\mathbf{A}^T \mathbf{A})^{-1} \mathbf{A}^T \mathbf{b}$ for this set of data. Also use available software to find the multiple regression equation. Compare the values of the coefficients from both calculations. Are they the same?

y	84.5	69.2	87.2	92.7
x_1	9	18	15	19
x_2	25	18	27	31

Table 6.20

12. The matrix method for calculating coefficients described in Exercise 11 above can be adapted to find the coefficients of a single-variable regression equation $y = \hat{m}x + b$. In this case, $\mathbf{m} = (\hat{m}, b)$ and $\mathbf{A}$ has only two columns, the first of which contains the observed values of the predictor variable, and the second contains all 1s. Use this approach to find the linear regression equation for the shoe length and height data given in Table 6.4. Give the matrix $\mathbf{A}$ and the vector $\mathbf{b}$. Compare this equation to that found in Example 6.3.1.

13. In Section 6.4, we defined a prediction interval for Y given a value of X for the case of a single predictor variable. This interval gives a range of possible values for Y when X has a given value. We can calculate a similar prediction interval for the case of multiple predictor variables. If we have k predictor variables $X_1, X_2, \ldots, X_k$ and we want to predict the value of Y when $X_1 = x_1, X_2 = x_2, \ldots, X_k = x_k$, then the prediction interval is $\hat{y} - E < Y < \hat{y} + E$ where the margin of error E is

$$E = s \cdot t_{\alpha/2} \sqrt{1 + \mathbf{x}^T (\mathbf{A}^T \mathbf{A})^{-1} \mathbf{x}}$$

and

- $\hat{y}$ is the predicted value of Y given by the multiple regression equation,
- $\mathbf{A}$ is as defined in Exercise 11 above,
- $\mathbf{x} = (x_1, x_2, \ldots, x_k, 1)$,
- $t_{\alpha/2}$ is a critical t-value with $[n - (k+1)]$ degrees of freedom, where n is the number of data points, and
- s is the standard error of the multiple regression equation fit to the data.

Consider the multiple regression equation fit to the data in Exercise 11 above. The standard error for this model is $s = 0.596927$. Calculate a 90% prediction interval estimate of Y when $X_1 = 14$ and $X_2 = 20$.

14. The matrix method for calculating a prediction interval described in Exercise 13 above can also be used in the case of a single predictor variable. Redo the calculations of the prediction interval in Example 6.4.2, using this matrix method, and verify that you get the same results.

15. A natural gas company would like to predict the amount of gas used by its customers during the winter months. The company believes that the average monthly consumption Y of its customers (measured in thousand cubic feet) is related to the average temperature X_1 (in degrees Fahrenheit) and price X_2 (in dollars). They also believe that there is an interaction between temperature and price, so they decide to model Y with the multiple regression model

$$Y = m_1 X_1 + m_2 X_2 + m_3 X_1 X_2 + b + \epsilon.$$

The table below gives the average consumption y, average temperature x_1, and price per 1000 ft^3 x_2 of several different months. Fit the multiple regression model to these data; then calculate the R^2 and adjusted R^2 values and SS$_{\text{Res}}$. Comment on how well this model fits the data.

y	9.68	7.26	12.39	13.98	11.32	9.87	14.65	5.24	12.53	16.7	10.76	4.92	2.06	6.73	1.16	5.08
x_1	34	37	26	27	26	30	26	38	25	26	28	37	39	34	39	36
x_2	9.1	9.1	9.1	9.1	9.75	9.75	9.75	9.75	10.2	10.2	10.2	10.2	10.9	10.9	10.9	10.9

16. Referring to the model for the consumption of natural gas in Exercise 15 above, one might wonder just how much the interaction term $m_3 X_1 X_2$ contributes to the quality of the model in terms of predicting values of Y. Informally, we could fit a "reduced" model of the form $Y = m_1 X_1 + m_2 X_2 + b + \epsilon$ to the data and compare the resulting R^2 values to those of the "complete" model considered in Exercise 15 above. If the R^2 values of the reduced model were not significantly less than those of the complete model, then we would conclude that the interaction term did not contribute much to the quality of the model.

A more formal way to compare the two models is to calculate the statistic

$$F = \frac{[\text{SS}_{\text{Res}}(R) - \text{SS}_{\text{Res}}(C)]/(k-g)}{\text{SS}_{\text{Res}}(C)/(n-k-1)}$$

where

- SS$_{\text{Res}}(R)$ and SS$_{\text{Res}}(C)$ are the residual sum of squares for the reduced and complete model, respectively,
- k is the number of nonconstant terms in the complete model,
- g is the number of nonconstant terms in the reduced model, and
- n is the number of data points.

If the complete model were significantly better than the reduced model, then SS$_{\text{Res}}(C)$ would be much smaller than SS$_{\text{Res}}(R)$, making F large. Thus if F were large, we would conclude that the terms *not* appearing in the reduced model do significantly contribute to the quality of the complete model. If F were small, we would conclude the opposite.

What we are really doing is testing the null hypothesis that in the complete model, the coefficients of the terms absent in the reduced model are 0. (In the model for natural gas consumption, we are testing the hypothesis H_0: $m_3 = 0$.) It can be shown that under this hypothesis, F is an observed value of a random variable with an F-distribution with $(k-g)$ numerator degrees of freedom and $(n-k-1)$ denominator degrees of freedom. Therefore, the appropriate critical value for the cutoff between a small and large value of F is $f_\alpha(k-g, n-k-1)$ found in Table C.5 (α is the desired significance level).

a. Fit the reduced model $Y = m_1 X_1 + m_2 X_2 + b + \epsilon$ to the data in Exercise 15, and calculate the R^2 and adjusted R^2 values and $SS_{Res}(R)$. Compare these to the values for the complete model in this exercise. Do the values appear to be significantly different? What does this mean about the usefulness of the interaction term?

b. Calculate the value of F for comparing the complete model to the reduced model. Find the appropriate critical value at the $\alpha = 0.05$ significance level. Does it appear that the interaction term is really necessary for predicting natural gas consumption?

CHAPTER 7

Nonparametric Statistics

Chapter Objectives

- Introduce nonparametric hypothesis tests
- Discuss similarities and differences with parametric tests
- Introduce the sign test
- Introduce the Wilcoxon tests
- Introduce the runs test for randomness

7.1 Introduction

Most of the hypothesis tests introduced in previous chapters require that the population has a certain type of distribution, such as a normal distribution. Such tests are called *parametric tests*. In this chapter, we introduce several tests that have no such requirement. These tests are called *nonparametric* or *distribution-free tests*.

Nonparametric tests have the advantage that no certain population distribution is needed. Thus they can be used in situations where parametric tests do not apply. Another advantage is that the required calculations are usually very simple. Many require only the sorting and counting of data values.

However, nonparametric tests have their disadvantages. They often "waste information" by ignoring the magnitude of values. For instance, in the sign test for a claim about a median, the only information used is whether each data value is above or below the claimed median. The magnitude of the differences is not considered. Also, nonparametric tests have less power compared to similar parametric tests.

7.2 The Sign Test

The *sign test* is a nonparametric test that can be used in a wide variety of situations. It can be used to test a claim about the median of a population, to compare matched pairs of data, or to test claims about qualitative data.

To perform the sign test, the sample data are divided into two categories; one is denoted by a $+$ sign, the other by a $-$ sign. We form a null hypothesis and an alternative hypothesis as with all other hypothesis tests. Assuming the null hypothesis is true, an unusually small number of one of these signs suggests that we reject H_0. We illustrate the procedure with the following example.

Example 7.2.1 Number of Children's Books The first row of Table 7.1 shows the number of children's books in 11 randomly selected homes with children in a town. Use the data to test the claim that the median number of books in homes with children in this town is greater than 12.

Number of Books	15	18	7	13	53	20	14	45	19	65	12
Sign	+	+	−	+	+	+	+	+	+	+	0

Table 7.1

We might be tempted to change the claim to state that the mean number of books is greater than 12 and then test it with a T-test. However, note that the sample size is less than 30 and a normal quantile plot indicates that the population is not normally distributed. Thus the requirements for the T-test are not met, so we cannot use this test.

Let m denote the median number of books. The claim is $m > 12$, so we test the hypotheses

$$H_0: m = 12, \quad H_1: m > 12.$$

We let a + sign denote a data value greater than 12, a − sign denote a value less than 12, and a 0 denote a value equal to 12, as shown in the second row of Table 7.1. We will simply ignore the 0. Thus we have a total of $n = 10$ signs, 9 of which are + signs, and 1 of which is a − sign.

If H_0: $m = 12$ is true, then the probability that a randomly selected data value is less than 12 is 0.5 (assuming the value is not equal to 12). Thus we would expect about one-half of the signs to be −. This is not the case in this set of data. To measure the significance of this small number of − signs, let the random variable X denote the number of − signs in the sample of $n = 10$. If H_0 were true, then X would be $b(10, 0.5)$.

In these data, we observed $x = 1$. To determine if this is an usually small value of X, we compute the probability

$$P(X \leq 1) = \binom{10}{0} 0.5^0 (1-0.5)^{10-0} + \binom{10}{1} 0.5^1 (1-0.5)^{10-1} \approx 0.0107.$$

This is the P-value for this test. Because this is less than 0.05, we reject H_0 and conclude that the data support the claim that the median is greater than 12. □

This example illustrates that we reject H_0 if x is "small enough." Specifically, for a one-tail test, we reject H_0 if $P(X \leq x) < \alpha$. Table C.6 gives the largest value of x such that this is true. These are the critical values for the sign test. From the table, we see that for a one-tail test with $n = 10$, the critical value is 1. This means in the previous example we would reject H_0 if $x \leq 1$ and do not reject it otherwise.

Note that the largest value of n in Table C.6 is 29. For larger values of n, we can convert the value of x to a z-score and use an appropriate critical z-value as follows: By the limit theorem of De Moivre and Laplace, the variable X is approximately

$$N(0.5n, 0.5(1-0.5)n) = N(0.5n, 0.25n)$$

so that

$$P(X \leq x) \approx P\left(Z \leq \frac{(x+0.5) - 0.5n}{\sqrt{0.25n}}\right)$$

where Z is $N(0, 1)$. Thus if $z = [(x+0.5)-0.5n]/\sqrt{0.25n}$ is less than $-z_{0.05}$, then $P(X \leq x)$ would be less than 0.05 and we reject H_0.

The procedures for the sign test are summarized in the box below.

The Sign Test

Purpose: To determine if a set of data consisting of $+$ and $-$ signs has an unusually small number of one sign or another. Let

- n be the total number of $+$ and $-$ signs in the data, and
- x be the number of times the *less* frequent sign appears.

The test statistic is

- x for $n \leq 29$, and
- $z = \dfrac{(x + 0.5) - 0.5n}{\sqrt{0.25n}}$ for $n > 29$.

The critical value is found in Table C.6 if $n \leq 29$. If $n > 29$, then the critical value is $-z_\alpha$ or $-z_{\alpha/2}$ for a one- or two-tail test, respectively, as found in Table C.2. The P-value is found as described below where X is $b(n, 0.5)$ and Z is $N(0, 1)$.

Type of Test	P-Value
One-tail	$P(X \leq x)$ or $P(Z \leq z)$
Two-tail	$2P(X \leq x)$ or $2P(Z \leq z)$

In any case, we reject H_0 if the test statistic is less than or equal to the critical value or the P-value $< \alpha$.

Requirement
1. The sample is random.

The next example illustrates how we can compare matched data by using the sign test.

Example 7.2.2 Comparing Test Scores At a large university, freshmen students are required to take an introduction to writing class. Students are given a survey on their attitudes toward writing at the beginning and end of the class. Each student receives a score between 0 and 100 (the higher the score, the more favorable the attitude toward writing). The scores of nine different students from the beginning and end of the class are shown in Table 7.2. Based on these data, do the attitudes toward writing appear to increase by the end of the class?

We could use these data to test the claim that the mean of the differences is greater than 0 with a paired T-test. But here we test the claim that the median of the differences is greater than 0. This claim is not equivalent to a claim that the mean is greater than 0,

Beginning	77.6	83.2	60.2	93.1	74.6	43.1	86.9	79.3	80.2
End	85.4	79.6	64.2	96.6	79.3	40.5	90.1	89.2	85.6
(End − Beginning)	+	−	+	+	+	−	+	+	+

Table 7.2

but the two claims are similar. Another difference between the two claims is that with the claim about the median, we do not need to assume the population of the differences is normally distributed.

We begin by recording the sign of the differences (end − beginning) as in the third row of Table 7.2. We have $n = 9$ and $x = 2$. The claim is $m > 0$ so we have the hypotheses

$$H_0: m = 0, \quad H_1: m > 0.$$

This is a one-tail test so that the critical value is at the 95% confidence level is 1. Because $x > 1$, we do not reject H_0 and we conclude that the data do not support the claim. Therefore, it appears that there is not a statistically significant change in the scores. Note that this is the opposite conclusion we would reach if we tested the claim that the mean difference is 0. □

The next example illustrates how we can use a sign test on qualitative data.

Example 7.2.3 High-Sugar Cereals A high-sugar cereal is defined as one in which the ratio of total carbohydrates to sugar is greater than 4:1. A student claims that one-half of all cereals are high-sugar. A sample of 120 types of cereal contains 70 that are high-sugar (data collected by Hannah McNeiley, 2009). Assuming that this is a random sample of all cereals, use these data to test the claim.

If we let p denote the proportion of all cereals that are high-sugar, then the claim is $p = 0.5$ so that the hypotheses are

$$H_0: p = 0.5, \quad H_1: p \neq 0.5.$$

If we let a + sign denote a cereal that is high-sugar and a − sign denote one that is not, then our data consist of 70 + signs and 50 − signs, so that $x = 50$ and $n = 120$. If H_0 were true, then we would expect to have about the same number of + and − signs. We can use the sign test to determine if $x = 50$ is unusually small. Because $n > 29$, the test statistic is

$$z = \frac{(50 + 0.5) - 0.5(120)}{\sqrt{0.25(120)}} \approx -1.83.$$

At the 95% confidence level, the critical value is $-z_{0.05/2} = -1.96$. Because $z > -1.96$, we do not reject H_0 and we conclude that there is not sufficient evidence to reject the claim. □

The next example illustrates that the sign test has significantly less power than a related Z-test.

Example 7.2.4 Calculating the Power of a Sign Test Consider a population that is $N(10, 1)$. Suppose that we know the variance but that we do not know the mean, we claim that the mean is less than 10.5, and we choose to test the claim with a Z-test, using a sample of size $n = 5$. This yields the hypotheses

$$H_0: \mu = 10.5, \quad H_1: \mu < 10.5.$$

Using the algorithm in Exercise 17 of Section 5.3, the power of this test is 0.2991 at the 95% confidence level. This power is the probability of rejecting the false null hypothesis.

Now suppose we use the sign test to test the claim that the median is less than 10.5. Because in a normal distribution the mean and median are equal, this claim is equivalent to the claim that $\mu < 10.5$. This claim about the median yields the hypotheses

$$H_0: m = 10.5, \quad H_1: m < 10.5.$$

This null hypothesis is a false statement. The power of this test is the probability that we reject H_0. If we use a sample of $n = 5$, from Table C.6 we see that at the 95% confidence level the critical value is 0. Thus we will reject H_0 only if there are no sample values greater than the claimed median of 10.5 (in other words, all the sample values must be less than 10.5).

To find this probability, let the random variable Y denote a randomly selected value from this population. Then

$$P(Y > 10.5) = P\left(Z > \frac{10.5 - 10}{1}\right) = P(Z > 0.5) = 0.3085.$$

Now let X denote the number of values that are greater than 10.5 in a sample of $n = 5$. Then X is $b(5, 0.3085)$. We will reject H_0 only if $X = 0$. The probability of this is

$$P(X = 0) = (1 - 0.3085)^5 = 0.1581.$$

Thus the power of the test is only 0.1581. This is quite a bit less than the power of the Z-test. To get the power of the sign test to be close to the power of the Z-test, we must choose a larger sample size. □

Software Calculations

To perform a sign test:

Minitab: If the data consist of a list of + and − signs, enter the data in a blank column by entering a 1 for a + sign, −1 for a − sign, and 0 for a 0. If the data consist of numeric values, enter the values in a blank column. Select **Stat** → **Nonparametrics** → **1-Sample Sign**. Enter the name of the column containing the data, select **Test Median**, enter the claimed median, and select the inequality in H_1.

R: To calculate the exact P-value, use the syntax pbinom$(x, n, 0.5)$ where x is the test statistic and n is the total number of + and − signs. For a two-tail test, double the result.

Excel: Excel does not have a built-in function for performing the sign test. However, the exact P-value can be calculated using the syntax BINOM.DIST$(x, n, 0.5, 1)$ where x is the test statistic and n is the total number of + and − signs. For a two-tail test, double the result.

TI-83/84 Calculators The calculators do not have a built-in function for performing the sign test. However, the exact P-value can easily be calculated by pressing **2ND** → **DISTR** → **B:binomcdf**. The syntax is binomcdf$(n, x, 0.5)$, where n is the total number of + and − signs and x is the test statistic. For a two-tail test, double the result.

Exercises

Directions: In each problem asking to test a claim, give (1) give the test statistic, (2) the critical value, and (3) the final conclusion. Unless otherwise specified, use a 95% confidence level.

1. A student claims that the median number of textbooks purchased by fellow students at his university is greater than 5. He surveys 10 students and records the data in the table below. Is there any point in performing a sign test on these data to test the claim? Explain.

Number of Textbooks	5	8	1	3	7	4	3	0	4	3

2. To determine if mice prefer a cage with a mirror or one without a mirror, a researcher set up 16 pairs of connected cages, one with a mirror and one without, and put a solitary mouse in each pair of cages for 30 min. She recorded the amount of time each mouse spent in each of its two cages (in minutes), as shown in the table below. Use a sign test to test the claim that mice spend more time in the cage with a mirror.

Without	8	9	12	18	2	10	13	4	11	26	12	11	14	15	9	10
With	22	21	18	12	28	20	17	26	19	4	18	19	16	15	21	20

3. To compare two different methods for measuring the percentage of starch in potatoes, call them methods A and B, 13 potatoes of various types were cut in half. Method A was used on one half and method B was used on the other half. The percentage of starch measured with each method is shown in the table below. Use the data to test the claim that the median of the differences between the two methods is 0. Based on these results, do the two methods appear to give significantly different results?

A	16.5	14.7	16.2	15.9	14.8	15.2	16.4	17.1	15.1	16.3	14.7	17.1	16.0
B	16.7	14.7	16.3	16.1	15.1	14.9	16.5	17.3	15.4	16.3	14.6	17.2	15.8

4. In a comparison of two brands of soda, call them A and B, 15 randomly selected consumers were given a sample of each. Out of these consumers, 12 preferred brand A while only 3 preferred B. Does there appear to be a significant difference between the preferences for the two brands?

5. To determine if there are more mosquitoes in his backyard or front yard, the author stood in each yard for 5 min on 12 randomly selected nights, swatted mosquitoes, and recorded which yard appeared to have mosquitoes. On 9 evenings the backyard appeared to have more, and on 3 evenings the front yard appeared to have more.

   a. Use a sign test to test the claim that the backyard has more mosquitoes on more than one-half of all evenings.

   b. Use a 1-proportion Z-test to test the same claim. Does this yield the same conclusion?

   c. Based on the results of parts a and b, which test requires stronger sample evidence to reject H_0, the sign test or the 1-proportion Z-test? Explain.

6. The table on the next page shows the number of piglets in 15 different litters and the number of females in each (data collected by Alexa Hopping and Brett Troyer, 2011). Use

the sign test to test the claim that a litter of piglets typically contains more males than females at the 90% confidence level. (**Hint**: Consider the difference of the number of males and females in each litter.)

Piglets in Litter	14	13	13	14	12	13	14	13	14	13	15	11	14	14	14
Number of Females	7	7	6	6	3	8	5	6	8	4	8	7	5	4	6

7. Find the critical value for a one-tail sign test in the case where $n = 50$ at the 99%, 95%, and 90% confidence levels. (**Suggestion**: Use software to calculate $P(X \leq x)$ for different values of x.)

8. Explain why a critical value for a one-tail test at the $\alpha = 0.025$ significance level is the same as a critical value for a two-tail test at the $\alpha = 0.05$ significance level.

9. Calculate the exact P-value for the sign test for a claim with the given alternative hypothesis and values of n and x.

   a. H_1: $m < 0$, $n = 10$, $x = 3$
   b. H_1: $m \neq 0$, $n = 25$, $x = 7$
   c. H_1: $m > 0$, $n = 15$, $x = 5$

10. Referring to the scenario in Example 7.2.4, suppose we use a sample of size $n = 6$ for the sign test. Noting that the critical value for this value of n is 0 at the 95% confidence level, calculate the power of the test. Give a conceptual explanation of why this has a lower power than in the case when $n = 5$. (**Hint**: Note that N, the number of values in the sample greater than 10.5, is $b(6, 0.3085)$. Find $P(N = 0)$.)

11. There are at least three approaches for handling ties in the sign test (when a data value is represented with a 0 rather than a + or − sign):

   1. Simply ignore the data value (this is what was done in Example 7.2.1).
   2. Treat one-half of the 0s as +'s and one-half as −'s (if the number of 0s is odd, ignore one).
   3. In a one-tail test, treat all the 0s as either +'s or −'s, whichever *contradicts* the alternative hypothesis.

Consider the claim that the median number of children's books in houses in a town is greater than 12, as described in Example 7.2.1. Suppose a sample of 95 houses yields 35 with fewer than 12 books, 55 with greater than 12 books, and 5 with exactly 12 books. Use the three approaches described above to test this claim at the 95% confidence level. Do the different approaches yield different conclusions?

7.3 The Wilcoxon Signed-Rank Test

The sign test discussed in the previous section is relatively easy to do, but it has a significant shortcoming in that it completely ignores the magnitudes of the values. In this section, we introduce a similar nonparametric test that takes the magnitudes into account, called the *Wilcoxon signed-rank test*, which was introduced by Frank Wilcoxon in 1945.

To introduce this test, consider again the scenario of determining if scores on a writing survey improve from the beginning to the end of the semester, as described in Example 7.2.2. The data are given in the first two rows of Table 7.3. The claim we tested is that the median of the population of the differences of the scores, denoted m, is positive. This yielded the hypotheses

$$H_0: m = 0, \quad H_1: m > 0.$$

To perform the Wilcoxon signed-rank test of these hypotheses, first we calculate the differences, $d = $ (end − beginning). Then we rank the absolute values of these differences from 1 to n, where the smallest absolute value is ranked 1 and the largest is ranked n. Next we calculate the signed ranks of each difference. If the difference is positive, the signed rank equals the rank. If the difference is negative, the signed ranked equals the negative of the rank. These calculations are shown Table 7.3.

End	85.4	79.6	64.2	96.6	79.3	40.5	90.1	89.2	85.6
Beginning	77.6	83.2	60.2	93.1	74.6	43.1	86.9	79.3	80.2
Difference d	7.8	−3.6	4	3.5	4.7	−2.6	3.2	9.9	5.4
Rank of $|d|$	8	4	5	3	6	1	2	9	7
Signed Rank	8	−4	5	3	6	−1	2	9	7

Table 7.3

To calculate the test statistic, we calculate the sum of the positive signed ranks and the sum of the negative signed ranks:

$$S^+ = 8 + 5 + 3 + 6 + 2 + 9 + 7 = 40 \quad \text{and} \quad S^- = 4 + 1 = 5.$$

The smaller of these two sums is the test statistic T. In this case, we have $T = 5$. Observe that the sum of all the ranks is 45. If $H_0: m = 0$ were true and we further assume that the distribution of the differences is symmetric around 0, then we would expect one-half the signed ranks to be positive, one-half to be negative, and that both S^+ and S^- should be close to $45/2 = 22.5$. Thus the test statistic T should also be close to 22.5. Because T is quite a bit smaller than 22.5, we have an indication that we should reject H_0.

To determine if T is small enough to reject H_0, we use the critical values given in Table C.7 (see Section A.7 of Appendix A for a discussion of how these critical values were calculated).

In this example, we have a one-tail test and $n = 9$, so the table gives us a critical value of 8 at the 95% confidence level. We reject H_0 if $T \leq$ the critical value. Because $T = 5 \leq 8$, we reject H_0 and conclude that the data support the claim. Note that this is the opposite conclusion reached from that using the sign test.

To approximate the P-value, we use the following theorem, whose proof is given in Appendix A.7. By symmetry, this theorem also holds for S^-.

Theorem 7.3.1 *Assuming the distribution of the differences is symmetric around 0, then the random variable S^+ is approximately normally distributed with mean and variance*

$$\mu = \frac{n(n+1)}{4} \quad \text{and} \quad \sigma^2 = \frac{n(n+1)(2n+1)}{24}$$

so that

$$P\text{-value} = P(S^+ \leq T) \approx P\left(Z \leq \frac{(T+0.5) - n(n+1)/4}{\sqrt{n(n+1)(2n+1)/24}}\right)$$

where Z is $N(0, 1)$. □

In this example,

$$\mu = \frac{9(9+1)}{4} = 22.5 \quad \text{and} \quad \sigma^2 = \frac{9(9+1)(2 \cdot 9 + 1)}{24} = 71.25$$

so that

$$P\text{-value} \approx P\left(Z \leq \frac{(5+0.5) - 22.5}{\sqrt{71.25}}\right) = P(Z \leq -2.01) = 0.0222.$$

Because the P-value ≤ 0.05, we reject H_0 as we did using the critical value. The procedures are summarized in box on the next page.

Notice that this test requires that the population of differences have a distribution that is symmetric. This means that the left half of the distribution is a mirror image of the right half. We are not requiring that the population have a certain type of distribution (such as a normal distribution), only that the distribution be symmetric. This is a weaker requirement than one of normality, as in other tests.

> ### The Wilcoxon Signed-Rank Test
>
> **Purpose**: To test the null hypothesis H_0: $m = 0$ where m is the population median of the differences of a set of paired data.
>
> **Procedures**: Let n denote the number of pairs of data values.
>
> 1. Calculate the difference d between each pair of data.
> 2. Rank the absolute values of the differences from 1 to n.
> 3. Calculate the signed rank of each difference, which equals the rank if the difference is positive and the negative of the rank if the difference is negative.
> 4. Calculate the sum of the positive signed ranks S^+ and the sum of the negative signed ranks S^-.
> 5. Let T equal the smaller of S^+ and S^-.
>
> The test statistic is
>
> - T for $n \leq 30$ and
> - $z = \dfrac{T - n(n+1)/4}{\sqrt{n(n+1)(2n+1)/24}}$ for $n > 30$.
>
> The critical value is found in Table C.7 for $n \leq 30$. If $n > 30$, then the critical value is $-z_\alpha$ or $-z_{\alpha/2}$ for a one- or two-tail test, respectively, as found in Table C.2. In any case, we reject H_0 if the test statistic is less than or equal to the critical value. For a one-tail test, the P-value is approximated with
>
> $$P\text{-value} \approx P\left(Z \leq \frac{(T+0.5) - n(n+1)/4}{\sqrt{n(n+1)(2n+1)/24}}\right).$$
>
> For a two-tail test, the P-value is twice this probability.
>
> **Requirements**
>
> 1. The sample is random.
> 2. The population of differences has a distribution that is symmetric.

To determine if this requirement is met, we construct a histogram of the differences. If this histogram is not "too far" from symmetric, then we conclude that it is reasonable to assume this requirement is met. Judging if the histogram is too far from symmetric is often very subjective, especially with a small set of data. The histogram of the differences for the survey scores is shown in the left half of Figure 7.1. This histogram does not look very symmetric, but because there are only nine data values, the difference between the left and right sides is not that great. Therefore we conclude that this requirement is met. If the histogram looked like the one in the right half of Figure 7.1, we would conclude that this requirement was not met.

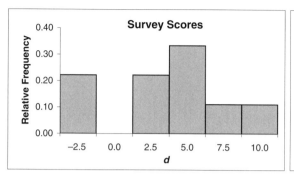
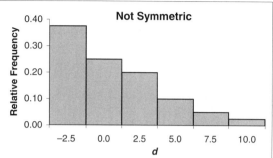

Figure 7.1

In some situations, we may have differences of 0 or absolute values of differences that are equal. Exercises 3 and 4 of this section describe simple procedures for dealing with these situations.

Software Calculations

To perform a Wilcoxon signed-rank test:

Minitab: Enter the two samples in separate columns. Calculate the differences by selecting **Calc → Calculator**. Enter an expression to calculate the difference of the columns, and store the result in a blank column. Then select **Stat → Nonparametrics → 1-Sample Wilcoxon**. Enter the name of the column containing the differences, select **Test Median**, enter the claimed median, and select the inequality in H_1. The output "Wilcoxon Statistic" is the value of S^+.

R: Enter the two samples of data $\{x_1, x_2, \ldots, x_n\}$ and $\{y_1, y_2, \ldots, y_n\}$ into vectors with the syntax x<−c($x_1, x_2, \ldots, x_n$) and y<−c($y_1, y_2, \ldots, y_n$). To perform a two-tail test where the P-value is calculated according to the approximation given in the text, enter the code

> wilcox.test(x, y, alternative="two-sided", paired=TRUE, exact=FALSE)

For a one-tail test, change "two.sided" to either "less" or "greater" depending on the inequality in H_1. R performs the test on the difference $(x - y)$. To calculate an exact P-value, change "exact=FALSE" to "exact=TRUE".

Exercises

Directions: In each exercise asking to test a claim, (1) give the test statistic, (2) give the critical value or the P-value, and (3) state the final conclusion. Unless otherwise specified, use a 95% confidence level.

1. Approximate the *P*-value for the Wilcoxon signed-rank test for a claim with the given alternative hypothesis and values of n and T.

   a. H_1: $m > 0$, $n = 10$, $T = 10$
   b. H_1: $m \neq 0$, $n = 25$, $T = 110$
   c. H_1: $m < 0$, $n = 50$, $T = 600$

2. A certain design of paper airplane was constructed from heavy paper and then from light paper. Ten different volunteers flew each airplane 5 times, and the distance flown was measured. The average distance (in feet) for each person and type of paper is shown in the table below. Test the claim that the median of the differences (heavy − light) is positive. Based on this result, does it appear that airplanes constructed from heavy paper fly farther?

Heavy	7.2	6.6	14.2	13.6	19.2	24.9	33.2	20.6	31.1	22.5
Light	6.5	7.1	10.3	13.9	20.5	21.6	25.7	24.9	28.9	34.8

3. The sign of the signed rank is the same as the sign of the difference. If the difference is 0, common practice is to simply ignore that pair of data and test the claim using the remaining pairs. To investigate whether college freshmen gain weight during their first semester, the weights of 12 randomly selected freshmen were recorded at the beginning and end of the semester, as shown in the table below. Use these data to test the claim that the median change in weight (end − beginning) is positive. What does this suggest about the commonly held belief that freshmen gain weight during their first semester?

End	136	163	179	156	184	138	103	119	140	156	195	137
Beginning	135	151	168	154	177	138	99	124	148	156	189	134

4. In the Wilcoxon signed-rank test, we rank the absolute values of the differences. If two or more of these absolute values are equal, then first we assign them the average of the ranks they would receive if they were a little bit different from each other, and then we calculate S^+ and S^- as usual. For instance, if three values of $|d|$ were 0.1, 0.1, 0.1, and 0.1 was the smallest value of $|d|$, then their ranks would each be the average of 1, 2, and 3, which is 2. Apply this procedure to the following scenario: A new processing method was tried on 12 different machines. The number of breakdowns experienced by each machine after one week was recorded. The number of breakdowns the previous week when using the

standard method was also recorded, as shown in the table below. Test the claim that the median of the difference in breakdowns (new − standard) is negative. Based on this result, does the new method appear to be better overall?

New	8	5	15	5	16	5	9	1	8	10	1	1
Standard	9	3	12	8	10	7	4	5	15	11	9	13

5. The table below shows the number of piglets in 15 different litters and the number of females in each (data collected by Alexa Hopping and Brett Troyer, 2011). Use the procedures in Exercises 3 and 4 above to test the claim that the median of the differences (number of males − number of females) is positive at the 90% confidence level. Based on this result, does it appear that a litter of piglets typically contains more males or females?

Piglets in Litter	14	13	13	14	12	13	14	13	14	13	15	11	14	14	14
Number of Females	7	7	6	6	3	8	5	6	8	4	8	7	5	4	6

6. To determine if mice prefer a cage with a mirror or one without a mirror, a researcher set up 16 pairs of connected cages, one with a mirror and one without, and put a solitary mouse in each pair of cages for 30 min. She recorded the amount of time each mouse spent in each of its two cages (in minutes) as shown in the table below. Use the procedures in Exercises 3 and 4 above to test the claim that mice spend more time in the cage with a mirror.

Without	8	9	12	18	2	10	13	4	11	26	12	11	14	15	9	10
With	22	21	18	12	28	20	17	26	19	4	18	19	16	15	21	20

7. We described the Wilcoxon signed-rank test as being used to compare data that come in pairs. However, the claim is really regarding the population of the differences. The fact that these population values are the differences of paired data is irrelevant. Therefore, this test can be used to test a claim about the median of *any* population with a symmetric distribution. To test the claim that a sample of data comes from a population with a specified median m_0, calculate the difference $d = x - m_0$ for each data value x, and then test the claim that the population of the differences has a median of 0, using the Wilcoxon signed-rank test. Use this approach to test the claim that the data in the table below come from a population with median $m_0 = 5$.

9.1	8.8	1.6	6.4	7.1	11.0	10.8	0.3	4.8	3.7

8. Consider the claim that the median number of books in homes with children in a town is greater than 12 as considered in Example 7.2.1. Given in that example are the data in the table below, which shows the number of children's books in 11 randomly selected homes from the town. Use these data to test the claim using the approach in Exercise 7 above. Do you come to a different conclusion from that in Example 7.2.1?

Number of Books	15	18	7	13	53	20	14	45	19	65	12

9. In this exercise, we explore the effects of an outlier on the results of a Wilcoxon signed-rank test. Consider the data in the table below.

−5	−2	1	9	12	15	20

   a. Use the data to test the claim that the data come from a population with median greater than 0.

   b. Now replace the data value of 20 with the outlier 40, and use the data to test the claim. Then replace −5 with the outlier −30 and test the claim. Does your final conclusion change?

   c. Based on the results of part b, is the Wilcoxon signed-rank test sensitive to outliers?

10. If a sample of data has n nonzero values and $S^+ = s$, find the value of S^- for a claim that the population median is 0. (**Hint**: What is the sum of all the ranks?)

11. For a one-tail test at the 95% confidence level, the critical value is defined to be the largest value of T for which the P-value is less than 0.05. One way to approximate this critical value is to approximate the P-value for different values of the test statistic T, then choose the largest value of T for which the P-value is less than 0.05. Use this approach to approximate the critical value when $n = 20$. Compare this approximation to the critical value given in Table C.7.

7.4 The Wilcoxon Rank-Sum Test

In this section, we present a method for comparing the medians of two independent populations, called the *Wilcoxon rank-sum test*. It is often used in place of the 2-sample T-test and can be shown to be nearly as powerful. This test is equivalent to another nonparametric test called the *Mann-Whitney U test* (see Exercise 5 of this section).

To illustrate the rank-sum test, consider the following scenario: There are two commonly used techniques for throwing a shot put: glide and rotation. The first and third rows of Table 7.4 give the maximum distances (in meters) of 12 different athletes using the glide method and 13 athletes using the rotation method at international competitions (data collected by David Meyer, 2010). Use these data to test the claim that there is not a significant difference between these two methods.

Glide	20.19	20.51	21.03	21.09	21.14	21.16	21.51	21.70	21.95	22.64	22.75	22.91	
Rank	1	2	4	5	6	7	10	11	14	21	23	25	
Rot.	21.00	21.26	21.29	21.87	21.89	22.00	22.09	22.22	22.43	22.51	22.54	22.67	22.86
Rank	3	8	9	12	13	15	16	17	18	19	20	22	24

Table 7.4

We might consider using the 2-sample T-test to compare the population means. However, both samples are small, and a normal quantile plot of each indicates that the populations may not be normally distributed. So the requirements of this test are not met. Also note that the data do not come in pairs, so the signed-rank test is not appropriate either.

The parameters of interest are

and
$$m_1 = \text{the median of all distances using the glide method}$$
$$m_2 = \text{the median of all distances using the rotation method}.$$

Our claim is $m_1 = m_2$ so that the hypotheses are

$$H_0\text{: } m_1 = m_2 \quad \text{and} \quad H_1\text{: } m_1 \neq m_2.$$

To calculate the test statistic, we combine both samples into one large sample and rank the values from 1 to 25. These ranks are shown in the second and fourth rows of Table 7.4. Next we add up the ranks from the first sample (by convention we let the first sample be the sample with the smaller sample size) and denote the sum R_1:

$$R_1 = 1 + 2 + 4 + \cdots + 25 = 129.$$

If R_1 is "large," it indicates that the values from the first sample are larger than the second. If R_1 is "small," it indicates the opposite. In either case, we have evidence to reject H_0. The following theorem, which we present without proof, is the key to determining if R_1 is large or small enough to reject H_0.

Theorem 7.4.1   If there are no ties and both populations have the same continuous distribution, then R_1 is an observed value of a random variable with mean and variance

$$\mu_R = \frac{n_1(n_1+n_2+1)}{2} \quad \text{and} \quad \sigma_R^2 = \frac{n_1 n_2 (n_1+n_2+1)}{12}.$$

The distribution of this variable is symmetric around the mean, and for $n_1, n_2 \geq 10$ it is approximately normal.   □

(For a proof of this theorem, see Jean Dickinson Gibbons, *Nonparametric Statistical Inference*, McGraw-Hill, New York, 1971, p. 165, or John A. Rice, *Mathematical Statistics and Data Analysis*, 3rd ed., Thomson Brooks/Cole, 2007, pp. 438–441.)

Describing the exact distribution of R_1 is rather tedious (see Exercise 6 of this section for one method of doing so). Table C.8 gives critical values for combined sample sizes of 35 or less. For larger samples, a normal approximation is used. The procedures for this test are outlined in the box on the next page.

For this example,

$$\frac{n_1(n_1+n_2+1)}{2} = \frac{12(12+13+1)}{2} = 156.$$

Because $R_1 = 129 \leq 156$, the test statistic is $R = 129$. From Table C.8, we see that the critical value is 119 for a two-tail test with $n_1 = 12$ and $n_2 = 13$ at the 0.05 significance level. Because $129 > 119$, we do not reject H_0 and we conclude that there is not a statistically significant difference between the medians of the two methods.

To find the *P*-value, we first find the mean and standard deviation of R:

$$\mu_R = \frac{12(12+13+1)}{2} = 156 \quad \text{and} \quad \sigma_R = \sqrt{\frac{12 \cdot 13(12+13+1)}{12}} = \sqrt{338}.$$

Thus the *P*-value is approximately

$$P\text{-value} \approx 2P\left(Z \leq \frac{(129+0.5)-156}{\sqrt{338}}\right) \approx 2P(Z \leq -1.44) = 0.1498.$$

> **The Wilcoxon Rank-Sum Test**
>
> **Purpose**: To test the null hypothesis H_0: $m_1 = m_2$, where m_1 and m_2 are the medians of two independent populations with continuous distributions.
>
> **Procedures**: Let n_1 and n_2 denote the sample sizes from the two populations where $n_1 \leq n_2$.
>
> 1. Combine the two samples into one large sample.
> 2. Rank the values from 1 to $n_1 + n_2$. In the case of a tie, assign the average rank to each tied value.
> 3. Calculate the sum of the ranks from the first sample. Call it R_1.
> 4. Let $\mu_R = n_1(n_1 + n_2 + 1)/2$.
>
> The test statistic is
>
> $$R = \begin{cases} R_1, & \text{if } R_1 \leq \mu_R \\ 2\mu_R - R_1, & \text{otherwise.} \end{cases}$$
>
> The critical value is found in Table C.8. We reject H_0 if $R \leq$ the critical value. The P-value for a one-tail test is approximated by
>
> $$P\text{-value} \approx P\left(Z \leq \frac{(R + 0.5) - \mu_R}{\sigma_R}\right) \quad \text{where} \quad \sigma_R = \sqrt{\frac{n_1 n_2 (n_1 + n_2 + 1)}{12}}.$$
>
> For a two-tail test, the P-value is twice this probability.
>
> **Requirement**
> 1. The samples are random and independent.

Confidence Interval Estimate of the Difference of the Population Medians

We can use the critical values from the rank-sum test to approximate a confidence interval estimate of the difference of the population medians $m_1 - m_2$, using the following algorithm:

1. Let x_i, $i = 1, \ldots, n_1$, and y_i, $i = 1, \ldots, n_2$, denote the sample values from the first and second samples, respectively. For each possible permutation of i and j, calculate the difference $x_i - y_j$. There are a total of $n_1 n_2$ differences.
2. Order these differences from smallest to largest according to their actual (not absolute) value. Denote them $d_{(1)}, d_{(2)}, \ldots, d_{(n_1 n_2)}$ so that $d_{(1)}$ denotes the smallest difference and $d_{(n_1 n_2)}$ denotes the largest.
3. Find the critical value for a two-tail rank-sum test at the desired significance level α. Denote this critical value c. Calculate

$$k = c - \frac{n_1}{2}(n_1 + 1).$$

4. An approximate $100(1 - \alpha)\%$ confidence interval interval estimate of $m_1 - m_2$ is

$$d_{(k+1)} < m_1 - m_2 < d_{(n_1 n_2 - k)}.$$

(See John A. Rice, *Mathematical Statistics and Data Analysis*, 3rd ed., Thomson Brooks/Cole, 2007, pp. 439–442, for a further discussion of this algorithm.)

Example 7.4.1 Calculating a Confidence Interval The tedious part of this algorithm lies in calculating the differences because there are so many of them. So to illustrate this algorithm, we consider two small samples of size $n_1 = 4$ and $n_1 = 5$ as shown along the top and left sides of Table 7.5. We calculate an approximate 95% confidence interval estimate of the respective population medians $m_1 - m_2$:

1. The differences $x_i - y_j$ are shown in the interior of Table 7.5.

		X			
		1.2	1.5	2.3	2.5
	0.9	0.3	0.6	1.4	1.6
	1.6	−0.4	−0.1	0.7	0.9
Y	1.9	−0.7	−0.4	0.4	0.6
	2.2	−1.0	−0.7	0.1	0.3
	2.7	−1.5	−1.2	−0.4	−0.2

Table 7.5

2. We order the differences:

−1.5 −1.2 −1 −0.7 −0.7 −0.4 −0.4 −0.4 −0.2 −0.1 0.1 0.3 0.3 0.4 0.6 0.6 0.7 0.9 1.4 1.6

3. From Table C.8 we find the critical value for a two-tail test where $\alpha = 0.05$ and $n_1 = 4$ and $n_1 = 5$ to be $c = 11$. This gives

$$k = 11 - \frac{4}{2}(4+1) = 1.$$

4. The lower and upper limits of the confidence interval are $d_{(1+1)} = -1.2$ and $d_{(4 \cdot 5 - 1)} = 1.4$, respectively, so that the confidence interval is

$$-1.2 < m_1 - m_2 < 1.4.$$

Because this confidence interval contains 0, we are approximately 95% confident that there is no difference between the population medians. □

Software Calculations

To perform a Wilcoxon rank-sum test:

Minitab: Enter the two samples in separate columns. Then select **Stat** → **Nonparametrics** → **Mann-Whitney**. Enter the names of the columns containing the samples, select **Test Median**, enter the claimed median, and select the inequality in H_1. Note that Minitab gives a different test statistic from that in the text, but the P-value is the same.

R: Enter the two samples of data $\{x_1, x_2, \ldots, x_n\}$ and $\{y_1, y_2, \ldots, y_n\}$ into vectors with the syntax x<−c($x_1, x_2, \ldots, x_n$) and y<−c($y_1, y_2, \ldots, y_n$). To perform a two-tail test where the P-value is calculated according to the approximation given in the text, and an approximate 95% confidence interval for the difference of the population medians is calculated, enter the code

> wilcox.test(x, y, alternative="two.side", paired=FALSE, exact=FALSE,
          conf.int=TRUE, conf.level=0.95)

For a one-tail test, change "two.sided" to either "less" or "greater" depending on the inequality in H_1. R performs the test on the difference $(x - y)$. To calculate an exact P-value, change "exact=FALSE" to "exact=TRUE". Note that R gives a different test statistic from that in the text, but the P-value is the same.

Exercises

Directions: For each exercise asking for a hypothesis test, (1) calculate the test statistic, (2) give the critical value or the P-value, and (3) state the final conclusion. Unless otherwise indicated, use a 95% confidence level.

1. To compare the prices of diapers at two chains of stores, call them A and B, a student randomly chose some packages of diapers from each chain and recorded the price per diaper as shown in the table below (data collected by Kay Zrust, 2012). Test the claim that the median price per diaper at chain A equals the median price at chain B.

Chain A	0.292	0.355	0.256	0.225	0.322	0.221
Chain B	0.296	0.370	0.238	0.242	0.380	0.200

2. To compare the weights of two brands of bottled water, call them A and B, two students selected four bottles of each brand and measured the mass (in grams) of water each, as shown in the table below (data collected by Brittany Singleton and Kelsie Elder, 2009). Test the claim that the median mass of brand A is higher than the median mass of brand B.

Brand A	512	509	511	511
Brand B	505	506	505	504

3. Twenty-one male and twenty-three female university students were each asked to draw a rectangle with an area of 6 in^2 without using a ruler. The dimensions of the rectangle were then measured, the actual area was calculated, and the error |actual area−6| was calculated. The results are given in the table below (data collected by Brandon Wood, 2011). Assuming that these students were randomly selected, test the claim that the median error of all male students at this university equals the median error of all female students. Based on this result, does there appear to be a significant difference between males and females in terms of ability to draw a rectangle with a given area?

Male Errors						Female Errors					
0.07	0.56	1.41	2.16	2.81	3.84	0.53	1.51	2.28	3.00	4.00	5.02
0.19	0.97	1.43	2.19	2.86		1.03	1.53	2.40	3.34	4.01	5.20
0.25	1.19	1.68	2.65	2.99		1.04	1.53	2.55	3.40	4.69	6.12
0.36	1.21	2.04	2.80	3.16		1.25	1.94	2.88	3.97	5.06	

4. The table below gives the run times of several movies rated G and PG (data collected by Meredith Hein and Rachel Dahlke, 2011). Assuming that these data are randomly selected, test the claim that the median run time of movies rated G is less the median of movies rated PG. Why might movies rated G be shorter than those rated PG?

G	63	68	74	75	76	82	83	88	98	106	117
PG	95	96	97	99	100	103	104	114	120	143	152

5. The *Mann-Whitney U test* is equivalent to the Wilcoxon signed-rank test in the sense that both tests lead to the same conclusion. To perform the U test, first we calculate the differences $x_i - y_j$ as in the first step of the algorithm for the confidence interval estimate of $m_1 - m_2$. Then the test statistic is

$$U = \text{the number of differences that are greater than 0.}$$

It can be shown that U is related to R_1 by

$$U = R_1 - \frac{n_1}{2}(n_1 + 1). \tag{7.1}$$

Calculate both U and R_1 for the data in Table 7.5, and show that Equation (7.1) holds for this set of data.

6. In this exercise we outline one method for calculating the critical value for a one-tail rank-sum test at the 0.15 significance level when $n_1 = 2$ and $n_2 = 4$.

   a. List all $\binom{6}{2} = 15$ combinations of the ranks of the values from the first sample. Let R_1 denote the sum of the ranks. Calculate the value of R_1 for each combination.
   b. Form a frequency distribution of R_1.
   c. Assuming the populations have the same distribution, so that the combinations from part a are equally likely, calculate the probability distribution of R_1.
   d. Find the largest number r_1 such that $P(R_1 \leq r_1) \leq 0.15$. The number r_1 is the critical value.

7. Calculate an approximate 90% confidence interval estimate of the difference of the two population medians, using the data in the table below. Does there appear to be a significant difference between the medians?

Sample 1	25.6	27.8	30.1	30.5	31.1
Sample 2	22.5	24.6	25.1	25.9	26.1

8. At the 95% confidence level, the critical value for a two-tail rank-sum test is a number c such that $2P(R_1 \leq c) = 0.05$. For given values of n_1 and n_2, it may not be possible to find an integer value of c for which this equation is true. Because of this, most of the critical values in Table C.8 satisfy this equation only approximately.

   One result of this is that a confidence interval such as that calculated in Example 7.4.1 is not exactly a 95% confidence interval. To find a more accurate measure of the confidence level, we use the following approximation:

   $$\text{Confidence level} \approx \left[1 - 2P\left(Z \leq \frac{(c+0.5) - \mu_R}{\sigma_R}\right)\right] \times 100\%$$

   where c is the critical value used in calculating the confidence interval and μ_R and σ_R are as given in Theorem 7.4.1. Use this to find a more accurate measure of the confidence level for the confidence interval in Example 7.4.1.

7.5 The Runs Test for Randomness

Suppose a group of eight males (M) and females (F) is seated in a row as MMMMFFFF and that another group is seated as MFMFMFMF. Does it appear that either group is arranged randomly? In this section, we present one very easy-to-use method for analytically answering such a question, called the *runs test for randomness*.

Most people would answer no to this question. Why? Notice that in the first group, there is a sequence of four males followed by a sequence of four females, while in the second group, there is no sequence of genders longer than one. Such sequences of data values are called *runs*.

> **Definition 7.5.1** A *run* is a sequence of data of the same type preceded and followed by data of a different type or by no data at all.

In the first group of people, there are only two runs, whereas in the second group there are eight runs. It seems reasonable to specify that an unusually large or small number of runs contradicts randomness. Theorem 7.5.1, which we present without proof, tells us about the distribution of the total number of runs in a sample where the data values are of two types. (For a proof of this theorem, see Jean Dickinson Gibbons, *Nonparametric Statistical Inference*, McGraw-Hill, New York, 1971, pp. 52–58. For a justification of part of this theorem, see Exercise 12 of this section.)

Theorem 7.5.1 *Let R represent the total number of runs in a sample of n_1 items of the first type and n_2 items of the second type. If the data values are chosen randomly, then R is an observed value of a random variable with pmf*

$$f(r) = \begin{cases} \dfrac{2\binom{n_1-1}{r/2-1}\binom{n_2-1}{r/2-1}}{\binom{n_1+n_2}{n_1}}, & \text{if } r \text{ is even} \\[2ex] \dfrac{\binom{n_1-1}{(r-1)/2}\binom{n_2-1}{(r-3)/2} + \binom{n_1-1}{(r-3)/2}\binom{n_2-1}{(r-1)/2}}{\binom{n_1+n_2}{n_1}}, & \text{if } r \text{ is odd} \end{cases}$$

for $r = 2, 3, \ldots, n_1 + n_2$. As $n_1 + n_2 \to \infty$, the distribution of R approaches a normal distribution with mean and variance, respectively, of

$$\mu_R = \frac{2n_1 n_2}{n_1 + n_2} + 1 \quad \text{and} \quad \sigma_R^2 = \frac{2n_1 n_2 (2n_1 n_2 - n_1 - n_2)}{(n_1 + n_2)^2 (n_1 + n_2 - 1)}. \qquad \Box$$

To apply this theorem, we use the hypotheses

H_0: the data are arranged randomly   and H_1: the data are not arranged randomly.

For a given set of data, we count the number of runs R and then use Theorem 7.5.1 to determine if this is unusually small or large, assuming H_0 is true. If it is large or small, then we reject H_0 and we conclude that the data are not arranged randomly. The procedures for the runs test are given in the box below. See Exercise 10 of this section for an explanation of how the critical values are calculated.

The Runs Test for Randomness

Purpose: To test the claim that a set of data with two types of values is arranged randomly. The null hypothesis is

H_0: The data are arranged randomly.

Procedures: Let
- n_1 = the number of data values of the first type,
- n_2 = the number of data values of the second type, and
- R = the total number of runs.

The test statistic is R. For $n_1, n_2 \leq 15$ the critical values at the 0.05 significance level are given in Table C.9. We reject H_0 if R is less than or equal to the smaller critical value or R is greater than or equal to the larger critical value.

The P-value is approximately twice the extreme region under the standard normal bell curve bounded by

$$z = \frac{R - \mu_R}{\sigma_R}$$

where μ_R and σ_R are as given in Theorem 7.5.1.

Requirements
1. The data values must be of two types.
2. The data need to be ordered in some meaningful way.

This test is very easy to perform, but we need to make two very important observations:

1. For this test to be appropriate, the data need to be ordered in some meaningful way. We are really testing if this order is random or not. For example, in the scenario of people sitting in a row, they are ordered according to the way they chose to sit. Determining if this order is random is interesting and meaningful. On the other

hand, if we were to randomly select several people from a crowd and write down their genders, the genders would be ordered according to the way we chose them. This order is not very meaningful.

2. We are *not* determining if one data type appears more or less frequently than the other. If we wanted to do this, we could use the 1-proportion Z-test. In the runs test, we are only testing for randomness.

The formula for the P-value gives an approximation. If n_1 and n_2 are large, this approximation is very accurate and the P-value will lead to the same technical conclusion as the critical values. However, if n_1 and n_2 are small, the P-value may not be very accurate and it may lead to a different technical conclusion. See Exercise 4 in this section for an example of this.

Example 7.5.1 Order of Songs A classical music fan has a collection of songs composed by Bach and Vivaldi on her MP3 player, which is supposed to randomly choose songs. The order of the composer of the songs played is shown in Table 7.6. Test the claim that the composers are arranged randomly.

$$\underline{B} \; \underline{V \; V} \; \underline{B \; B \; B} \; \underline{V \; V} \; \underline{B} \; \underline{V} \; \underline{B} \; \underline{V} \; \underline{B \; B} \; \underline{V \; V \; V} \; \underline{B} \; \underline{V \; V \; V} \; \underline{B}$$

Table 7.6

The hypotheses for this claim are

H_0: the composers are arranged randomly

and $\qquad H_1$: the composers are not arranged randomly.

From the data we see that there are $n_1 = 10$ songs by Bach and $n_2 = 12$ songs by Vivaldi. The individual runs are underlined in the table. There are a total of $R = 13$ runs. From Table C.9 we find that the critical values are 7 and 17. Because R is between these critical values, we do not reject H_0 and we conclude that the composers are arranged randomly. Thus it appears that the songs are indeed randomly chosen.

To calculate the P-value for this claim, we first find μ_R and σ_R^2:

$$\mu_R = \frac{2 \cdot 10 \cdot 12}{10 + 12} + 1 \approx 11.91 \quad \text{and} \quad \sigma_R^2 = \frac{2 \cdot 10 \cdot 12 \left(2 \cdot 10 \cdot 12 - 10 - 12\right)}{(10 + 12)^2 (10 + 12 - 1)} \approx 5.15.$$

The z-score for calculating the P-value is then

$$z = \frac{13 - 11.91}{\sqrt{5.15}} \approx 0.48.$$

Because this z-score is positive, the extreme region is to the right so that the P-value is

$$P\text{-value} = 2P(Z > 0.48) = 2(1 - 0.6844) = 0.6312.$$

Because this P-value is greater than 0.05, we do not reject H_0 as we did when using critical values. □

The previous example dealt with qualitative data. The runs test for randomness can also be applied to quantitative data by first classifying each data value into one of two categories. Then we can do a test of randomness on the sequence of categories. The next example illustrates this.

Example 7.5.2 Temperature Trend The first and third rows of Table 7.7 show the daily high temperature (in degrees Fahrenheit) near the author's home during the 31 days of May in a recent year. The median of these temperatures is 74°F. Test for randomness above and below the median.

64	61	63	80	69	82	83	87	96	95	83	75	52	55	63	68
B	B	B	A	B	A	A	A	A	A	A	B	B	B	B	B
71	68	70	67	77	79	82	74	64	68	60	76	80	88	82	
B	B	B	B	A	A	A	B	B	B	B	A	A	A	A	

Table 7.7

We first classify each data value as being above or below the median as in the second and fourth rows of the table. Note that by convention we count the data value of 74 in the second row as below the median. There are $n_1 = 15$ values above the median, $n_2 = 16$ values below, and a total of $R = 8$ runs. (Note that we should think of these data as being one long sequence, not two short ones, so that the run of B's that begins on the first row and ends on the second is counted as only one run.) From Table C.9 we find that the critical values are 10 and 23. Because $8 \leq 10$, we reject H_0 and conclude that the temperatures do not randomly fall above and below the median. Thus there appears to be a pattern. Examining the data, we see that there were alternating cool and warm spells with two long spells in the middle of the month. □

Software Calculations

To perform a runs test for randomness:

Minitab: If the data are qualitative of exactly two types, enter them in a blank column by entering a 0 for data of the first type and a 1 for data of the second type. Then select **Stat** → **Nonparametrics** → **Runs Test**. Enter the names of the columns containing the data, select **Above and below**, and enter the number 0.5. If the data are quantitative, enter them in a column as they are.

R: First install and load the package **lawstat**. To test the claim that a set of quantitative data randomly falls above and below its median, first enter the sample data $\{x_1, x_2, \ldots, x_n\}$ into a vector with the syntax x<-c($x_1, x_2, \ldots, x_n$). Then enter the code

> runs.test(x, plot.it=TRUE)

In the output, the statistic z described in the hypothesis test in the box seen earlier in this section is labeled "Standardized Runs Statistic." The graph given in the output illustrates whether each data value is above or below the median.

Exercises

Directions: In each exercise asking to test a claim, (1) give the test statistic, (2) give the critical value, and (3) the final conclusion. For all data given in tables, the data are ordered from left to right, top to bottom.

1. Approximate the P-value for the runs test for a set of data with the following values of n_1, n_2, and R.

   a. $n_1 = 8$, $n_2 = 12$, $R = 8$

   b. $n_1 = 25$, $n_2 = 16$, $R = 30$

   c. $n_1 = 30$, $n_2 = 15$, $R = 21$

2. The table below gives the average daily high temperatures (in degrees Fahrenheit) for the month of November in the city of Lincoln, Nebraska, for the years 1995 to 2009 (data collected by Brandon Metcalf, 2009). The mean of these temperatures is 40.19°F.

Test for randomness above and below the mean. Do these data suggest any trend in the temperatures?

37.7	33.1	36.0	43.2	46.5	32.6	48.7	38.1
36.8	42.3	43.0	40.1	39.9	40.2	44.6	

3. The table below gives the answers to 40 true/false questions on an exam. Test for randomness in the sequence of answers. Did the professor who wrote this exam do a good job in mixing up the answers?

T	T	F	T	F	F	F	T	T	T	F	T	F	F	F	T	T	T	T	
F	F	T	T	F	F	T	T	F	F	T	F	T	T	F	F	T	F	T	T

4. A manufacturer of baseball bats selects every 10th bat that comes off the production line and weighs it. The weights (in ounces) from one day's production, in order of time, are shown in the table below. The mean of these weights is 32.56.

31.4	31.7	31.5	31.8	31.6	32.2	32.8	33.1
32.1	32.3	33.1	33.3	33.5	33.6	33.5	33.5

a. Test for randomness above and below the mean, using critical values.

b. Now test for randomness above and below the mean, using the *P*-value. Do you come to the same conclusion as in part a? (Remember, the critical values are given for the 95% confidence level.)

c. Which conclusion do you trust more, part a or b? Explain.

5. Consider the temperature trend data in Example 7.5.2.

a. Calculate the *P*-value for this test.

b. In this set of data, there was a data value that equals the median of 74. We counted that data value as being below the median. Explain why, in this particular set of data, the value of R and the *P*-value would not change if we counted the data value of 74 as being above the median rather than below the median.

c. When there is a data value equal to the median, as in this set of data, some textbooks simply ignore the data value and adjust n_1, n_1, and R accordingly. Calculate the P-value for the claim that the values fall randomly above and below 74, using this strategy. Does this strategy lead to a different final conclusion?

6. The table below gives the first 75 decimal places of the number π. Test for randomness of odd and even digits.

1 4 1 5 9 2 6 5 3 5 8 9 7 9 3 2 3 8 4 6 2 6 4 3 3 8 3 2 7 9 5 0 2 8 8 4 1 9
7 1 6 9 3 9 9 3 7 5 1 0 5 8 2 0 9 7 4 9 4 4 5 9 2 3 0 7 8 1 6 4 0 6 2 8 6

7. The table below gives the first 75 decimal places of the number e. Test for randomness of odd and even digits.

7 1 8 2 8 1 8 2 8 4 5 9 0 4 5 2 3 5 3 6 0 2 8 7 4 7 1 3 5 2 6 6 2 4 9 7 7 5
7 2 4 7 0 9 3 6 9 9 9 5 9 5 7 4 9 6 6 9 6 7 6 2 7 7 2 4 0 7 6 6 3 0 3 5 3

8. The table below gives the sequence of the genders of 99 students entering a university cafeteria (data collected by Nick Thill and Bryan Pick, 2009). Test for randomness in the sequence. Do students appear to randomly enter the cafeteria according to gender?

M M F F F F M M M F M F F F M M M F M M M M M F F
F F M M M M M M M M F F F F F F F M M M F M M M
F M M M M M M F F F F F F M M M M M M M M M M
F F F F F M F M F M M M M F F M M F F M M M F F F

9. In Example 7.5.1, we concluded that the songs are arranged randomly. This does *not* mean that the proportions of songs by each composer in the population of all songs chosen by the MP3 player are equal. Use a 1-proportion Z-test and the data in Table 7.6 to test the claim that these proportions are indeed equal (treat these data as if they were a random sample of all songs chosen by the MP3 player, even though the data are not random). (**Hint**: If both proportions are equal, what is the value of each?)

10. In this exercise, we outline the method for finding the critical values for the runs test at the 0.05 significance level. Let $n_1 = 8$ and $n_2 = 7$.

a. Use Theorem 7.5.1 to calculate $P(R = r) = f(r)$ for $r = 2, \ldots, 15$.
b. Calculate $P(R \leq r)$ and $P(R \geq r)$ for each value of r.

7.5 The Runs Test for Randomness

c. This is a two-tail test, so the critical values are numbers r_1 and r_2 such that $P(r_1 \leq R \leq r_2) = 1 - \alpha = 0.95$. Thus $P(R \leq r_1) = \alpha/2 = 0.025$ and $P(R \geq r_2) = 0.025$. There may not be numbers r_1 and r_2 such that these two equalities hold exactly, so r_1 is taken to be the largest value r such that $P(R \leq r) \leq 0.025$ and r_2 is taken to be the smallest value r such that $P(R \geq r) \leq 0.025$. Use the results of part b to find r_1 and r_2.

11. Use the procedure described in Exercise 10 to find the critical values for the runs test at the 0.10 significance level when $n_1 = 4$ and $n_2 = 7$.

12. In this exercise, we examine the pmf of R given in Theorem 7.5.1. We consider only the case where $R = r$ is even. By definition, $f(r)$ is the probability of exactly r runs. The numerator is the number of ways this can happen while the denominator is the total number of ways of arranging the $n_1 + n_2$ items.

a. Explain why this denominator is $\binom{n_1+n_2}{n_1}$.

b. If there are a total of r runs, explain why the number of runs of each type of item is $r/2$.

c. If there are a total of r runs, explain why these runs can be arranged in two different ways.

d. Each run of items could be of length 1 item, 2 items, up to some maximum length. Explain why the lengths of the runs of the first type of item can be chosen in $\binom{n_1-1}{r/2-1}$ ways and the lengths of the runs of the second type of item can be chosen in $\binom{n_2-1}{r/2-1}$ ways. (**Hint:** See Exercise 24 of Section 1.3.)

e. Explain why the numerator in $f(r)$ is $2\binom{n_1-1}{r/2-1}\binom{n_2-1}{r/2-1}$.

13. The runs test for randomness introduced in this section is based on the total number of runs. Another test for randomness is based on the length of the longest run. Let $n = n_1 + n_2$ denote the total number of items in the sequence, and let S denote the length of a run. It can be shown that if the sequence is random, then when $S \geq n/2$,

$$P(\text{at least one run of length } S \text{ or more}) = \frac{2 + 2(n-S)(S+1)}{(S+2)!}. \tag{7.2}$$

When $S < n/2$, the fraction on the right provides an overestimate of the probability on the left, but if the value of the fraction is small (that is, ≤ 0.05), then it is probably a good approximation. (See James V. Bradley, *Distribution-Free Statistical Tests*, Prentice-Hall,

Englewood Cliffs, N.J., 1968, pp. 274–277 and 280–281, for a proof of this.) This result leads to the following hypothesis test:

1. Find the length S of the longest run in the sequence and the total number of items n.
2. Calculate the probability of observing a run at least as long as S, using Equation (7.2).
3. If this probability is less than the desired significance level, then reject the claim of randomness.

In parts a–c, test the claim of randomness at the 0.05 significance level for each sequence with the given values of n and S.

a. $n = 9$, $S = 4$
b. $n = 40$, $S = 5$
c. $n = 1000$, $S = 7$
d. Apply this test to Example 7.5.2. Do you come to the same conclusion?
e. Try applying this test to Example 7.5.1. Does Equation (7.2) give a good estimate of the probability in this case? Explain.
f. If $n = 25$, find the largest value of S that would *not* contradict randomness at the 0.05 significance level.
g. Show that if $n \leq 1000$, then any run of length 7 or greater would contradict randomness at the 0.05 significance level.

APPENDIX A

Proofs of Selected Theorems

A.1 A Proof of Theorem 3.7.5

Let X and Y be continuous or discrete random variables. Define the joint cdf of X and Y to be

$$F_{XY}(x, y) = P(X \leq x, Y \leq y)$$

and the marginal cdfs to be

$$F_X(x) = P(X \leq x) \quad \text{and} \quad F_Y(y) = P(Y \leq y).$$

Then X and Y are independent if and only if

$$F_{XY}(x, y) = F_X(x) \cdot F_Y(y).$$

We use this idea to prove Theorem 3.7.5.

Theorem A.1.1 *Let X and Y be independent discrete or continuous random variables, and let $g(x)$ and $h(y)$ be any functions. Then the random variables $Z = g(X)$ and $W = h(Y)$ are independent.*

Proof. Let z and w be any elements in the ranges of Z and W, respectively. Also let R_X and R_Y denote the ranges of X and Y, respectively. Define the sets

$$A_z = \{x \in R_X : g(x) \leq z\} \quad \text{and} \quad B_w = \{y \in R_Y : h(x) \leq w\}.$$

Then the joint cdf of Z and W is

$$F_{ZW}(z, w) = P(Z \leq z, W \leq w) = P(g(X) \leq z, h(Y) \leq w) = P(X \in A_z, Y \in B_w).$$

But X and Y are independent, so

$$\begin{aligned} F_{ZW}(z,w) &= P(X \in A_z, Y \in B_w) = P(X \in A_z) \cdot P(Y \in B_w) \\ &= P(g(X) \le z) \cdot P(h(Y) \le w) \\ &= P(Z \le z) \cdot P(W \le w) \\ &= F_Z(z) \cdot F_W(w) \end{aligned}$$

which proves that Z and W are independent, as desired. □

A.2 A Proof of the Central Limit Theorem

A proof of the central limit theorem in its full generality is beyond the scope of the text. Below we give a proof assuming that the moment-generating function exists, which, as shown in Example 2.7.4, may not always be true. Other proofs use the *characteristic function*, whose existence is guaranteed. The proof we give relies on the following lemma from advanced probability theory.

Lemma A.2.1 *Let $S_1, S_2, \ldots$ be a sequence of random variables. Let $M_{S_i}(t)$ denote the moment-generating function and $F_{S_i}(s)$ denote the cumulative distribution function of the ith variable. Also, let S be another random variable with mgf $M_S(t)$ and cdf $F_S(s)$. If*

$$\lim_{n \to \infty} M_{S_n}(t) = M_S(t)$$

for all t in an open interval containing 0, then

$$\lim_{n \to \infty} F_{S_i}(t) = F_S(t)$$

for all $-\infty < s < \infty$. □

Informally, this lemma says that if the mgfs of the variables in the sequence approach the mgf of S, then the distributions of the variables in the sequence approach the distribution of S. This means that for any number b

$$\lim_{n \to \infty} P(S_n < b) = F_S(b).$$

The proof also uses the following results from elementary calculus and probability theory:

1. If the random variable X has mean μ, variance σ^2, and mgf $M_X(t)$, then the random variable $Y = (X - \mu)/\sigma$ has mean 0, variance 1, and mgf

$$M_Y(t) = e^{-\mu/(\sigma t)} M_X\left(\frac{t}{\sigma}\right).$$

2. By Theorem 3.7.3, if $Y_1, \ldots, Y_n$ are mutually independent random variables with a common mgf $M(t)$, then the mgf of

$$S_n = aY_1 + \cdots + aY_n$$

is $[M(at)]^n$.

3. By Taylor's theorem, if the first and second derivatives of a function $f(t)$ exist on an open interval containing 0, then for each t in this interval, there exists a number r between 0 and t such that

$$f(t) = f(0) + f'(0)t + \frac{f''(r)}{2} t^2.$$

4. If a random variable has the mgf $M(t)$, then all the derivatives of $M(t)$ exist at 0. Specifically, $M'''(0)$ exists. By results from elementary calculus, this implies that $M''(t)$ is continuous at $t = 0$.

5. If $\lim_{n \to \infty} b_n = b$, then

$$\lim_{n \to \infty} \left(1 + \frac{b_n}{n}\right)^n = e^b.$$

The central limit theorem says that if μ and σ^2 are the population mean and variance, respectively, then the distribution of the sample mean $\bar{X}_n$ approaches $N(\mu, \sigma^2/n)$ as $n \to \infty$. By result 1 from the above list, this is equivalent to saying that the distribution of

$$S_n = \frac{\bar{X}_n - \mu}{\sigma/\sqrt{n}}$$

approaches $N(0, 1)$. By Lemma A.2.1, we can do this by showing that the mgf of S_n approaches $e^{t^2/2}$, which is the mgf of a random variable that is $N(0, 1)$.

Theorem A.2.1   The Central Limit Theorem *If $X_1, \ldots, X_n$ are mutually independent random variables with a common mgf $M_X(t)$, mean μ, and variance σ^2, then as $n \to \infty$, the distribution of the sample mean*

$$\bar{X}_n = \frac{X_1 + \cdots + X_n}{n} \quad \text{approaches} \quad N\left(\mu, \frac{\sigma^2}{n}\right).$$

Proof. Using elementary algebra, we may write

$$S_n = \frac{\bar{X}_n - \mu}{\sigma/\sqrt{n}} = \sqrt{n}\left[\frac{(X_1 + \cdots + X_n)/n - \mu}{\sigma}\right]$$

$$= \frac{\sqrt{n}}{n}\left(\frac{X_1 + \cdots + X_n - n\mu}{\sigma}\right)$$

$$= \frac{1}{\sqrt{n}}\left(\frac{X_1 - \mu}{\sigma} + \cdots + \frac{X_n - \mu}{\sigma}\right)$$

$$= \frac{1}{\sqrt{n}}Y_1 + \cdots + \frac{1}{\sqrt{n}}Y_n$$

where $Y_i = (X_i - \mu)/\sigma$ for $i = 1, 2, \ldots n$. By result 1 from the list above, each Y_i has mean 0, variance 1, and the common mgf

$$M(t) = e^{-\mu/(\sigma t)} M_X\left(\frac{t}{\sigma}\right).$$

By result 2, the mgf of S_n is

$$M_{S_n}(t) = \left[M\left(\frac{t}{\sqrt{n}}\right)\right]^n.$$

Next note that for each i, $0 = E(Y_i) = M'(0)$ and

$$1 = \text{Var}(Y_i) = M''(0) - [M'(0)]^2 = M''(0).$$

Now, applying result 3 to $M(t)$, we get that, for each t in some open interval containing 0, there exists a number r between 0 and t such that

$$M(t) = M(0) + M'(0)t + \frac{M''(r)}{2}t^2 = 1 + \frac{t^2}{2}M''(r).$$

From this we get

$$M\left(\frac{t}{\sqrt{n}}\right) = 1 + \frac{(t/\sqrt{n})^2}{2}M''(s) = 1 + \frac{(t^2/2)M''(s_n)}{n}$$

for some number s_n between 0 and $t/\sqrt{n}$.

Thus we have

$$\lim_{n\to\infty} M_{S_n}(t) = \lim_{n\to\infty}\left[M\left(\frac{t}{\sqrt{n}}\right)\right]^n = \lim_{n\to\infty}\left[1 + \frac{(t^2/2)M''(s_n)}{n}\right]^n.$$

Now for each t, s_n is between 0 and $t/\sqrt{n}$. Thus $-|t|/\sqrt{n} < s_n < |t|/\sqrt{n}$, so by the squeeze theorem from elementary calculus,

$$\lim_{n\to\infty} s_n = \lim_{n\to\infty} \pm \frac{|t|}{\sqrt{n}} = 0.$$

Furthermore, by result 4, $M''(t)$ is continuous at 0. This means that

$$\lim_{n\to\infty} M''(s_n) = M''\left(\lim_{n\to\infty} s_n\right) = M''(0) = 1.$$

Thus we have

$$\lim_{n\to\infty} \left(\frac{t^2}{2}\right) M''(s_n) = \frac{t^2}{2}$$

so that by result 5,

$$\lim_{n\to\infty} M_{S_n}(t) = \lim_{n\to\infty} \left[1 + \frac{(t^2/2)\, M''(s)}{n}\right]^n = e^{t^2/2}$$

as desired. □

A.3 A Proof of the Limit Theorem of De Moivre and Laplace

This theorem states that if X is $b(n, p)$, then as $n \to \infty$, its distribution approaches $N(np, npq)$, where $q = 1 - p$. This is equivalent to saying that the distribution of

$$\frac{X - np}{\sqrt{npq}}$$

approaches $N(0, 1)$.

Theorem A.3.1 Limit Theorem of De Moivre and Laplace Let X be $b(n, p)$. Then as $n \to \infty$, the distribution of X approaches

$$N(np, npq)$$

where $q = 1 - p$.

Proof. By definition, X denotes the number of successes in a sequence of n Bernoulli trials. For $i = 1, 2, \ldots n$ define the random variables

$$X_i = \begin{cases} 1, & \text{if the } i\text{th trial is a success} \\ 0, & \text{otherwise.} \end{cases}$$

Then the number of successes in the n trials is the sum of these variables. That is,

$$X = X_1 + \cdots + X_n.$$

Note that for each i, X_i is $b(1, p)$, $E(X_i) = p$, $\text{Var}(X_i) = pq$, and by Example 2.7.3 each X_i has the same mgf, $M(t) = (q + pe^t)$. Furthermore, because the trials are independent, these variables are mutually independent. Thus by the central limit theorem, as $n \to \infty$, the distribution of the sample mean $\bar{X}_n = X_1 + \cdots + X_n$ approaches $N(p, pq/n)$. Stated another way, the distribution of

$$\frac{(X_1 + \cdots + X_n)/n - p}{\sqrt{pq}/\sqrt{n}}$$

approaches $N(0, 1)$. However, note that

$$\frac{(X_1 + \cdots + X_n)/n - p}{\sqrt{pq}/\sqrt{n}} = \frac{1}{n} \cdot \frac{(X_1 + \cdots + X_n) - np}{\sqrt{pq}/\sqrt{n}} = \frac{X - np}{\sqrt{npq}}$$

so that the distribution of $(X - np)/\sqrt{npq}$ also approaches $N(0, 1)$, as desired. $\square$

Note that in this proof we showed that the distribution of

$$\frac{(X_1 + \cdots + X_n)/n - p}{\sqrt{pq}/\sqrt{n}} = \frac{X/n - p}{\sqrt{pq/n}}$$

approaches $N(0, 1)$. This means that as $n \to \infty$, the distribution of the variable

$$\hat{P} = \frac{X}{n},$$

which is called the *sample proportion*, approaches $N(p, pq/n)$. This proves Theorem 4.4.1.

A.4 A Proof of Theorem 4.6.1

Theorem A.4.1 Let $\bar{X}$ denote the mean and S denote the standard deviation of a sample of size n taken from a normally distributed population with mean μ. The random variable

$$T = \frac{\bar{X} - \mu}{S/\sqrt{n}}$$

has a Student-t distribution with $(n-1)$ degrees of freedom.

Proof. From Theorem 3.8.1 we know that $\bar{X}$ is $N(\mu, \sigma^2/n)$ where σ^2 is the population variance. Thus the variable

$$Z = \frac{\bar{X} - \mu}{\sigma/\sqrt{n}}$$

is $N(0,1)$. From Theorem 4.7.1 we know that the variable

$$C = \frac{(n-1)S^2}{\sigma^2}$$

is $\chi^2(n-1)$. It can be shown that $\bar{X}$ and S are independent (the proof of this is beyond the scope of this text) so that Z and C are independent. Thus by Definition 3.9.5, the variable

$$T = \frac{Z}{\sqrt{C/(n-1)}}$$

has a Student-t distribution with $(n-1)$ degrees of freedom. Substituting and rewriting, we get

$$T = Z \cdot \sqrt{\frac{n-1}{C}} = \frac{\bar{X} - \mu}{\sigma/\sqrt{n}} \cdot \sqrt{\frac{n-1}{(n-1)S^2/\sigma^2}} = \frac{\bar{X} - \mu}{\sigma/\sqrt{n}} \cdot \frac{\sigma}{S} = \frac{\bar{X} - \mu}{S/\sqrt{n}}.$$

This proves the theorem. $\square$

A.5 Confidence Intervals for the Difference of Two Means

In this section, we discuss the theory behind the confidence intervals for the difference of two means described in the box titled *Small Sample Confidence Intervals for the Difference of two Means* in Section 4.8. The random variables, parameters, and statistics referenced below are the same as those described in the discussion preceding this box.

We begin with the case where the two population variances are assumed to be equal. Define $\sigma = \sigma_1^2 = \sigma_2^2$ to be the common variance. The random variable

$$Z = \frac{(\bar{X}_1 - \bar{X}_2) - (\mu_1 - \mu_2)}{\sqrt{\sigma^2/n_1 + \sigma^2/n_2}}$$

is $N(0, 1)$. By Theorem 4.7.1, the random variables

$$\frac{(n_1 - 1) S_1^2}{\sigma^2} \quad \text{and} \quad \frac{(n_2 - 1) S_2^2}{\sigma^2},$$

where S_1^2 and S_2^2 denote the sample variances of the samples taken from the first and second populations, respectively, are $\chi^2 (n_1 - 1)$ and $\chi^2 (n_2 - 1)$. Because the two samples are independent, these random variables are independent. Thus by Exercise 12 of Section 3.9, the variable

$$C = \frac{(n_1 - 1) S_1^2}{\sigma^2} + \frac{(n_2 - 1) S_2^2}{\sigma^2}$$

is $\chi^2 (n_1 + n_2 - 2)$. By Definition 3.9.5, the random variable

$$T = \frac{Z}{\sqrt{C/(n_1 + n_2 - 2)}}$$

has a Student-t distribution with $(n_1 + n_2 - 2)$ degrees of freedom. Replacing Z and C with their definitions yields

$$T = \frac{\dfrac{(\bar{X}_1 - \bar{X}_2) - (\mu_1 - \mu_2)}{\sqrt{\sigma^2/n_1 + \sigma^2/n_2}}}{\sqrt{\left[\dfrac{(n_1 - 1) S_1^2}{\sigma^2} + \dfrac{(n_2 - 1) S_2^2}{\sigma^2}\right] / (n_1 + n_2 - 2)}}.$$

Algebraically rearranging the right-hand side yields

$$T = \frac{(\bar{X}_1 - \bar{X}_2) - (\mu_1 - \mu_2)}{\sqrt{\left[\dfrac{(n_1 - 1) S_1^2 + (n_2 - 1) S_2^2}{n_1 + n_2 - 2}\right] \cdot \left[\dfrac{1}{n_1} + \dfrac{1}{n_2}\right]}}. \tag{A.1}$$

If $t_{\alpha/2}$ is a critical t-value with $(n_1 + n_2 - 2)$ degrees of freedom, we have

$$P(-t_{\alpha/2} \leq T \leq t_{\alpha/2}) = 1 - \alpha. \tag{A.2}$$

A.5 Confidence Intervals for the Difference of Two Means

Substituting Equation (A.1) into inequality (A.2) and rewriting the inequality so $\mu_1 - \mu_2$ is in the middle yields

$$P\left[\left(\bar{X}_1 - \bar{X}_2\right) - t_{\alpha/2}S_p\sqrt{\frac{1}{n_1} + \frac{1}{n_2}} \leq \mu_1 - \mu_2 \leq \left(\bar{X}_1 - \bar{X}_2\right) + t_{\alpha/2}S_p\sqrt{\frac{1}{n_1} + \frac{1}{n_2}}\right] = 1 - \alpha$$

where

$$S_p = \sqrt{\frac{(n_1 - 1)S_1^2 + (n_2 - 1)S_2^2}{n_1 + n_2 - 2}}.$$

This gives the margin of error as described in part 1 of the box.

If the two population variances are *not* assumed to be equal, things are much more complicated. If we knew their values, then the variable

$$Z = \frac{\left(\bar{X}_1 - \bar{X}_2\right) - (\mu_1 - \mu_2)}{\sqrt{\sigma_1^2/n_1 + \sigma_2^2/n_2}}$$

would be $N(0,1)$. We might be tempted to simply replace the population variances σ_1^2 and σ_2^2 with their respective sample variances S_1^2 and S_2^2, forming the variable

$$Y = \frac{\left(\bar{X}_1 - \bar{X}_2\right) - (\mu_1 - \mu_2)}{\sqrt{S_1^2/n_1 + S_2^2/n_2}}.$$

If we knew a critical value $c_{\alpha/2}$ such that

$$P\left(-c_{\alpha/2} \leq Y \leq c_{\alpha/2}\right) = 1 - \alpha,$$

then we could form a confidence interval by solving the inequality for $(\mu_1 - \mu_2)$, giving us

$$P\left[\left(\bar{X}_1 - \bar{X}_2\right) - c_{\alpha/2}\sigma \leq \mu_1 - \mu_2 \leq \left(\bar{X}_1 - \bar{X}_2\right) + c_{\alpha/2}\sigma\right] = 1 - \alpha$$

where $\sigma = \sqrt{S_1^2/n_1 + S_2^2/n_2}$. The problem is that we do not know the distribution of Y, so we do not know the appropriate critical values. In the paper "Tables for Use in Comparisons Whose Accuracy Involves Two Variances, Separately Estimated" (Alice A. Aspin, *Biometrika*, issue 36, 1949, pp. 290–296), Aspin gives a table of appropriate critical values. In an appendix to the paper, B. L. Welch gives a way of approximating these critical values by using critical t-values with r degrees of freedom, where r is

$$r = \frac{\left(s_1^2/n_1 + s_2^2/n_2\right)^2}{[1/(n_1 - 1)]\left(s_1^2/n_1\right)^2 + [1/(n_2 - 1)]\left(s_2^2/n_2\right)^2}.$$

A.6 Coefficients in the Linear Regression Equation

In this section we derive the following formulas for the least-squares estimates of m and b in the simple linear model $Y = mX + b + \epsilon$:

$$\hat{m} = \frac{n \sum x_i y_i - \sum x_i \sum y_i}{n \sum x_i^2 - (\sum x_i)^2} \quad \text{and} \quad \hat{b} = \bar{y} - \hat{m}\bar{x}.$$

We want $\hat{m}$ and $\hat{b}$ to minimize the quantity

$$S = \sum_{i=1}^{n}(y_i - \hat{y}_i)^2 = \sum_{i=1}^{n}\left[y_i - \left(\hat{m}x_i + \hat{b}\right)\right]^2.$$

A necessary condition for optimality is that the partial derivatives with respect to $\hat{m}$ and to $\hat{b}$ be zero. This gives the equations

$$\frac{\partial S}{\partial \hat{m}} = \sum 2\left(y_i - \hat{m}x_i - \hat{b}\right)(-x_i) = 0$$

$$\frac{\partial S}{\partial \hat{b}} = \sum 2\left(y_i - \hat{m}x_i - \hat{b}\right)(-1) = 0.$$

Dividing both these equations by -2 and rewriting, using the fact that $\hat{m}$ and $\hat{b}$ are both constants, yield the following two equations:

$$\sum x_i y_i - \hat{m}\sum x_i^2 - \hat{b}\sum x_i = 0 \qquad (A.3)$$

$$\sum y_i - \hat{m}\sum x_i - n \cdot \hat{b} = 0. \qquad (A.4)$$

Solving Equation (A.3) for $\hat{b}$ yields

$$\hat{b} = \frac{1}{n}\left(\sum y_i - \hat{m}\sum x_i\right) = \bar{y} - \hat{m}\bar{x}.$$

This gives the formula for $\hat{b}$. Multiplying Equation (A.3) by $-n$ and Equation (A.4) by $\sum x_i$ (assuming $\sum x_i \neq 0$) yields

$$-n\sum x_i y_i + n\cdot\hat{m}\sum x_i^2 + n\cdot\hat{b}\sum x_i = 0$$

$$\sum x_i \sum y_i - \hat{m}\left(\sum x_i\right)^2 - n\cdot\hat{b}\sum x_i = 0.$$

Adding these two equations together yields

$$-n\sum x_i y_i + \sum x_i \sum y_i + \hat{m}\left(n\sum x_i^2 - \left(\sum x_i\right)^2\right) = 0.$$

Solving this equation for $\hat{m}$ yields

$$\hat{m} = \frac{n\sum x_i y_i - \sum x_i \sum y_i}{n\sum x_i^2 - (\sum x_i)^2}.$$

This gives the formula for $\hat{m}$.

A.7 Wilcoxon Signed-Rank Test Distribution

Below we prove Theorem 7.3.1, and we also provide a method of finding the exact P-values and the critical values for the Wilcoxon signed-rank test.

First we note that the ranks take values between 1 and n. The value of S^+ is the sum of the positive signed ranks. Without loss of generality, assume the differences $d_1, \ldots, d_n$ are arranged in such a way that $|d_1| < \cdots < |d_n|$. For $i = 1, \ldots, n$, define the random variable

$$X_i = \begin{cases} i, & \text{if } d_i > 0 \\ 0, & \text{otherwise.} \end{cases}$$

Then

$$S^+ = X_1 + X_2 + \cdots + X_n. \tag{A.5}$$

Assuming that the population of differences is distributed symmetrically about the median $m = 0$, each of the signed-ranks is positive with probability $\frac{1}{2}$ and negative with probability $\frac{1}{2}$, and the sign of each is independent of the signs of others. Thus $P(X_i = i) = P(X_i = 0) = \frac{1}{2}$ for each i, and these variables are mutually independent.

The mean of X_i is

$$\mu_i = E(X_i) = i \cdot \frac{1}{2} + 0 \cdot \frac{1}{2} = \frac{i}{2}$$

and the variance of X_i is

$$\sigma_i^2 = \mathrm{Var}(X_i) = E\left[\left(X_i - \frac{i}{2}\right)^2\right] = \left(i - \frac{i}{2}\right)^2 \cdot \frac{1}{2} + \left(0 - \frac{i}{2}\right)^2 \cdot \frac{1}{2} = \frac{i^2}{4}. \tag{A.6}$$

Now, applying Theorems 3.7.1 and 3.7.2 to Equation (A.5) and using the relations $\sum_{i=1}^{n} i = n(n+1)/2$ and $\sum_{i=1}^{n} i^2 = n(n+1)(2n+1)/6$, we see that

$$E(S^+) = \sum_{i=1}^{n} \mu_i = \sum_{i=1}^{n} \frac{i}{2} = \frac{n(n+1)}{4}$$

and
$$\text{Var}\left(S^+\right) = \sum_{i=1}^{n} \sigma_i^2 = \sum_{i=1}^{n} \frac{i^2}{4} = \frac{n(n+1)(2n+1)}{24}. \quad (A.7)$$

Now to show that S^+ is approximately normally distributed, we use the following theorem attributed to Lyapunov (see R. V. Hogg and A. T. Craig, *Introduction to Mathematical Statistics*, 3rd ed., The Macmillan Company, New York, 1970, p. 362).

Theorem A.7.1 *Let $X_1, X_2, \ldots, X_n$ be a sequence of mutually independent random variables with respective means and variances μ_i and σ_i^2. If $E\left(|X_i - \mu_i|^3\right)$ is finite for each i and if*

$$\lim_{n \to \infty} \frac{\sum_{i=1}^{n} E\left(|X_i - \mu_i|^3\right)}{\left(\sum_{i=1}^{n} \sigma_i^2\right)^{3/2}} = 0$$

then the sequence of random variables

$$Z_n = \frac{\sum_{i=1}^{n} X_i - \sum_{i=1}^{n} \mu_i}{\sqrt{\sum_{i=1}^{n} \sigma_i^2}}$$

has a limiting distribution that is $N(0, 1)$. □

Informally, this theorem says that Z_n is approximately $N(0, 1)$ for each n, and this approximation becomes better as n gets larger. Applying this theorem to the variables X_i defined above, we see that

$$E\left(|X_i - \mu_i|^3\right) = \left|i - \frac{i}{2}\right|^3 \cdot \frac{1}{2} + \left|0 - \frac{i}{2}\right|^3 \cdot \frac{1}{2} = \frac{i^3}{8},$$

which is finite for each i. Using the relation $\sum_{i=1}^{n} i^3 = n^2(n+1)^2/4$, we see that

$$\sum_{i=1}^{n} E\left(|X_i - \mu_i|^3\right) = \sum_{i=1}^{n} \frac{i^3}{8} = \frac{n^2(n+1)^2}{32}.$$

Combining this with Equation (A.7), we see that

$$\lim_{n \to \infty} \frac{\sum_{i=1}^{n} E\left(|X_i - \mu_i|^3\right)}{\left(\sum_{i=1}^{n} \sigma_i^2\right)^{3/2}} = \lim_{n \to \infty} \frac{n^2(n+1)^2/32}{[n(n+1)(2n+1)/24]^{3/2}} = 0$$

because the numerator is of degree 4 and the denominator is of degree 9/2. Therefore, we conclude that

$$Z_n = \frac{\sum_{i=1}^n X_i - \sum_{i=1}^n \mu_i}{\sqrt{\sum_{i=1}^n \sigma_i^2}} = \frac{S^+ - n(n+1)/4}{\sqrt{n(n+1)(2n+1)/24}}$$

is approximately $N(0,1)$. Now, S^+ is a discrete random variable while the normal distribution is continuous. So to use this result to approximate the probabilities of S^+, we use a continuity correction as in Section 3.10.

$$P(S^+ \leq T) \approx P\left(Z \leq \frac{T + 0.5 - n(n+1)/4}{\sqrt{n(n+1)(2n+1)/24}}\right).$$

This establishes Theorem 7.3.1.

To find the exact P-value and the critical values for this test, we use moment-generating functions. By Theorem 2.7.1, the mgf of X_i is

$$\frac{1}{2}e^{0t} + \frac{1}{2}e^{it} = \frac{1}{2}\left(1 + e^{it}\right).$$

By Equation (A.5) and Theorem 3.7.3, the mgf of S^+ is

$$M_{S^+}(t) = \prod_{i=1}^n \frac{1}{2}\left(1 + e^{it}\right) = \frac{1}{2^n}\prod_{i=1}^n \left(1 + e^{it}\right).$$

To understand how to get a P-value from this mgf, suppose $n = 4$. Then

$$M_{S^+}(t) = \frac{1}{2^4}\prod_{i=1}^4 \left(1 + e^{it}\right)$$
$$= \frac{1}{2^4}\left(1 + e^{1t}\right)\left(1 + e^{2t}\right)\left(1 + e^{3t}\right)\left(1 + e^{4t}\right)$$
$$= \frac{1}{16}\left(1 + e^t + e^{2t} + 2e^{3t} + 2e^{4t} + 2e^{5t} + 2e^{6t} + 2e^{7t} + e^{8t} + e^{8t} + e^{9t} + e^{10t}\right).$$

From the coefficient of e^{3t} we see that $P(S^+ = 3) = \frac{2}{16} = \frac{1}{8}$. The probabilities of other values of S^+ are found similarly. Thus if the test statistic is $T = S^+ = 3$, for instance, then the P-value is

$$P\text{-value} = P(S^+ \leq 2) = \frac{1}{16} + \frac{1}{16} + \frac{1}{16} + \frac{1}{8} = \frac{5}{16} = 0.3125.$$

The critical value is the largest value of S^+ such that the corresponding P-value is less than the significance level.

APPENDIX B

Software Basics

B.1 Minitab

Minitab is a graphical-interface statistics package originally developed at Pennsylvania State University. All Minitab commands described in this text were tested in release 14.20.

The Minitab window is shown in Figure B.1. There are two main parts to the window. The top part is called the **session window**. The session window displays many of the outputs. The bottom part is called the **worksheet**. Data are entered into the worksheet in **rows** and **columns**. Each row is labeled with a number, and each column has a default name (C1, C2, ...), but columns can be labeled with user-specified names as well. To illustrate how to use Minitab, suppose we want to find the mean of the data $\{1, 2, 4, 5\}$. We first enter the data in a column, labeling it x, as in Figure B.1.

Then select **Calc** $\to$ **Column Statistics**, format the window as in Figure B.2, and press **OK**. The **Input variable** can be selected by clicking in the blank box and then double-clicking on the desired column in the box to the left. The result is displayed in the session window.

B.2 R

R is a programming language for doing statistical analysis. It can be freely downloaded from the website http://cran.r-project.org/bin/windows/base/. This is a very powerful programming language. The calculations we do in this text only begin to scratch the surface of R's capabilities. All R syntax described in this book was tested in version 2.13.1.

APPENDIX B Software Basics

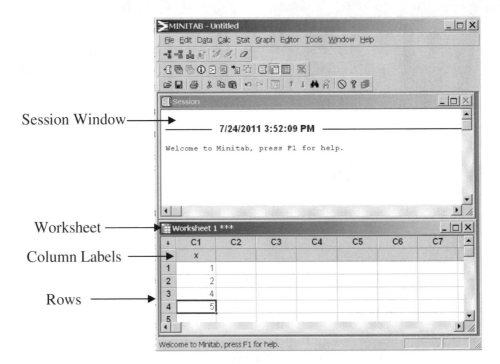

Figure B.1

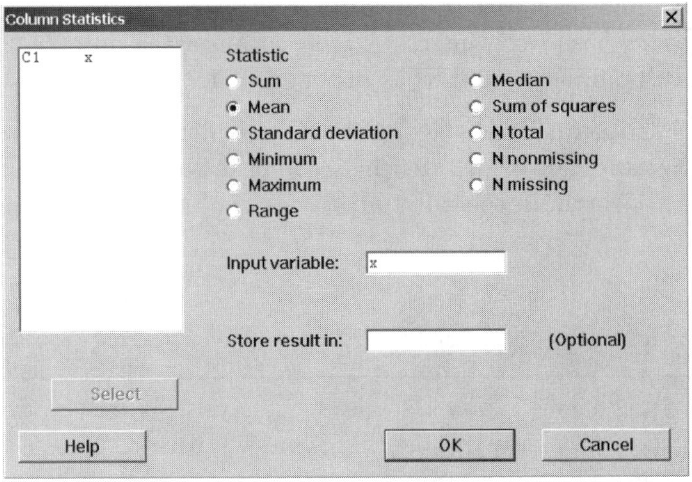

Figure B.2

All commands in R are entered next to the > prompt. Below we describe basic ways of entering data into R.

Vectors

Single lists of data are entered into a vector with the c command. For instance, to enter the data $\{1, 2, 4, 5\}$ into the vector x, we use the code

> x<−c(1,2,4,5)

We can then refer to this vector in functions. For instance, to find the mean of this set of data, we use the code

> mean(x)

To view the vector x, simply enter the code

> x

Information on a particular command or function can be found by typing

> help("*function*")

where *function* is the name of the command or function.

Data Frames

A *data frame* is a structure in which a set of data consisting of several different components can be stored. To illustrate a data frame, suppose we select five people and record the gender, shoe length, and height of each person as shown in Table B.1.

Gender	Shoe Length	Height
M	9	62
M	10	64
F	10.5	64.5
M	11	69
F	11.5	70

Table B.1

To enter these data in the data frame ShoeData, we begin by entering the code

> ShoeData<−data.frame(Gender=factor(1:5), ShoeLength=c(1:5), Height=c(1:5))

Then open the **data editor** by entering the code

> fix("ShoeData")

Then enter the data in the data editor as shown in Figure B.3. Close the data editor window, and ignore any warnings about added factor levels.

	Gender	ShoeLength	Height
1	M	9	62
2	M	10	64
3	F	10.5	64.5
4	M	11	69
5	F	11.5	70

Figure B.3

To view the entire data frame ShoeData, enter the code

> ShoeData

Individual components of the data frame can be viewed or referenced using the $ command. For instance, to view the height data and then calculate the mean, enter the code lines

> ShoeData$Height

> mean(ShoeData$Height)

B.3 Excel

Microsoft® Excel® has many built-in functions for doing statistical analysis. Most of the calculations in this text can be done using these functions. The capabilities of Excel can be extended with add-ins such as Data Desk XL. All Excel commands in this book were tested on Excel 2010, but they should work in other versions as well.

When you first open Excel 2010, you should get a window that looks similar to Figure B.4. (Your window may not look exactly like that in the figure due to customization of the ribbons.)

A **worksheet** consists of an array of **rows** and **columns**. The intersection of a row and column is called a **cell**, where data and functions are entered. An array of cells is called a **range**. To illustrate how to enter data and functions in a worksheet, suppose we have the data values $\{1, 2, 4, 5\}$ and we want to square each value and then find the mean of these squares. To do this, format a blank worksheet as in Figure B.5. The result is displayed in cell **C2** when Enter is pressed.

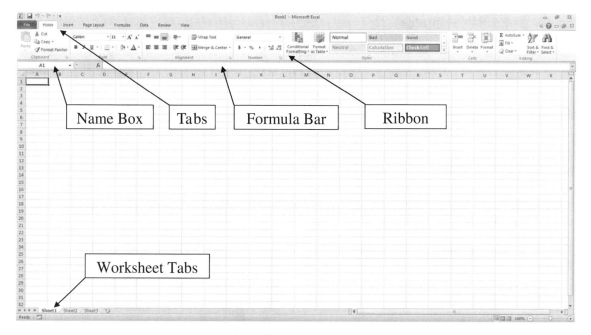

Figure B.4

Figure B.5

Note that each cell is referred to by its column and then row position, a range is referred to by (upper left):(lower right), and each function begins with an = sign. Whenever Excel syntax is described in this book, enter it in a blank cell starting with an = sign.

Some of the Excel commands described in this book are tools in the **Data Analysis** add-in. To install this add-in, select **File** → **Options** → **Add-Ins**. Near the bottom of the window, select **Manage: Excel Add-ins** and press **Go…**. Check the box beside **Analysis ToolPak** and press **OK**. In the **Data** tab, there should now be an **Analysis** section with a **Data Analysis** command.

B.4 TI-83/84 Calculators

The TI-83 and TI-84 calculators have almost identical statistical analysis functions. A few functions available in the TI-84 are not available in the TI-83. The capabilities of these calculators can be extended by installing various programs available on the Internet.

Most of the statistical functions are listed in the **STAT** menu. To illustrate how to use some of these functions, suppose we want to find the mean of the data $\{1, 2, 4, 5\}$. We enter the data in a **list** by pressing **STAT** and then selecting **1:EDIT...** under the **EDIT** option. This opens the list editor. Lists are labeled L1, L2, ..., L6 (if not all of these lists show in the editor, press **STAT** $\to$ **EDIT** $\to$ **5:SetUpEditor** and then press **ENTER** twice). The entries of a list can be cleared by highlighting the name of the list and then pressing **CLEAR**.

Enter the data as in the left half of Figure B.6. Then press **STAT** $\to$ **CALC** $\to$ **1:1-Var Stats**. Press Enter. This pastes the 1-Var Stats command at the home screen. Then enter the name of the list containing the data by pressing **2ND** and then **1** for list L1, **2** for list L2, etc. The home screen should look like the right half of Figure B.6. The output contains many results. The mean is labeled $\bar{x}$.

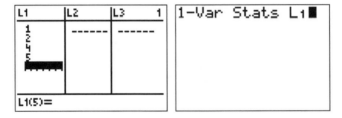

Figure B.6

APPENDIX C

Tables

Note: All tables were calculated using Microsoft Excel.

	Standard Normal Distribution $\Phi(z) = P(Z \leq z)$									
z	.00	.01	.02	.03	.04	.05	.06	.07	.08	.09
≤ -3.50	.0001									
−3.4	.0003	.0003	.0003	.0003	.0003	.0003	.0003	.0003	.0003	.0002
−3.3	.0005	.0005	.0005	.0004	.0004	.0004	.0004	.0004	.0004	.0003
−3.2	.0007	.0007	.0006	.0006	.0006	.0006	.0006	.0005	.0005	.0005
−3.1	.0010	.0009	.0009	.0009	.0008	.0008	.0008	.0008	.0007	.0007
−3.0	.0013	.0013	.0013	.0012	.0012	.0011	.0011	.0011	.0010	.0010
−2.9	.0019	.0018	.0018	.0017	.0016	.0016	.0015	.0015	.0014	.0014
−2.8	.0026	.0025	.0024	.0023	.0023	.0022	.0021	.0021	.0020	.0019
−2.7	.0035	.0034	.0033	.0032	.0031	.0030	.0029	.0028	.0027	.0026
−2.6	.0047	.0045	.0044	.0043	.0041	.0040	.0039	.0038	.0037	.0036
−2.5	.0062	.0060	.0059	.0057	.0055	.0054	.0052	.0051	.0049	.0048
−2.4	.0082	.0080	.0078	.0075	.0073	.0071	.0069	.0068	.0066	.0064
−2.3	.0107	.0104	.0102	.0099	.0096	.0094	.0091	.0089	.0087	.0084
−2.2	.0139	.0136	.0132	.0129	.0125	.0122	.0119	.0116	.0113	.0110
−2.1	.0179	.0174	.0170	.0166	.0162	.0158	.0154	.0150	.0146	.0143
−2.0	.0228	.0222	.0217	.0212	.0207	.0202	.0197	.0192	.0188	.0183
−1.9	.0287	.0281	.0274	.0268	.0262	.0256	.0250	.0244	.0239	.0233
−1.8	.0359	.0351	.0344	.0336	.0329	.0322	.0314	.0307	.0301	.0294
−1.7	.0446	.0436	.0427	.0418	.0409	.0401	.0392	.0384	.0375	.0367
−1.6	.0548	.0537	.0526	.0516	.0505	.0495	.0485	.0475	.0465	.0455
−1.5	.0668	.0655	.0643	.0630	.0618	.0606	.0594	.0582	.0571	.0559
−1.4	.0808	.0793	.0778	.0764	.0749	.0735	.0721	.0708	.0694	.0681
−1.3	.0968	.0951	.0934	.0918	.0901	.0885	.0869	.0853	.0838	.0823
−1.2	.1151	.1131	.1112	.1093	.1075	.1056	.1038	.1020	.1003	.0985
−1.1	.1357	.1335	.1314	.1292	.1271	.1251	.1230	.1210	.1190	.1170
−1.0	.1587	.1562	.1539	.1515	.1492	.1469	.1446	.1423	.1401	.1379
−0.9	.1841	.1814	.1788	.1762	.1736	.1711	.1685	.1660	.1635	.1611
−0.8	.2119	.2090	.2061	.2033	.2005	.1977	.1949	.1922	.1894	.1867
−0.7	.2420	.2389	.2358	.2327	.2296	.2266	.2236	.2206	.2177	.2148
−0.6	.2743	.2709	.2676	.2643	.2611	.2578	.2546	.2514	.2483	.2451
−0.5	.3085	.3050	.3015	.2981	.2946	.2912	.2877	.2843	.2810	.2776
−0.4	.3446	.3409	.3372	.3336	.3300	.3264	.3228	.3192	.3156	.3121
−0.3	.3821	.3783	.3745	.3707	.3669	.3632	.3594	.3557	.3520	.3483
−0.2	.4207	.4168	.4129	.4090	.4052	.4013	.3974	.3936	.3897	.3859
−0.1	.4602	.4562	.4522	.4483	.4443	.4404	.4364	.4325	.4286	.4247
−0.0	.5000	.4960	.4920	.4880	.4840	.4801	.4761	.4721	.4681	.4641

Table C.1 (continues)

	Standard Normal Distribution $\Phi(z) = P(Z \leq z)$									
	.00	.01	.02	.03	.04	.05	.06	.07	.08	.09
0.0	.5000	.5040	.5080	.5120	.5160	.5199	.5239	.5279	.5319	.5359
0.1	.5398	.5438	.5478	.5517	.5557	.5596	.5636	.5675	.5714	.5753
0.2	.5793	.5832	.5871	.5910	.5948	.5987	.6026	.6064	.6103	.6141
0.3	.6179	.6217	.6255	.6293	.6331	.6368	.6406	.6443	.6480	.6517
0.4	.6554	.6591	.6628	.6664	.6700	.6736	.6772	.6808	.6844	.6879
0.5	.6915	.6950	.6985	.7019	.7054	.7088	.7123	.7157	.7190	.7224
0.6	.7257	.7291	.7324	.7357	.7389	.7422	.7454	.7486	.7517	.7549
0.7	.7580	.7611	.7642	.7673	.7704	.7734	.7764	.7794	.7823	.7852
0.8	.7881	.7910	.7939	.7967	.7995	.8023	.8051	.8078	.8106	.8133
0.9	.8159	.8186	.8212	.8238	.8264	.8289	.8315	.8340	.8365	.8389
1.0	.8413	.8438	.8461	.8485	.8508	.8531	.8554	.8577	.8599	.8621
1.1	.8643	.8665	.8686	.8708	.8729	.8749	.8770	.8790	.8810	.8830
1.2	.8849	.8869	.8888	.8907	.8925	.8944	.8962	.8980	.8997	.9015
1.3	.9032	.9049	.9066	.9082	.9099	.9115	.9131	.9147	.9162	.9177
1.4	.9192	.9207	.9222	.9236	.9251	.9265	.9279	.9292	.9306	.9319
1.5	.9332	.9345	.9357	.9370	.9382	.9394	.9406	.9418	.9429	.9441
1.6	.9452	.9463	.9474	.9484	.9495	.9505	.9515	.9525	.9535	.9545
1.7	.9554	.9564	.9573	.9582	.9591	.9599	.9608	.9616	.9625	.9633
1.8	.9641	.9649	.9656	.9664	.9671	.9678	.9686	.9693	.9699	.9706
1.9	.9713	.9719	.9726	.9732	.9738	.9744	.9750	.9756	.9761	.9767
2.0	.9772	.9778	.9783	.9788	.9793	.9798	.9803	.9808	.9812	.9817
2.1	.9821	.9826	.9830	.9834	.9838	.9842	.9846	.9850	.9854	.9857
2.2	.9861	.9864	.9868	.9871	.9875	.9878	.9881	.9884	.9887	.9890
2.3	.9893	.9896	.9898	.9901	.9904	.9906	.9909	.9911	.9913	.9916
2.4	.9918	.9920	.9922	.9925	.9927	.9929	.9931	.9932	.9934	.9936
2.5	.9938	.9940	.9941	.9943	.9945	.9946	.9948	.9949	.9951	.9952
2.6	.9953	.9955	.9956	.9957	.9959	.9960	.9961	.9962	.9963	.9964
2.7	.9965	.9966	.9967	.9968	.9969	.9970	.9971	.9972	.9973	.9974
2.8	.9974	.9975	.9976	.9977	.9977	.9978	.9979	.9979	.9980	.9981
2.9	.9981	.9982	.9982	.9983	.9984	.9984	.9985	.9985	.9986	.9986
3.0	.9987	.9987	.9987	.9988	.9988	.9989	.9989	.9989	.9990	.9990
3.1	.9990	.9991	.9991	.9991	.9992	.9992	.9992	.9992	.9993	.9993
3.2	.9993	.9993	.9994	.9994	.9994	.9994	.9994	.9995	.9995	.9995
3.3	.9995	.9995	.9995	.9996	.9996	.9996	.9996	.9996	.9996	.9997
3.4	.9997	.9997	.9997	.9997	.9997	.9997	.9997	.9997	.9997	.9998
> 3.5	.9999									

Table C.1 (continued)

Critical z-Values

p	0.200	0.100	0.050	0.025	0.010	0.005
z_p	0.842	1.282	1.645	1.960	2.326	2.576

Table C.2

Critical T-Values $t_p(r)$

r	\multicolumn{5}{c}{p}	r	\multicolumn{5}{c}{p}
	0.005	0.01	0.025	0.05	0.1		0.005	0.01	0.025	0.05	0.1
1	63.657	31.821	12.706	6.314	3.078	40	2.704	2.423	2.021	1.684	1.303
2	9.925	6.965	4.303	2.920	1.886	45	2.690	2.412	2.014	1.679	1.301
3	5.841	4.541	3.182	2.353	1.638	50	2.678	2.403	2.009	1.676	1.299
4	4.604	3.747	2.776	2.132	1.533	55	2.668	2.396	2.004	1.673	1.297
5	4.032	3.365	2.571	2.015	1.476	60	2.660	2.390	2.000	1.671	1.296
6	3.707	3.143	2.447	1.943	1.440	65	2.654	2.385	1.997	1.669	1.295
7	3.499	2.998	2.365	1.895	1.415	70	2.648	2.381	1.994	1.667	1.294
8	3.355	2.896	2.306	1.860	1.397	75	2.643	2.377	1.992	1.665	1.293
9	3.250	2.821	2.262	1.833	1.383	80	2.639	2.374	1.990	1.664	1.292
10	3.169	2.764	2.228	1.812	1.372	85	2.635	2.371	1.988	1.663	1.292
11	3.106	2.718	2.201	1.796	1.363	90	2.632	2.368	1.987	1.662	1.291
12	3.055	2.681	2.179	1.782	1.356	95	2.629	2.366	1.985	1.661	1.291
13	3.012	2.650	2.160	1.771	1.350	100	2.626	2.364	1.984	1.660	1.290
14	2.977	2.624	2.145	1.761	1.345	110	2.621	2.361	1.982	1.659	1.289
15	2.947	2.602	2.131	1.753	1.341	120	2.617	2.358	1.980	1.658	1.289
16	2.921	2.583	2.120	1.746	1.337	130	2.614	2.355	1.978	1.657	1.288
17	2.898	2.567	2.110	1.740	1.333	140	2.611	2.353	1.977	1.656	1.288
18	2.878	2.552	2.101	1.734	1.330	150	2.609	2.351	1.976	1.655	1.287
19	2.861	2.539	2.093	1.729	1.328	160	2.607	2.350	1.975	1.654	1.287
20	2.845	2.528	2.086	1.725	1.325	170	2.605	2.348	1.974	1.654	1.287
21	2.831	2.518	2.080	1.721	1.323	180	2.603	2.347	1.973	1.653	1.286
22	2.819	2.508	2.074	1.717	1.321	190	2.602	2.346	1.973	1.653	1.286
23	2.807	2.500	2.069	1.714	1.319	200	2.601	2.345	1.972	1.653	1.286
24	2.797	2.492	2.064	1.711	1.318	250	2.596	2.341	1.969	1.651	1.285
25	2.787	2.485	2.060	1.708	1.316	300	2.592	2.339	1.968	1.650	1.284
26	2.779	2.479	2.056	1.706	1.315	350	2.590	2.337	1.967	1.649	1.284
27	2.771	2.473	2.052	1.703	1.314	400	2.588	2.336	1.966	1.649	1.284
28	2.763	2.467	2.048	1.701	1.313	450	2.587	2.335	1.965	1.648	1.283
29	2.756	2.462	2.045	1.699	1.311	1000	2.581	2.330	1.962	1.646	1.282
30	2.750	2.457	2.042	1.697	1.310	2000	2.578	2.328	1.961	1.646	1.282
35	2.724	2.438	2.030	1.690	1.306	Large	2.576	2.326	1.960	1.645	1.282

Table C.3

Chi-Square Critical Values $\chi^2_p(r)$

d.f. r	\multicolumn{9}{c}{p}								
	0.995	0.990	0.975	0.950	0.900	0.100	0.050	0.025	0.010	0.005
1	0.000	0.000	0.001	0.004	0.016	2.706	3.841	5.024	6.635	7.879
2	0.010	0.020	0.051	0.103	0.211	4.605	5.991	7.378	9.210	10.60
3	0.072	0.115	0.216	0.352	0.584	6.251	7.815	9.348	11.34	12.84
4	0.207	0.297	0.484	0.711	1.064	7.779	9.488	11.14	13.28	14.86
5	0.412	0.554	0.831	1.145	1.610	9.236	11.07	12.83	15.09	16.75
6	0.676	0.872	1.237	1.635	2.204	10.64	12.59	14.45	16.81	18.55
7	0.989	1.239	1.690	2.167	2.833	12.02	14.07	16.01	18.48	20.28
8	1.344	1.646	2.180	2.733	3.490	13.36	15.51	17.53	20.09	21.95
9	1.735	2.088	2.700	3.325	4.168	14.68	16.92	19.02	21.67	23.59
10	2.156	2.558	3.247	3.940	4.865	15.99	18.31	20.48	23.21	25.19
11	2.603	3.053	3.816	4.575	5.578	17.28	19.68	21.92	24.72	26.76
12	3.074	3.571	4.404	5.226	6.304	18.55	21.03	23.34	26.22	28.30
13	3.565	4.107	5.009	5.892	7.042	19.81	22.36	24.74	27.69	29.82
14	4.075	4.660	5.629	6.571	7.790	21.06	23.68	26.12	29.14	31.32
15	4.601	5.229	6.262	7.261	8.547	22.31	25.00	27.49	30.58	32.80
16	5.142	5.812	6.908	7.962	9.312	23.54	26.30	28.85	32.00	34.27
17	5.697	6.408	7.564	8.672	10.09	24.77	27.59	30.19	33.41	35.72
18	6.265	7.015	8.231	9.390	10.86	25.99	28.87	31.53	34.81	37.16
19	6.844	7.633	8.907	10.12	11.65	27.20	30.14	32.85	36.19	38.58
20	7.434	8.260	9.591	10.85	12.44	28.41	31.41	34.17	37.57	40.00
25	10.52	11.52	13.12	14.61	16.47	34.38	37.65	40.65	44.31	46.93
30	13.79	14.95	16.79	18.49	20.60	40.26	43.77	46.98	50.89	53.67
35	17.19	18.51	20.57	22.47	24.80	46.06	49.80	53.20	57.34	60.27
40	20.71	22.16	24.43	26.51	29.05	51.81	55.76	59.34	63.69	66.77
45	24.31	25.90	28.37	30.61	33.35	57.51	61.66	65.41	69.96	73.17
50	27.99	29.71	32.36	34.76	37.69	63.17	67.50	71.42	76.15	79.49
60	35.53	37.48	40.48	43.19	46.46	74.40	79.08	83.30	88.38	91.95
70	43.28	45.44	48.76	51.74	55.33	85.53	90.53	95.02	100.4	104.2
80	51.17	53.54	57.15	60.39	64.28	96.58	101.9	106.6	112.3	116.3
90	59.20	61.75	65.65	69.13	73.29	107.6	113.1	118.1	124.1	128.3
100	67.33	70.06	74.22	77.93	82.36	118.5	124.3	129.6	135.8	140.2

Table C.4

Critical F-Values $f_p(n,d)$

p	d	n=1	2	3	4	5	6	7	8	9	10
0.05	1	161.4	199.5	215.7	224.6	230.2	234.0	236.8	238.9	240.5	241.9
0.025		647.8	799.5	864.2	899.6	921.8	937.1	948.2	956.7	963.3	968.6
0.01		4052	4999	5403	5625	5764	5859	5928	5981	6022	6056
0.05	2	18.51	19.00	19.16	19.25	19.30	19.33	19.35	19.37	19.38	19.40
0.025		38.51	39.00	39.17	39.25	39.30	39.33	39.36	39.37	39.39	39.40
0.01		98.50	99.00	99.17	99.25	99.30	99.33	99.36	99.37	99.39	99.40
0.05	3	10.13	9.55	9.28	9.12	9.01	8.94	8.89	8.85	8.81	8.79
0.025		17.44	16.04	15.44	15.10	14.88	14.73	14.62	14.54	14.47	14.42
0.01		34.12	30.82	29.46	28.71	28.24	27.91	27.67	27.49	27.35	27.23
0.05	4	7.71	6.94	6.59	6.39	6.26	6.16	6.09	6.04	6.00	5.96
0.025		12.22	10.65	9.98	9.60	9.36	9.20	9.07	8.98	8.90	8.84
0.01		21.20	18.00	16.69	15.98	15.52	15.21	14.98	14.80	14.66	14.55
0.05	5	6.61	5.79	5.41	5.19	5.05	4.95	4.88	4.82	4.77	4.74
0.025		10.01	8.43	7.76	7.39	7.15	6.98	6.85	6.76	6.68	6.62
0.01		16.26	13.27	12.06	11.39	10.97	10.67	10.46	10.29	10.16	10.05
0.05	6	5.99	5.14	4.76	4.53	4.39	4.28	4.21	4.15	4.10	4.06
0.025		8.81	7.26	6.60	6.23	5.99	5.82	5.70	5.60	5.52	5.46
0.01		13.75	10.92	9.78	9.15	8.75	8.47	8.26	8.10	7.98	7.87
0.05	7	5.59	4.74	4.35	4.12	3.97	3.87	3.79	3.73	3.68	3.64
0.025		8.07	6.54	5.89	5.52	5.29	5.12	4.99	4.90	4.82	4.76
0.01		12.25	9.55	8.45	7.85	7.46	7.19	6.99	6.84	6.72	6.62
0.05	8	5.32	4.46	4.07	3.84	3.69	3.58	3.50	3.44	3.39	3.35
0.025		7.57	6.06	5.42	5.05	4.82	4.65	4.53	4.43	4.36	4.30
0.01		11.26	8.65	7.59	7.01	6.63	6.37	6.18	6.03	5.91	5.81
0.05	9	5.12	4.26	3.86	3.63	3.48	3.37	3.29	3.23	3.18	3.14
0.025		7.21	5.71	5.08	4.72	4.48	4.32	4.20	4.10	4.03	3.96
0.01		10.56	8.02	6.99	6.42	6.06	5.80	5.61	5.47	5.35	5.26
0.05	10	4.96	4.10	3.71	3.48	3.33	3.22	3.14	3.07	3.02	2.98
0.025		6.94	5.46	4.83	4.47	4.24	4.07	3.95	3.85	3.78	3.72
0.01		10.04	7.56	6.55	5.99	5.64	5.39	5.20	5.06	4.94	4.85

Table C.5 (continues)

Critical F-Values $f_p(n, d)$

p	d	n=1	2	3	4	5	6	7	8	9	10
0.05	12	4.75	3.89	3.49	3.26	3.11	3.00	2.91	2.85	2.80	2.75
0.025		6.55	5.10	4.47	4.12	3.89	3.73	3.61	3.51	3.44	3.37
0.01		9.33	6.93	5.95	5.41	5.06	4.82	4.64	4.50	4.39	4.30
0.05	15	4.54	3.68	3.29	3.06	2.90	2.79	2.71	2.64	2.59	2.54
0.025		6.20	4.77	4.15	3.80	3.58	3.41	3.29	3.20	3.12	3.06
0.01		8.68	6.36	5.42	4.89	4.56	4.32	4.14	4.00	3.89	3.80
0.05	20	4.35	3.49	3.10	2.87	2.71	2.60	2.51	2.45	2.39	2.35
0.025		5.87	4.46	3.86	3.51	3.29	3.13	3.01	2.91	2.84	2.77
0.01		8.10	5.85	4.94	4.43	4.10	3.87	3.70	3.56	3.46	3.37
0.05	24	4.26	3.40	3.01	2.78	2.62	2.51	2.42	2.36	2.30	2.25
0.025		5.72	4.32	3.72	3.38	3.15	2.99	2.87	2.78	2.70	2.64
0.01		7.82	5.61	4.72	4.22	3.90	3.67	3.50	3.36	3.26	3.17
0.05	30	4.17	3.32	2.92	2.69	2.53	2.42	2.33	2.27	2.21	2.16
0.025		5.57	4.18	3.59	3.25	3.03	2.87	2.75	2.65	2.57	2.51
0.01		7.56	5.39	4.51	4.02	3.70	3.47	3.30	3.17	3.07	2.98
0.05	40	4.08	3.23	2.84	2.61	2.45	2.34	2.25	2.18	2.12	2.08
0.025		5.42	4.05	3.46	3.13	2.90	2.74	2.62	2.53	2.45	2.39
0.01		7.31	5.18	4.31	3.83	3.51	3.29	3.12	2.99	2.89	2.80
0.05	60	4.00	3.15	2.76	2.53	2.37	2.25	2.17	2.10	2.04	1.99
0.025		5.29	3.93	3.34	3.01	2.79	2.63	2.51	2.41	2.33	2.27
0.01		7.08	4.98	4.13	3.65	3.34	3.12	2.95	2.82	2.72	2.63
0.05	120	3.92	3.07	2.68	2.45	2.29	2.18	2.09	2.02	1.96	1.91
0.025		5.15	3.80	3.23	2.89	2.67	2.52	2.39	2.30	2.22	2.16
0.01		6.85	4.79	3.95	3.48	3.17	2.96	2.79	2.66	2.56	2.47
0.05	200	3.89	3.04	2.65	2.42	2.26	2.14	2.06	1.98	1.93	1.88
0.025		5.10	3.76	3.18	2.85	2.63	2.47	2.35	2.26	2.18	2.11
0.01		6.76	4.71	3.88	3.41	3.11	2.89	2.73	2.60	2.50	2.41
0.05	∞	3.84	3.00	2.60	2.37	2.21	2.10	2.01	1.94	1.88	1.83
0.025		5.02	3.69	3.12	2.79	2.57	2.41	2.29	2.19	2.11	2.05
0.01		6.63	4.61	3.78	3.32	3.02	2.80	2.64	2.51	2.41	2.32

Table C.5 (continued)

		Critical F-Values $f_p(n,d)$									
		\multicolumn{10}{c}{n}
p	d	12	15	20	24	30	40	60	120	200	∞
0.05	1	243.9	245.9	248.0	249.1	250.1	251.1	252.2	253.3	253.7	254.3
0.025		976.7	984.9	993.1	997.2	1001	1006	1010	1014	1016	1018
0.01		6106	6157	6209	6235	6261	6287	6313	6339	6350	6366
0.05	2	19.41	19.43	19.45	19.45	19.46	19.47	19.48	19.49	19.49	19.50
0.025		39.41	39.43	39.45	39.46	39.46	39.47	39.48	39.49	39.49	39.50
0.01		99.42	99.43	99.45	99.46	99.47	99.47	99.48	99.49	99.49	99.50
0.05	3	8.74	8.70	8.66	8.64	8.62	8.59	8.57	8.55	8.54	8.53
0.025		14.34	14.25	14.17	14.12	14.08	14.04	13.99	13.95	13.93	13.90
0.01		27.05	26.87	26.69	26.60	26.50	26.41	26.32	26.22	26.18	26.13
0.05	4	5.91	5.86	5.80	5.77	5.75	5.72	5.69	5.66	5.65	5.63
0.025		8.75	8.66	8.56	8.51	8.46	8.41	8.36	8.31	8.29	8.26
0.01		14.37	14.20	14.02	13.93	13.84	13.75	13.65	13.56	13.52	13.46
0.05	5	4.68	4.62	4.56	4.53	4.50	4.46	4.43	4.40	4.39	4.37
0.025		6.52	6.43	6.33	6.28	6.23	6.18	6.12	6.07	6.05	6.02
0.01		9.89	9.72	9.55	9.47	9.38	9.29	9.20	9.11	9.08	9.02
0.05	6	4.00	3.94	3.87	3.84	3.81	3.77	3.74	3.70	3.69	3.67
0.025		5.37	5.27	5.17	5.12	5.07	5.01	4.96	4.90	4.88	4.85
0.01		7.72	7.56	7.40	7.31	7.23	7.14	7.06	6.97	6.93	6.88
0.05	7	3.57	3.51	3.44	3.41	3.38	3.34	3.30	3.27	3.25	3.23
0.025		4.67	4.57	4.47	4.41	4.36	4.31	4.25	4.20	4.18	4.14
0.01		6.47	6.31	6.16	6.07	5.99	5.91	5.82	5.74	5.70	5.65
0.05	8	3.28	3.22	3.15	3.12	3.08	3.04	3.01	2.97	2.95	2.93
0.025		4.20	4.10	4.00	3.95	3.89	3.84	3.78	3.73	3.70	3.67
0.01		5.67	5.52	5.36	5.28	5.20	5.12	5.03	4.95	4.91	4.86
0.05	9	3.07	3.01	2.94	2.90	2.86	2.83	2.79	2.75	2.73	2.71
0.025		3.87	3.77	3.67	3.61	3.56	3.51	3.45	3.39	3.37	3.33
0.01		5.11	4.96	4.81	4.73	4.65	4.57	4.48	4.40	4.36	4.31
0.05	10	2.91	2.85	2.77	2.74	2.70	2.66	2.62	2.58	2.56	2.54
0.025		3.62	3.52	3.42	3.37	3.31	3.26	3.20	3.14	3.12	3.08
0.01		4.71	4.56	4.41	4.33	4.25	4.17	4.08	4.00	3.96	3.91

Table C.5 (continued)

		Critical F-Values $f_p(n,d)$									
		\multicolumn{10}{c}{n}
p	d	12	15	20	24	30	40	60	120	200	∞
0.05	12	2.69	2.62	2.54	2.51	2.47	2.43	2.38	2.34	2.32	2.30
0.025		3.28	3.18	3.07	3.02	2.96	2.91	2.85	2.79	2.76	2.72
0.01		4.16	4.01	3.86	3.78	3.70	3.62	3.54	3.45	3.41	3.36
0.05	15	2.48	2.40	2.33	2.29	2.25	2.20	2.16	2.11	2.10	2.07
0.025		2.96	2.86	2.76	2.70	2.64	2.59	2.52	2.46	2.44	2.40
0.01		3.67	3.52	3.37	3.29	3.21	3.13	3.05	2.96	2.92	2.87
0.05	20	2.28	2.20	2.12	2.08	2.04	1.99	1.95	1.90	1.88	1.84
0.025		2.68	2.57	2.46	2.41	2.35	2.29	2.22	2.16	2.13	2.09
0.01		3.23	3.09	2.94	2.86	2.78	2.69	2.61	2.52	2.48	2.42
0.05	24	2.18	2.11	2.03	1.98	1.94	1.89	1.84	1.79	1.77	1.73
0.025		2.54	2.44	2.33	2.27	2.21	2.15	2.08	2.01	1.98	1.94
0.01		3.03	2.89	2.74	2.66	2.58	2.49	2.40	2.31	2.27	2.21
0.05	30	2.09	2.01	1.93	1.89	1.84	1.79	1.74	1.68	1.66	1.62
0.025		2.41	2.31	2.20	2.14	2.07	2.01	1.94	1.87	1.84	1.79
0.01		2.84	2.70	2.55	2.47	2.39	2.30	2.21	2.11	2.07	2.01
0.05	40	2.00	1.92	1.84	1.79	1.74	1.69	1.64	1.58	1.55	1.51
0.025		2.29	2.18	2.07	2.01	1.94	1.88	1.80	1.72	1.69	1.64
0.01		2.66	2.52	2.37	2.29	2.20	2.11	2.02	1.92	1.87	1.80
0.05	60	1.92	1.84	1.75	1.70	1.65	1.59	1.53	1.47	1.44	1.39
0.025		2.17	2.06	1.94	1.88	1.82	1.74	1.67	1.58	1.54	1.48
0.01		2.50	2.35	2.20	2.12	2.03	1.94	1.84	1.73	1.68	1.60
0.05	120	1.83	1.75	1.66	1.61	1.55	1.50	1.43	1.35	1.32	1.25
0.025		2.05	1.94	1.82	1.76	1.69	1.61	1.53	1.43	1.39	1.31
0.01		2.34	2.19	2.03	1.95	1.86	1.76	1.66	1.53	1.48	1.38
0.05	200	1.80	1.72	1.62	1.57	1.52	1.46	1.39	1.30	1.26	1.19
0.025		2.01	1.90	1.78	1.71	1.64	1.56	1.47	1.37	1.32	1.23
0.01		2.27	2.13	1.97	1.89	1.79	1.69	1.58	1.45	1.39	1.28
0.05	∞	1.75	1.67	1.57	1.52	1.46	1.39	1.32	1.22	1.17	1.00
0.025		1.94	1.83	1.71	1.64	1.57	1.48	1.39	1.27	1.21	1.00
0.01		2.18	2.04	1.88	1.79	1.70	1.59	1.47	1.32	1.25	1.00

Table C.5 (continued)

Sign Test Critical Values

n	0.005 (one-tail) 0.01 (two-tail)	0.01 (one-tail) 0.02 (two-tail)	0.02 (one-tail) 0.04 (two-tail)	0.025 (one-tail) 0.05 (two-tail)	0.05 (one-tail) 0.10 (two-tail)	0.10 (one-tail) 0.20 (two-tail)
4	★	★	★	★	★	0
5	★	★	★	★	0	0
6	★	★	0	0	0	0
7	★	0	0	0	0	1
8	0	0	0	0	1	1
9	0	0	1	1	1	2
10	0	0	1	1	1	2
11	0	1	1	1	2	2
12	1	1	2	2	2	3
13	1	1	2	2	3	3
14	1	2	2	2	3	4
15	2	2	3	3	3	4
16	2	2	3	3	4	4
17	2	3	3	4	4	5
18	3	3	4	4	5	5
19	3	4	4	4	5	6
20	3	4	4	5	5	6
21	4	4	5	5	6	7
22	4	5	5	5	6	7
23	4	5	6	6	7	7
24	5	5	6	6	7	8
25	5	6	6	7	7	8
26	6	6	7	7	8	9
27	6	7	7	7	8	9
28	6	7	8	8	9	10
29	7	7	8	8	9	10

Note: Reject H_0 if $x \leq$ the critical value. A ★ indicates there is no critical value.

Table C.6

Wilcoxon Signed-Rank Test Critical Values

	α				
	0.005 (one-tail)	0.010 (one-tail)	0.025 (one-tail)	0.050 (one-tail)	0.100 (one-tail)
n	0.010 (two-tail)	0.020 (two-tail)	0.050 (two-tail)	0.100 (two-tail)	0.200 (two-tail)
5	★	★	★	0	2
6	★	★	0	2	3
7	★	0	2	3	5
8	0	1	3	5	8
9	1	3	5	8	10
10	3	5	8	10	14
11	5	7	10	13	17
12	7	9	13	17	21
13	9	12	17	21	26
14	12	15	21	25	31
15	15	19	25	30	36
16	19	23	29	35	42
17	23	27	34	41	48
18	27	32	40	47	55
19	32	37	46	53	62
20	37	43	52	60	69
21	42	49	58	67	77
22	48	55	65	75	86
23	54	62	73	83	94
24	61	69	81	91	104
25	68	76	89	100	113
26	75	84	98	110	124
27	83	92	107	119	134
28	91	101	116	130	145
29	100	110	126	140	157
30	109	120	137	151	169

Note: Reject H_0 if $T \leq$ the critical value. A ★ indicates there is no critical value.

Table C.7

Wilcoxon Rank-Sum Test Critical Values

n_1	α one-tail	α two-tail	n_2=20	19	18	17	16	15	14	13	12	11	10	9	8	7	6	5	4	3
2	0.005	0.01	3	3																
2	0.025	0.05	5	5	5	5	4	4	4	4	4	3	3	3	3					
2	0.05	0.10	7	7	7	6	6	6	6	5	5	4	4	4	4	3	3	3		
3	0.005	0.01	9	9	8	8	8	8	7	7	7	6	6	6						
3	0.025	0.05	14	13	13	12	12	11	11	10	10	9	9	8	8	7	7	6		
3	0.05	0.10	17	16	15	15	14	13	13	12	11	11	10	10	9	8	8	7	6	6
4	0.005	0.01	18	17	16	16	15	15	14	13	13	12	12	11	11	10	10			
4	0.025	0.05	24	23	22	21	21	20	19	18	17	16	15	14	14	13	12	11	10	
4	0.05	0.10	28	27	26	25	24	22	21	20	19	18	17	16	15	14	13	12	11	
5	0.005	0.01	28	27	26	25	24	23	22	22	21	20	19	18	17	16	16	15		
5	0.025	0.05	35	34	33	32	30	29	28	27	26	24	23	22	21	20	18	17		
5	0.05	0.10	40	38	37	35	34	33	31	30	28	27	26	24	23	21	20	19		
6	0.005	0.01	39	38	37	36	34	33	32	31	30	28	27	26	25	24	23			
6	0.025	0.05	48	46	45	43	42	40	38	37	35	34	32	31	29	27	26			
6	0.05	0.10	53	51	49	47	46	44	42	40	38	37	35	33	31	29	28			
7	0.005	0.01	52	50	49	47	46	44	43	41	40	38	37	35	34	32				
7	0.025	0.05	62	60	58	56	54	52	50	48	46	44	42	40	38	36				
7	0.05	0.10	67	65	63	61	58	56	54	52	49	47	45	43	41	39				
8	0.005	0.01	66	64	62	60	58	56	54	53	51	49	47	45	43					
8	0.025	0.05	77	74	72	70	67	65	62	60	58	55	53	51	49					
8	0.05	0.10	83	80	77	75	72	69	67	64	62	59	56	54	51					
9	0.005	0.01	81	78	76	74	72	69	67	65	63	61	58	56						
9	0.025	0.05	93	90	87	84	82	79	76	73	71	68	65	62						
9	0.05	0.10	99	96	93	90	87	84	81	78	75	72	69	66						
10	0.005	0.01	97	94	92	89	86	84	81	79	76	73	71							
10	0.025	0.05	110	107	103	100	97	94	91	88	84	81	78							
10	0.05	0.10	117	113	110	106	103	99	96	92	89	86	82							

Table C.8 (continues)

Wilcoxon Rank-Sum Test Critical Values																				
	α		n_2																	
n_1	one-tail	two-tail	20	19	18	17	16	15	14	13	12	11	10	9	8	7	6	5	4	3
11	0.005	0.01	114	111	108	105	102	99	96	93	90	87								
	0.025	0.05	128	124	121	117	113	110	106	103	99	96								
	0.05	0.10	135	131	127	123	120	116	112	108	104	100								
12	0.005	0.01	132	129	125	122	119	115	112	109	105									
	0.025	0.05	147	143	139	135	131	127	123	119	115									
	0.05	0.10	155	150	146	142	138	133	129	125	120									
13	0.005	0.01	151	148	144	140	136	133	129	125										
	0.025	0.05	167	163	158	154	150	145	141	136										
	0.05	0.10	175	171	166	161	156	152	147	142										
14	0.005	0.01	172	168	163	159	155	151	147											
	0.025	0.05	188	183	179	174	169	164	160											
	0.05	0.10	197	192	187	182	176	171	166											
15	0.005	0.01	193	189	184	180	175	171												
	0.025	0.05	210	205	200	195	190	184												
	0.05	0.10	220	214	208	203	197	192												

Note: Reject H_0 if $R \leq$ critical value.

Table C.8 (continued)

Critical Values for the Runs Test for Randomness

n_1	n_2=2	3	4	5	6	7	8	9	10	11	12	13	14	15	16	17	18	19	20	21	22	23	24	25
2	1 5	1 6	1 6	1 6	1 6	1 6	1 6	1 6	1 6	1 6	2 6	2 6	2 6	2 6	2 6	2 6	2 6	2 6	2 6	2 6	2 6	2 6	2 6	2 6
3	1 6	1 7	1 8	1 8	2 8	2 8	2 8	2 8	2 8	2 8	2 8	3 8	3 8	3 8	3 8	3 8	3 8	3 8	3 8	3 8	3 8	3 8	3 8	3 8
4	1 6	1 8	1 9	2 9	2 9	2 10	3 10	3 10	3 10	3 10	3 10	3 10	3 10	4 10	4 10	4 10	4 10	4 10	4 10	4 10	4 10	4 10	4 10	4 10
5	1 6	1 8	2 9	2 10	3 10	3 11	3 11	3 12	3 12	4 12	4 12	4 12	4 12	4 12	4 12	5 12	5 12	5 12	5 12	5 12	5 12	5 12	5 12	5 12
6	1 6	2 8	2 9	3 10	3 11	3 12	3 12	4 13	4 13	4 13	4 13	5 14	5 14	5 14	5 14	5 14	5 14	6 14	6 14	6 14	6 14	6 14	6 14	6 14
7	1 6	2 8	2 10	3 11	3 12	3 13	4 13	4 14	5 14	5 14	5 14	5 15	5 15	6 15	6 16	6 16	6 16	6 16	6 16	7 16	7 16	7 16	7 16	7 16
8	1 6	2 8	3 10	3 11	3 12	4 13	4 14	5 14	5 15	5 15	6 16	6 16	6 16	6 16	7 17	7 17	7 17	7 17	7 17	8 18	8 18	8 18	8 18	8 18
9	1 6	2 8	3 10	3 12	4 13	4 14	5 14	5 15	5 16	6 16	6 16	6 17	7 17	7 18	7 18	7 18	8 18	8 18	8 19	8 19	8 19	9 19	9 19	9 19
10	1 6	2 8	3 10	3 12	4 13	5 14	5 15	5 16	6 16	6 17	7 17	7 18	7 18	7 18	8 19	8 19	8 19	8 20	9 20	9 20	9 20	9 20	9 20	10 20
11	1 6	2 8	3 10	4 12	4 13	5 14	5 15	6 16	6 17	7 17	7 18	7 19	8 19	8 19	9 20	9 20	9 20	9 21	10 21	10 21	10 22	10 22	10 22	10 22
12	2 6	2 8	3 10	4 12	4 13	5 14	6 16	6 16	7 17	7 18	7 19	8 19	8 20	8 20	9 21	9 21	9 21	10 22	10 22	10 22	10 22	11 23	11 23	11 23
13	2 6	2 8	3 10	4 12	5 14	5 15	6 16	6 17	7 18	7 19	8 19	8 20	9 20	9 21	9 21	10 22	10 22	10 23	10 23	11 23	11 24	11 24	11 24	12 24
14	2 6	3 8	3 10	4 12	5 14	5 15	6 16	7 17	7 18	8 19	8 20	9 20	9 21	9 22	10 22	10 23	10 23	11 23	11 24	11 24	12 24	12 25	12 25	12 25
15	2 6	3 8	3 10	4 12	5 14	6 15	6 16	7 18	7 18	8 19	8 20	9 21	9 22	10 22	10 23	11 23	11 24	11 24	12 25	12 25	12 25	13 26	13 26	13 26

Table C.9 (continues)

Critical Values for the Runs Test for Randomness

n_1	n_2 = 2	3	4	5	6	7	8	9	10	11	12	13	14	15	16	17	18	19	20	21	22	23	24	25
16	2	3	4	4	5	6	6	7	8	8	9	9	10	10	11	11	11	12	12	12	13	13	13	14
	6	8	10	12	14	16	17	18	19	20	21	21	22	23	23	24	25	25	25	26	26	27	27	27
17	2	3	4	4	5	6	7	7	8	9	9	10	10	11	11	11	12	12	13	13	13	14	14	14
	6	8	10	12	14	16	17	18	19	20	21	22	23	23	24	25	25	26	26	27	27	27	28	28
18	2	3	4	5	5	6	7	8	8	9	9	10	10	11	11	12	12	13	13	13	14	14	14	15
	6	8	10	12	14	16	17	18	19	20	21	22	23	24	25	25	26	26	27	27	28	28	29	29
19	2	3	4	5	6	6	7	8	8	9	10	10	11	11	12	12	13	13	13	14	14	15	15	15
	6	8	10	12	14	16	17	18	20	21	22	23	23	24	25	26	26	27	27	28	29	29	29	30
20	2	3	4	5	6	6	7	8	9	9	10	10	11	12	12	13	13	13	14	14	15	15	15	16
	6	8	10	12	14	16	17	18	20	21	22	23	24	25	25	26	27	27	28	29	29	30	30	31
21	2	3	4	5	6	7	7	8	9	10	10	11	11	12	12	13	13	14	14	15	15	16	16	16
	6	8	10	12	14	16	18	19	20	21	22	23	24	25	26	27	27	28	29	29	30	30	31	31
22	2	3	4	5	6	7	8	8	9	10	10	11	12	12	13	13	14	14	15	15	16	16	16	17
	6	8	10	12	14	16	18	19	20	22	22	24	24	25	26	27	28	29	29	30	30	31	31	32
23	2	3	4	5	6	7	8	8	9	10	11	11	12	12	13	14	14	15	15	16	16	16	17	17
	6	8	10	12	14	16	18	19	20	22	23	24	25	26	27	27	28	29	30	30	31	32	32	33
24	2	3	4	5	6	7	8	9	9	10	11	11	12	13	13	14	14	15	15	16	16	17	17	18
	6	8	10	12	14	16	18	19	20	22	23	24	25	26	27	28	29	29	30	31	31	32	33	33
25	2	3	4	5	6	7	8	9	10	10	11	12	12	13	14	14	15	15	16	16	17	17	18	18
	6	8	10	12	14	16	18	19	20	22	23	24	25	26	27	28	29	30	31	31	32	33	33	34

Notes:
1. Reject H_0 if the total number of runs R is less than or equal to the smaller critical value or greater than or equal to the larger critical value.
2. Critical values are for the 0.05 significance level only.

Table C.9 (continued)

APPENDIX D

Answers to Selected Exercises

Section 1.2

1. $P(A) = \frac{7}{8}$

3. a. $\frac{1}{365}$, b. $\frac{1}{12}$

5. a. $\frac{5}{7}$ b. $\frac{6.5}{7}$ c. $\frac{5}{7}$ d. $\frac{5}{7}$ e. $\frac{1}{7}$
f. 0

11. $\frac{8}{81}, \frac{16}{81}, \frac{24}{81}, \frac{32}{81}$, $P(\text{Bullseye}) = \frac{1}{81}$

Section 1.3

1. a. 17,576,000 b. 11,232,000 c. 15,210,000

3. a. 80 b. 20 c. 64 d. 5120 e. 3

5. a. 2730 b. 455

7. 56

9. 90

13. 2^n

15. a. 0.00024 b. 0.00144 c. 0.0211

17. a. $\frac{9}{91}$ b. $\frac{3}{364}$ c. $\frac{1}{364}$ d. $\frac{3}{182}$

19. 24, 12, 4

21. a. 1000 b. $\frac{3}{500}$ c. $\frac{3}{1000}$ d. $\frac{1}{1000}$

23. $2/n$

Section 1.4

9. 0.4

13. 0.6181

17. 0.25

Section 1.5

1. a. 0.335, 0.525, 0.494 b. 0.208, 0.208

3. a. $\frac{44}{91}$ b. $\frac{1}{455}$ c. $\frac{66}{455}$ d. $\frac{198}{455}$

5. $\frac{3}{1225}$

7. a. $\frac{5}{33}$ b. $\frac{35}{66}$ c. $\frac{7}{22}$

9. a. $\frac{1}{5}$ b. $\frac{1}{5}$ c. $\frac{1}{20}$

Section 1.6

3. 0.16

5. $\frac{1}{4}$

7. 0.1199

9. $\frac{1}{6}$

11. $\frac{81}{82}$

13. 128, 1

Section 1.7

3. a. 0.4437 b. 9

5. 4

7. a. 0.00879 b. 0.00879 c. 10 d. 0.08789

9. b. 0.9035 c. 0.9039

11. 4

13. $p\left(4p^2 - 6p + 3\right)$

Section 2.2

1. $\frac{1}{4}, \frac{1}{2}, \frac{29}{36}, \frac{1}{36}$

3. a. 6 b. $\frac{2}{3}$ c. 0.44

7. $F(x) = \lfloor x \rfloor / k$ for $0 \leq x < k+1$; 0 for $x < 0$, 1 for $x \geq k+1$

15. $f(x) = \left(\frac{1}{2}\right)^x$

Section 2.3

1. a. 0.274 b. 0.930 c. 0.0055

3. 0.472

5. 0.1757

7. 0.15

9. 0.918

11. a. 0.798 b. 0.0000

13. b. 0.0142 c. 0.9971

15. 0.0330

17. a. 0.147 b. 0.002

Section 2.4

3. 0.0067

5. a. 0.224 b. 0.0446 c. 0.0446

7. 0.393

11. a. 0.271 b. 0.138 c. 0.677

13. 0.0768

Section 2.5

1. a. 2.3, 1.21, 1.1 b. 6, $\frac{112}{3}$, 6.11 c. 1, 0, 0 d. 1, $\frac{1}{2}$, 0.707

3. 22.5

5. 4

7. $\frac{1}{2}$, $\frac{3}{8}$

11. −$0.25

13. a. $\frac{200}{3}$ c. 204

15. 10

Section 2.6

1. $\frac{13}{8}$, $\frac{31}{8}$, $\frac{79}{64}$, $\frac{39}{4}$, 2.564

3. a. $\frac{5}{2}$ b. $\frac{5}{2} - n$ c. $\frac{5}{2}$

9. 0, 1

11. 0

15. a. 16.1 b. 23.1, 23.6, 20.2 c. 3

Section 2.7

1. a. $\frac{13}{5}$, $\frac{86}{25}$ b. 3, 2

11. a. 4.78 b. 97.8 d. $p < 0.206$

13. r/p

19. $1/p$

Section 3.2

1. d. 0.969, 0, 0.237

3. d. $\frac{1}{8}$, 0, $\frac{15}{16}$

5. a. 5, $\frac{25}{3}$ b. 2, $\frac{1}{5}$ c. 5, 25 d. $\frac{3}{2}$, $\frac{3}{4}$

7. a. $(x-5)/x$ b. $\frac{50}{9}$, 500

9. b. 2000, 2000 c. 0.06075 d. 2608.4

11. a. $\frac{9}{16}$ b. $\frac{7}{8}$ c. $\frac{41}{64}$ d. $\frac{7}{9}$

13. $\left(e^{4t} - e^t\right)/(3t)$ for $t \neq 0$, 1 for $t = 0$

Section 3.3

3. a. 20 b. $\frac{25}{3}$ c. $\frac{1}{2}$ d. $\frac{3}{10}$

7. a. $d = 0.75b + 0.25a$

9. a. $\frac{1}{4}$ b. $\frac{1}{2}$

11. 0.264

15. a. 1.15 b. 29.96 c. $\pi_p = \ln(1-p)/(-\lambda)$

17. $(-\ln 0.5)/\lambda$, $0.5(a+b)$

Section 3.4

1. a. 0.0374 b. 0.7747 c. 0.0124 d. 0.7924

3. a. 50.88% b. 168.4 c. 219.7

7. a. 1.645 b. 1.96 c. 2.575

9. 6.69%

11. 0.1311

19. 0.54917

Section 3.5

3. $1/8 e^{\sqrt{y}/2}$

5. $\sqrt{6} - \sqrt{5}$

7. $1/(9\sqrt{y})$ for $0 \le y \le 16$, $1/(18\sqrt{y})$ for $16 < y \le 25$

11. $1/(y\sigma\sqrt{2\pi}) e^{-(\ln y - \mu)^2/(2\sigma^2)}$

Section 3.6

1. c. 1, $\frac{7}{10}$

3. b. $f_X(x) = (9x+6)/39$, $f_Y(y) = (9+2y)/39$ c. $\frac{32}{39}, \frac{20}{39}, \frac{32}{39}$

5. a. $\frac{4}{9}$ b. $\frac{1}{18}$ c. $\frac{1}{2}$ d. $\frac{25}{72}$

7. $f_X(x) = x + 0.5$, $f_Y(y) = y + 0.5$

9. a. $f_X(x) = 2x$, $f_Y(y) = 3y^2$ c. $\frac{1}{10}$

15. a. 0.03823 b. 0.9797

Section 3.7

1. a. $\frac{17}{12}$ b. $\frac{2}{3}$ c. $\frac{3}{4}$ d. $\frac{5}{4}$

3. 5

5. Does not exist

9. c. $\frac{6}{5}$

13. b. 0.0225, 0.0075

15. 5

Section 3.8

1. a. 0.3745 b. 0.1772 c. no
3. 0.0003
5. b. The probability gets larger
7. a. 0
9. 0.2434
11. c. 0.2642 d. 0.2389
13. 0.2821

Section 3.9

1. a. 12, 72 b. 0.7127 c. 0.4754
3. 1.225, 0.919, 1.608
5. 0.975
9. 0.96875
15. a. 0.975 b. 0.01 c. 0.04

Section 3.10

3. b. 0.5561
5. 0
7. 243

Section 4.2

3. a. $\bar{x} = 0.89037$, $s = 0.04562$
13. b. 1.095, 1.67, 1.21
17. a. Medians all equal 2.5
21. $\bar{x} \approx 2.1$, $s \approx 1.73$

Section 4.3

1. $\frac{14}{3}$
3. $\hat{\theta} = 1.749$

Section 4.4

1. a. 0.592 b. 0
5. a. 0.0808 b. 0.0808, 0.0239, 0.0026, 0.0000
7. a. 7, 3 c. 3.11, 4.667

Section 4.5

1. a. 0.255, 0.84 b. 0.525, 1.035 c. 2.515, 2.81

7. [0.593, 0.827]

9. [0.546, 0.734], [0.573, 0.707], [0.598, 0.682]

11. a. [0.537, 0.607]

13. a. [0.496, 0.584] b. [0.509, 0.571] c. [0.5186, 0.56184]

15. a. [0.780, 1]

17. 0.431

19. a. [0.276, 0.539] b. [0.276, 0.538]

Section 4.6

1. a. [68.86, 92.34]

3. a. [213.7, 256.3], [219.4, 250.6]

5. a. [1.629, 1.67]

7. a. [29.4, 134.1], [−0.87, 59.9]

11. a. [2.081, 2.520] b. [9.88, 10.12], [9.83, 10.17]

Section 4.7

1. [1.600, 4.124]

3. [0.0348, 0.0898]

5. a. [21.60, 68.71] b. [9.15, 59.40]

9. a. 73.77, 129.07 b. 3.949, 31.682 c. [1.961, 2.873]

Section 4.8

1. [−0.146, 0.0059]

3. [−0.0863, 0.0223]

7. [7.795, 21.545]

9. a. [−0.0744, 0.2344] b. [−0.079, 0.239], [−0.0764, 0.23638]

11. [−3.92, 0.024]

13. [−2.084, 2.806]

15. [−0.169, 1.369]

19. [21.46, ∞)

Section 4.9

3. 601

5. 59

7. a. 0.325 b. 71

11. a. 16,590

13. 21

15. a. $n = (E/z)^2 \left[p_1 (1 - p_1) + p_2 (1 - p_2) \right]$

APPENDIX D Answers to Selected Exercises

Section 5.1

5. a. 0.1075 b. 0.0076 c. 0.1922

7. P-value $= 0.0028$

9. a. 0.283, -0.566, 2.828, -3.677, 0

Section 5.2

1. P-value $= 0.042$

3. P-value $= 0.2327$

5. P-value $= 0.1190$

7. P-value $= 0.0004$

9. P-value $= 0.0307$

11. P-value $= 0.1814$

15. $[0.644, 0.727]$, $[0, 0.236]$

17. 0.8078

Section 5.3

1. a. Exact $= 0.0302$ b. Exact $= 0.0647$ c. Exact $= 0.1081$

3. $z = 1.04$

5. $z = -5.29$

7. $t = 12.25$

9. b. $t = 2.30$

17. 0.6141

Section 5.4

3. $z = -1.15$

5. $z = 0.41$

7. a. $z = 1.81$ b. $z = 1.47$ c. $z = -0.94$

9. a. Females $(0.534, 0.666)$, Males $(0.433, 0.567)$ b. $z = 1.74$

11. $z = 2.03$

Section 5.5

1. $f = 5.27$

3. $f = 1.79$, P-value $= 0.0236$

5. a. $F_R = 3.61$, $f_L = 0.214$
b. $f_R = 2.84$, $f_L = 0.272$

7. a. $[18.43, 78.97]$ b. $[28.90, 50.36]$

9. $[0.795, 6.500]$

11. $[0.518, 3.748]$

Section 5.6

1. $z = 1.02$
3. $t = 0.053$, $s_p = 2.185$
5. $t = -2.28$
7. P-values $= 0.2076, 0.2100$
9. $t = 4.061$
11. $t = -2.86$, $r = 44$
13. Yes
15. $t = 1.77$
17. $t = -2.17$
19. b. $f = 1.39$ c. $s_p = 0.0207$, $t = 11.24$

Section 5.7

1. $c = 25.22$
3. $c = 5.36$
5. $c = 8.12$
7. a. $c = 25.225$
9. $c = 1.917$
11. a. P-value $= 0.4238$

Section 5.8

3. $c = 79.43$
5. $c = 4.72$
7. $c = 5.97$
9. $c = 4.30$
11. P-value $= 0.0472$

Section 5.9

1. $F = 0.942$
3. $F = 11.72$
7. $F = 1.355$
9. $[70.4, 91.2]$, $[96.3, 117.5]$, $[112.8, 134.0]$, $[100.6, 112.8]$
11. d. $F = 9.33$

Section 5.10

1. $F_1 = 16.1$, $F_2 = 17.3$
3. $F_1 = 0.035$, $F_2 = 19.86$
5. $F_B = 8.93$, $F_A = 1.63$, $F_{AB} = 0.93$
7. $F_B = 0.024$, $F_A = 1.180$, $F_{AB} = 0.023$

Section 6.2

3. b. $E(XY) = E(X) = 0$, $E(Y) = 4/3$, $\text{Cov}(X, Y) = 0$

7. b. 0.6025

9. $\rho = 0$

13. $r = 0.855$

15. $r = \pm 1$

17. $0.668 \leq \rho \leq 0.811$

Section 6.3

1. $y = -191.74 + 10.89x$, $\hat{y} = 91.5$

3. $\hat{y} = 73.265$

5. $\hat{y} = 27.17$

7. b. $\hat{m} = 3.32$

11. a. 2.059 **b.** 1.75

Section 6.4

3. $(114.3, 191.3)$

5. $(9.05, 12.74)$

9. $(17.414, 34.264)$

11. a. $3.878x + 25.839 \pm 2.825\sqrt{1.1 + (x - 11.4)^2/14.775}$

Section 6.5

1. $\text{SS}_{\text{Reg}} = 105.4$, $\text{SS}_{\text{Res}} = 7.297$, $\text{SS}_{\text{Tot}} = 112.75$, $r^2 = 0.935$, $f = 28.9$

3. a. $y = 0.8084x + 4.3927$ **b.** $r^2 = 0.0824$ **c.** $f = 0.808$

Section 6.6

1. a. $a = 2.0042$, $b = 15.173$, $R^2 = 0.9797$
b. $a = 3.261$, $b = -1.3905$, $R^2 = 0.9997$
c. $a = 1.2313$, $b = 5.5803$, $R^2 = 1$

3. $y = 103.4 - 5.2876 \ln x$, $R^2 = 0.8255$

7. $a = 4.0686$, $b = -0.5211$, $R^2 = 0.992$

Section 6.7

1. Area: $R^2 = 0.2232$, Acres: $R^2 = 0.1249$, Bedrooms: $R^2 = 0.0036$

11. $y = -0.28x_1 + 1.84x_2 + 41.15$

13. $(69.338, 78.813)$

15. $y = 1.49x_1 + 6.03x_2 - 0.22x_1x_2 - 28.27$, $R^2 = 0.92$, adjusted $R^2 = 0.90$

Section 7.2

3. $n = 10$, $x = 3$

5. a. $x = 3$, $n = 12$ b. P-value $= 0.042$

7. 16, 18, 19

9. a. 0.172 b. 0.043 c. 0.151

11. $z = -2.00, -1.96, -1.44$

Section 7.3

1. a. 0.042, b. 0.162 c. 0.360

3. $T = 13$

5. $T = 28$

7. $T = 16$

9. a. $T = 5$ b. $T = 5, 8$

11. 60

Section 7.4

1. $R = 37$, P-value $= 0.8104$

3. $R = 362$, P-value $= 0.0096$

5. d. $a = 4$

7. $(1.7, 6)$

Section 7.5

1. a. 0.2112 b. 0.0016 c. 1

3. $R = 21$, critical values $= 14, 28$

5. a. 0.001916

7. P-value $= 0.4238$

9. P-value $= 0.670$

11. 3, 9

13. a. 0.0722 b. 0.0837 c. 0.0438 d. yes f. 4

Index

Φ
    definition, 152
    evaluating, 161

absolute zero, 476
acceptance sampling, 62
addition rule, 32
adjusted R^2, 495
adjusted Wald method, 279
alternative hypothesis, 335
ANOVA
    definition, 419
    one-way, 420
arrangements, 15
assessing normality, 324
average. *See* expected value
axioms of probability, 27

Bayes' theorem, 44, 48
Bayesian statistics, 50
Bernoulli distribution, 121
Bernoulli experiment, 83
Bernoulli random variable, 109
Bernoulli trials, 83
Bernoulli's law of large numbers, 214
beta distribution, 210
biased estimator, 255
biased sample, 191
bimodal, 74, 243
bin width, 126

binomial distribution, 78
    definition, 83
    mean, 124
    mean and variance, 119
    probability-generating function, 124
binomial theorem, 20
bins, 126
birthday problem, 30
bivariate normal distribution, 451
block, 225
block effect, 432
Bonferroni multiple comparison method, 425
box-plot, 238
Buffon's needle problem, 174

carrying capacity, 487
census, 221
central limit theorem, 194, 542
    finite population correction, 197
characteristic function, 542
Chebyshev's criterion, 467
Chevyshev's inequality, 216
chi-square critical values approximation, 299
chi-square distribution
    critical values, 203
    definition, 202
    mean, 211
    mean and variance, 203
    sum, 211

chi-square goodness-of-fit test. *See* goodness-of-fit test
classes. *See* bins
Clopper-Pearson "exact" method, 280
cluster sample, 226
coefficient of determination, 480
coefficient of variation, 248
combinations, 16
complement of an event, 29
  independence, 55
conditional probability, 37, 38
confidence bands, 478
confidence interval
  difference of medians, 528
  difference of population means, known variances, 304
  difference of population means, unknown variances, 306
  difference of population medians, 527
  difference of population proportions, 302
  population mean, known variance, 282
  population mean, one-sided, 291
  population mean, unknown variance, 285
  population proportion, 272
  population proportion, one-sided, 274
  population variance, 295
  population variance, one-sided, 297
  ratio of variances, 379
  simple linear model, 475
  single population mean using ANOVA, 429
  slope of simple linear model, 472
  y-intercept of simple linear model, 476

confidence interval method comparing proportions, 373
confidence level, 531
confounding, 225
confusion of the inverse, 45
continuous data, 222
continuous random variable definition, 129
control chart, 293
convenience sample, 226
correlation, 444
correlation coefficient ρ, definition, 448
correlation coefficient ρ
  alternate form, 454
  confidence interval for, 458
  of independent random variables, 452, 454
counting problems, 13
covariance, 183, 446
critical z-values, 270
cross-sectional study, 224
cumulative distribution function, 130
  definition, 72

data, 220
data analysis add-in, 559
De Moivre and Laplace limit theorem, 213, 545
degrees of freedom, 244
density curve, 128
dependent events, 55
discrete data, 222
discrete random variable, definition, 67
disjoint events, 27
distribution function. *See* cumulative distribution function
distribution of $\overline{X} - \mu$, 196
distribution of a random variable
  definition, 70

distribution of difference of sample
        means, 197
distribution-free tests. *See* nonparmetric
        tests
double-counting outcomes, 32

empirical rule, 154
empty set, 30
equally likely events, 28
estimator, 255
event, 2
exact binomial test, 354
expected value
        continuous definition, 132
        discrete definition, 101
        function of a random variable, 112
        properties, 113
expected value of a function of two
        variables, 180
expected value of a game, 102
experiment, 224
experimental error, 262
exponential distribution
        cumulative distribution function,
                143
        definition, 141
        kurtosis, 150
        mean and variance, 142
        median, 149
        MLE, 252, 259
        moment-generating function, 142
        skewness, 149
exponential model, 485

F-distribution
        critical values, 208
        definition, 207
        reciprocal, 212

F-test
        factorial experiment, 436
        randomized block design, 431
        slope of a simple linear model, 481
        for variance, 376
factor, 224. *See also* treatment
factorial, 14
factorial experiment, 434
failure, 83
final conclusion, 337
finite population correction
        central limit theorem, 197
        confidence interval for mean, 291
Fisher's exact test, 417
Fisher's least significant difference
        method, 425
5-number summary, 237
5% guideline, 59
functions of a random variable, 111
fundamental counting principle, 13
fuzzy central limit theorem, 328

gamma distribution
        definition, 199
        general definition, 202
        mean and variance, 200
        moment-generating function, 211
gamma function, 201
generalized factorial. *See* gamma
        function
generating values of a random variable,
        164
geometric distribution, 91, 257
        mean, 124
        MLE, 257
        moment-generating function, 122
        probability-generating function, 124
golden ratio, 178

hypergeometric distribution, 78
    approximation with binomial, 85
    definition, 80
    mean, 188
hypothesis testing
    confidence interval method, 354
    definition, 334

independent events, 54, 55
independent random variables, definition, 173
interaction effect, 436
interpretation of P-value, 345
intersection symbol, 28
interval level of measurement, 223

joint distribution, 168
    discrete definition, 169
joint probability density function,
    definition, 171

Kolmogorov, 26
kurtosis
    definition, 150
    sample, 246

large-sample confidence interval, 283
law of large numbers, 5
least-squares criterion, 460, 492
levels of measurement, 223
Levene's test, 396
    more than two variances, 430
likelihood function, 251
linear operator, 114
linear regression, 459
linear regression equation, 459
    formulas for coefficients, 460
    matrix formula for coefficients, 505
logarithmic model, 485
logistic distribution, 166

logistic model, 485
lose a degree of freedom, 402
lurking variables, 225

Mann-Whitney U test, 524, 530
marginal probability density function,
    definition, 171
marginal probability mass function, 170
matched pairs, 388
mathematical expectation
    continuous definition, 132
    discrete definition, 112
mathematical statistics, 228
maximum-likelihood estimate, 250, 251
mean. See expected value
mean absolute deviation (MAD), 269
mean squares, 422
measures of center, 247
measures of variation, 236
median
    continuous definition, 134
    discrete definition, 74
memoryless, 148
mid-P method, 280
midrange, 244
MLE. See maximum-likelihood estimate
mode, discrete definition, 74
modified Levene's test. See Levene's test
moment-generating function, 117
    continuous definition, 132
    discrete definition, 117
    linear combination, 121
    sum of independent variables, 184
    uses, 119
Monty Hall problem, 9
multimodal, 74
multiple coefficient of determination,
    495, 503
multiple R, 495

multiple regression, 492
    determining significance of terms, 506
    formula for coefficients, 504
multiplication rule, 37, 40
multistage sample, 226
mutually independent events, 56
mutually independent random variables, definition, 173

natural frequencies, 49
negative binomial distribution, 91
    moment-generating function, 123
nominal level of measurement, 223
nonlinear models, 485
nonparametric tests, 510
nonsampling error, 262
normal distribution
    definition, 153
    kurtosis, 160
    moment-generating function, 153
    skewness, 160
normal quantile plot, 324
nuisance factors, 225
null hypothesis, 335

observational study, 224
odds against an event, 36
1-proportion Z-test, 343
ordinal level of measurement, 223
outcome, 2
outlier
    definition, 246
    effect on mean and standard deviation, 246
    effect on median, 247

P-value method, 339, 345
paired data, 308
paired T-test, 387

pairwise independent events, 56
parameter, 221
partition, 47, 48
Pascal's equation, 26
Pearson correlation coefficient. *See* sample correlation coefficient r
Pearson product-moment correlation coefficient. *See* sample correlation coefficient r
percentiles, 134, 237
permutation
    different objects, 15
    some identical objects, 18
point estimate, 270
Poisson distribution, 92
    approximation with binomial, 97
    definition, 93
    mean, 124
    mean and variance, 106, 123
    mode, 100
    moment-generating function, 123
    probability-generating function, 124
poker hands, 19, 23
polynomial model, 498
pooled standard deviation, 305, 384
pooled variance, 300
population, 191, 220
posterior probability, 50
power model, 485
power of a test, 347
    1-proportion Z-test, 348
    Z-test, 365
prediction bands
    simple linear model, 477
prediction interval
    multiple regression, 505
    simple linear model, 471
predictor variable, 459
prior probability, 50

probability density function, 69, 129
probability function. *See* probability mass function
probability generating function, 123
probability histogram, 70
probability mass function, 67
    definition, 69
probability plot, 324, 331
probability space, 27
prospective study, 224

Q-Q plot, 324. *See also* probability plot
qualitative data, 222
quantiles, 134
quantitative data, 222
quartiles, 134, 237

R squared. *See* multiple coefficient of determination
random experiments, 2
random number generator, 167
random sample, 191, 227
random variable, 7
    definition, 66
randomized block design, 225, 431
range of a random variable, 66
range of values, 269
range rule of thumb, 123
ratio level of measurement, 223
rectangular distribution. *See* uniform distribution
redundancy, 55
regression
    Chebyshev's criterion, 467
    definition, 444
    minimizing absolute values, 467
    multiple. *See* multiple regression
    $Y = mX$, 466
    $Y = mX + b_0$, 466

relative frequency approximation, 3
relative frequency distribution, 127
relative frequency histogram, 86
residuals, 469
response variable, 459
retrospective study, 224
run, 532
runs test for randomness, 533
    critical values, 538
    length of run, 539

sample, 68, 191, 220
sample correlation coefficient r
    alternate form, 483
    critical values, 458
    definition, 449
sample distribution, 127, 235
sample mean, 192
sample proportion, 258–260, 546
sample size, 191, 315
    approximation, 317
    difference of means, 323
    difference of proportions, 323
    proportion, 316
    small population, 322
    variance, 319
sample space, 2
sample statistics, 236
sample variance, 257
    alternate formula, 249
sampling distribution of the mean, 264
sampling distribution of the proportion, 261
sampling error, 262
sampling techniques, 226
sampling variability, 4
scatterplot, 444
sign test, 512
    handling ties, 517

simple linear model, 468
    definition, 468
simple random sample, 227
simulations, 163
68-95-99.7 rule. *See* empirical rule
skewness
    definition, 149
    sample, 245
small sample confidence interval, 283
St. Petersburg paradox, 111
standard deviation
    continuous definition, 132
    discrete definition, 103
standard error of estimate, 470, 496
standard error of the coefficient, 472
standard normal distribution
    definition, 150
    moment-generating function, 152
statistic, 221
statistical control, 293
statistical study, 220, 224
statistics, definition, 220
stratified sample, 226
Student-t distribution
    critical values, 206
    definition, 205
Student-t ratio, 205
subjective probabilities, 8, 50
subset of an event, 31
success, 83
sum of squares, 419, 479
summary statistics, 236
support of a random variable. *See* range
    of a random variable
systematic sample, 226

T-test, 358
T-test of the Slope of a Simple Linear
    Model, 472

technical conclusion, 337
test of homogeneity, 411
theoretical approach, 5, 28, 29
transforming data, 484
transitivity, 373
treatment, 224, 420
tree diagram, 46
trial, 3, 57
trimmed mean, 247
Tukey's honestly significant difference
    method, 425
2-proportion Z-test, 367
    claimed difference, 374
2-sample T-interval, 305
2-sample T-test, 383
2-sample Z-interval, 304
2-sample Z-test, 383
type II error, 347

unbiased estimator, 255
uniform distribution
    continuous random variable, 139
    discrete random variable, 70
    mean and variance, 106
    median, 149
    skewness, 150
union symbol, 28
unusual event, 12
    guideline, 87
    principle, 87

variance
    alternate formula, 105
    of aX, 114
    continuous definition, 132
    discrete definition, 103
    estimation, 318
    in terms of expected value, 113
voluntary response sample, 226

Wald method, 279
Weibull distribution, 210
Wilcoxon rank-sum test critical values, 531
Wilcoxon signed-rank test
    approximating critical values, 524
    effects of outliers, 524
    exact critical values, 553
    handling ties, 522

Wilson score method, 279

$\bar{x}$ chart, 292

Yates' correction for continuity, 416

Z-interval, 282
z-score, 156
    of a data value, 244
Z-test, 356